HUNTERS ON THE TRACK

HUNTERS ON THE TRACK

WILLIAM PENNY AND THE SEARCH FOR FRANKLIN

W. GILLIES ROSS

McGill-Queen's University Press
Montreal & Kingston • London • Chicago

ISBN 978-0-7735-5283-8 (cloth)
ISBN 978-0-7735-5832-8 (ePDF)
ISBN 978-0-7735-5833-5 (ePUB)

Legal deposit second quarter 2019
Bibliothèque nationale du Québec

Printed in Canada on acid-free paper that is 100% ancient forest free (100% post-consumer recycled), processed chlorine free

This book has been published with the help of a Bishop's University Senate Research Committee Publication Grant.

Funded by the Government of Canada | Financé par le gouvernement du Canada | Canada

Canada Council for the Arts | Conseil des arts du Canada

We acknowledge the support of the Canada Council for the Arts.
Nous remercions le Conseil des arts du Canada de son soutien.

Library and Archives Canada Cataloguing in Publication

Title: Hunters on the track : William Penny and the search for Franklin / W. Gillies Ross.
Names: Ross, W. Gillies (William Gillies), 1931– author.
Description: Includes bibliographical references and index.
Identifiers: Canadiana (print) 2019008314x | Canadiana (ebook) 20190083220 | ISBN 9780773552838 (hardcover) | ISBN 9780773558328 (ePDF) | ISBN 9780773558335 (ePUB)
Subjects: LCSH: Penny, William, 1809–1892. | LCSH: Franklin, John, 1786–1847. | LCSH: John Franklin Arctic Expedition (1845–1851) | LCSH: Whalers (Persons) Scotland—Biography. | LCSH: Whaling—Arctic regions—History—19th century. | LCSH: Arctic regions—Discovery and exploration. | LCSH: Northwest Passage.
Classification: LCC G660 .R67 2019 | DDC 919.804—dc23

This book was typeset in 10.5/13 Sabon.

Contents

Acknowledgments

I extend sincere thanks to the helpful staffs of many libraries, archives, and other institutions, especially the following: Aboriginal Affairs and Northern Development Canada (Gatineau); American Philosophical Society (Philadelphia); British Newspaper Library (Colindale, London); Hudson's Bay Company Archives (Winnipeg); Library and Archives Canada: Ottawa); Library of Congress (Washington); National Portrait Gallery (London), New-York Historical Society (New York); National Archives of the United Kingdom (formerly Public Record Office: Kew); Rauner Special Collections Library (Dartmouth College); Scott Polar Research Institute (Cambridge).

Many individuals have contributed in valuable ways to this book, for which I am very grateful. Foremost among them is Liz Tregonning, a descendant of one of William Penny's uncles. She has carried out extensive research into Penny's career and life, with support from the Crowther Fund of the Open University, and has given me the immense benefit of her knowledge, as well as encouragement in the project

I also thank the following: Andrew Lambert, who provided valuable details about the British Admiralty, Tim Morton, who showed Liz and me Stephen Pearce's 1853 portrait of Penny at the Wimbleton storehouse of the National Portrait Gallery; Charles Broadhurst, who provided several illustrations from John B. Cator's arctic journal; Cathy Haycock, who sent a painting of Penny's boat by her father, the geologist Maurice Haycock; Mary L. Robertson, who sent excerpts from the Beaufort Papers in the Huntington (San Marino, California); Douglas Wamsley, who supplied a rare image of HMS *Lady Franklin*, and Allan Rowell, for sending a satellite image of Devon Island. For the final maps, I thank Tim Doherty of Visimage. Last in the sequence of things, but of special importance, is Holly McMillan, whose formidable computer skills and sharp eye for inconsistencies have been invaluable.

Thanks also to the anonymous reviewers of the manuscript for McGill-Queen's University Press, and to copy editor Maureen Garvie.

William Penny, by Stephen Pearce, 1853. Oil on canvas © National Portrait Gallery, London.

Aerial photo showing Beechey Island and Erebus and Terror Bay, off the coast of North Devon. Just beyond the island, the north shore of Low Shingle Beach and Port Riley can be seen. Photo T432L-234 © Her Majesty the Queen in right of Canada, Department of Energy, Mines and Resources.

Wintering at Assistance Bay, 6 April 1851, from the journal of John B. Cator. Photograph courtesy of C. Broadhurst. Collection of the author.

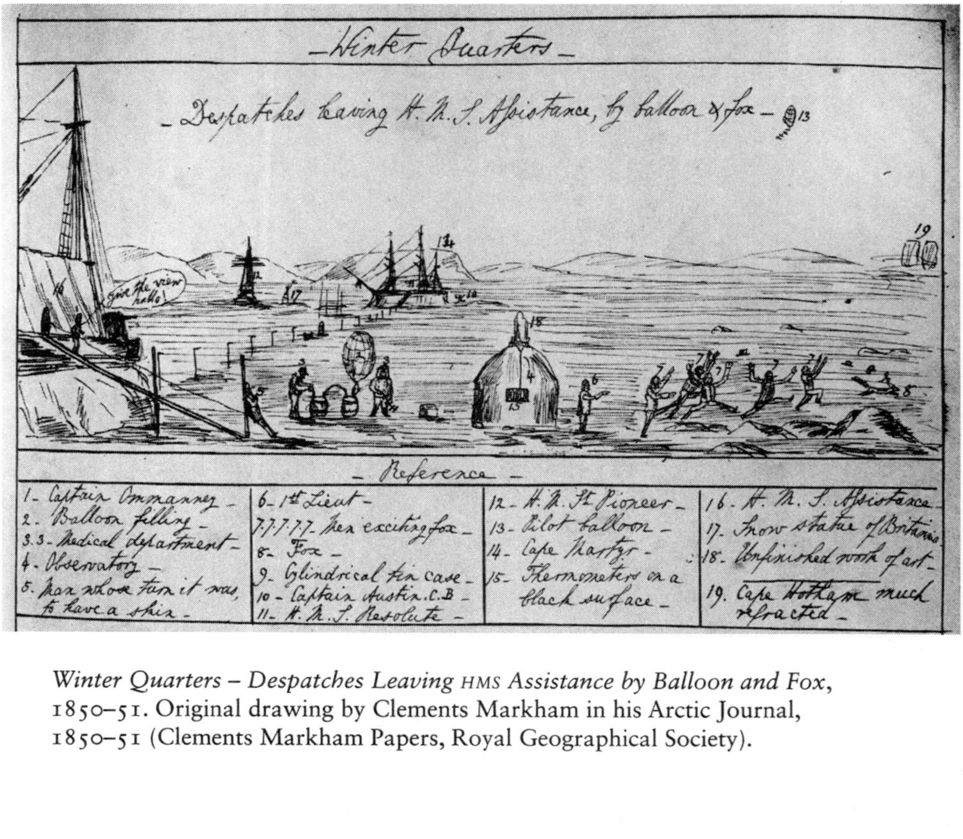

Winter Quarters – Despatches Leaving HMS *Assistance by Balloon and Fox*, 1850–51. Original drawing by Clements Markham in his Arctic Journal, 1850–51 (Clements Markham Papers, Royal Geographical Society).

Penny's Boat at Abandon Bay, Cornwallis Island, Maurice Hall Haycock, 1974. Reprinted with permission from *On-Site with Maurice Haycock, Artist of the Arctic* (Toronto: Dundurn 2007).

Four Beechey Island Graves, Maurice Hall Haycock, 1975. Reprinted with permission, from *On-Site with Maurice Haycock, Artist of the Arctic* (Toronto: Dundurn 2007).

Assistance Bay, 24 February 1851, the Coldest Day, Mercury Frozen, by M. & N. Hanhart, 1852. The illustration shows Penny's two ships and Ross's *Felix* wintering at Assistance Bay, the crews playing rounders. Frontispiece, *Journal of a Voyage in Baffin's Bay and Barrow Straits in the Years 1850–1851 ....*, vol. 1, by Peter Sutherland (London: Longman, Brown, Green, and Longmans 1852).

Assistance in the Ice, by Thomas Sewell Robins, 1853. National Maritime Museum, Greenwich, London, BHC4239.

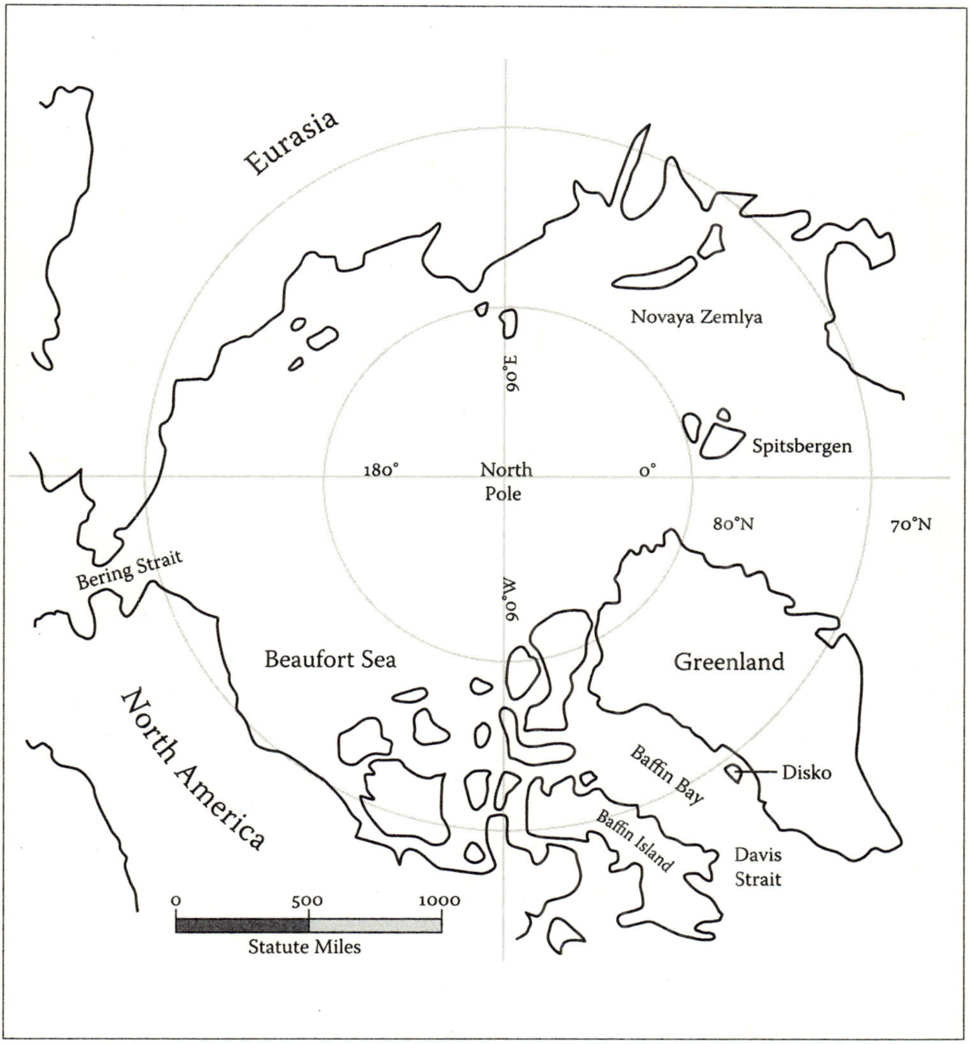
Eurasia
Novaya Zemlya
90°E
Spitsbergen
180°
North
Pole
0°
80°N
70°N
Bering Strait
90°W
Beaufort Sea
Greenland
North America
Baffin Bay
Disko
Baffin Island
Davis
Strait
0
500
1000
Statute Miles

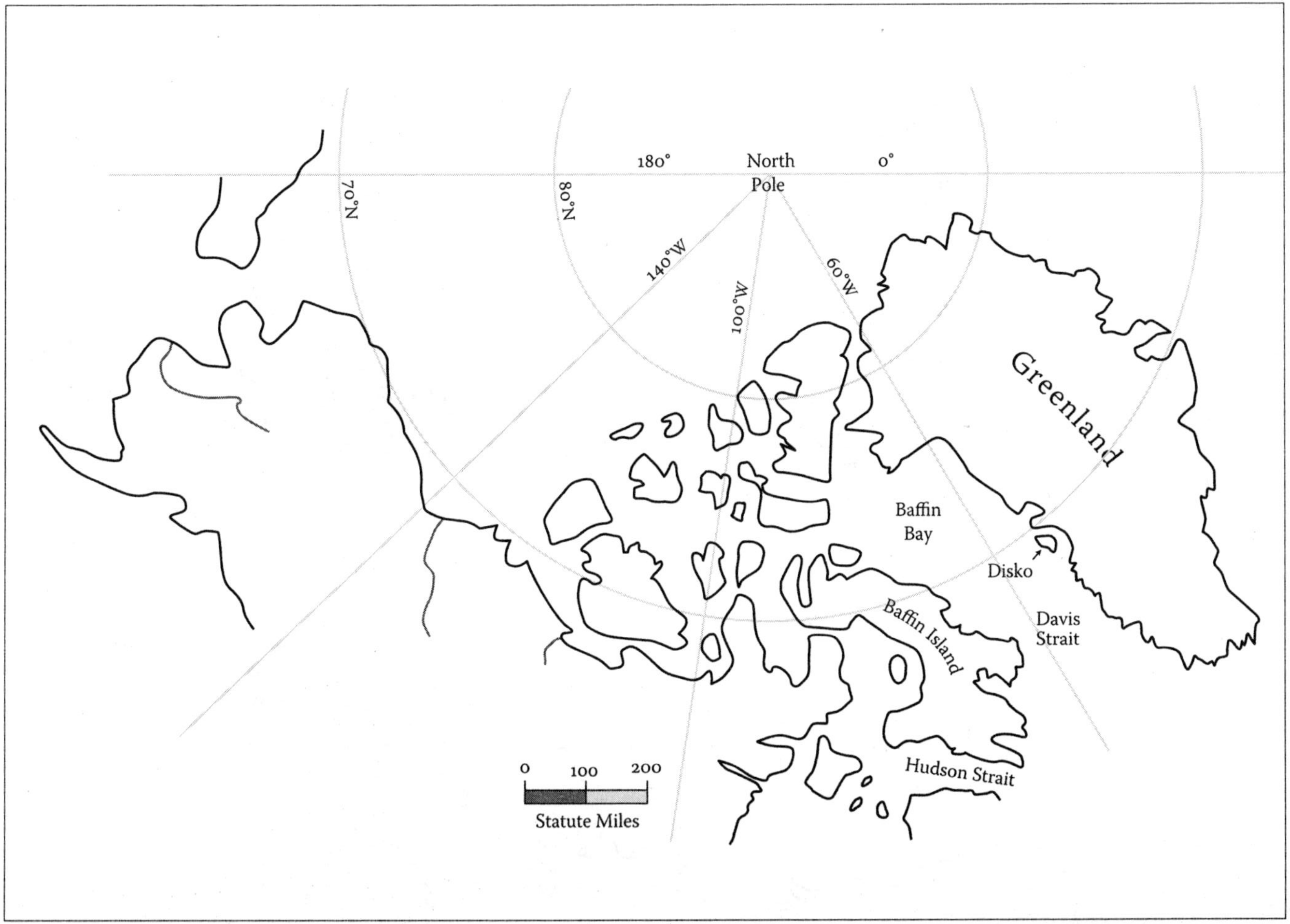
180°
North Pole
0°
70°N
80°N
140°W
100°W
60°W
Greenland
Baffin Bay
Disko
Davis Strait
Baffin Island
Hudson Strait
0
100
200
Statute Miles

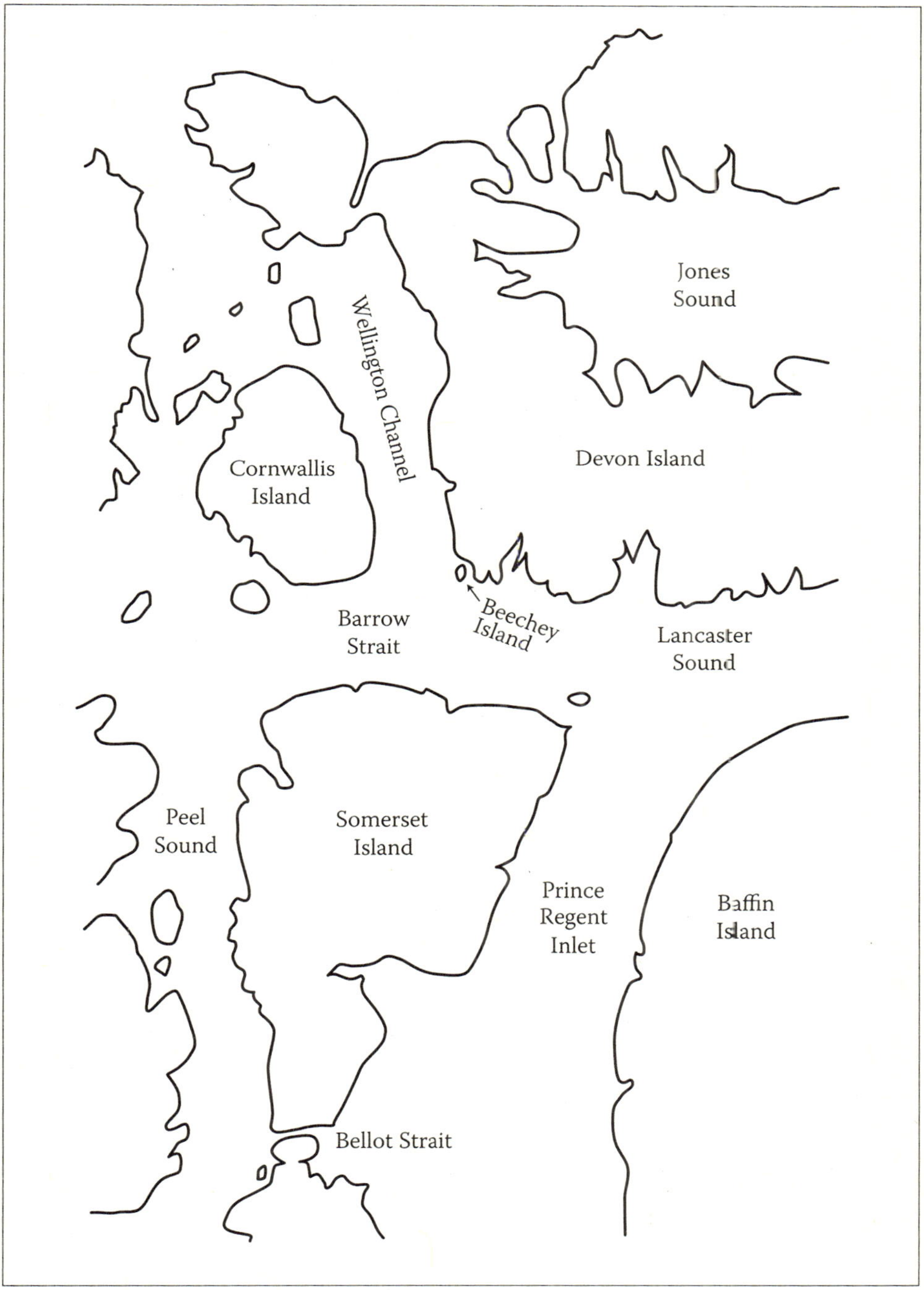
Jones Sound
Wellington Channel
Cornwallis Island
Devon Island
Beechey Island
Barrow Strait
Lancaster Sound
Peel Sound
Somerset Island
Prince Regent Inlet
Baffin Island
Bellot Strait

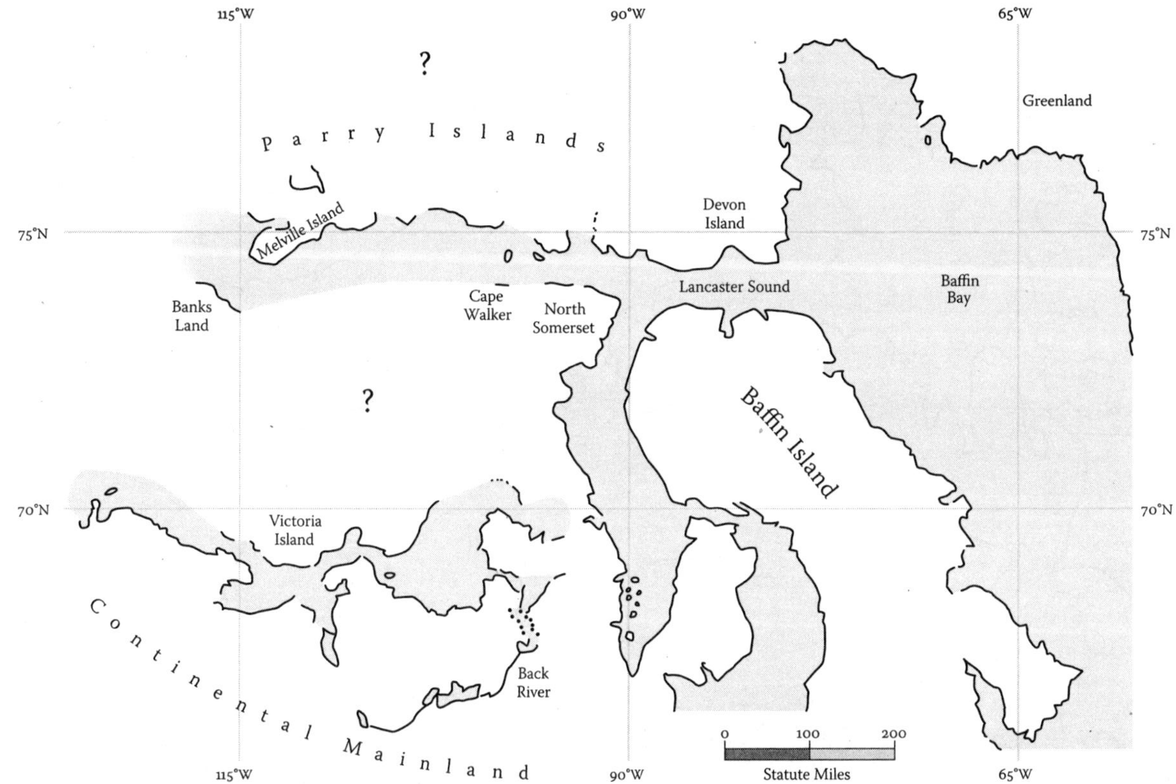

Sketch of "Chart of Part of the Arctic Regions as Known in 1845, Being a Copy of the Chart Supplied to the Franklin Expedition," in Brown, *The North-West Passage*, 3–4.

HUNTERS ON THE TRACK

Prologue

Aberdeen, Scotland, 13 April 1850. Two small sailing vessels are shortly about to set off in search of Captain Sir John Franklin and his 128 companions on board HMS *Erebus* and *Terror*. Five years ago, Franklin's ships departed to find a Northwest Passage from the Atlantic to the Pacific through the maze of islands lying between the North American continent and the North Pole. They have not been heard from since.

These little brigs are not the first to search for Franklin. Two years ago, a British Admiralty expedition sailed through Baffin Bay and Lancaster Sound into the Arctic Archipelago, the largely uncharted labyrinth of islands lying between the North American continent and the Pole. Other ships have sailed halfway round the world by way of Cape Horn to reach Alaska. Overland parties have descended rivers to examine the arctic coast. So far, however, all efforts have been in vain. Rumours and hoaxes have kept the Admiralty busy investigating potential leads and have cruelly manipulated the emotions of relatives and friends, but not a single trustworthy clue to the whereabouts of the missing expedition has come to light.

Although this new search expedition is an official venture of the British government, comprising two of "Her Majesty's Ships" and proceeding under orders issued in London by the Admiralty, it does not have the appearance of a Royal Navy enterprise. In contrast to the large naval ships preferred by the Admiralty, these brigs, newly built in Scotland, are small; and instead of bearing the sort of awe-inspiring names common in the Navy – *Courageous*, *Indefatigable*, *Valiant*, and the like – they are named for two women, Franklin's wife and niece. Each ship has a crew of only two dozen men or so, none of whom are in the navy. They are whalemen and merchant seamen.

Perhaps the most unusual thing about this Admiralty expedition is that its leader is not a naval officer but a whaling captain – a *mere* whaling captain, as snobs like to say. William Penny, a Scotsman, has been sailing on arctic whaling voyages since the age of twelve and has been in command of ships for the past fifteen years. He is part of a large brotherhood of seafaring men – whalemen, fishermen, and merchant seamen – whose activities have not been adequately recorded and whose exploits (unlike those of the *gentlemen* officers of the Royal Navy) have not been glorified in published narratives.

There are fervent hopes that Penny's little vessels from Aberdeen will solve the great mystery of Franklin's disappearance. They will be far from alone in the quest, however. Four other maritime expeditions – three from Britain (one of them, like Penny's, sailing under Admiralty orders), and one from the United States – also intend to follow Franklin's presumed path through Baffin Bay and into the Arctic Archipelago, while others will approach overland or by way of the Pacific and Bering Strait.

Penny's brigs have been inspected and declared seaworthy. The last supplies and bonded stores have been brought on board and stowed. The Blue Peter has been hoisted to recall the crews. HMS *Lady Franklin* and *Sophia* are now moored to buoys in the lower dock awaiting the harbour pilot. The steam tug *Victory* idles nearby, ready to pass them a towing hawser and haul them out to sea.

1

Franklin Departs

A desert waste of waters lies before –
Behind, the anxious hospitable shore,
Which like a parent bird sees ye depart,
Bold winged messenger of daring Art![1]

Mid-morning on 19 May 1845, Lady Jane Franklin stood on the quay at Greenhithe with her husband's daughter Eleanor and his niece Sophia Cracroft, as HMS *Erebus* and *Terror* and their supply ship *Baretto Junior* cast off and were taken in tow by the steam-tugs *Rattler* and *Monkey*. Thousands of spectators cheered and waved as the flotilla headed down the Thames toward the North Sea.

From the deck of the *Erebus*, Sir John Franklin waved a handkerchief and watched the figures of the three women slowly recede into the distance until they merged indistinctly with the assembled throng and finally disappeared behind a bend in the river. His view through the telescope, framed dramatically within the circular cropping of the glass, was the last he would ever have of them.

More than a week earlier, when Lady Franklin had draped a flag over a reclining Sir John in their London lodgings, he had remarked that this was customarily done after death: was her gesture an omen? She was already worried about the uncertainties and risks that would inevitably confront the expedition in remote and uncharted territory, and she may have wondered whether she would ever see her husband again. She recalled the near-fatal difficulties he had encountered in his overland voyages of the 1820s; she knew that he had sailed on only one maritime expedition to the Arctic before (twenty-seven years earlier!); and she doubtless felt that a man in his sixtieth year whose girth had expanded from a surfeit of mess dinners should not be sailing away on a dangerous mission like a schoolboy on a class picnic. Even if the ships did succeed in finding a sea passage to the Pacific, it would likely be years before he returned to England.

She had stayed with him on the *Erebus* before departure and remained on board while the ship was towed downriver from Woolwich to Greenhithe on 12 May to be swung for compass corrections. On Sunday, the day before the ships were to leave, the three ladies attended church on the deck of the ship, the service conducted by Sir John himself.

The departure of the ships was in many ways a joyous celebration, the manifestation of a fervent national pride that Great Britain would at last solve the geographical problem of the Northwest Passage. It was also a celebration of human courage in the inevitable confrontation with the powerful forces of nature. In the vocabulary of the times, Franklin and his "gallant" companions were setting out to overcome the "hostility" of the arctic environment, and a triumphal "conquest" of the elements was confidently predicted.

According to Franklin's orders, the primary objective of the expedition was to find a navigable passage to the Pacific. While attempting to do so, he was to record observations in geography, geology, oceanography, and natural history. Of special importance to the scientific community in England were daily measurements of the earth's magnetism, to supply data for a growing network of observations that would enhance the accuracy of ship navigation around the world.

Elaborate preparations had been made for Franklin's struggle against the arctic ice. The ships' hulls were doubled and strengthened inside by cross timbers; their bows were masses of timber eight feet thick, protected on the exterior by plates of sheet iron extending twenty feet back from the stems. Utility, not beauty, was the key: even the figureheads had been replaced by thick wooden blocks. To assist the square-rigged sailing ships in calms and among floes, they were powered by twenty-five-horsepower, coal-burning locomotive engines that had previously operated on railway lines emanating from London. The engines were mounted thwart-ship in the afterholds, the rotation of their driving wheels transmitted by an extended axle to propellers sheltered within enclosed wells at the stern. Steam-powered screw propulsion had never been used in the Arctic before.

With the extra supplies carried on board the transport *Baretto Junior* for transfer to the discovery ships at Disko Bay on Greenland's west coast, the expedition carried enough food for three years, including ten live oxen. With a generous supply of guns and ammunition, they could procure more fresh meat by hunting when they reached the Arctic.

For the 134 officers and men on board the ships, the occasion was a long-anticipated moment, at once an end to the long and hectic preparations for the voyage and a beginning to the challenging task before them.

The high spirits of the crews were evident, and one can imagine the pet monkey that Lady Franklin had donated to her husband's ship leaping nimbly about the mast and rigging – the closest approximation to a live tree that the poor animal could find. Perhaps it was attired in the ridiculous outfit that it allegedly wore for festive occasions in the officers' mess – "trousers, frock and blanket."[2]

After leaving the Thames, the little squadron headed north toward the Orkney Islands, following the usual route of whaling ships bound for Davis Strait and Baffin Bay. While battling headwinds and heavy seas off East Anglia, the ships being towed by the steam tugs broke their hawsers more than once and finally had to drop them and proceed under short sail. The Admiralty sent the steamer *Blazer* to overtake them and advise Franklin that if the headwinds persisted, he could turn back and go by way of the English Channel instead, but he decided to maintain his course. Off Aberdeen, he sent a message back to the lords commissioners of the Admiralty in London, stating that he had sent the tug *Monkey* back to Woolwich and the *Rattler* had gone into port to disentangle a towing hawser from the propeller and get more coal. The *Erebus*, *Terror*, and *Baretto Junior* were continuing as planned, accompanied by the *Blazer* to help tow if necessary. By the time they made Stromness, Orkney, at the beginning of June, four of the ten bullocks carried on the transport to provide fresh meat were dead. It was a disquieting beginning.

* * *

At the Whalefish Islands, midway up the west coast of Greenland, the crews worked feverishly for six days to transfer supplies from the transport to the already crowded expedition ships. Then *Baretto Junior* returned to Britain bearing letters for families and friends – the last they would ever receive from HMS *Erebus* and *Terror*. The messages were reassuringly optimistic. One man had been sent home from Orkney and four men from Greenland (two for health reasons and two for incompetence), but the remaining 129 were said to be in good health. The ice conditions were considered favourable and the prospects of a swift voyage excellent.

The assistant-surgeon on HMS *Terror* was Alexander McDonald, a young man who had already made several arctic voyages as surgeon on whaling ships commanded by William Penny. British whaling ships were required to carry a "surgeon" – in many cases, not qualified physicians or surgeons but merely medical students such as McDonald, who had accompanied Penny between university terms. The presence of surgeons on the whaling ships had an important bearing not only on the health of crews but also on science; medical men of the day tended to be intensely curious about the natural world, and they often recorded and published valuable observations on a wide spectrum of phenomena.

From Greenland, McDonald wrote to his brother James, "I am inclined to think that this is what is called an open season. Should this prove correct we will make great and rapid progress: indeed some among us are so sanguine as to believe that we will reach the Pacific without wintering."[3] He confidently advised his brother to write to him in care of the Russian ambassador in London, who would then forward the letters to Moscow. From there they would somehow be carried thousands of miles eastward across Siberia to meet the ships at Petropavlovsk on the Kamchatka Peninsula, which Franklin planned to visit after completing the Northwest Passage. Or – if James preferred – he could send his letters to the Sandwich Islands (Hawaii) or to Panama, both of which were on Franklin's itinerary. The possibility that the ships might be held up for a time in the Arctic did not worry him: "You may continue to do this for several years should you not hear from or of me at that time,"[4] he advised his brother. McDonald's optimistic outlook was typical of those held by officers and men on the discovery ships. The expedition was the best ever sent out: how could it possibly fail?

A few months later, there was more encouraging news: a whaling ship had spoken Franklin's ships in northern Baffin Bay in late July. They were on course for Lancaster Sound, and all was well.

The year came to a close without further word of the expedition, but few in Britain saw cause to worry. The ships had plenty of food, and wintering in the ice should not be life-threatening. Had not Parry's crews in Winter Harbour amused themselves with hearty games and jolly theatricals while their two ships remained encased in ice six feet thick? Had he not returned to the Arctic to spend two more winters frozen up in Foxe Basin, and then gone back a third time to winter in Prince Regent Inlet? And what about Sir John Ross, who had survived no fewer than four successive arctic winters on the voyage that earned him a knighthood? Their experiences showed that winter need not be feared.

Lady Franklin may have had some cause for anxiety, but there were plenty of distractions in the months following Sir John's departure. She was an active woman, certainly not the sort to sit around and mope. She travelled to France and Madeira and in the following year to the West Indies and the United States, where she hiked up Mount Washington, New Hampshire, the tallest mountain in New England. But Sir John was never far from her thoughts. Back in London again, she wrote to Sir James Ross, "I dare not be sanguine as to their success – indeed the very thought seems to me presumptuous, so entirely absorbed is my soul in aspirations for their *safety* only."[5] If the expedition failed to turn up, she said, it would give her great comfort if he would "be the man to go in search of them." As it happened, he was.

2

William Penny and Arctic Whaling

One of the most arduous of all maritime adventures.[1]

During the early months of 1845, while the attention of the reading public in Britain was focused sharply on the imminent departure of Franklin's widely publicized expedition in search of a Northwest Passage, the whaling captain William Penny was in Aberdeen, making preparations for another whaling voyage.

Like most whalemen, he was interested in the progress of the discovery expedition because it might reveal new whaling grounds. According to a widely held opinion, the declining returns from whaling were attributable not to overhunting but rather to the whales' retreat to more remote haunts to escape persecution. Exactly where these refuges might be was naturally a subject of great interest, but whaleship owners discouraged their captains from leaving the customary places in which whales were certainly *known to exist* in order to seek them out in regions where they *might possibly exist*. Their attitude was that exploration was not their responsibility but that of the government, because it might lead to the discovery of new marine resources and ultimately benefit the nation as a whole. The whaling industry and the nation had profited immensely from the Admiralty voyages of John Ross in 1818 and William Edward Parry in 1819–20, which had shown that the west side of Baffin Bay and Davis Strait was within the summer territory of the whales that British and Dutch whalers had been hunting along the coast of Greenland for more than a century. Perhaps Franklin's voyage would yield important information too.

Penny may have had another reason for being interested in the progress of HMS *Erebus* and *Terror*. As a boy, he had probably been in Peterhead in late October 1820 when Parry, returning from his remarkable transit westward from Baffin Bay to the margin of the Beaufort Sea, disembarked at the Scottish port to speed to London and report to the Admiralty. Penny, although only a dozen years of age, had returned three weeks earlier from his first arctic whaling voyage, on the whaleship *Alert*, commanded by his father, and it seems unlikely that they would have missed the opportunity of getting a peek at Parry as he stepped triumphantly ashore from HMS *Hecla*.

For the next three decades, Penny had sailed to the Arctic on whaling ships, hunting Greenland whales (*Balaena mysticetus*, also called polar or bowhead whales) on their summer feeding grounds. For six years, he served as apprentice under his father's command, voyaging to the "Greenland whale fishery" in the region between Greenland and Spitsbergen. But returns from this region had declined, and many British whaling captains were shifting to the "Davis Strait whale fishery" between Greenland and the arctic flank of North America, where yields were generally higher. When Penny accompanied his father in 1827 on the Peterhead whaler *Eclipse*, they sailed for their first time across the Atlantic, south of Iceland, around Cape Farewell – the southern tip of Greenland – and north through Davis Strait into Baffin Bay. After exploiting this whaling ground for five seasons, his father retired and Penny found a berth as mate on the Peterhead whaler *Traveller*, commanded by Captain George Simpson. They sailed to the Davis Strait fishery for two highly successful seasons, after which Penny accepted an offer to join the *Hecla*, one of Aberdeen's whalers. If this was the same ship that Parry had sailed through the Arctic Archipelago to Melville Island in 1819 and had taken on two subsequent expeditions into arctic waters – the ship from which he had landed at Peterhead when Penny was a boy – it may have been a factor in Penny's decision to leave the *Traveller* and move from Peterhead to Aberdeen.

Penny attained his first command in 1835, and for the next four seasons he sailed to Davis Strait in charge of the Aberdeen whaler *Neptune*. These years were characterized by unusually severe ice conditions and extraordinarily high ship losses among the whaling fleet, but Penny exhibited a talent for interpreting ice movements and a flair for ship-handling among floes and bergs, and the *Neptune* avoided disaster. When the ship was sold in 1840, Penny took command of the *Bon Accord*. After she was put on the auction block a year later, he commanded the *Saint Andrew* on trading voyages to North America and one whaling voyage to Davis Strait. In 1845, the year in which Sir John Franklin set off to complete the first-ever Northwest Passage, Penny sailed again to the Davis Strait fishery. By this time he had made at least twenty-one voyages to the Arctic, ten of them to the very region through which Franklin was to pass.

The dramatic story of the British search for a maritime route from the Atlantic to the Pacific northward of the intervening continent is often told without adequate reference to the whaling industry, as if the discovery expeditions were penetrating an unknown part of the Arctic. In fact, the discovery ships that headed for Lancaster Sound, the principal entrance into the archipelago and a potential route to Bering Strait, were traversing a region that had being attracting European whaling ships every season for two centuries.

Records of whaling activity reveal that from 1719 (when annual statistics of voyages and catch began to be compiled in Britain) up to the beginning of 1845 (the year that Franklin sailed) more than 5,300 voyages were made to the Davis Strait whale fishery by ships from Holland, Britain, Germany, and the United States.[2] As several serious gaps exist in the statistical record, the real total must be significantly higher. Until the 1790s, Dutch ships dominated the fishery, along with some whalers from Britain, Germany, France, and a few other countries, but after 1800, Dutch ships rarely sailed to the region, and British whaleships were by far the most numerous until the demise of the arctic whaling industry early in the twentieth century.

Before 1820, the whaling effort was restricted mainly to the southern part of Davis Strait and the coast of Greenland, but after two British whaling ships reached the ice-free polynya of the "North Water" at the head of Baffin Bay in 1817, and the discovery expeditions of John Ross and William Parry drew attention to the presence of whales off the coast of Baffin Island, British whalers concentrated their efforts on the "West Side" and adopted a counter-clockwise circuit of Baffin Bay to exploit the region. They would work northward along the coast of Greenland in May and June, cross the dreaded Melville Bay to Cape York (ice conditions permitting), sail westward in the North Water, pursue whales off Pond Inlet and the entrance to Lancaster Sound, and then cruise southeastward along the flank of Baffin Island to Davis Strait before heading homeward in September or October.

In the quarter-century between the initiation of whaling on the West Side in 1820 and the departure of Franklin's expedition in 1845, British whaling ships made more than twelve hundred whaling voyages to the Davis Strait/Baffin Bay region. In peak years, more than eighty ships were involved in this enterprise. The region was familiar to thousands of whalemen.

The idea of sailing thousands of miles to the margin of polar ice, lowering small open boats, and attempting to harpoon and lance creatures half as long as the whaling ships themselves may seem like madness, but there were good reasons to undertake such long and dangerous voyages. In the era before petroleum and electricity, Greenland whales supplied resources vital to the economy. The oil rendered from their foot-thick overcoat of blubber was used as an illuminant in street lighting systems – either directly or in the form of oil-gas – as well as in factories, offices, business establishments, homes, and churches. Oil was used in the manufacture of paints, varnishes, soaps, and cordage, served as a fine lubricant for watches, clocks, and machinery, and provided a softening agent for flax and jute fibres. The arctic whaling ships brought home chopped-up pieces of blubber in casks, to be rendered into oil in boiling houses ashore.

In addition to blubber, each whale had a mouthful of valuable baleen – more than six hundred plastic-like slabs attached to the upper jaw to filter out their planktonic food from sea water. Called "whalebone" or simply "bone" by the whalemen, baleen was springy and durable, ideal for making articles such as corset stays, umbrella ribs, riding crops, buggy whips, brushes, and combs. The hair-like fringe along the trailing edge of baleen plates was used to stuff upholstery. The slabs of bone, some of which could exceed twelve feet in length in a large whale, were tied together on board ship and later dried and processed in home port.

The economic benefits from whaling extended in cascading fashion through the industrial and commercial fabric of each port – from the ships that returned with blubber and bone, to the boiling houses that extracted the oil, the town and city utilities that provided street lighting, the small factories that made consumer products, and the stores that sold the items. The men and women who worked in such places and the customers who bought the products all benefited from the whaling trade. The industry was important in a national sense too, bringing wealth to several countries in Europe and North America and constituting a factor in international rivalries and conflicts. In Britain, the whale fisheries were regarded as a "nursery for seamen," because they provided a reservoir of experienced seamen in case of war – sailors who had been trained in the harshest marine environment on earth. Even when the trade was unprofitable, the government paid bounties to encourage whaleship owners to send out ships because it was in the national interest.

* * *

Penny had earned a reputation in the whaling fraternity for being in the vanguard, getting through obstacles, overcoming difficulties, and avoiding disaster. He was also insatiably curious about the geography of northern lands and seas and was willing to turn aside from the beaten path and strike off into unknown territory if he felt that some useful information could be gained. In addition, he had a great respect for the culture and geographical knowledge of the Indigenous inhabitants of the Arctic and saw them as valuable sources of information as well as potential assistants in securing whales.

While serving under Captain Simpson on the *Traveller* in 1832, Penny came into contact with Inuit at Durban Harbour on the west coast of Davis Strait, where whalers usually stopped before setting out on their homeward voyages. The natives' stories of a large body of water somewhere to the south, in which whales abounded, excited his imagination, and he endeavoured to reach it by whaleboat from Exeter Bay (or possibly Exeter Sound) with several companions. The waterway they penetrated turned out to be a fiord rather than a strait, so they were obliged to turn back, but his quest for the large gulf was not over. When he commanded the *Neptune* in 1839, he returned to Durban, where he met a young Inuk named Eenoolooapik (variously spelled) who had been born on the south coast of the great bay that they called Tenudiakbeek (variously spelled). At the age of ten, he had travelled with his family all the way to Durban Harbour, a coasting voyage of several hundred miles, in an umiak, an open boat made of sea-mammal skins stretched over a driftwood framework. Now, ten years later, Eenoo was able to draw from memory a rough map showing in remarkable detail the entire coast along which they had journeyed. It gave a reasonable idea of the location, size, and shape of the sought-after gulf. It appeared to be about two hundred miles long and fifty miles wide, and its entrance was approximately two hundred miles south of Durban Island. By placing the Inuit map within the framework of parallels and meridians provided on the Admiralty chart, Penny could see how Tenudiackbeek related to the comparatively well-known coastline to the north.

The Inuit were a nomadic people whose seasonal migrations were synchronized with those of the animals they hunted. Observations of the natural world were absolutely vital to their survival in the harsh and unforgiving environment. While travelling over land and ice, they noted useful details so that they could retrace their paths later or tell others about them. Without telescopes, chronometers, or surveying instruments, without training in cartography or drawing, without knowledge of the magnetic compass or the concept of latitude and longitude, they could somehow imagine the layout of the land as it would appear from above and thus form a mental map of the region. Although they lacked the tradition and materials for writing, they could communicate oral information about routeways, landmarks, and obstacles. On occasion they drew sketch maps in sand or snow, sometimes using mounds and hollows to indicate relief, but of course these could not be preserved or duplicated for use later. Many of the sketch maps they drew on paper for explorers, however, have been preserved and carefully analyzed.[3]

Inuit maps were usually not aligned in any specific compass direction, and they tended to distort scale in response to varying travel speeds. In the sense that European explorers also followed linear paths and made observations that were implemented on maps, they were similar to the Inuit, but their relatively sophisticated technology for position-finding enabled them to produce more accurate representations of a landscape or a coastline, and their technique of printing on paper made it possible to reproduce and distribute charts to a large number of users.

On the British Admiralty's charts of the world's oceans, latitude was marked along the sides and longitude along the top and bottom. Whatever distribution of land and sea these charts indicated was therefore set within a global grid and could easily be related to other locations anywhere in the world, making it possible to navigate from one place to another. As their accuracy was only as good as the data that went into their compilation, the charts were frequently updated and reissued as corrections were brought to the attention of the Hydrographic Office by the captains of whaling or trading vessels or by subsequent government expeditions.

Each chart reflected the state of European knowledge about a region at a particular time, which in turn reflected the extent and intensity of European maritime activity there. In 1839 the cartography of the arctic regions of North American was only in its infancy. Some coasts were depicted in great detail, but others were shown sketchily or not at all. Fifteen years earlier, the Admiralty had issued Chart No. 262, entitled "Chart of Baffin's Bay, with Davis and Barrow Straits; by Captn Ross & Lieut^{t} Parry, R.N in 1818, 19 & 20. and the Discoveries of Captn Parry in 1822 and 1823; & Captn Lyon in 1824." On it, the coasts of Lancaster Sound, Hudson Strait, northern Hudson Bay, and Melville Peninsula, which had been traversed by a few expeditions in search of a Northwest Passage or by supply ships of the Hudson's Bay Company, were depicted in the greatest detail and adorned with English place names, with depths indicated wherever soundings had been taken.

The coast of Baffin Island facing Baffin Bay had not attracted as much attention and exhibited a more generalized outline. Several indentations were shown incompletely, their unexplored inner recesses mere dotted lines, and place names were much less common than in the more travelled regions. At various places, the chart contained brief descriptions of prominent features to aid the navigator. In addition to notations such as "Apparent Harbour" and "Low land," there were no fewer than seven landmarks deemed to be *remarkable*, including "Remarkable Rock," "Remarkable perpendicular dark cliff," "Remarkable peaked hill," "Remarkable round topped hill," and "A very remarkable black bluff headland." The chart was least accurate in the southeastern part of Baffin Island between Cape Dyer (opposite the narrowest part of Davis Strait) and Resolution Island (at the entrance to Hudson Strait). In this section of nearly four hundred miles, the distribution of land and sea were fanciful and misleading. Both "Cumberland Strait" and "Frobisher's Strait" were vaguely delineated, but as no Europeans had visited these indentations since their discovery by Martin Frobisher and John Davis in the sixteenth century, the Admiralty's cartographers could not portray them with any confidence. In fact, neither was a strait.

When Penny's *Neptune* departed for Scotland in 1839, Eenoolooapik went with him. Speculating on Penny's purpose in bringing the young man back, the *Aberdeen Journal* (13 November) wrote, "Capt. Penny has found that he [Eenoolooapik] has a correct notion of the geography of his country, and is inclined to think he may be able to impart some important information regarding the places where whales 'most do congregate,' now that they seem to have forsaken their old haunts." The Inuk passed the winter in Aberdeen and amused the populace by giving a kayak demonstration in the harbour. He attended dinner parties and balls, visited stores and country lodges, received many gifts, and had his portrait painted by James Cassie, a promising young artist. In the following spring, he looked forward to accompanying Penny back to the Arctic on the whaler *Bon Accord*.

Scottish shipowners were alert to the possibility that the newly discovered gulf (now Cumberland Sound) might provide a tonic for the declining whaling industry, and they were anxious that it be explored and mapped. In November 1839, shortly after Penny and Eenoolooapik arrived in Aberdeen, Kirkcaldy shipowners petitioned the Admiralty to send an expedition to the region to determine whether it would be suitable for commercial whaling. It is a measure of Penny's standing within the whaling community that he was sent a copy of the petition, with the message, "If you have not obtained success you have acquired fame – the country is ringing with the wonders of 'Bobbie' [Eenoolooapik] – his growing acuteness, and with the enterprise of the spirited Commander to whom he is indebted for a peep at the *huts* of Britain."

Whaling interests in Aberdeen also asked the government to send out an expedition to follow up Eenoo's information and make a survey of the bay. The Admiralty, however, decided against sending an expedition, or even interviewing Eenoolooapik. They preferred to leave the entire matter to the whalemen themselves. Shortly before the departure of the *Bon Accord*, they donated the sum of twenty pounds for the purchase of guns, ammunition, and other useful articles for the Inuk to take back to his people. If the attitude of the Admiralty was disappointing, that of the whaleship owners was hardly less so. Penny was instructed to adhere to the usual whaling itinerary and leave the search for Tenudiakbeek until the end of the season.

Penny had good reason to be discouraged. He had obtained important geographical information from Inuit at Durban. He had attempted to reach the alleged bay by boat from the Davis Strait coast. He had interviewed Inuit again and had obtained a rough map of the indentation, drawn by memory by Eenoolooapik. He had brought the Inuk back to Scotland and taken care of him over the winter. He was now about to take him back to the Arctic, to act as guide and interpreter. Yet the Admiralty was unwilling to carry out a survey of the region, and *Bon Accord*'s owners insisted that any exploration be crammed into the tail end of the navigation season. The discovery of a new whaling ground might save the whaling industry from disaster, but neither government nor the whaling interests were willing to take any meaningful action.

The Admiralty had taken one important measure, however. On arrival home in the previous autumn, Penny had submitted the manuscript chart that he had compiled with Eenoolooapik, which depicted Tenudiakbeek and showed its probable relationship to the known coast to the north. On 12 February 1840, the Admiralty's Hydrographic Office published an official version entitled "Davis Strait Cumberland Isle from the Observations of Captain Penny of the Greenland ship Neptune of Aberdeen, and from the Information of Enoolooapeek, an Intelligent Esquimaux 1839."

With the new chart on hand and Eenoo as guide, Penny found the entrance to the inlet, entered it in early August, closely followed by three other ships, and spent almost two months exploring and looking for whales. When the *Bon Accord* and a few other whaling ships entered the sound, they were the first Europeans to enter the gulf since John Davis some two and a half centuries before in 1585. Davis had named it Cumberland Sound after the Earl of Cumberland; Penny named it Hogarth's Sound; and a Newcastle captain called it Northumberland Inlet. It is now once again called Cumberland Sound.

Penny's rediscovery of the sound led to a new and improved Admiralty chart and an expansion of the frontier of arctic whaling. Although the returns were initially disappointing because the whalers were unable to enter the bay at the time when whales were most abundant, Cumberland Sound would come to play a vital role in British whaling a decade later, after Penny introduced the techniques of wintering on board ship and establishing year-round shore stations.

3

The Summer of 1845

A quixotic search for a useless waterway.[1]

Penny had a special reason to be interested in Franklin's expedition. He had received a letter from his former shipmate, Alexander McDonald, proudly announcing that he had been appointed to HMS *Terror* and would soon be traversing the Northwest Passage. A close relationship had developed between the two men since 1835, when McDonald sailed as surgeon with Penny on his first whaling command. They had been shipmates again in 1837 and also in 1840, when Penny took Eenoolooapik back to Baffin Island after his winter sojourn in Aberdeen. McDonald's popular book about Eenoo's experiences in Scotland and the voyage back to his homeland put Penny before the public eye for the first time.

During his years as a medical student, McDonald had relied on summer whaling voyages to earn money for university life. Hailing from Laurencekirk, an inland village several miles north of Montrose, he had been obliged to live in Edinburgh and pay for his lodging, food, tuition, and books. Evidently his financial resources were barely adequate, and in hard times he put the touch on Penny. Writing in December 1836, more than a year after their first voyage together, McDonald confessed that he was a "curious correspondent" who wrote only when he needed something. He was sorry to bother Penny so often, he said, but he needed a loan of a one or two pounds.[2] After Penny sent two pounds in January, McDonald wrote back, "You have added another favour to those already bestowed."[3] Two months later, he appealed for additional help. "I am ashamed," he wrote, "seeing the extent to which you have already assisted me, but I hope the day will yet arrive when I will be able to repay you."[4] That McDonald felt comfortable in asking for money, and that Penny cheerfully granted his requests, says a good deal about the relationship between the two men. Penny was more dutiful and considerate than the shipowner William Hogarth, who neglected even to answer McDonald's letters "concerning some pecuniary affairs" – probably some back wages he was still owed for his services as surgeon.[5]

After winning the John Argyle Robertson silver medal at the University of Edinburgh, McDonald went on to become a licentiate of the Royal College of Surgeons of Edinburgh. Following his voyage with Penny and Eenoolooapik in 1840, he took up medical practice in Laurencekirk, living with his parents and a younger sister. Village life may have seemed humdrum after the excitement of navigating arctic seas, and as he prepared the narrative of his voyage for publication, his thoughts turned to the possibility of a career at sea. After his book was published in 1841, he applied to the Royal Navy, and on 2 September was confirmed at the rank of assistant-surgeon and appointed to HMS *Belvidere*. He was then twenty-three years of age. A month later, he wrote to his father from London, where he was rooming temporarily in a house "full of navy and army surgeons, all from Scotland."[6] He had just purchased his uniforms and other items on credit, and complained that the sum of a hundred pounds would not cover his expenses during the month before payday. On board ship at Spithead, he wrote again. He and his fellow officers had been excited at the prospect of sailing with a flotilla bound for China, where British forces were besieging several coastal ports to protect Britain's drug trade in the so-called Opium War, but to their disappointment the *Belvidere* was being dispatched to the Mediterranean instead. Cigar smoking was discouraged on the ship, he said, but "I am glad of that for I'll soon conquer the habit."[7]

Now, four years later, writing to Penny on 24 March 1845, he said that he had been discharged from HMS *Belvidere* and was about to write the final medical exams that would make him eligible for promotion. And then his most exciting news: "I am already appointed to a ship. I go out in the expedition under Sir John Franklin. The object is to discover the northwest passage. So if you happen to be in Davis Strait this year we may meet. I belong to the 'Terror' Captain Crosier [*sic*]. The appointment is a good one."[8]

The appointment is a good one. How many keen young men thought thus when they left the Thames on board the *Erebus* or *Terror*, fully confident of becoming the discoverers of the long-sought passage through the attic of North America to the Pacific Ocean? A dozen years later, when Francis Leopold McClintock finally located the grim evidence of the expedition's failure and the death of its officers and men, it would become clear that the optimism of the Admiralty, the public, and the expedition members themselves had not been justified. But in the hopeful spring of 1845, McDonald was thrilled to take his place among the confident history-makers and eager to get on with the job. Penny must have been proud of his protégé, and possibly envious.

When Franklin's ships left the Thames in mid-May 1845, Penny's *Saint Andrew* and the other Davis Strait whaling ships were already off the west coast of Greenland, having set off in April with much less fanfare. As HMS *Erebus* and *Terror* were battling headwinds and rough seas off Aberdeen, the *Saint Andrew* was passing Upernavik, more than 2,500 miles ahead.

Penny knew that an early arrival on the Greenland coast did not guarantee an early arrival at the principal whaling grounds off northern Baffin Island, but a late arrival could mean missing an opportunity of getting through the troublesome barriers of ice and crossing Baffin Bay in time to exploit the run of whales on the "West Side." Most of the British whaling fleet (comprising about sixteen vessels in that year) were now held up by impassable ice at the margin of Melville Bay, the most dangerous place of all. The *Saint Andrew* had already secured two whales, a good start for the season, but time was passing. In the short arctic summer, every moment was precious. For the next four weeks, the various captains worked their ships laboriously through the narrow leads (channels in an ice field) and small pools of water that opened briefly among the floes. Rarely was there enough space to sail; most of the progress was achieved by dragging the ships ahead by brute strength, either from whaleboats or from the surface of the landfast ice. The latter technique, known as tracking, made a vivid impression on the surgeon of the Dundee ship *Alexander*: "It was truly a novel scene to behold five companies of forty men each march along the ice singing cheerfully some Scottish lays and marching [in] time to the simple music." But when soft snow lay deep on the ice, he described it as "a truly horrid business."[9]

On one occasion, the power of the pack almost spelled disaster for Penny's *Saint Andrew*. The ship was caught by a berg and lifted onto the ice. Twenty-two of its timbers were broken and twenty-six feet of its keel cut out. Before the crew's clothes could be got out, the ship settled down again, and made water very rapidly. When, afterwards, this vessel was put on the slips at Aberdeen, everyone who saw it was surprised that Penny had been able to bring it home. The ground-tier casks were broken from side to side, involving a loss of twenty tons of oil.[10]

After many severe squeezes in the pack ice and much desperate and back-breaking work by the crews, the whaling ships left the dangers of Melville Bay behind, passed Cape York, and entered the ice-free region of the North Water. On 23 June the surgeon of the *Alexander* wrote, "When we had proceeded as far north as near seventy-seven degrees north we sought away to the west-ward for a passage to the west side. We were now in an ocean of water to all appearances[,] no solid ice seen to the northward."[11] They coasted southward along the mountains of the West Land – "a most splendid landscape, beggaring description."[12] Old-timers in the crew could not recall such an early crossing of the North Water and took it as a good omen, both for their own success in whaling and for Franklin's completion of the Northwest Passage. In the former supposition they were correct but in the latter they could hardly have been more wrong.

By 26 June, all the ships were off Pond Inlet near the northern tip of Baffin Island. The deep inlet itself was still firmly frozen, and the landfast ice extending out beyond its entrance provided a convenient natural quay. The *Saint Andrew* slanted in to the floe edge, and men leaped down to plant the ice anchors. Soon there were ten ships moored there, lined up in a row like horses at a trough. In addition to the *Saint Andrew*, there were the *Alfred* from Bo'ness, *Advice*, *Alexander*, and *Princess Charlotte* from Dundee, *Abram* from Hull, *Caledonia* and *Chieftain* from Kirkcaldy, *Lady Jane* from Newcastle, and *Joseph Green* from Peterhead. From this position, for as long as weather and sea permitted, their whaleboats would venture forth to harpoon whales, kill them with lances, and tow them back to the ships for flensing. A few days later, when the wind increased and a heavy swell made their exposed position dangerous, Penny and the other captains cast off and worked their vessels away from the ice to dodge back and forth in the open water until things improved.

Meanwhile, Franklin's discovery ships were a thousand miles behind, crossing the Arctic Circle abreast of Holsteinborg on the southwest coast of Greenland – much too late to take full advantage of the unusually open ice conditions in northern Baffin Bay.

British ships had been whaling off Pond Inlet for a quarter-century. In the heyday of the 1820s, one whaling surgeon had written, "Saw 41 sail made fast to the land floe about 8 miles to the N of Pond Bay. Every ship spoken was found to have been very successful. Some had caught 14 [whales], others 12, many 8, & only one believed to be clean."[13] Whenever a ship hove into sight and moored to the floe edge, Inuit inhabitants of nearby coastal settlements and hunting camps would travel across the intervening landfast ice to meet the strangers. Although the magnitude of the whaling effort (and the size of the whale population) had declined drastically since the 1820s, ten ships and several hundred men were enough to provide the Inuit with plenty of opportunities for trade and amusement. By the time the ships had moored, the surgeon of the *Alexander* noted that the ice was "litterally [*sic*] covered with natives, men women and children with their dogs and sledges,"[14] all raising a prolonged cry of welcome that ended with a yell so loud that commands given on deck were almost inaudible.

Lots of visiting and good fun followed. The doctors – probably regarded by the Inuit as powerful shamans of the European race – were much in demand to work their magic for various ailments. When one native visitor to the *Advice* took a fit and went into a coma, his companions urged the ship's surgeon to bleed the man. After the blood-letting the patient arose, apparently cured, and in gratitude offered the surgeon the temporary use of his wife.

Differences in cultural attitudes and modes of behaviour are often evident in such early encounters. In the eyes of the surgeon, the Inuit men were tough-looking and inclined to pilfer objects, while the women were "very lewd and immoral, making the most disgusting capers to our sailors, some of who encouraged their abominable maneuvers."[15] One can only wonder what the Inuit said about the sailors. Some of the Inuit were adept at imitation, slyly mimicking excessively stern or arrogant captains behind their backs. Most were keen on music, and some ships' officers assumed the role of entertainers, playing lively tunes on their fiddles. On this occasion, the amateurish bowing of one ship's surgeon produced among the native audience grimaces of such a "truly frightful aspect" that he desisted for fear of driving more men into comas. Later, he brought out the fiddle again for the benefit of two women who were taking a cup of tea in the cabin. "Such a tea party I daresay was never witnessed," he wrote. "After tea I performed on the violin a few reels when they instantly assumed their fierce & savage stance, running their fingers along their head & breasts."[16]

Although the Inuit of Baffin Island responded enthusiastically to the sprightly music of the Scots and Englishmen, they had not yet adopted European instruments. On the Greenland coast, however, where whalemen, explorers, and colonizers had been in contact with the native population for more than a century, the situation was quite different. When a native Greenlander visiting the whaler *Alexander* expressed a desire to inspect the surgeon's violin, the surgeon had handed it over with great reluctance, worried that the man might damage the delicate instrument. He later wrote in his journal, "I indulged his wish by giving him it in his hands, when to my astonishment he put it in tune and play[ed] in capital style the popular hornpipe, 'The King's Head,' and a good many like it[,] all our most popular reels & hornpipes."[17]

Through July and into August 1845, until the fast ice became unstable, the whaling ships remained at the floe edge near Pond Inlet. On 16

August, the Hull whaler *Prince of Wales*, under Captain Dannett (variously spelled), caught up with them, having sailed first to Spitsbergen for seals and then all the way across the Atlantic to Davis Strait and Baffin Bay for whales. As the ship had made its way up the Greenland coast toward Melville Bay in late July, it had encountered HMS *Erebus* and *Terror* working through the ice. Some of Franklin's officers had come on board bearing an invitation from Sir John to join him for dinner on the following day. But what would have been a memorable experience for Dannett did not occur. When an opportunity presented itself to hoist sails and move ahead through the ice, he put business ahead of pleasure, passed up the invitation, and pressed on. Dannett's account stimulated a lot of table talk among the whalemen at Pond Inlet. It was now mid-August, and the general feeling was that if the discovery ships had sailed as early as the whalers, they could have arrived on the West Side by late June, leaving at least two months to proceed through the Arctic Archipelago to Bering Strait.

More than six years later, Penny would be told by Robert Martin, who had commanded the whaler *Enterprise* in 1845, that he too had met up with Franklin's ships in Baffin Bay in late July and been invited to dine on HMS *Erebus*. Like Dannett, he had instead seized a sudden opportunity to go ahead, leaving Franklin's ships moored to an iceberg.

One wonders why Franklin, already late in reaching Baffin Bay, had lingered alongside an iceberg issuing dinner invitations while the whaling captains pressed on for the West Side. The reason may have been inexperience. Whereas British whalers had made more than a thousand voyages to the region since 1818, only three discovery expeditions, comprising a mere five vessels, had sailed through these waters during the same period. Thousands of officers and men in the whaling ports were familiar with the geography of Davis Strait and Baffin Bay, accustomed to the challenges of ice navigation there, and cognizant of the need to press on without delay. No comparable reserve of experienced men existed within naval ranks.

Seven officers of HMS *Erebus* and *Terror* had previous arctic experience, but only four had been in Baffin Bay before. Two of these, however, were the former whaling captains James Reid and Thomas Blanky, who were serving as ice masters. A third was the assistant-surgeon Alexander McDonald, who had sailed there on whaling voyages under William Penny. Aside from McDonald, the sole naval officer on Franklin's expedition with previous experience in the region was Francis Crozier, the commander of HMS *Terror*, who had sailed through Baffin Bay and Lancaster Sound with Parry in 1824–25. Franklin's only experience in arctic waters had been north of Spitsbergen, commanding HMS *Trent* in 1818. He had never been in waters west of Greenland.

Crozier, at least, appears to have been conscious of the need to get past the Middle Pack and reach Lancaster Sound as early as possible. From Disko he wrote to his brother James in early July, "We are, I fear, sadly late."[18] It is conceivable that Franklin, making his first voyage to Baffin Bay, did not fully appreciate the shortness of the navigation season and the need to make strong efforts to get through or around the Middle Ice, and it appears that he was not inclined to follow the example of the whaling masters Dannett and Martin, or to take seriously the advice of Crozier or his ice masters from the whaling trade.

After Captain Dannett reported his contact with Franklin to the Admiralty in the autumn, the news was released to the newspapers. Martin also submitted a report of his encounter, but the Admiralty did not see fit to inform the press about it, possibly believing it would not add anything significant. (Not even Lady Franklin knew about it until six years later, when Penny made it public.) Nothing further was heard of Franklin's expedition, but that was neither unexpected nor alarming. It was assumed that his ships had advanced well beyond the region exploited in summer by British whalers and would not have an opportunity to get a message out until they reached Alaska. They might be out of touch for some time.

Penny and his men enjoyed a successful season. The *Saint Andrew* returned to Aberdeen in the autumn with the blubber of eighteen or nineteen whales, enough to yield 150 tons of oil.

4

Whaling Interlude

We leave our sweethearts and our wives
All weeping on the pier;
Cheer up, my dears, we'll soon return,
'Tis only half a year.[1]

As a veil of silence descended over Franklin's Northwest Passage expedition, the whalers went about their usual work. In the summer of 1846, Penny took the *Saint Andrew* back to the Davis Strait whale fishery, with Peter Cormack Sutherland serving for a second season as ship's surgeon. Sutherland's experience on a whaler ignited his interest in the Arctic, and he would later participate in other whaling voyages and two Franklin searches, including Penny's official search expedition of 1850–51.

By the first week of May, the *Saint Andrew* was off Disko with several other ships, including the ancient *Truelove* from Hull (commanded by Penny's friend John Parker), the *Traveller* of Peterhead (on which Penny had twice served as mate in the 1830s), and the Dundee ship *Advice* (which Penny would command a few years later). Other ships soon caught up, including the *Pacific* from Aberdeen, *Alfred* from Bo'ness, *Caledonia* and *Regalia* from Kirkcaldy, *Abram*, *Lord Gambier*, and *Prince of Wales* from Hull, *Alexander*, *Horn*, and *Princess Charlotte* from Dundee, and *Lady Jane* from Newcastle.

The whaling fleet made slow progress northward through closely packed floes. Three weeks later, when a field of ice was driven in upon them, the *Saint Andrew* and seven other ships embedded their ice anchors in the solid buttress of an iceberg to wait out the pressure. When the ice slackened, they cast off their lines and worked northward past Black Hook, lowering boats for whales on one occasion but without success. In early June, after hazardous sailing and strenuous tracking, they moored to the floe edge for a day or two to wait out a gale thick with blowing snow. When the storm subsided, they advanced into a dangerous region in which

Andrew Hamilton, surgeon of the *Pacific*, observed icebergs "as numerous as trees in the forest."[2] This formidable barrier remained solid during the entire month, preventing the pack ice from dispersing. On the first of July, "there being no probability of getting through Melville Bay,"[3] the captains resolved to turn back and seek a crossing further south.

After they had crossed to the west side of Davis Strait, Penny turned south and entered Cumberland Sound. His old friend Eenoolooapik came on board to assist in the whaling operations. The ship's surgeon, Peter Sutherland, was an enthusiastic and talented naturalist and ethnographer, and during the two-month cruise in Cumberland Sound he undertook detailed researches on the Inuit who resided at Eenoo's birthplace, Kingmiksoke (variously spelled). Visiting each of the sixteen huts in the settlement, he enumerated the names and family relationships of 111 individuals and recorded their ages, heights, and many facts of a highly personal nature. Sutherland was a stranger among the people and almost certainly had little or no fluency in Inuktitut. Probably Eenoo made the rounds with him, introduced him to each family, explained the purpose of the survey, posed the questions, and translated the responses into English. Yet when Sutherland published the results a decade later in the *Journal of the Ethnological Society of London*, he did not acknowledge the help of the young Inuk.

His census-taking, the first at any Inuit settlement north of Labrador, revealed the salient characteristics of a community of nomadic hunters and their families. It was a young population, all under fifty-one, with an average age of only twenty-two. Owing to the mortality associated with hunting, males were in a distinct minority (twenty-three adult males to thirty females). Watercraft were essential to their existence. There were twenty-two kayaks for hunting sea mammals; almost every male aged nineteen or older possessed one. In addition, the community contained six umiaks – usually called "luggage boats" or "women's boats" by explorers. These were mainly used for shifting camp or transporting heavy loads of meat from the site of a kill.

Like Penny, Sutherland admired the Inuit. "We have too long judged races in heathen darkness ... by our own laws," he wrote; "it is worthy of our intelligence now to make ourselves thoroughly acquainted with their mental constitution and, having done so, we shall be better able to return a just verdict when points in their character are disputed."[4] He had been greatly impressed by the musical talents demonstrated by Greenland natives under the tutelage of their Danish administrators: some men played the violin "with great taste and correctness,"[5] while the women sang choral works "with a degree of harmony superior to what we find among other communities, with far greater pretension, in our own country."[6] He also praised Inuit powers of observation, their courage and skill in hunting, and their ability to anticipate weather changes. Sutherland's attitude toward the Inuit was much more tolerant and benevolent than that of Penny's former surgeon Alexander McDonald, now on his way through the Northwest Passage with Franklin; McDonald's limited vision discerned only "dismal solitudes of snow . . . where mental cultivation is unknown."[7] Sutherland expressed a contrary view: "We have reason to hold this people in esteem, on account of their mental endowment."[8]

The *Saint Andrew* returned to Aberdeen with the blubber of six whales, yielding of sixty tons of oil. Although the routine voyages of British whaleships had appeared to offer some prospect of obtaining news of Franklin's ships, these hopes had not been realized. Ice had blocked their customary route and forced them to return southward and cross to the West Side at a lower latitude. They had failed to reach the mouth of Lancaster Sound, where messages left by the missing expedition would probably be found.

* * *

Concerns about HMS *Erebus* and *Terror*, and suggestions that a relief expedition be organized, were first voiced by Sir John Ross, who claimed he had told his friend Franklin that if nothing was heard after two years he would set out to look for him. He tried to get official wheels creaking into motion in the fall of 1846, writing to the Admiralty and offering to lead a search party in the following summer. But the lords commissioners of the Admiralty, influenced by the strong opinions of its permanent secretary, Sir John Barrow Sr, who knew perfectly well how to hold a grudge, had not looked with favour upon Ross for thirty years, nor did they now. They rejected Ross's generous offer of service, and when he repeated it more than once in the new year, they stood by their earlier decision. Ross later blamed his failure to obtain a command on "a malevolent influence at the Admiralty," by which, of course, he meant Barrow.[9] The lords commissioners also turned down a plan proposed by Dr Richard King, another *persona non grata* in official circles. But these and other search proposals may have stirred them to an overdue contemplation of relief measures. In February 1847, Parry advised them to offer incentives to whaling ships to look for cairns and flagstaffs where Franklin might have left messages. As twenty whalers were about to start for Davis Strait, the Admiralty's secretary wrote to shipowners and captains in seven ports:

> My Lords being desirous to trace as far as possible the progress of the Expedition under Sir John Franklins [*sic*] orders since its

> departure from Disco . . . are anxious to engage the attention and secure the assistance of the masters of the several whaleships in their enquiry, and for their purpose are disposed to offer a reward to such masters of whaleships and others assisting in the search who may succeed in obtaining any information or record of the progress of the Erebus and Terror through Lancaster Sound, and to the westward.[10]

Penny, commanding the *Saint Andrew* again, crossed the Atlantic and headed up the west Greenland coast. Somewhere north of Disko, Captain Dring of the *Mary* from Peterhead delivered a letter from Sutherland, Penny's surgeon during the previous two seasons. Writing in Scotland, Sutherland reported that his father had died a month before and that he intended to spend the summer in Aberdeen, after which he would take the degree of MD at King's College and then a diploma in surgery at the University of Edinburgh. He had become interested in scurvy, he told Penny. One hospital had admitted seventy cases in a week, and his own hospital had more than twelve on hand. Doctors had found that an ounce of lime juice and sugar daily with a "nourishing diet" brought the patients back to health rapidly. The disease was "evidently produced by want of proper food," and he made particular mention of "that invaluable antiscorbutic vegetable the *Potato*."[11]

British whalers normally included potatoes among their provisions. Potatoes kept well, and the crews considered them essential companions to meat and to the fresh cod they often obtained in Davis Strait by hook and line or through barter with commercial fishing vessels. Unfortunately, however, airborne spores of *Phytophthora infestans*, the potato fungus that had spread through Europe and devastated crops in Ireland, had now reached whaling ports on the east coast of Scotland, and the supply of healthy potatoes had been greatly reduced, with disastrous effects. When the whaling ships returned from the whaling grounds in the autumn, the *Aberdeen Journal* (3 November 1847) – after asserting that "the tar at sea, like Paddy ashore, declares that there is nothing like the potato" – wrote, "The scurvy broke out in all the ships at Davis Strait this year, and Captain Reid [of the *Pacific*] says he can trace it to no other cause than the want of potatoes." The ravages of scurvy were apparent in all the ports. Most of the crew of the *Advice* came down with the disease, and two men died before reaching Orkney. The *Princess Charlotte* reached Orkney with five men ill, but fresh food obtained on the island brought them back to health. The *Chieftain* reached Dundee with twenty-five to thirty cases, several of whom were so weak that they had to be carried home in carts.

The *Saint Andrew* did not reach Pond Inlet until 23 July. Penny had not forgotten the Admiralty's request for information about Franklin. As he later told John Barrow Jr, "I determined to proceed to Lancaster Sound, both with a view to the capture of whales and in search of H.M.S."[12] Unfortunately the weather turned against him: "I contended for a week against adverse winds and a strong swell." Failing to make any westward progress into Lancaster Sound, Penny coasted southward along Bylot Island to Pond Inlet. "Being a little acquainted with the 'Esquimaux' language," he reported, "I made every inquiry of the various tribes met with in 'Ponds Bay' but could procure no information respecting Sir John Franklin who I think must have attained a very high longitude [west]"[13]

Reaching Cumberland Sound in late August, Penny sought out Eenoolooapik but was saddened to learn that his Inuit friend had died, apparently of consumption, leaving a wife and one child. Eenoo had become a skilful entrepreneur among his countrymen, collecting whalebone and then trading it to Penny when he arrived in late summer. According to the *Dundee Advertiser* (19 November 1847), he had always negotiated "a pretty stiff bargain" with the whaling captain who had taken him across the sea to Scotland.

Penny spent seven weeks cruising for whales in Cumberland Sound before wintry blasts and the formation of young ice heralded the end of the navigation season. In mid-October, the men happily hoisted sail for home, satisfied with their catch of six whales. The blubber would make about seventy-five tons of oil. The *Saint Andrew* came into port with the tail of a whale mounted over the stern to indicate its success.

Penny's brief attempt to find traces of Franklin in Lancaster Sound in 1847 was the first of many British and American search efforts. It was a spontaneous venture, neither sanctioned by the owners of his vessel nor supported by the Admiralty. Although it did not involve any substantial departure from the usual whaling grounds, his search did delay his pursuit of the whales, which had already started on their southward migration. Wind and waves conspired against him, rendering his efforts futile, but at least he had tried. From this moment on, the challenge of finding Franklin would not be far from his mind.

Many people, Lady Franklin among them, had hoped that whaling ships sailing to the Davis Strait fishery would obtain some news during the summer of 1847 because their principal fishing ground was off Pond Inlet, close to the mouth of Lancaster Sound, to which Sir John Franklin had first been directed. But when the ships returned in the fall, their captains had nothing to report. They had seen no evidence of Franklin's ships having passed that way.

* * *

The year 1848 marks the beginning of the long series of government and private expeditions sent out specifically to search for Franklin. Complacency had by now been replaced by anxiety, and after considering several search proposals, the Admiralty initiated the first measures.

The problem of finding Franklin was basically one of geography. From Greenland in the east to Bering Strait in the west along the 70th parallel is a distance of almost three thousand miles, and from the continent northward through the labyrinth of the Arctic Archipelago to the tip of Ellesmere Island is about one thousand miles. As water bodies were frozen for most of the year, HMS *Erebus* and *Terror* might have been trapped in the ice, crushed, or driven ashore almost anywhere in that immense territory. Conceivably, they might even have attained a further point before being beset, perhaps somewhere in the vast circumpolar Arctic Ocean, which many believed to be navigable. Most likely, however, they were still within the Arctic Archipelago, a complex region bisected by the Parry Channel – the east-west corridor that Parry had used in 1819 to travel some eight hundred miles west from Baffin Bay to Melville Island.

What geographical information did the Admiralty have in 1845 when they sent Franklin out to complete the Northwest Passage? The contemporary historian John Brown included in his 1858 book, *The North-West Passage*, a copy of the chart that Franklin carried with him.[14] It illustrates the state of geographical knowledge about the Arctic Archipelago at the time and reveals the magnitude of the challenge faced by Franklin and those who would search for him.

On the chart, two parallel east-west waterways are represented in reasonable detail – the area adjacent to the continental mainland, and the Parry Channel approximately four hundred miles to its north. Although Parry had seen islands further north during his voyage (including parts of Cornwallis, Bathurst, Byam Martin, Sabine, and Melville Islands), he had observed almost nothing to the south, only Cape Walker and a small portion of Banks Island. With the exception of these two widely separated spots, the entire space between the two latitudinal passages and west of Prince Regent Inlet is *entirely blank* on Franklin's chart. No western boundary is indicated; some contemporary maps even showed an extension of the "Polar Sea" reaching in all the way to Boothia Peninsula.

Did this "unexplored quadrilateral"[15] really consist of open sea? Or did it contain islands? If there were islands, were any of the passages among them navigable by deep-draft ships? The Admiralty had assumed that some navigable passages would be found and that Franklin would find a way through.

Unless Franklin had met disaster in Baffin Bay after separating from the two whaling ships in late July 1845, or taken a detour northward into Smith Sound or westward into Jones Sound, he had almost certainly obeyed his instructions and proceeded through Lancaster Sound to Cape Walker, and then turned southward into some channel that would lead him to the continent. Once there, his intention was to turn west and follow the handrail of the mainland coast (much of which he himself had explored and mapped more than two decades earlier) all the way to Bering Strait, the door into the Pacific.

Despite the lack of knowledge about the geography of the "unexplored quadrilateral," the Admiralty and Franklin considered the task straightforward, and their optimism was reflected in the press. There were concerns about the ice, of course. Franklin himself had stated, "I conceive that the greatest impediments from the ice will probably be between the 95th & 125th degrees of longitude."[16] As his statement embraced a territory fully six hundred miles wide, it was somewhat vague, yet it was correct. His ships were destined to be inextricably trapped in the ice at a longitude of approximately 98° W – between the two meridians he had mentioned.

Not having provided Franklin with explicit instructions on how to proceed if he encountered an impassable barrier of land, shallow water, or impenetrable ice, or if his ships happened to be wrecked, the Admiralty was faced by unanswerable questions. Would he turn back to Baffin Bay or probe in another direction? (They had mentioned Wellington Channel, leading northward, as an alternative if all other routes were blocked.) If his crews were forced to abandon their ships and make their own way out of the Arctic, would they gravitate toward the supply depot left by Parry at Fury Beach? Would they attempt to reach whaling ships near Pond Inlet? Would they trudge toward distant fur-trade posts south of the tree line? Nobody knew.

In an attempt to cover most eventualities, the Admiralty adopted a three-pronged search strategy in 1848. Some searching ships would follow Franklin's intended route through Lancaster Sound into the Arctic Archipelago. Others would sail around Cape Horn and north to Alaska in case Franklin had emerged from the passage. Small overland parties would travel down rivers to search the continental coast in between.

The eastern division consisted of HMS *Enterprise* and *Investigator* under the overall command of James Clark Ross, Sir John's nephew and an experienced polar explorer in his own right – the man Lady Franklin had wanted to lead the search. James Ross left England in mid-May (late, like Franklin), passed Disko, crossed Baffin Bay, entered Lancaster Sound, and took up winter quarters at Port Leopold at the northern tip of what is now called Somerset Island on 11 September 1948. On that day, HMS *Herald* was approaching the north coast of Alaska, HMS *Plover* was in the North Pacific heading toward the Aleutian Islands, and a party under Sir John Richardson and Dr John Rae, having already searched the central mainland coast, was retreating to Fort Confidence on Great Bear Lake; from there Rae intended to extend the coastal search in the following summer. As Lady Franklin's biographer Frances Woodward has written, "The macabre game of hide-and-seek had begun."[17]

The Admiralty, the press, and the general public were confident that such an ambitious and extensive search could not fail to reveal the whereabouts of the missing explorers. If Franklin and his men needed any help – which most people still doubted – one of the expeditions in the field would surely be able to provide it. Lady Franklin, however, was not so sure. Although the Admiralty measures were encouraging, she was not content. In February she had written to William Scoresby, the extraordinary whaling captain who had written the definitive book on the arctic regions, studied at two universities, been elected to the Royal Society for his scientific achievements, and then entered the church. "I beg to apologise for the liberty I am taking in addressing you," she wrote, before confessing to feeling a "great uneasiness" about the forthcoming search expeditions.[18] If they failed to locate the missing ships during the coming summer, she told him, the search would have to be postponed for a further year, by which time her husband and his men would be at the end of their three-year supply of provisions. It was imperative, therefore, to do as much as possible in the short navigation season ahead. She continued, "It appears to me that my private resources might be made available for this purpose, and I venture to ask of you who have so much knowledge in these matters how I could best employ a couple of thousand pounds or more, if necessary, in the relief or rescue of the missing ships or any portion of their crews – Is there any scale of rewards that could be offered with advantage to the Captains, Crews or Owners of Whaling Vessels?"[19] She had considered two possible courses of action, and asked Scoresby which would be more feasible. "Do you think that a vessel could be induced to go on purpose, looking only into those parts which will not interfere with the course of Sir James Ross's ships, or would it be worth while to place extra Boats & Crews under the command perhaps of Lieutenants in the Navy who would doubtless volunteer for the purpose, on board the Whalers, whose sole business should be to explore?"[20]

The letter was the first of many between Jane Franklin and William Scoresby and marks the beginning of her active involvement in the search for her husband. She had come to realize that her own ideas, strengthened by the practical advice of experienced whaling masters and supported by her own money, might play a useful role by supplementing the efforts of the Admiralty. She saw clearly the many advantages of a private venture that could operate independently, unfettered by time-consuming and restrictive governmental procedures, free to set its own objectives and to adopt quickly whatever means were necessary to attain them. Her growing commitment to the search, her suggestion of using whaling ships, and her willingness to sponsor private searching efforts were setting her on a course that would steadily converge with that of William Penny. While she was thinking that whalers could take a more active part in the search, he was looking for ways to undertake a search without the frustrating restraints imposed by his shipowners. Their common objective – to find Sir John and his companions – would soon bring them together in a powerful partnership.

As a result of her correspondence with Scoresby, Lady Franklin decided to offer financial incentives to encourage whaling ships to sail beyond their usual grounds in search of the missing ships and crews. Any ship that discovered, communicated with, and afforded relief to the missing expedition, in Prince Regent Inlet, Admiralty Inlet, Jones Sound, or Smith Sound, would receive the sum of £1,000, the money to be divided among the owner, captain, and crew. Another £1,000 would reward the men of any ship that made special efforts early in the season to locate Franklin. Unfortunately, however, her announcement was published in late March and failed to catch the whalers before they left for Davis Strait.

Lady Franklin found some comfort in the departure of James Ross's search expedition in May. She liked the man, had supported his application at the Admiralty, and may have perceived great significance in the fact that he had once commanded a highly successful expedition to south polar regions with the same two ships in which her husband was now sailing through the Northwest Passage. On that occasion, Ross had been away for four years, during which he circumnavigated the continent of Antarctica, and after returning home safely he had been knighted for his achievement. James Ross was a cheerfully optimistic man, just the sort one would like to have on such a mission, although Lady Franklin may have considered him a trifle overconfident when he glibly promised that he would return from the Arctic with Sir John in four months. Four months passed after he left England, then six, then eight. James Ross still did not reappear. News came back with the whaling ships that he had not managed to reach Lancaster Sound until 28 August, which would have left almost no time for searching before freeze-up.

When no further news arrived, it seemed fairly certain that HMS *Enterprise* and *Investigator*, like the *Erebus* and *Terror* before them, had entered the Arctic Archipelago much too late and were now enclosed somewhere within its thickening winter ice cover. Yet Ross might be in a good position to search effectively from winter quarters by dispatching man-hauling sledge parties in late winter. It was also conceivable that the expeditions sent to Alaska and the Mackenzie River had already encountered the missing ships. News travelled so slowly.

* * *

In the 1848 season, while James Ross headed for Lancaster Sound, Penny took the *Saint Andrew* to the Greenland whale fishery in late winter. The waters off Spitsbergen, where he had begun his seafaring career at the age of twelve, had experienced a resurgence of popularity and were again attracting more whaleships than the Davis Strait fishery. Penny is also said to have sailed to Davis Strait in 1848, but whether he visited the two whaling grounds in succession that year or made two separate voyages is uncertain.

He may have been back home in Aberdeen in time to share the great excitement surrounding the visit of Queen Victoria and her German-born consort Prince Albert, who passed through the city on their way to Balmoral Castle in early September. A week before their arrival, the newspapers were already talking of little else. News of foreign wars, financial matters, crime, cholera, and the search for Franklin all faded into the background as plans for the gala occasion were outlined in detail. Six royal carriages, eight magnificent horses and their grooms, some luggage vans, and members of the household staff had left Windsor Castle for London to travel north on a steamer (appropriately named the *Earl of Aberdeen*) and would arrive on Monday. On Tuesday they would proceed to Balmoral to prepare for the arrival of the queen. A hundred soldiers of the 93rd Highlanders were already there.

Meanwhile, a solid but temporary wooden jetty and a triumphal arch were under construction at Waterloo Quay, on the north side of what became known as Victoria Dock, and a stage that could accommodate upwards of fifteen hundred persons was approaching completion nearby. From Victoria Dock a railing had been erected along the quay as far as Marischal Street. The queen and her party would be travelling in the royal yacht *Victoria and Albert*, escorted by several other ships – the Admiralty's yacht *Black Eagle*, the steam-sloop *Virago*, the screw schooner *Sharpshooter*, the Dover steam packet *Vivid*, and the steam yacht *Undine*. This impressive flotilla was scheduled to arrive in time to catch the high tide about 0900 on Friday to enter port. Its approach would be observed by a lookout in the lighthouse on Girdleness, the hilly point just south of the harbour entrance, and telegraphed visually to the city, where three canons would be fired and city bells rung for half an hour. When the ships were secured in Victoria Dock, the queen and her party would step ashore, receive a royal salute, and be welcomed by a delegation of civic officials, after which they would proceed through the main part of the city and on to Balmoral.

The *Aberdeen Herald* of 2 September 1848 wrote that Aberdeen was determined to show by the warmth of its welcome that "Scotland has *not* degenerated from the loyal and patriotic feeling which has ever characterized her as a nation." Bitter memories of battles against invading English armies a century earlier were set aside. The premature arrival of Queen Victoria's squadron almost scuppered the careful preparations, but after an initial panic, some hectic last-minute work made everything ready in time. The queen and Albert landed, were welcomed with speeches, and set off on their procession through the town, accompanied by civic officials and a guard of honour. From the quay they proceeded up Marischal Street to Union Street, and westward through the main business district to Wellington Place, then southwestward as far as Cuparston Place, where the official escort bid the queen adieu as she continued at last into the countryside towards Balmoral and a little peace and quiet. Attempting to estimate the number of spectators, the *Aberdeen Herald* (9 September) concluded that from eighty thousand to a hundred thousand people had seen the queen, which it found rather surprising considering that the city contained a total of only seventy thousand residents. It concluded that "the enormous crowd was to be accounted for by a most unprecedented immigration from the country parishes."

Riding up Marischal Street in their carriage, Victoria and Albert passed right in front of the Penny home. One can imagine Margaret, and perhaps William as well, leaning out an upper window for a glimpse of the procession – the royal couple, the guard of honour, and the city officials. Penny was a staunch nationalist, not merely in a narrow Scottish sense but in a wider British context. A few years later, he would speak out strongly in defence of British interests in the North American Arctic. If he was in Aberdeen that day, he would have counted it a great honour to witness the queen's passage.

In November, as the *Saint Andrew* lay alongside the quay in Aberdeen, a young boy was visiting the ship. Like Penny during his boyhood, he was fascinated by everything he saw, especially the forest-like canopy of masts, yards, and rigging rising skyward, but he should have been watching where he would next place his foot, for suddenly he tripped and went headlong over the side. Penny, who was on board at the time, either saw the incident or heard the splash. According to the *Aberdeen Journal* (8 November 1848), he "immediately threw off his coat, leaped into the tide, and succeeded in keeping himself and the poor little fellow afloat, until a boat came to their assistance." His rapid response in the situation was characteristic of the man.

* * *

After the end of the 1848 whaling season, Penny's relationship with the owners of the *Saint Andrew* deteriorated. Some dispute over policy or conflict of personalities led to a serious falling out. The nature of the matter is unknown, but it is possible that Penny indicated that in his next whaling voyage he would make another effort to find traces of the Franklin expedition, and the owners objected. He decided to seek employment elsewhere and in November applied for command of the Dundee whaling ship *Advice*, whose captain had just been discharged by the Tay Whale Fishing Company. The company's manager, John Hume, told Penny that the directors would like to hire him but wished to have a personal interview with him first. Penny travelled to Dundee for the meeting, and on 1 December the minutes of the company's committee of management recorded the following:

> The manager stated to the meeting that he had as requested written Capt. Penny of Aberdeen (whose letter offering his services as master of the ship Advice was before them) to come here to have a personal interview with the committee, and that he was now ready to wait upon them.
>
> Capt. Penny, when introduced, stated that he had been fourteen years as a shipmaster chiefly in the whale fishery, and that he was ready to take command of the "Advice," on the same terms as the other masters from this port, engaged in the whale fishery, namely £7 per month, 3 Guineas for each measurable fish, and 3 Guineas for each ton of whale fins [slabs of baleen], & six percent on the divisible profits, exclusive of premiums of insurance.
>
> The meeting after conversing with Capt. Penny, unanimously agreed to appoint him master of the Advice, and recommended him to be on the outlook for harpooners and other officers, for that vessel, the time of sailing to be fixed at a future meeting of the committee.[21]

Having secured a command at Dundee, Penny notified the manager of the *Saint Andrew* that he was resigning: "In consequence of various communications which I have received and from reports generally current in Aberdeen of measures to be adopted by some of the owners of the 'St. Andrew' and of their expressed feeling toward myself – I have deemed it necessary to accept a situation elsewhere. I will thank you therefore to intimate to the owners that I continue in their employ only until the 1st Janry 1849."[22]

On 11 January 1849, the managing committee of the Tay Whale Fishing Company, in consultation with Penny and Captain Deuchars of the company's whaler *Princess Charlotte*, established Davis Strait as the destination of their two ships and set 10 March as their sailing date. Two weeks later, the committee took out £4,000 insurance on each vessel, and "the masters of the two ships (who were present) were instructed that in the event of their meeting with any of the government discovery ships, to give them every assistance in their power, *but not to deviate from their usual fishing stations.*"[23] In the eyes of the company there was no doubt whatever about the priorities of its captains and ships. Their first responsibility was to seek whales, not missing explorers. Any searching effort on the part of its captains could detract from whaling operations and jeopardize the commercial success of the venture.

To the Admiralty and to Lady Franklin, however, the regular voyages of whaling ships to Baffin Bay, where Sir John had last been seen, appeared to be an excellent means of obtaining news. Both had announced financial incentives for whalers to depart from their usual whaling grounds to seek information about the missing expedition. In 1848 none of the whaling ships had found any evidence of the missing mariners, however. Late announcements of the rewards were partly to blame, and perhaps the financial incentives were not attractive enough. For the 1849 season, the Admiralty decided to reward any private parties that provided assistance to Franklin's ships and crews with an amount up to the immense sum of £20,000. Lady Franklin increased her offer as well: "With the view of inducing any whaling ships which shall resort to Davis's Strait and Baffin's Bay to make special efforts in search of the expedition under the command of Sir John Franklin, I hereby offer the sum of three thousand pounds (3,000 £.), or a proportion thereof, according to the services rendered, to such ship or ships as, departing from the usual fishing-grounds, shall discover, and, if needed, afford effectual relief to the above expedition, or to any portion of it."[24]

Despite such substantial incentives by the Admiralty and Lady Franklin, the whaling companies were directly opposed to their captains sailing and searching beyond their usual whaling grounds. The shipowners had to proceed cautiously, however, because they did not want to be publicly perceived as greedy investors who placed their pocketbooks ahead of humanitarian concerns, or as uncooperative businessmen who undermined the government's emergency measures. Therefore Penny's Dundee owners agreed to let him carry Admiralty messages, Captain Reid's owners in Aberdeen allowed him to carry coal, and Captain Parker's owners in Hull permitted him to take coal and provisions. The messages, coal, and provisions were to be deposited somewhere on the south coast of Lancaster Sound in case the *Erebus* or *Terror*, or any of their men, happened to pass that way.

* * *

Nobody was very surprised that since 1845 the whalers had not seen HMS *Erebus* and *Terror*; by this time they should have progressed beyond the limits of the whaling grounds, deep into the complex archipelago lying poleward of the North American continent. Even if the expedition had succeeded in reaching the western end of the passage, it would take a long time for the news to reach England. But it did seem strange that none of the whaling ships operating in the mouth of Lancaster Sound and off Pond Inlet had retrieved any cairn messages left by Franklin. Beyond the range of human vision the only way to communicate, other than by message balloon or drift bottle (neither of which had much chance of success), was to deposit a written document in a cairn where someone would be likely to find it.

Throughout the world's oceans, specific locations were used as letter-boxes by whaling ships and trading vessels. On a ship away from port for several years, the captain and crew, eager to send news home, would anchor at a frequently visited island and leave messages in a prominent cairn or other structure. Subsequent visitors would sort through the letters at the spot, pick up ones they could deliver or carry to another distribution point closer to their destinations, and leave the remainder for ships that would follow. Such drop points were normally established at uninhabited islands where local people would not disturb their contents. The use of Lancaster Sound by whaling ships and discovery expeditions, and the absence of permanent human settlements, clearly favoured the use of message cairns. The ease of spotting unnatural elements in the treeless landscape meant that prearranged drop sites were not necessary. A man-made structure of rocks tall enough to be clearly visible from the sea, surmounted by a pole, flag, or other marker and containing a message within a waterproof bottle, can, or metal cylinder, could be erected on any prominent headland or other conspicuous topographical feature. This communication system had weaknesses, certainly. Cairns could easily be overlooked if darkness, fog, rain, or snow obscured the visibility; mirages could deceive searchers; inuksuit erected by Inuit hunters could be mistaken for cairns built by Europeans. Furthermore, a cairn message could never provide real-time communication; if the message found inside had been deposited months or even years earlier, it might be too late to take action appropriate to the note's content.

As concern for the safety of the Franklin expedition intensified, one particular cairn came to play a significant role in stimulating searching activities. Thomas Lee Sr, commanding the Hull whaleship *Prince of Wales* (one of the last ships to have had contact with HMS *Erebus* and *Terror*, under a different commander), reported that during the summer of 1848 he had entered Jones Sound and sailed west for more than a hundred miles. As no Europeans were known to have penetrated that waterway before, his achievement was of great interest, and it raised the possibility that Franklin had taken that route westward instead of Lancaster Sound. Lee provided what appeared to be evidence: he claimed to have sighted a cairn on the shore.

Early in 1849, Lady Franklin heard about the episode while visiting Hull, and she managed to meet and interview Lee and his son. Their testimony on this vital matter was very disappointing. In a letter to William Scoresby, Sophia Cracroft wrote:

> The Lees, father & son, could give no satisfactory account of where they had been, tho' they were positive it was in some place *north* of Lancaster Sound, & that they ran some 150 miles westward f^m the entrance in Baffin's Bay. As to the cairn, they did not seem to have, or wish to have anything to say ab^t it. Some of their men *thought* they had seen one, & that was all they admitted. They probably, & with much reason, expected that to admit its existence wd be to incur blame for not examining it.[25]

Lee's logbook contained no reference to the cairn and no details about the ship's alleged course into Jones Sound. His information was so vague that it was impossible to tell where the ship had been and whether or not a cairn existed. But until its existence could be proved or disproved, Jones Sound would remain a region of considerable interest to search expeditions.

5

Concern Mounts

A cry for help from 130 men has come to us on the northern gale now blowing.[1]

By the beginning of 1849, concern for the missing expedition had become widespread and intense. Special prayers were read in churches for the missing men; official dispatches, articles, and letters from readers appeared in newspapers. Franklin's overland expedition in 1819–22 had been marked by starvation, murder, and cannibalism, and he had emerged as something of a national hero. Was an even greater catastrophe now looming? Was he about to become a martyr in the cause of geographical discovery? Sensing the possibility of a newsworthy disaster unfolding, and eager to exploit the British fascination with heroic failures, the press laid the groundwork for such an outcome by giving more space to the searching efforts.

More and more, the search for Franklin was perceived as a military operation in which *gallant* men sought *gallant* icebound explorers whose survival time was running out in the *desolate*, *lifeless*, and *hostile* arctic region. So frequently were the officers (rather than the sailors) described as *gallant* that the adjective became a tiresome cliché. Although whalemen, fishermen, and merchant seamen also came face to face with the fury of the oceans during their frequent voyages, only the naval officers seemed to warrant the term, and they were accorded the honour whether they had done anything to deserve it or not. "We were called heroes long before we had earned our laurels," admitted Sherard Osborn, a naval officer who would participate in two search expeditions.[2] A reader calling himself "Wide Awake" wrote to the editor of the *Nautical Standard* (10 January 1850): "We want a new vocabulary, Mr. Editor – Lord bless you, what would they think of me down here did I call Jok Jubbins, commanding the Collier . . . as the 'Gallant Captain' or 'Gallant Master!' What the devil is gallantry? . . . Now we have various sorts of gallantry and daring, but what disgusts me is the constant harping on the same note; – the gallant officer[,] the gallant captain and merely because he has a parchment commission and is entitled to a pair of swabs [epaulets]."

In early March 1849, as Penny was making preparations for his forthcoming whaling voyage on the *Advice*, he received a letter from Lady Franklin. She knew that in 1847 he had attempted on his own initiative to enter Lancaster Sound and look for evidence of the missing ships, and she was aware that he was about to carry Admiralty messages to deposit in Lancaster Sound. She enclosed a letter for her husband and asked if Penny could take it with him in case he encountered Sir John during the summer.

Penny immediately wrote back, "Rest assured that every effort in my power shall be used to gain intelligence of him who I have no doubt will be restored to you by the end of October."[3] It was much too confident, too optimistic a statement. He meant to encourage Lady Franklin, but one doubts that she was deceived by his unwarranted assurance. James Ross had made a similar promise before he departed for the Arctic, but nothing had been heard of him for nearly a year. Undoubtedly she realized that on a commercial whaling voyage Penny would not be not free to depart widely from his customary itinerary or spend much time looking for a missing naval expedition. Nonetheless, she respected his wide experience in icy seas and sensed that he would give his best effort in the cause. This was the beginning of a close relationship between Jane Franklin and William Penny, two headstrong and impassioned individuals of boundless energy who, once they had decided upon a goal, pursued it with unrelenting determination. For Lady Franklin, Penny would turn out to be an effective tool in the quest for her husband. For Penny, she would prove to be the means by which he could assume a major role in the search.

As Lady Franklin became increasingly involved in the search, she became a symbol of the anxiety felt by all the relatives and friends of the missing men. Her actions were reported in the newspapers, and she was widely applauded for the forceful and eloquent appeals she made to the Admiralty, to influential men in government, and to whaleship owners and captains. Acutely conscious of the value of publicity, she used it to excellent advantage. Sophia Cracroft told William Scoresby that her aunt believed that influencing public opinion was vital. When she badgered people to enlist their support, she often sent copies of her letters to the newspapers to put matters on record in plain view. When she took the extraordinary step of writing to the American president Zachary Taylor, entreating his "great and kindred nation to take up the cause of humanity which I plead,"[4] her letter – which her biographer has called "an epistolary masterpiece"[5] – quickly reached the House of Commons and was reprinted in newspapers on both sides of the Atlantic.

She was tireless in suggesting ways to make the search more effective and in imploring people to act on her suggestions. Both her stepdaughter, Eleanor, and Sophia Cracroft helped her in this work. At times the house must have resembled a secretarial school. In January – "mama" being too busy to write – Eleanor had penned a letter to Scoresby, telling him that a committee was about "to meet at the Admiralty, for considering Arctic matters" and that he could attend if he were in town.[6] In late April she wrote again, thanking him for agreeing to give one or more public lectures to raise funds for a private expedition. Admiral Beaufort, hydrographer of the navy and a confidant of Lady Franklin, sent Scoresby some charts and advised him to draw sketch maps on a blackboard with chalk to illustrate his lectures. The gesture not only reveals a revolutionary advance in the techniques of lecturing with visual aids but shows the extent to which certain members of the "establishment" were willing to go in order to please Lady Franklin and assist her cause. Sir Francis Beaufort at the Hydrographic Office and John Barrow Jr and W.A. Baillie Hamilton at the Admiralty were influential contacts, and she did not hesitate to ask for their help to sway opinion in the corridors of power.

Influenced by Scoresby, Lady Franklin had come to support the concept of small ships of shallow draft rather than the large, deep-draft vessels normally used by the navy (such as HMS *Hecla*, *Fury*, *Erebus*, *Terror*, *Enterprise*, and *Investigator*). In May, thinking of organizing her own search expedition, she asked the Admiralty if it could either lend or sell her two small ships, perhaps dockyard lighters. Their answer was no; they had already arranged for HMS *North Star* to take supplies to Lancaster Sound to enable James Ross to extend the duration of his search, and they considered that to be enough.

In fact, it was not enough. The *North Star* sailed on 18 May 1849 under command of James Saunders but was trapped in the ice off the northwest coast of Greenland in the autumn and forced to remain in Wolstenholme Sound over the winter – unknown to the Admiralty or to James Ross, who was wintering almost five hundred miles further west and had no idea that help was on its way. By October 1849, the voracious tentacles of arctic ice had ensnared five ships in the eastern Arctic – Franklin's two discovery vessels, James Ross's two ships searching for Franklin, and Saunders's transport carrying supplies so that Ross could prolong his search for Franklin.

Having had no success begging dockyard lighters from the navy, Lady Franklin considered other possibilities for the 1849 season, including whalers. She wrote to Scoresby in mid-May: "I have almost decided on purchasing the Abram of Hull . . . [which] is really I am told a sound & good ship & could be ready for sea in a fortnight. I have sent a person down to look at her & expect the result immediately."[7] In the end, she could not purchase the whaler *Abram*; Captain John Gravill, whose ship *William Ward* had been lost in a severe gale at the Greenland whale fishery, returned to Hull and, as she told Scoresby at the end of the month, "immediately on landing at Hull he set about chartering the Abram for a voyage to Davis Strait."[8] It was by now too late for her to find another ship and organize a searching expedition, but as she knew Gravill and admired his enterprising spirit, she invited him down to London to discuss possibilities. They decided on a compromise whereby Gravill would combine searching with whaling operations, paying particular attention to Jones Sound, where Thomas Lee had reported sighting a cairn. In the end she agreed to pay the *Abram*'s owner, Mr Barkworth, £500 to have Gravill undertake a search during his voyage. Barkworth, however, turned out to be very "cunning," she confided to Scoresby.[9] In addition to the exorbitant fee, he required her to pay for extra insurance to cover the ship being at greater risk than usual and to purchase a quarter-share in the ship as well! "I am become a whale-fisher," she wrote.[10] With these expenses and the gratuity she had promised Gravill, she reckoned she would be paying £1,100 to have Jones Sound searched, but there was some consolation in the fact that as a shareholder she would be entitled to a part of the oil money if the whaling were successful.

Owing to Gravill's late arrival from the Greenland fishery, he was unable to depart on Lady Franklin's mission until 12 June. Most of the Davis Strait whaling fleet had left three months earlier and were already far up the Greenland coast, approaching Melville Bay. Among them was the *Advice*, commanded by Penny.

* * *

Despite the strictures of the Tay Whale Fishing Company, Penny intended to make a greater effort to find evidence of Franklin than he had in 1847. As surgeon for the voyage, he hired Robert Anstruther Goodsir, whose brother Harry had sailed as assistant-surgeon on HMS *Erebus* in 1845. Anxious to search for his missing brother, Robert had offered to serve on the *Advice*. The Goodsirs were a Scottish family in which the practice of various forms of medicine and the study of natural history were held to be important. Robert's father and grandfather were both medical men. Another brother, John, a professor of anatomy at Edinburgh University, made a number of significant contributions to medical knowledge, was elected to the Royal Society, and was closely associated with men of science including the naturalist and pioneer oceanographer Edward Forbes. Before joining Franklin's expedition, Harry had served as curator of the Edinburgh Museum and had written several chapters for a book on anatomy and physiology edited by John. Robert Goodsir later produced a well-written narrative of his voyage with Penny.

The *Advice* left Dundee, cleared Stromness by 17 March 1849 (a week before the rewards were announced), and headed west towards Greenland. Goodsir found the first ten days on the Atlantic "rather tedious" but when the wind and waves increased on the twenty-seventh, he had an opportunity to observe the sea in a decidedly more boisterous mood. "I ventured on deck," he wrote, "and, for the first time in my life, saw what the ocean *looks like* in a storm. I could see nothing all around but heaving mountains of water; each succeeding wave seemed as if it would swallow up the labouring vessel, but it always appeared to melt away gently under us, except when one more rapid, or 'cross,' would send water and spray washing over her decks and high up into the rigging."[11]

After he retreated below, there was "a report like thunder, mingled with the rending and crashing of timber; sudden and complete darkness, with a rush of water through the skylight, and the ship thrown on her beam-ends." He rushed on deck and saw destruction everywhere. "The quarter-deck was literally swept of everything, rails and bulwarks, almost all the stanchions, the binnacle, compasses, dog's couch, and nothing could be seen of the wheel but the nave."[12] Two men had been carried overboard, one already dead from severe blows and the other, still alive, clinging desperately to the floating binnacle in the angry seas, rapidly disappearing astern in the frothy wake of the vessel. They could do nothing to save him.

In mid-April they encountered a great many icebergs and small floes. Penny posted two men aloft to keep a sharp lookout. After rounding Cape Farewell, the ship sailed north through Davis Strait and reached Disko about the end of April. Beyond this point, pack ice slowed their progress, and it was not until 8 June that they reached the Devil's Thumb, the unofficial entrance to Melville Bay. In this dangerous region, there was always the risk of wind-driven floes crushing a ship against the solid edge of the landfast ice. According to Goodsir's information, fourteen ships had been "nipped" and lost in this way in 1819, eleven in 1821, seven in 1822, and nineteen in 1830. Earlier he had noticed with some apprehension that emergency supplies were being brought up and stacked on deck in case the ship was destroyed, and he had been advised to have a bag of clothes ready to throw onto the ice. A ship stove by ice could go down in minutes, and when the crisis came, no one wanted to be caught down below frantically packing a bag.

The power of the moving pack was fearsome. From the relative security of a protective dock that the crews sawed out of the margin of the land floe, Goodsir watched the floes "grinding and crushing against one another with the most irresistible force": "It seemed like a trial of strength between the opposing floes, the hollow grinding noise under one's feet booming lower and lower in the distance. It was as if one was standing over the site of an earthquake. The ponderous ice, trembling and slowly rising, would rend and rift with a sullen roar, and huge masses, hundreds of tons in weight, would be heaved up, one above the other, until, where it was before a level, an immense rampart of angular blocks became piled."[13]

The ships, a dozen in number, then worked northward in small increments through the ice-encumbered coastal waters, employing the laborious techniques of tracking, warping, towing, and taking shelter from time to time in docks cut into the land ice. Four of the ships were falling behind the others and appeared to be in trouble. Two were listing badly, and their crews had already moved onto the fast ice with their boats, tents, and emergency provisions. Penny and the others were powerless to go back through the closely packed floes to assist; they could only hope that the ships would survive the pressure and eventually reach the West Land. They continued on, crossed Melville Bay to Cape York, entered the North Water on 4 July, and moored to the floe edge off Pond Inlet a few days later.

Goodsir had hopes that the *Advice* would proceed directly west to look for signs of his brother Harry and the other missing explorers, but Lancaster Sound was still blocked by ice. For three weeks the ship remained near Pond Inlet, its seven boats busily pursuing whales. Penny suggested to Goodsir that during this extended period of floe-edge whaling, he might be able to hire a local Inuk named Toonick as guide and travel westward by dog sledge towards Navy Board Inlet. However, Toonick and other able-bodied men were away salmon fishing, so the plan had to be dropped. Goodsir would have to be patient until the run of whales was over. The waiting was frustrating but not altogether boring, for there were many periods of hectic activity on board the ship, and the whaling and flensing operations were fascinating to watch. On one occasion, when boats were hurriedly being lowered for whales, Goodsir impulsively volunteered to help. Moments later he found himself straining at an oar in a twenty-five-foot boat on the heaving ocean. Pulling an oar in a whaleboat was harsh treatment for the delicate hands of a surgeon, but the exhilaration of the chase was an effective analgesic. After labouring for several miles, they came to the assistance of another boat that had harpooned a large whale, and Goodsir was able to participate in the most dangerous part of the hunt. The animal dived; the line ran out swiftly through the bow chocks; a few turns around the bollard gradually increased the drag; smoke arose from the friction of the line on the wood; the whale grew tired; it surfaced again and in time was lanced and killed, but not before lifting its powerful flukes high above the boat in a final gesture of defiance and smashing them down upon the water nearby – alarmingly close.

The exhausted men summoned enough energy to raise a feeble cheer before taking up the pursuit of other whales. They had already pulled fifteen miles from the ship, and Goodsir sank to the bottom of the boat for a short nap before they encountered the next quarry. On and on they went, exploiting the cetacean bonanza to the limit of their endurance, eating where they sat, sleeping doubled over their oars, unconscious of time in the continuous summer daylight. When the boat at last returned to the ship, Goodsir and his companions had been seated on the thwarts or curled up on the floorboards for almost forty-eight hours. It felt marvellous to stretch their legs on deck.

The experience did not dampen his ardour. A few days later he was off again in a whaleboat, participating in the capture of a large female whale, which excited his scientific interest. He carefully measured twenty different dimensions of its body. Its length "from the fork of the tail along the abdomen to the tip of the lower jaw" was sixty-five feet, half as long as the ship. The girth behind the fins was thirty feet. The tail flukes were twenty-four feet across. On each side of the mouth there were about 360 baleen plates, the longest of which measured ten and a half feet in length.[14]

Meanwhile, the short Arctic summer was passing. "As fish after fish added to the extent and value of our cargo," Goodsir wrote, "my hopes of being able to get up Lancaster Sound, began to wax fainter and fainter."[15] This was all the more frustrating because there was a report of Inuit origin circulating among the whaling ships to the effect that Franklin and his men were safe, their ships frozen in somewhere further west. The report, however, seemed to become more embellished as it passed from one vessel to another, which aroused the suspicions of Penny and Goodsir.

On 1 August, while off Cape McCulloch just south of the inlet, the *Advice* spoke the *Truelove* of Hull, and Penny invited the captain, John

Parker, on board for breakfast. As both were carrying supplies or messages for the missing expedition they decided to continue together into Lancaster Sound. Parker told Penny what he knew about the Inuit report. He said that two weeks earlier, while Captain Kerr's ship *Chieftain* was moored to the ice off the entrance to Pond Inlet, visiting Inuit had reported seeing four ships frozen into the ice some distance to the west. Two ships were said to be in one locality and two in another. One pair had been there over three winters. The Inuit said they had actually visited all the vessels three months before, apparently around the beginning of May. The ships had then appeared to be undamaged, and their crews seemed in good health. The Inuit drew a sketch map showing the ships and stated that the two expeditions had been in contact with each other. From their story and sketch, Kerr understood the ships to be Franklin's *Erebus* and *Terror* and James Ross's searching ships *Enterprise* and *Investigator*, the former apparently at Whaler Point (enclosing Port Leopold at the northwest corner of Prince Regent Inlet) and the latter at Port Jackson (on the inlet's eastern side). These two places were only about fifty miles apart. As Kerr planned to remain on the whaling grounds until late in the season, he had asked Parker to carry the Inuit sketch back to England and send it immediately to the Admiralty.

At last, after four anxious years, there was news of Franklin! And he was safe! Evidently James Ross's searching expedition had found him but had itself become trapped in ice. "Both the seeker and the sought have become involved in one common calamity," *The Athenaeum* would later (13 October) report. If the four ships were still beset, there was yet time for a favourable alteration of weather and ice to release them before winter set in again. Or, if they could not break free, their crews could probably be rescued by whalers or relief expeditions entering Lancaster Sound.

The mystery of Franklin's disappearance now appeared to be explained. He had simply had the misfortune to be caught in the unpredictable arctic ice, an event not uncommon in the whaling trade. While it was disappointing that he had been unable to complete the passage to the Pacific, it was an immense relief to learn that he and his men were still alive and within range of assistance.

Penny and Goodsir desperately wanted to believe the story, yet they thought it strange that the Inuit who claimed to have visited the missing ships had not produced any tangible evidence of having done so. They had not delivered messages from either Franklin or Ross, and they had not displayed any tools, materials, or trinkets obtained from the Europeans. Furthermore, it did not seem plausible that Sir John could have spent three or four winters trapped in the ice near Lancaster Sound without sending boats eastward to contact whalers, or dispatching sled parties in winter to deposit notes in cairns for them to pick up, or using Inuit as messengers. James Ross, whose instructions had specifically mentioned sending information to Pond Inlet by steam launch, had sent no word at all, even though Prince Regent Inlet was less than three hundred miles away.

After listening to Parker's account, Goodsir wrote a summary addressed to the Admiralty and placed it in Parker's hands, assuming that the *Truelove* would reach England before the *Advice*. A version of his letter was later published in *The Athenaeum* (13 October 1849). In it Goodsir was candid in his analysis of the Inuit report. "All this was very pleasing intelligence," he wrote, "but when I began to consider, I saw much to throw doubt upon the correctness and authenticity." He spoke of "the extreme difficulty of extracting correct information of any kind from the Esquimaux even by those best acquainted with their habits and language" and asserted that "a leading question they are sure to answer in the affirmative." As Goodsir had never encountered Inuit before joining the *Advice* and had no knowledge of their language, his remarks probably reflected the opinions of Penny. When he published his book in 1850, Goodsir would take aim at Parker, who had shared the scepticism of the others about the story while in the Arctic, "yet on his return home spread it throughout the length and breadth of the land, raising high hopes in the breasts of hundreds, which, he was fully aware, would in a few days be dashed aside."[16]

Kerr's report of the safety of Franklin's expedition required investigation. It might contain some element of truth, perhaps distorted during its transmission by memory lapses or linguistic misunderstandings. Goodsir and Penny were still eager to reach Lancaster Sound, scour its coasts for signs of Franklin and Ross, and deposit the Admiralty messages ashore. If ice conditions were favourable, they might be able to get into Prince Regent Inlet, where the four missing ships were said to be frozen in. The hold of the *Advice* already contained enough casks of blubber to make 140 tons of oil and the *Truelove* had slightly more stowed away, and this presented Penny and Parker with a dilemma. With large cargoes of blubber and bone already on hand, they could afford the time to undertake some searching in Lancaster Sound without their owners accusing them of wasting whaling opportunities. But as they would be entering more dangerous waters in which ice had already trapped four ships, they would be putting their valuable cargoes at risk, and that would certainly displease the owners.

On 2 August Penny and Parker ordered ice anchors picked up, sails hoisted, and northward courses set. On their port side, Pond Inlet led off into the mountainous coast. Owing to the enduring cover of landfast ice, its inner recesses had never been explored by Europeans; the whalers' familiarity with the place was limited to the seaward margin of the landfast ice where they moored their ships in July to pursue whales. From the decks and rigging, they could look through the inlet's entrance – twenty miles wide and framed on each side by glacier-capped rock walls rising more than two thousand feet from the sea. The inlet obviously extended for a considerable distance, but they were unable to see how far.

From the sea, Pond Inlet and Navy Board Inlet looked like separate fiords incised into the mountainous flank of Baffin Island. It was not yet known that a hundred miles in from Baffin Bay, Pond Inlet turned north for an equal distance through Navy Board Inlet and joined Lancaster Sound, thus enclosing what became known as Bylot Island.

With Pond Inlet receding on the port quarter, the *Advice* and *Truelove* passed the southeastern corner of Bylot Island. Penny and Parker took their ships as close to the coast as they dared, with lookouts posted at the mastheads and officers on deck keenly scanning the land with telescopes for cairns or other man-made features. They continued northward past a succession of bold, black headlands and broad intervening bays and lowlands with glacier-topped mountains rising behind them.

Suddenly Goodsir, scanning the coast by telescope, sighted an upright post with a flag flying from it! Another man on deck confirmed the sighting, but before they could notify the captain, the flagstaff dissolved into a dancing, disjointed column, then metamorphosed into an inverted pyramid. Finally, it vanished altogether. It had been nothing but a tantalizing mirage.

Gradually the coast of Bylot Island began to curve westward into the spacious entrance to Lancaster Sound. The lofty white-tipped mountains on the ship's port side receded behind an extensive coastal plain. Off Cape Hay at the island's northwest corner, Penny cast an Admiralty message adrift in an official metal cylinder enclosed within a ballasted cask and surmounted by a pole and a red vane. As the marker fell steadily astern in the ship's wake, it was easily visible for a long time, but it must have been clear to everyone that in the vast expanse of the Arctic this object had little chance of washing up at the feet of Franklin or Ross – or of reaching the Admiralty in London.

The *Advice* then crossed the opening of Navy Board Inlet, the back door to Pond Inlet. Late on Friday, 3 August, clouds began to gather and thicken. By four the next morning, when Goodsir went on deck, a strong easterly wind was whipping up big waves. Penny ordered topgallant sails stowed and topsails reefed, but the wind continued to rise and visibility deteriorated. Two hours later, he decided to heave to under a close-reefed main topsail only. When a ship heaves to, it becomes a passive servant of the wind, the power of its sails no longer used to advance but only to survive. With a small area of canvas aloft and a rudder adroitly positioned, it meets the dangerous seas almost head-on so that the waves expend their energy rushing along the hull rather than striking it broadside and damaging or capsizing the ship. But in waiting out a storm in this way, the ship is steadily pushed downwind. This is of little consequence as long as there is plenty of sea room, but is extremely dangerous if a rocky coast, or a body of ice, lies to leeward.

At ten o'clock the next morning, the *Advice*'s downwind drift was taking it directly towards an area of ice floes over which the seas were breaking in an alarming manner. Penny ordered the men aloft to set enough sail to enable the ship to work slowly away from danger, but near the north side of Lancaster Sound they had to heave to again and begin to drift back. By ten that evening, the ice was once more in sight. As Goodsir later described it, "The sea was here breaking with the greatest violence and magnificence upon the heavy masses of ice, and upon a solitary berg which was in sight."[17]

They hoisted storm sails and clawed their way into the wind, moving slowly away from the danger. When at last the fury of the gale diminished, the *Truelove* was no longer in sight. As the sea moderated and the visibility improved, Penny made sail again, attempting to advance westward, but before long the same heavy pack ice that had threatened their safety during the preceding days barred their progress. They were brought to a halt somewhere near the entrance to Prince Regent Inlet. From the crow's nest, Penny scanned the area to the westward with his telescope. He could just make out the top of a fortress-like island, its encircling cliffs rising a thousand feet out of the sea. It was Prince Leopold Island, off the northeastern tip of North Somerset (now Somerset Island). Although Penny could not know it at the time, the two ships of James Ross were in a snug harbour about twenty miles beyond the island, barricaded in by pack ice.

Meanwhile Captain Parker, whose ship had become separated from Penny's during the gale, had also encountered impenetrable pack ice and had turned homeward. He took the *Truelove* in toward Cape Hay on Bylot Island, landed thirty bags of coal, one cask of preserved meat, and a cask containing letters, for the men of HMS *Erebus* and *Terror*. He then shaped a course for Britain.

Months later, when Penny and Parker both reached Britain, they would give different reports about the nature and distribution of the ice in

Lancaster Sound. These were matters of the utmost importance to the Admiralty. If the ice barrier was a solid and continuous sheet, and if it remained in place during the rest of the summer, it would prevent the icebound expeditions of Franklin and James Ross from breaking out to safety. If, on the other hand, the ice was a mass of unconsolidated floes, any dispersal or dissolution brought about by wind and temperature changes might enable them to cut their ships out of harbour ice and get under way.

After Parker's arrival in Hull on 3 October, the *Truelove*'s owner, Thomas Ward, wrote to the Admiralty and described the ice in Barrow Strait as solid with no water visible to the westward. His letter was published in many newspapers, and the perception of the ice barrier being impenetrable was thus promulgated to the public. But later, when Parker appeared in person before the Admiralty Lords in London and presented his ship's logbook for examination, a different picture emerged. An Admiralty memo about the interview reads in part: "In reply to the question put to Capt. Parker as to the ice described as 'solid Ice,' he observes – The Land Ice extended a few miles offshore, but the main body barring the strait, was not a continuous frozen mass, similar to the land ice, but a packed mass of drift-Ice which would readily break up with a change of wind, and be moved to the eastward."[18] This view of the situation – a decidedly more optimistic one – agreed with the observations of Goodsir and Penny, who had seen a water sky (low-lying dark clouds indicating open water below) to the west.

Despite their doubts about the authenticity of the Inuit story at Pond Inlet, Penny and Goodsir must have found it exhilarating to be so close to the alleged location of the ships of Franklin and James Ross. Soon they might have the opportunity to either make contact with the expeditions and provide whatever assistance was required or prove the native story false. Unfortunately, further progress was not feasible. The *Advice* was lying so deep in the water that its exterior doubling and interior fortifications were mostly below the water-line, rendering the upper part of the hull vulnerable to ice pressure. It would be foolhardy to force a full ship deeper into the pack through narrow leads, risking entrapment and destruction. Penny decided to retreat from Lancaster Sound. At Pond Inlet he learned that the ships they had seen beset on the Greenland coast in June (*Lady Jane*, *Superior*, and *Prince of Wales*) had all been crushed by ice.

All ideas of finding the ships of Franklin and James Ross were reluctantly abandoned. Goodsir, although desperately keen to rescue his brother, later acknowledged that Penny "had scarcely another course open to him."[19] At the very moment that Penny gave up his quest in

Lancaster Sound, James Ross's men were busily cutting channels through the harbour ice to extricate their ships, never suspecting there were whaling vessels within a day's sail. A few weeks later the *Enterprise* and *Investigator* succeeded in breaking out, but they were then beset in drifting pack ice for a month and were fortunate to be released in Baffin Bay.

Penny took the *Advice* out of Lancaster Sound on 9 August. As Goodsir related, "an Admiralty cylinder was now got ready and enclosed in a small cask, along with some of the latest newspapers which we had on board."[20] Off the mouth of Navy Board Inlet, two boats were lowered, and Goodsir took his place in one of them. The men pulled for two hours to reach a small island in the Wollaston group. On its summit they dug a hole, buried the cask containing the messages and some newspapers, erected a cairn of stones, and marked it with a pole surmounted by a black ball.

Meanwhile, the weather had been deteriorating. Penny hoisted the recall signal and brought the ship in as close to the land as he dared. The shore party pulled out from the island and clambered on board, the boats were quickly hoisted, and the *Advice* got under way. Thick weather soon enveloped them, and they sailed southward with only occasional glimpses of the formidable mountain ramparts of Bylot and Baffin Islands. They crossed Home Bay, passed Cape Searle and Durban Island, and rounded Cape Dyer, the sentinel overlooking the narrowest part of Davis Strait. In Exeter Harbour, they dropped anchor for the first time since leaving Stromness five months before. Several other ships were lying there, their boats out "rock-nosing" for whales migrating close inshore towards their wintering area further south.

During the next ten days in the fiord, Goodsir found time to go out exploring and shooting. In an excursion in a dinghy with a ship's boy, he came upon two large polar bears feasting on a whale carcass. When one of the animals charged the intruders, Goodsir and the boy retreated hastily. Back at the ship they related their experience, and Penny, whom the doctor described as an enthusiastic bear-hunter, decided to go after the bears. He rounded up some men and a mighty arsenal of muskets, rifles, and lances and set off eagerly in a whaleboat, but the animals had moved on.

After the sojourn in Exeter Harbour, Penny weighed anchor and continued toward Cumberland Sound. He found it defended by a formidable phalanx of large bergs and floes, and so decided to head immediately for Scotland rather than put his cargo at risk again. Turning homeward after a long arctic voyage was always a joyous moment for the crew, but especially so on this occasion because the ship was carrying almost ten tons of whalebone and the blubber of seventeen whales, enough to render into 159 tons of oil, a rich bounty.

By the end of the 1849 voyage, the Franklin search was assuming a prominent position in Penny's mind. He was now fully conscious of the difficulty of combining whaling and searching in one voyage and also clear about the motives of the shipowners. In 1835 he had been outraged when Mr Hogarth, principal owner of the *Neptune*, opposed sending aid to icebound whalers in Davis Strait, and his new employers, the Tay Whale Fishing Company, had instructed him not to go out of his way to search for Franklin. They were interested in blubber, not overdue government expeditions. Penny was developing ideas on how further searching should proceed, but he was frustrated by the hard-hearted attitude of the whaleship owners on whom he was dependent for a livelihood. What he really needed was a search expedition of his own.

* * *

Meanwhile the whaler *Abram*, under the command of John Gravill, had proceeded up the Greenland coast, broken out of the ice into open water on 21 August, and sailed across to Cape Clarence on what was later named Ellesmere Island. Gravill then coasted south to the mouth of Jones Sound, where he would attempt to find the cairn allegedly seen by Lee. He later reported that after sailing round a prominent island a mile off "Cape Hardwicke" (probably King Edward Point, the southern tip of what is now called the Paddy Hamilton Peninsula), he "entered the Sound," went ashore, and climbed a slope. From an elevation of more than three hundred feet, he could look deep into Jones Sound with his glass. Although he could see enough open water for ships "to make their way out without difficulty,"[21] he did not attempt to go any further into the sound, possibly because he thought it would be contrary to the shipowner's wishes (although Lady Franklin had paid for extra insurance to cover the additional risk of entering the sound). Gravill was therefore unable to confirm or refute Lee's story of a cairn.

It was a great disappointment to Lady Franklin, who felt that the whaling captain whose spirit of discovery she had greatly admired had not done his utmost. Writing to Edmund Gratton, the British consul in Boston, she said, "The 'Abram' whaler has been a failure."[22] But perhaps the whaling captain was not to blame. She learned much later that Mr Ward, the managing director, had sent Gravill a letter "which made it impossible to explore Jones Sound under the circumstances in which he was placed."[23] Although she had paid handsomely for Gravill to examine Jones Sound with the *Abram*, his owners had apparently prohibited him from doing so. For more than a year she withheld her full payment, insisting that the terms of her agreement with Barkworth had not been met, while Ward stubbornly refused to hand over her share of the oil money – the produce of four whales – and threatened legal action against her. Penny backed her up in the matter, telling her, "You have a claim upon them and not them upon you[.] Do not pay them a farthing."[24]

Writing to Lady Franklin more than a year later, Gravill said he was sorry about the "unpleasantness between the shareholders and yourself regarding the proceedings of 1849, but I did what I could under the trying circumstances & if the wind had been up the Sound, the 'Abram' certainly would have run up, let the consequences have been what they might, but as it was I could do no more."[25]

* * *

During the summer of 1849, there was little for Jane Franklin to do but try to get on with life and wait for news. It was difficult not to be impatient. News from Baffin Bay would not reach home until the whalers returned in October. News from the central arctic coast would travel only as fast as the men who carried mail over snow and ice by dog sledge or on snowshoes, along rivers by canoe, over roads and trails by horse or sleigh, and across the Atlantic by ship. News from Alaska would require a sea voyage of half a year or more by way of Cape Horn, or slightly less if mail was carried across the isthmus of Panama or via the St Lawrence River. Even if one of the search parties did encounter the missing crews, Lady Franklin would not hear of their rescue until long after the event. It was conceivable that news of her husband's safety was already on a circuitous, snail-slow journey homeward. She had no way of knowing.

In her heart she did not believe that her husband had completed the Northwest Passage. Her intuition and the paranormal vision of Ellen Dawson, a clairvoyant she had consulted in the spring, led her to believe that he had been brought to a halt somewhere within the Arctic Archipelago. She was confident that he would do his utmost to follow his instructions, and if he had gone southward from Cape Walker as ordered, he should have been able to reach the continental coast whose configuration he had done so much to delineate. But some unforeseen obstacle could easily have intervened: shallow waters or an uncharted barrier of land might have turned a supposed channel into a cul-de-sac; the heavy multi-year ice known to drift southeastward toward King William Island might have proved to be impenetrable; pack ice might have crushed the ships.

To get whatever news the returning whalers might bring back from Davis Strait at the end of the season, Lady Franklin and Sophia Cracroft travelled north to Scotland by steamer in July. In Edinburgh they met with William Scoresby to discuss ideas on future searching strategies, should they be necessary. Then they continued to the Orkney Islands, where returning Davis Strait whalers usually stopped to discharge some of their men. At Stromness they visited the mother of Dr John Rae, the hardy Hudson's Bay Company explorer who had mapped part of the arctic coast on foot during previous summers. Cracroft wrote in her journal, "My aunt and I were perfectly astonished at the beauty of the picture before us – for Mrs. Rae in her widows cap, black gown, & delicate knitted shawl falling fm. her head & forming as it were a background to a most beautiful face, was no ordinary sight. There is the fine Grecian nose, lovely hazel eyes, a mouth wh. even the want of teeth has not spoiled, coral lips, & even yet a brilliant complexion . . . our first sight of Mrs. Rae we shall not readily forget."[26] When they called on her again two days later, "Mrs. Rae looked even more beautiful than before."[27] Five years later, her son John would uncover evidence of the death of a large party of Franklin's men, but when he dared to report that some had resorted to cannibalism in their last days, Lady Franklin and her niece would turn against him.

After proceeding further north to Shetland for a holiday, the two women returned to Orkney to be on hand for the return of the whaling ships. In Kirkwall they received their first piece of news. A Danish ship arrived carrying two whaling captains and about thirty men who had been shipwrecked on the Greenland coast. One of them, Captain Paterson of the *Lady Jane* (a Newcastle ship that had entered Cumberland Sound in company with Penny's *Bon Accord* and others in 1840), had brought with him a metal cylinder addressed to the Admiralty and containing a document from the Franklin expedition! It had been found by Greenland hunters along the shore near Egedesminde on 14 July and taken to the Danish inspector of the district, who in turn had entrusted it to Paterson, then awaiting transportation to Britain after the loss of his ship.

In five languages the note requested the finder to send it to the Admiralty or to a British consul, with details on where and when it had been retrieved. To Lady Franklin's deep disappointment, however, the message had been written and cast adrift four years earlier, while her husband was still in Baffin Bay. It was nothing more than a note indicating his position, and it was not even the last news received, for Dannett's whaling ship *Prince of Wales* had spoken Franklin's ships a month later.

On the heels of this cruel disappointment, another piece of news reached her ears, this one exciting beyond belief. On 29 September, Parker's whaling ship *Truelove* reached Orkney and one of her crew, James Donaldson, disembarked in Longhope Bay. There he recounted the Inuit story of four ships frozen in to the west of Pond Inlet and reported that Sir John and his men were alive and had been in contact with James Ross's search expedition! In Kirkwall, Deputy Sheriff James Robertson recorded Donaldson's statement and then sent the man to talk with Lady Franklin at Skaill, where she and Sophia Cracroft were staying.

It was impossible for Lady Franklin to verify the details with Parker, because he had already left for London to report to the Admiralty. She was annoyed that he had not taken the time to contact her and relate the news in person. To Scoresby she wrote, "You are perhaps aware that we received while at Stromness the *earliest* intelligence brought by Captn. Parker of the Truelove, who, however, did not attempt to see, nor even to write to me, in which I think he failed in what I might reasonably have expected from him."[28] She had purchased the provisions and coal that Parker had deposited in Lancaster Sound during the voyage; she had made the journey to Orkney for the express purpose of getting news firsthand; and, as Sir John Franklin's wife, she of all people deserved to be informed of any information concerning his safety! Parker later said that he had been unaware that she was in the islands, although a man named Jobson at Longhope claimed to have told Parker of her presence.

The news of Franklin's safety put everyone in high spirits, caused church bells to be rung in Kirkwall, and spread rapidly throughout Britain by way of the telegraph system. In London the Admiralty lords, reluctant to accept the report as completely reliable before they interviewed Parker, released a cautious statement which read, according to the *Historic Times* (12 October 1849), "Some hopes are entertained that the news brought by Captain Parker ... of Sir John Franklin's ships having been seen by the natives as late as March last ... is not without foundation." That restraint did nothing to dampen the joyful feelings, however. The *Athenaeum* (13 October) wrote that "as news generally travels with the properties of an avalanche, swelling in importance with every step, many hours had not elapsed before the return of the Erebus and Terror in the course of the present month was spoken of as an almost certain event."

Desperately wanting to believe the report yet hesitant to put all their trust in it, Jane Franklin and Sophia Cracroft crossed to the mainland in mid-October to obtain more information at the whaling ports and interview Captain Kerr when the *Chieftain* arrived in Kirkcaldy. There they talked with George Turnbull, a manager of whaling ships, and with Captain Kinnear who, according to the *Fifeshire Advertiser* (20 October), "for a long period of his life has been familiar with that desolate country." Three years earlier, Kinnear had brought a young Inuk back to Scotland on board the whaler *Caledonia*. Although his name was Aukutook Zininnuck (variously spelled), the Scots called him Kookie-Eekie, even though he always protested "No, no." The newspaper now thought it unfortunate that the young Inuk, who had learned a good deal of the English language during his winter in Kirkcaldy, had not been at Pond Inlet to interpret when the Inuit of that place, by means of pidgin English, sign language, and a crude sketch, had described to Captain Kerr the four ships they had claimed to have seen in the ice. At the time Aukutook had been at Cape Searle, a few hundred miles to the south. While awaiting the arrival of the *Chieftain*, the women visited the studio of an artist named Nairn and admired the portrait he had painted of Aukutook during his sojourn in Scotland. At last the *Chieftain* arrived, and they were able to talk directly with Kerr. He assured them that he considered the report reliable and a few days later set off for London to report to the Admiralty in person.

The two women continued on to Dundee. They were particularly anxious to see Penny, who, like Parker of the *Truelove*, had intended to enter Lancaster Sound to search and leave messages. His ship *Advice* had arrived from Davis Straits only a few days before and was now anchored offshore. Having alerted the *Advice*'s managing owner, Mr Hume, before arriving, Lady Franklin and Sophia got the royal treatment. They were taken out to Penny's ship on the steam-powered *Mercury*, hired specially for the occasion. In honour of the visit, Penny had his ship dressed with flags.

One can imagine the scene: the small, brightly painted steamboat slicing through the waves in a smart breeze, black smoke streaming downwind from the sooty funnel, as the two ladies, neatly turned out in long dresses, gloves, stylish hats, and boots, take shelter from smoke and spray inside the small cabin. The *Mercury* comes alongside the ice-battered hull of the *Advice*, fenders are rigged, lines thrown and secured, and the two ships come gently together. On the steamer, the ladies, holding their fancy bonnets in the brisk wind, their long skirts below waists fashionably pinched in with whalebone busks, look nervously between the rising and falling hulls at the gap now opening, now closing. Seamen remove a section of the bulwarks and run a gangway between the two vessels, and the ladies are helped over, not looking down at the seething water below. They arrive breathless on the oil-saturated deck of the veteran whaleship and are at once enveloped by the smell of blubber, a pervasive odour their clothing will long retain.

Captain Penny welcomes them on board, shows them an authentic Inuit kayak, and invites them to watch a demonstration. The second mate, wearing an Inuit parka borrowed from Dr Goodsir, lowers himself into the slender craft and shows how the double-bladed paddle is used. Penny then tries to explain in landlubbers' terms the spiderweb of halyards, braces, sheets, stays, shrouds, masts, yards, and booms rising above the deck. He leads them forward to admire the foredeck equipment, and aft to see the helm and binnacle. Then, gathering their skirts, his guests descend the near-vertical companionway leading to the interior spaces of the ship. Finally, seated together on a settee in the captain's small day cabin, they get down to serious discussion.

Unfortunately, Penny's news is discouraging, a blow after Kerr's report. Although he has succeeded in leaving messages for Sir John in Lancaster Sound as planned, he has seen no evidence of him or his ships, and both he and Goodsir are sceptical of the Inuit report from Pond Inlet. This puts a damper on Lady Franklin's hopes, but she prefers to hear honest opinions and plain truth rather than sugar-coated flattery, to be in touch with reality rather than illusion. Penny's opinions are not ambiguous or camouflaged to please or exaggerated to inflate his own achievements. A forthright individual herself, Lady Franklin is favourably impressed.

The discussion then turns on the prospects of finding her husband, and they compare ideas on how a more effective search could be undertaken. Lady Franklin asks Penny whether he would be interested in commanding a search expedition, either one sent out by the government or one sponsored by her. This is the chance Penny has been waiting for, and he assures her of his willingness to serve. After an hour's conversation, she gives Penny some money to enable the men to drink Sir John's health, and she and her niece return to shore on the steamboat. The smell of blubber persists in their nostrils for days.

This first face-to-face meeting between two of the most dedicated individuals in the search for Franklin was loaded with significance. Lady Franklin was seeking news of her husband, which Penny, despite the strong efforts he had made during three whaling voyages, was unable to provide. Disappointed, yet impressed by Penny's sincerity and resolution, she became convinced that he was a man who could play an effective role in future searching activities. At the same time, Penny, whose own searching ambitions had been reined in by his shipowners, saw Lady Franklin as a person of great influence and a potential sponsor whose support might enable him to cast off the shackles of commercial whaling and devote himself wholeheartedly to the search. On many points, including the desirability of using small vessels rather than the large ones customarily sent out by the Admiralty, they found themselves of like mind.

Soon afterward, Lady Franklin wrote to Robert Goodsir, asking if he would provide a summary of the voyage of the *Advice*. This he gladly did. She forwarded the piece to Barrow at the Admiralty for critical comments and then submitted it to the *Morning Herald* for publication. She also asked Goodsir for his frank opinion of Penny's character and reliability. The doctor spoke highly of him, describing him as a leader of the whaling fleet, cool under pressure, passionate in a cause, "goodhearted & good natured" – but surprisingly awkward when dealing with women or first lords of the Admiralty.[29] Penny had passed Lady Franklin's tests with flying colours, and she was now prepared to support and promote his candidacy for leadership of an official search expedition.

Any elation she might still have felt after hearing the story of the four ships frozen into the ice in the vicinity of Prince Regent Inlet faded early in early November when James Ross arrived back in Britain. Although he had in fact wintered near the entrance to Prince Regent Inlet, he said, he had seen no Inuit and no other ships. Writing to Edmund Gratton, the British consul in Boston, Lady Franklin confessed that Ross's return after only one winter without any positive news "threw us all into dismay, & destroyed amongst other hopes that which we founded on the whalers' report."[30] To Silas Burrows, another American, she said that it has "put an almost fatal extinguisher on my hopes."[31] She wrote in similar vein to Sir George Simpson and Dr John Rae of the Hudson's Bay Company, urging Rae to "sweep all these unknown seas & channels" on his next expedition.[32] It was impossible to know whether the alleged Inuit report, as understood by Captain Kerr, had been entirely false or merely incorrect in some of its details. To answer this question and search further west than Ross had sailed would require a new expedition in 1850. Lady Franklin's hopes now rested on Penny.

In her attempts to learn where her husband was, she continued to consult clairvoyants. On 19 November she described to her friend Admiral Beaufort, the Admiralty's hydrographer, a recent session with Ellen Dawson, and on the following day, Beaufort accompanied her on a visit to another clairvoyant, who, however, was not available. Like Lady Franklin, Beaufort was fully conscious of the need to continue the search with vigour and to follow up the Kerr story, but he was finding it difficult to keep the official wheels turning. After an interview with the first lord on 23 November, he drew up a detailed paper for him, but the next day he wrote in his private journal with evident frustration, "Nothing done 1st Lord went out of town!" A few days later he wrote, "Admy moving very torpidly about Franklin."[33] His impatience was fully justified: HMS *Erebus* and *Terror* had now been gone for four and a half years.

6

A Silver Penny

It would be wise to let loose his energy.[1]

A month after meeting Lady Franklin on board the *Advice*, Penny eagerly opened a letter from her:

> My letter relates to your employment by the Govt. as commander of a small vessel of search for the missing expedition, or at any rate your being engaged by the Admiralty for some important post in any such renewed Expedition. I have no official authority for addressing you on this subject, but am anxious as soon as possible to know your views & inclinations in order that I may represent them in the proper quarter. I have expressed my own opinion of your capacity and enterprise, derived from Mr. Goodsir & others & shall be glad to learn that you continue to entertain the same sentiments as to being engaged by Govt. in search of the missing ships, which you expressed to me on board the "Advice."[2]

She was willing, she added, to send any "books, papers, or Admiralty Instructions & Charts which may assist you in understanding all that is necessary to be known of the Arctic Expedition." In closing, she asked Penny to consider the matter "strictly confidential."[3]

Penny replied promptly. "My views with regard to the interesting subject we conversed upon on board the Advice can never change. I am both ready and willing to enter upon such a cause . . . but I would like it were a command for unless it is your Ladyship or those immediately connected few could feel as I do and I fear my enthusiastic nature would brook no control in matters which is to me a second nature."[4] The prose was awkward but the sentiments were clear; he wished to have sole command because he was used to it and needed it. Without it, his enthusiasm might be curbed by a more cautious or less motivated superior, which he would resent. In arctic navigation, a moment's hesitation could spell disaster, so he had to be empowered to take whatever course of action he thought advisable without wasting time consulting someone else.

In early December he wrote again to Lady Franklin: "I send you an outline of my ideas what I think the most adviseable [*sic*] plan to proceed in search of Sir John with any probability of success." He now offered (reluctantly, one senses) to serve under another commander if necessary, but expressed the hope that through Lady Franklin's influence he would be appointed sole commander, or at least co-commander.[5] He was pessimistic about the prospects of securing sole command, however, with good reason. Three decades earlier, the renowned whaling captain William Scoresby Jr, after observing unusually open ice conditions at high latitude, had suggested that the time was ripe for a renewal of the search for a Northwest Passage. But his offer to lead an expedition was refused, largely through the influence of the permanent secretary of the Admiralty, Sir John Barrow Sr, whom Scoresby's biographers have called "a mean-spirited sycophant who had wormed his way upward by devious means and . . . was determined not to give way to anyone, least of all a whaling captain."[6] Since that time, although whaling masters had served on most of the arctic expeditions sent out by the British government, they had always been appointed in the advisory capacity of ice master, never in command.

Penny returned to this crucial question of command a few weeks later. "I am perfectly aware of the almost insurmountable difficulty of a person in my situation being employed except in a very insubordinate [*sic*] capacity in such an undertaking therefore I fear my application to their Lordships would only meet with a refusal however I will see by your next what your Ladyship proposes."[7] Fearing that he was about to back out of the project, she implored him not to be discouraged, to which he replied, "Nothing can discourage me from an object when once resolved upon."[8] Serving under a naval officer could lead to problems, Penny said, but he would be willing to accept a joint command, and he specifically mentioned Commander Charles Codrington Forsyth, a friend of the Franklins. Penny and Forsyth were not destined to sail together as co-commanders, however; each would obtain a command of his own.

In early December Lady Franklin expressed her views about Penny to Francis Beaufort. He wrote back, "I think the best thing that your (silver) Penny could do would be a direct proposal to the Admiralty [to command an expedition himself]," giving particulars about the desired ship's tonnage and draft, the number of crew, the provisions required for a certain number of months, and his own monthly salary.[9] "Let him send the letter to you though addressed to the Secretary," Beaufort cannily advised, "and then if you will summon me I will go to your bureau & discuss with you the best way of presenting it."[10]

Lady Franklin sent some books and a chart to Penny. Thanking her by return mail, he confessed that he was still worried that the Admiralty might reject his application. Nevertheless, in response to more urging, he prepared a formal letter to the Admiralty and sent it for her perusal a few days before Christmas. She submitted it to the Admiralty, probably after discussing its contents with Beaufort and editing it. By 27 December, Penny's application to lead a search expedition under the auspices of the Admiralty was on the table.

Penny's letter began:

> My Lords,
> It is my most anxious desire to be employed in the search for Sir John Franklin, and I beg to offer you my services, which if you are pleased to accept, all that man can do to save that gallant officer and his companions shall be done by me. I have been engaged 28 years, since the age of 12, in the whale trade, and in command of vessels for 14 years, and your Lordships may command my testimonials, if you think proper . . .
>
> I now beg to make the following proposition. It is, to be put in command of one or two small vessels, of from 150 to 200 tons; the crews to be of my own selection, picked men, whom I can trust, and with these to proceed in company with the other whalers in March next, or not later than 1st of April, to Davis Straits.[11]

He then outlined his intended plan of search on the West Side of Baffin Bay. He would first examine Jones Sound because in his experience it was usually open – at least at its entrance – when the whalers arrived, and because it might connect with the top of Wellington Channel (which in fact it does, in a roundabout way). If he found such a passage to be navigable, he would proceed through it and examine the northern part of Wellington Channel, searching the coasts for evidence of Franklin, who might have attempted to go north of the Parry Islands, the several islands lying north of Parry's 1819 route westward through Lancaster Sound to Melville Island. If it was impossible to enter Jones Sound, or if it turned out to be a dead end, he would continue to Lancaster Sound and try to enter Wellington Channel from its southern end. If he found Wellington Channel blocked by ice, he would carry on westward through Barrow

Strait and look for messages at Cape Walker, the place to which Franklin had been first directed in his official instructions.

In his application, Penny placed great emphasis on Wellington Channel, which Franklin had been ordered to try if he failed to find a navigable route southward to the continental mainland. In view of the brief navigation season, Penny was prepared to winter in the region and renew the search in the following year, both by ship and by sledge. If he had two ships, they might winter in separate localities in order to enlarge the area of search (a measure he did not implement later). If the Admiralty decided to employ other ships as well, he said, "I would beg leave still to offer my services for any branch of the searching Expedition from the eastward, so that I might have the command of my own vessel, and am allowed to conduct the particular search committed to my charge, according to the best of my ability, and with full power to act according to the circumstances."[12] He was willing to be part of a larger Admiralty expedition, but only if given command of a ship and freedom to act as he saw fit.

At the Admiralty, Beaufort strongly supported the proposal, pointing to Penny's "daring but prudent conduct" and his long experience in ice navigation: "It would be wise to let loose his energy, and to give him the opportunity for which he begs."[13] He recommended that Penny be encouraged to look for a suitable vessel and submit cost estimates and other particulars, such as the number of men required. Penny required a decision as soon as possible because he would have to tell his employers whether or not he would be available to command a whaling ship in the approaching season. During his initial correspondence with Lady Franklin in November, Penny had said nothing to his shipowners, waiting to see what would develop, but as time went on the need to inform them that he might be employed in the search rather than in whaling, grew stronger. This he did in late December after he had made his formal application to the Admiralty. The directors of the Tay Whale Fishing Company were, as Penny reported to Lady Franklin, "very much displeased."[14]

The next several weeks were stressful for Penny. While he and Lady Franklin were discreetly urging the Admiralty to make a prompt decision on his proposal, the Tay Whale Fishing Company was pressuring him for a firm commitment on a whaling command. If the Admiralty delayed, Penny would either have to withdraw his search application and commit himself to the whaling, or decline the whaling command in the mere hope of getting a favourable response later from the Admiralty. As he and Lady Franklin were well aware, there was a strong possibility that he could end up with no employment at all.

On 1 January 1850, Lady Franklin wrote to Barrow at the Admiralty, saying that she had told Penny, "I must consider myself responsible for his coming years income if the delay occasioned his losing his usual employ^{t} with his owners."[15] Penny had replied that he could not accept money from her unless he did something to earn it, and he could not remain unemployed during the forthcoming season because "idleness accords ill with my nature."[16] At the end of the month, Lady Franklin wrote to Hume, the managing director of the whaling company, offering to deposit £200 in case Penny was not employed by either the Admiralty or the whaling company, but Hume declined too, probably at Penny's insistence.

Time was very short. Despite the uncertainty of the situation, Penny had to begin preliminary preparations immediately and assume that the Admiralty would decide in his favour. The most important task was to find suitable ships and see how much modification they would need for sailing among ice floes. The size of the ships would allow estimates to be made of crew size and in turn the wages, provisions, and equipment required for them. In early January, Lady Franklin sent him estimates made for her by a whaling captain named Davidson a year earlier when she was thinking of sponsoring an expedition for the 1849 season. Penny was to use these as a guide.

In his application he had mentioned "picked men, whom I can trust." In letters to Lady Franklin, he was more specific. Orkney men who had worked for the Hudson's Bay Company should comprise at least a third of the crew of each ship, he told her in December. He had employed such men before, and "Fatigue had no effect upon them and for food nothing came wrong."[17] The rest would consist of whalemen from mainland ports. Men with naval backgrounds were not on his wish list: "As for men of wars men, I have never found them to suit our whaling trade."[18] He had received applications from experienced men capable of leading sledging parties, had invited Robert Goodsir to serve again in a medical capacity, and had written to Stromness to get ten good sailors.

Precious weeks passed and still no word came from the Admiralty. At last a letter arrived on 14 January. The secretary of the Admiralty (according to Penny's handwritten copy) wrote:

> I am commanded to acquaint you that my Lords are well assured of your ability and zeal, but with respect to your questions as to a more extensive search and the offer of your services for any branch of the searching Expedition from the Eastward my Lords are not prepaired [*sic*] to sanction such search or to accede to your proposals[.]

> My Lords therefore desire it may be distinctly understood by you that they in no way hold themselves bound to employ you yet at the same time as your views cannot fail to be of service in the event of any expedition being decided upon they would be much obliged to you for whatever details you might be able to furnish them with relation to any such expedition[,] such as the number [*illegible word*] cost and description of the the [*sic*] vessels you would propose to employ[,] their equipment and number of their crews and such other observations and remarks as may occur to you to offer[.][19]

Instead of offering him command of an expedition, or turning him down, they were declaring that they were not obliged to hire him but wanted more information!

This opaque Admiralty-speak was a wet blanket thrown over Penny's aspirations after what had already been a frustrating and unnecessary delay. The response backed him into a corner, for the Tay Whale Fishing Company was pressing him for a decision. In late January, John Hume reported to the committee of management that "Capt. Penny has informed him that he has offered his services to Government to go out on an expedition in search of Sir John Franklin, but had not got a positive reply from the Lords of the Admiralty if his offer was accepted, and asked for a week to get a positive reply to his application before appointing another master to the 'Advice' – the meeting agreed to allow him till the 30 Inst. to give his final answer."[20]

Thinking of the plight of Franklin's missing men, Penny acted with his heart and notified Hume at the end of the month that he had committed himself to the search and would have to relinquish command of the *Advice*. Regretting Penny's decision, Hume replied, "I have a meeting of my Committee tomorrow & your letter will be laid before them, at this late period we shall be rather at a loss to get a master for the 'Advice.'"[21] On 30 January the minutes of the committee of management recorded, "There was laid before the meeting a letter from Capt. Penny resigning the command of the ship Advice, in consequence of receiving an appointment from the Admiralty to go out in an expedition, in search of Sir John Franklin – which the meeting very much regret."[22] In fact, Penny had not yet received an Admiralty appointment. But the next day his contract with the whaling company expired and he was unemployed, facing an uncertain future.

Having destroyed the bridge behind him, Penny sent the Admiralty the information they required. "I beg to state," he wrote, "that for the conduct of that part of the Expedition mentioned in my former letter, viz. Jones' Sound, Wellington and north shore of the Parry Islands, I would propose to your Lordships that two small square-rigged vessels of from 150 to 200 tons, with decked boats for each, be employed."[23] He suggested that twenty-five men would man each ship, and that the expedition would plan to stay out for eighteen months but would carry provisions for two years. He estimated the total costs as £12,238, but, he added, if the purchase and outfitting of two ships were too costly or too time-consuming, naval dockyard lighters might be adapted for the search. They were strong and light in draft, two essential qualities for the task, and he had been told that some were available. This suggestion, and a few other aspects of his plan, suggest something of the influence of William Scoresby, who on the suggestion of Lady Franklin had set forth some of his own ideas in a series of articles in the *Morning Herald* in the autumn of 1849 and then gathered them into a book, *The Franklin Expedition, or, Considerations of Measures for the Discovery and Relief of Our Absent Adventurers in the Arctic Regions*. The book, evidently published in January 1850, specifically recommended dockyard lighters or small coastal merchant ships for exploring westward from a base at Port Leopold or Cape Walker. As Scoresby had never sailed beyond Cape Farewell, he had to rely on information provided by Penny and several other whaling captains in writing about ice conditions and ship movements in Davis Strait and Baffin Bay.

As both Penny and Scoresby corresponded with Lady Franklin, there was a three-cornered exchange of information and a stimulating cross-fertilization of ideas between two capable arctic navigators, both very experienced in the Arctic, and the woman who played the central role in the search. While Lady Franklin was doing her best to influence men at the Admiralty, she remained in close touch with those who through long experience knew most about ice navigation in arctic regions – the whaling masters. Scoresby had made sixteen arctic voyages and Penny twenty-eight, but James Clark Ross, by far the most experienced of the Royal Navy's "arctic worthies," had only made eight. Some of the naval men had wintered in the Arctic, however, which neither Scoresby nor Penny had done.

* * *

The lords commissioners of the Admiralty, who bore the responsibility for finding Franklin, had been badly shaken when James Ross returned in the fall of 1849 after a voyage of a year and a half without finding a single trace of Sir John's passage. Disappointment and concern were widespread among the general public as well. On the first day of 1850, the *Times* printed a letter from a reader signing himself "An Observer": "A cry for help from 130 gallant men comes to us on the northern gale now blowing. Oh! Let them not cry in vain."

No word of any shipwrecks in Baffin Bay had been received from the Greenland settlements or whaling ships. No news had come from fur-trade posts in the vast subarctic region that stretched across the continent. No reliable reports had arrived from Alaska, where HMS *Plover* had been waiting hopefully for the overdue expedition at the far end of the Northwest Passage. In order to draw up a search strategy for the approaching season, the Admiralty sought the opinions of officers experienced in arctic navigation – Back, Beaufort, Beechey, Parry, Richardson, James Ross, and Sabine. Did they believe the men to be alive or dead? Where did they think the missing expedition was? What direction should searching ships take?

The results were published in the *Athenaeum* and summarized in the *Aberdeen Herald* on 6 April 1850. A surprising unanimity was revealed. (No doubt the various officers had discussed the matter among themselves.) They all believed that Franklin and his men were alive and would be found either in the vicinity of Melville and Banks Islands, or even further west. Some were under the impression that the climate became milder in the west and were confident that the abundance of animal life in that region would provide both food and fuel.

A quarter-century earlier, when Parry had encountered unusually open ice conditions, he made the passage from Baffin Bay to Melville Island look easy. Had he retained an unreasonably optimistic view of the so-called Parry Channel, one that unduly influenced his colleagues?

Nobody at this time considered that Franklin might have discovered a route northward into a navigable polar ocean and could be working his way his way toward Bering Strait and the Pacific Ocean – or even back across the roof of the world toward Europe. And not one of the arctic experts dared to suggest that after five years of effort Franklin might still be beset within the Arctic Archipelago, on the very path that his orders had specified. That was unthinkable.

The Admiralty planned to send another expedition in 1850 along part of the course that Franklin had been ordered to follow – across Baffin Bay, into the Arctic Archipelago by way of Lancaster Sound, and then westward past Cape Walker, possibly to Melville Island. To some extent this would be a repetition of James Ross's voyage, but he had gotten no further west than Somerset Island and had left some gaps to be filled. He was a competent commander, more experienced in polar navigation, wintering, and sledge travel than any other Royal Navy officer, but the unpredictable and unforgiving arctic ice had treated him harshly. It had denied his ships *Enterprise* and *Investigator* access to the main objectives, forced them into winter quarters in early September, kept them frozen up for almost a year, and then, after the men had finally managed to cut a channel two miles long to open water the following spring, it had encircled the ships and carried them out of Lancaster Sound and into Baffin Bay like passive pieces of driftwood.

If ice was the problem, the Admiralty lords believed they now had the solution – steam power. They would send two large sailing ships as before, but accompanied by two steam auxiliary tenders to force their reinforced prows through thin ice, wedge floes apart to make channels, and take their wind-driven consorts in tow during calms. Although Franklin's ships had been equipped with steam engines and screws, the placing of railway locomotives athwartship below deck had been awkward and space-consuming. On the new search expedition, the two tenders would be equipped with dedicated marine engines, much more compact and three times as powerful.

To command the new search expedition, they selected Captain Horatio Thomas Austin, RN. Although he had served as a lieutenant on Parry's third expedition, he had not been back to the Arctic in twenty-five years. Austin's chief qualification was that he had commanded the Navy's first engined ship, *Salamander*, in 1832 and gained considerable experience with steam power, which was now to be the Admiralty's secret weapon. He would fly his flag on one of the two sailing ships, however, and give his orders from there.

Second-in-command of the squadron, and in charge of the other sailing ship, was Captain Erasmus Ommanney. His only previous arctic experience had been on HMS *Cove* in 1836, serving under James Ross on a desperate mid-winter voyage to Davis Strait and Labrador to assist whaling ships trapped in pack ice the previous autumn. Like Austin, he had attained some practical knowledge of steam power, having commanded the steam-sloop *Vesuvius* for three years in the Mediterranean.

The steam-powered ships were to be under the command of two young lieutenants, John Bertie Cator and Sherard Osborn. Cator had served on two steam vessels and Osborn on one, but neither had handled engined ships among ice floes and bergs or even been in arctic waters before. Austin and Ommanney, the two most senior officers, would direct the operations of the steam tenders that were expected to lead them through the ice, an arrangement that would prove awkward.

In mid-February Penny travelled to London for an interview at the Admiralty. It was probably at this meeting, after outlining his views in person, that he was at last offered command of the search expedition that he had proposed, which would sail under Admiralty orders. He also met Austin, who had been appointed commander of the four-ship naval squadron. Somehow Penny and Austin did not see eye to eye. What actually transpired is not a matter of record, but it did not bode well for the future. Sometime after their meeting, a man named James Peake wrote Penny from Woolwich to say that he had heard "a sad tale of the treatment you have received . . . from Captain A." He advised Penny to "keep him at a distance and go your own road."[24] A month later, Sherard Osborn, who would soon command one of Austin's steamers, wrote to Penny, "You must not think ill of Naval Officers, from the little disagreeable conversation you had with my leader – he is a hot-headed – but still a warm hearted sailor – blunt of speech perhaps but nevertheless take my word he will not be a bit more unwilling than I should be to claim you as a brother workman in a good and glorious cause and go hand in hand with you, wherever duty may call us."[25]

The disagreement between Penny and Austin was an inauspicious beginning for the two official search operations, and despite Osborn's assurances that Austin would "go hand in hand" with Penny in the "good and glorious cause," the relationship between the two commanders in the Arctic would be marred by friction and misunderstanding.

Penny returned to Aberdeen to focus his energy on the search expedition. The lateness of the Admiralty decision made it necessary to compress preparations into a very short time and carry out all the operations at hectic speed, with the attendant risks of confusion and careless mistakes. Penny had talked about getting away by the beginning of March but that deadline was a mere two weeks away and he did not even have any ships yet!

In late February he found two clipper brigs of 225 and 120 tons under construction – one in Aberdeen (later named *Lady Franklin*), owned by Walter Hood & Company and the other in Dundee (later named *Sophia*), owned by Thomas Adamson. Penny purchased them, and after an Admiralty shipwright reported favourably on both, the work of getting them ready for an arctic voyage proceeded with frantic haste. The hulls had to be double-planked and their interiors strongly fortified with cross beams. Cabins had to be rearranged.

On 13 March, after being christened by Margaret Penny, *Lady Franklin* "glided into her future element in first-rate style, amidst the cheers of a numerous crowd," the *Aberdeen Herald* reported (16 March). The loading of supplies and equipment, some purchased locally and others sent from London, began immediately. Two weeks later the *Sophia* was launched in Dundee and towed north to Aberdeen by steamer to commence the outfitting.

The wide variety of items required for a wintering voyage to the Arctic included cork lifebuoys (two per ship), brooms (twelve dozen), a barrel organ with twenty tunes, housing cloths to be suspended above deck while wintering, leather to make boot soles, a warming apparatus for each ship, rockets, blue lights, Inuit dictionaries, and food. The "Account of Provisions required by Fifty Men for two years"[26] lists impressive quantities of beef, pork, biscuit, flour, oatmeal, barley, rice, fish, molasses, sugar, butter, cheese, coffee, lime juice, and vinegar. Although no vegetables appear in the list, another document, "Scale of Provisioning the Arctic Expedition,"[27] includes potatoes, turnips, carrots, and peas. Penny wrote to Barrow, "I am taking vegetables in every shape I can think of such as salted in casks leeks onions and cabbages."[28] Spirituous beverages were not overlooked: there were sixty-nine gallons of rum, gin, and brandy and 209 bottles of wine. And of course there was tobacco for the men's clay pipes – more than half a ton of it.

Nine of Green's message balloons – the latest in communication methods – arrived for the expedition. They would be used to inform Franklin and his men of the location of the wintering ships. Green, a well-known aeronaut, had earlier carried out a test in front of Lady Franklin and others, with "a balloon containing 32 packets, consisting of 3000 slips of silk and paper, which were to be detached at the rate of one every five minutes." The balloon, filled with hydrogen, lifted off and headed southwest, shedding message slips as it drifted away "toward the Channel," the *Aberdeen Herald* reported on 23 March.

To Londoners it must have seemed sometimes that it was snowing message slips. George Shepherd, another manufacturer, had sent up experimental balloons in late February, using slow matches to release the messages at regular intervals. Slips of paper five by one-and-a-half inches in size were addressed to Franklin with space to insert the name and location of the sending vessel. Finders were asked to send them back to the Admiralty in return for a small reward. Message slips had been returned "from Colchester, from Sussex, and various parts of Kent, and from Ryde in the Isle of Wight" (*Times*, 1 March 1850). In a subsequent test, Shepherd released four balloons from the roof of the Admiralty building, the largest carrying 2,500 slips of paper, which began to drop off at 3,000 feet of elevation. "Some of the slips fell in Lower Thames-street," reported the *Times* (7 March 1850), "some in Bethnal-green cemetery; others were found on the Surrey side of the water, and down the river side to near Greenwich, affording evidence that the first lot were scattered over an area of several miles." By the time the balloon sailed over Dungeness and headed across the Channel, it had dropped half of its slips, leaving 1,300 to bemuse residents of France and nearby countries. One drifted as far as Hamburg, four hundred miles away.

Shepherd had agreed to supply message balloons for Collinson's expedition to Alaska, but the *Enterprise* and *Investigator* had departed in early January before his tests were complete. In March he sent the balloons out with the West Indian mail to be taken across Central America and forwarded to Hawaii in time to intercept the ships.

Message balloons were not the only potential method of communicating with Franklin. Sound could be employed as well. A correspondent signing himself "F.A.C." wrote to the *Nautical Standard* (9 March 1850) to suggest the use of gongs and bugles. Even if they were not heard by Franklin's men, he said, they might lead to contact with Inuit who possessed relevant information. He recommended trials to see which would be heard over the longer distance. A few months later, as Austin's four ships advanced north along the Greenland coast, Austin would issue a memo advising his commanders that later, when the ships were secure in winter quarters, they should employ sounds as well as rockets and lights to advertise their presence during fog or darkness. In addition to firearms, he mentioned "drums, gongs, bells, and whistles."[29]

* * *

Miscellaneous work of various sorts had to be performed on Penny's ships before departure, such as gilding and painting, carving and fitting figureheads, sawing 1,300 linear feet of lumber to take along, and adding ships' names to their lifebuoys and boats. As expenditures had to be approved by the Admiralty, Penny was engaged in a constant exchange of messages, an unfamiliar and frustrating burden. He confessed to Lady Franklin, "The many delays I meet with from the many forms I have to go through is like an eternity to me."[30]

As the work of securing and stowing provisions and gear was proceeding, Penny concerned himself with the important business of hiring crews. Many men, motivated by a desire to find Franklin, to rescue missing friends, to sail under Penny, or perhaps simply to earn double pay for a voyage into the ice zone, had applied for positions. It was vital to find good officers, and in particular a reliable commander for the *Sophia*. Early in March, after a trip to Peterhead, Penny wrote to Lady Franklin, "I have engaged a fine young fellow[,] a second Osborn."[31] Lieutenant Sherard Osborn, who would command HMS *Pioneer* in the forthcoming search expedition under Captain Austin, was a capable officer who lacked the look-down-the-nose superiority exhibited by some naval officers. He was approachable, open-minded, fair, enthusiastic, and cheerfully optimistic, qualities that had endeared him to both Penny and Lady Franklin. To attribute similar qualities to Alexander Stewart, Penny's choice as second-in-command, was therefore a great compliment. In the previous year, Stewart had sailed as mate of the Peterhead whaler *Joseph Green* under his father's command. Penny had recommended the young man to replace him as captain of the whaler *Advice*, but Stewart found the prospect of joining the search expedition more attractive. Penny was glad to have him.

For a voyage extending into arctic regions thousands of miles from home and lasting a year and a half and possibly more, it was important to have a reliable medical man on board, not merely a competent surgeon but a man who was also an agreeable companion, prepared to accept the discomfort and privations of shipboard life, willing to help organize cultural activities among the men during the winter, fit enough to march over ice and snow on extended man-hauling sled trips, and capable of finding his way across vast expanses of unfamiliar snow-covered terrain. Penny had been in touch with Robert Goodsir since their voyage together on *Advice* and discussed with him the possibility that he would obtain command of a search expedition. In mid-January, when the government's intentions were still unknown, Goodsir had written to Lady Franklin: "I need not say how anxious I am that I may have an opportunity of devoting my services to the cause during the ensuing season; and I hope I am not asking too much in requesting your good offices in obtaining me an appointment of any kind in any of Expeditions that may be sent out during the next season."[32] He wondered whether the Admiralty would send naval ships or whalers and told her he believed that whalers would be best, "particularly if the Commander is unrestricted by unnecessary instructions, and left to the guidance of his own good sense and experience"[33] – an opinion that Penny would have applauded. Lady Franklin forwarded Goodsir's letter to the Admiralty, and he was later able to join *Lady Franklin* as surgeon.

Another doctor who wanted to accompany Penny was Peter Cormack Sutherland. Already an experienced mariner, he had served as surgeon under Penny on the whaler *Saint Andrew* in 1845 and again in 1846. As Penny wished to have a medical man on each ship in case they wintered in different localities, he invited him to join the *Sophia*. Sutherland replied immediately, "I accept the appointment which you have kindly and with no small trouble obtained for me."[34] Another man, John Stuart, was appointed as assistant-surgeon to the expedition. He was assigned to the *Sophia* under Captain Stewart and assumed the duties of third mate as well.

The opportunity to embark on a search for Franklin brought offers of service from sometimes unlikely people. Among those who wrote directly to Lady Franklin (who forwarded applications to the Admiralty) was a curate in Somerset, the Reverend Joseph Wolff, a man of German birth whose geography was as bad as his English. He wrote, "I . . . offer my services by accompanying the expedition to the North Pole, or go by land to the Esquimaux, and make inquiry after Sir John Franklin." Despite his crimes against grammar and syntax, he claimed that his proficiency in languages would be an asset to a search expedition. "I learn any language in two months,"[35] he stated (without specifying how well).

Wolff was undeniably a courageous and enterprising adventurer. As relevant experience, he referred to a mission he had once taken to Bokhara on behalf of Lieutenant-Colonel Charles Stoddart and Captain Arthur Conolly, two British officers who in 1842 had paid the ultimate price for meddling in the touchy politics of central Asia, where Britain and Russia were constantly competing for the allegiance of local tribes and shifty rulers. The emir of Bokhara imprisoned the Englishmen in a vermin-ridden dungeon for months and finally had them beheaded. Rumours of the men's deaths were doubted by many in Britain, and Wolff was sent to Bokhara to ascertain the situation. Somehow he managed to get to the dangerous region, confirm the rumours of the men's execution, and return unscathed to write his *Narrative of a Mission to Bokhara*.

Now this intrepid voyager was volunteering for a mission to the polar regions. There were a few strings attached, however. He told Lady Franklin that if the Admiralty could see fit to find a curate for his parish, pay his replacement a hundred pounds per year, and provide Mrs Wolff with a like amount for upkeep during his absence, he would be willing to put his talents at the disposal of a search expedition. Not surprisingly, neither Lady Franklin nor the Admiralty felt inclined to meet his terms. It was not merely that his demands were outrageous. Lady Franklin and Sophia Cracroft, who had met the man, did not like him. The fact that he had jumped from Judaism through Catholicism into Anglicanism may have given an impression of insincerity or flightiness, and his appearance and manner did not meet their standards. Sophia described him as "very – corpulent is the only word – & by a reputation of many years standing – unclean . . . He has a thick snuffling intonation & is very nearsighted."[36]

In the hierarchy of a ship's command, the first mate occupied an important place. When the captain gave the order to weigh anchor, shorten sail, alter course, or perform any other of the numerous operations that enabled a ship to leave port, fulfill its functions and reach its destination, it was the mate's job to translate the order into action and ensure the work was carried out promptly and correctly. All of this required an experienced man, competent in all aspects of seamanship, cool in times of crisis, capable of getting the best out of the crew, and reliable at making entries in the logbook. It helped if he was good natured and sociable, so that he could ease the captain's burden of isolation and consult frankly with him if his advice were sought. For *Lady Franklin*, Penny hired as first mate John Marshall, a veteran whaleman who had made thirty-one voyages to the arctic regions. John Leiper, another experienced whaleman, was appointed second mate. To serve as mate on the *Sophia* under the young and relatively inexperienced Alexander Stewart, Penny was fortunate in getting Donald Manson, who had had a long association with arctic whaling and the Penny family. In 1817, he had commanded the *Superior* on a voyage to the Greenland whale fishery with Peterhead's nine other vessels, including the *Alert*, then under the command of Penny Sr. In 1821, when young Penny made his first voyage to the Arctic with his father on the *Alert*, Manson had been in charge of the Peterhead whaling ship *Pacific*. By 1850 he had made a total of forty-two arctic whaling voyages.

After commanding the *Pacific* for years without mishap, Manson had experienced his first shipwreck the previous year while serving as mate under Captain Arbuthnot on the *Superior*. In early June the ship was nipped in Melville Bay, along with *Lady Jane* and *Prince of Wales*. Penny's *Advice* and other ships were slightly further ahead: their commanders could see the crisis developing but were powerless to assist. The event illustrates the precarious nature of navigation along the Greenland coast. When strong onshore winds blew pack ice against the solid rampart of the land floe, any ships caught in between were at serious risk. Unable to move away, they were at the mercy of the wind-driven battering rams of ice. Unless the men had time to cut docks into the land ice and haul their ships in, they could do nothing but wait and hope that the weather gods were in a benevolent mood.

The *Superior* was lying tight against the abrupt face of the landfast ice with ice floes packed solidly around the rest of the hull, relentlessly grinding, crunching, twisting, crumbling, rafting one over the other, some forcing their way below water level and others sliding up higher than the gunwales. After a few days of intense pressure, ice pierced the hull below the water line and then, in addition to the frightful din made by the floes themselves, there was a dreadful sound of splintering planks, breaking timbers, and in-rushing water. Racing to save what they needed for survival, the men managed to get all seven boats onto the surface of the landfast ice, with some food, water, and extra clothing. From an improvised camp they watched the death throes of their ship. The nearby *Lady Jane* and *Prince of Wales* were destroyed in similar manner.

When the ice pressure diminished and some leads of water appeared, the survivors – numbering probably about 150 men – left the smashed debris of their ships and travelled about 500 miles south along the Greenland coast in whaleboats – a remarkable voyage. At the settlement of Lievely, they were given shelter and food until they were able to take passage home on a Danish ship. Despite this harrowing experience in the twilight of his long career, Manson was eager to set off again for the Arctic, this time in search of Franklin instead of whales, and happy to serve under Penny, the son of his former colleague.

The man hired as *Sophia*'s second mate was James Reid, from a well-known whaling family. His father, also named James, had joined Franklin's expedition in 1845 as ice master on HMS *Erebus*. His uncle Charles, a veteran of thirty-four arctic voyages, had offered his services to Lady Franklin in late January in the hope of joining the search for his brother, but she declined his services after Penny told her that Charles Reid couldn't command himself, let alone a ship. (He was later hired to command Penny's former ship *Advice*). Penny had no doubts about James Reid, however. He was confident that James Reid and Manson would both provide valuable support for Alexander Stewart.

As for men of lower ranks, there had been a surfeit of volunteers ever since the news of Penny's appointment reached the whaling fraternity. It was said that he could man an entire fleet if he so desired. The crews of his two ships would number forty-five all told. Their average age was just under thirty. One man was under twenty; twenty-seven were in their twenties, twelve in their thirties, four (including Penny himself) in their forties, and only one older – the well-seasoned Manson, fifty-eight years old. Of the forty-five men, at least thirty-six stated that they had been born in towns and villages on the Scottish mainland, four were English, and the rest of the birthplaces are unclear. More than half were from the ports of Aberdeen and Peterhead, from which Penny had sailed for most of his life. From the crew lists it is impossible to know how many had sailed on arctic whalers before because only their last ship is named. Ten, however, named a whaler as their last ship, and of these, five had sailed under Penny's command before.

7

Under Way

The opinion your Ladyship has formed of me if in my power
shall not be disappointed.

Penny to Lady Franklin[1]

Shortly before the *Lady Franklin* and *Sophia*'s departure, six hundred people attended the Mariners' Church for a special service "to commend to the care and guidance of the Lord the brave men who are about to set out on the benevolent and perilous search for Sir John Franklin" (*Aberdeen Herald*, 13 April 1850). Penny and his men were in attendance. Reverend Longmuir addressed the congregation, urging the two captains to include Bibles among their stores and to "establish and maintain worship on board their vessel." He asked the men to be "respectful and obedient to their superior officers; and kindly affectioned one towards another." Borrowing from Shakespeare, he expressed the hope that they would find Franklin in the "thick-ribbed ice" and return safely.

Penny's two ships left Aberdeen on 13 April. The harbour pilot came on board, and at 2 p.m. the crews took on board a hawser from the steam tug *Victory*, slipped their moorings, and were towed single file across the harbour, through its narrow mouth, and out to sea.

On board *Lady Franklin*, John Stuart, the third mate and assistant-surgeon, saw "the whole length of the spacious quays and piers . . . covered with spectators cheering us heartily."[2] From Regent Bridge all the way to the breakwater and the pier bracketing its mouth, both sides of the harbour were lined with people. The *Aberdeen Journal* (17 April) called the assemblage "one of the most numerous ever seen on our Quay, equalling even that which congregated to welcome our gracious Sovereign on her landing among us." The *Herald* (20 April) was optimistic about the outcome of the expedition: "There cannot be a shadow of doubt, that all that human energy, directed by extensive knowledge and experience, can achieve, will be accomplished by Captain Penny, unless unforeseen and unavoidable accidents intervene." Penny himself was confident of success. In a letter to Barrow from Pentland Firth a few days later, he said, "I hope to surprise you coming back with Sir John and his crews by way of Cape Horn."[3] If he found the missing ships somewhere within the arctic islands, he said, they might together manage to complete the Northwest Passage, circumnavigate the American continents, and return home in triumph.

Yet amongst the crowds of Aberdonian well-wishers, however, the reporter from *The Herald* (20 April) discerned a sombre undertone. "The prevailing sensation seemed to be one of deep, and, to some extent, depressing interest," he wrote. "As the flotilla passed the quays and north pier, attempts were made to get up a cheer, but they were chiefly confined to the juveniles, and almost entirely failed, apparently from a conviction that the occasion was not one on which exuberant manifestations of feeling could be considered appropriate." In addition to the sorrow and anxiety experienced by relatives and friends on seeing loved ones leave on a long, dangerous voyage, there was apprehension among the spectators that whatever calamity had befallen Franklin's large, well-equipped ships might overwhelm Penny's small vessels as well.

Outside the harbour it was discovered that two sailors were missing. To retrieve them caused a delay that must have strained everyone's patience, so eager were they finally to embark. The tug went back into the port and brought the men out, but it was not until six in the evening that the ships finally hoisted sail and put to sea. At the outset of the voyage, they were accompanied by William Hogarth on the yacht *Sylph*. The prominent merchant and shipowner who had been largely responsible for completing the rapid refitting and provisioning of the ships provided a fatherly escort all night. The following day, after a farewell dinner on board *Lady Franklin* with captains Penny and Stewart, he started back for Aberdeen while the ships continued northward to round the tip of Scotland.

Near Duncansby Head, where the coast turned westward, the two ships encountered an adverse tide running out of Pentland Firth and were forced to dodge back and forth in a bay under shortened sail to await a change. When the tide slackened, they continued into the firth, which, living up to its reputation, kicked up a confused sea that amputated one arm of the *Sophia*'s figurehead – the first casualty of the voyage. Pentland Firth was notorious among seamen, especially those manning sailing vessels. Currents, tides, and winds sometimes seemed to be engaged in a free-for-all. Three weeks later, after Austin's ships had forced a passage through, Sherard Osborn, commander of HMS *Pioneer*, would remark, "They who want excitement had better go and beat a vessel up the Pentland Firth, against both wind and tide. I tried it, but shall not repeat the experiment."[4] Osborn's ship had the advantage of steam power, but Penny and Stewart, with sail power alone, had to wait patiently for favourable conditions.

In view of his late departure and his desire to reach Baffin Bay as soon as possible, Penny had decided to bypass the Orkney Islands; he had hired all his men and taken on his provisions at Aberdeen. A stop in Stromness would not only delay their progress but would expose his crew to the charms of easy women and the dangers of strong drink. (Later, when Austin's ships stopped at the port on their way to the Arctic, Osborn would write, "The cheapness of whiskey appeared likely to be the cause of our going to sea with a crew in a lamentable state of drunkenness, and rather prejudiced me against Stromness."[5]) Two days out from Aberdeen, *Lady Franklin* and *Sophia* passed Cape Wrath, the mainland's northwest corner, and fifty miles further west, off the Butt of Lewis, they "took their departure," noting on the chart the last position fixed by bearings taken on landmarks.

Although it was mid-April, the North Atlantic was still in a boisterous mood. The logbooks recorded a succession of "brisk gales" and "strong gales" in which the crews often had to stow mainsails and jibs, reef foresails, and double-reef topsails. Whenever the weather moderated, they shook out reefs and hoisted the principal sails again. Conflicts between winds and currents sometimes created chaotic sea conditions. On one occasion, the *Sophia*'s log recorded "a high tumbling Sea[,] ship roling [*sic*] about a good deal," and on another, a "high confused sea running." Sutherland, who had sailed to Davis Strait with Penny before, wrote that "the sea was sometimes very rough and stormy, and its foaming waves raged and roared angrily around us, and appeared disposed to threaten our safety, [but] like true 'blue-jackets' we feared nothing and cared not, so long as the wind continued fair, and we could hold canvas to it."[6] Stuart observed that even after the oilskin-clad sailors had worked precariously aloft to take double reefs in the topsails, "squalls frequently heeled her over to an extent rather disagreeable to a landsman."[7] *Lady Franklin* proved to be a faster sailer than *Sophia*, and to keep the two ships in company through the variable conditions required constant attention by the watch officers and frequent sail changes by the crews.

On 17 April, they passed within a dozen miles of Rockall, the rugged summit of an isolated seamount. As it was shown on the Admiralty charts, the captains had another chance to check their longitudes and adjust their chronometers. Stuart described the uninhabited outpost of

British sovereignty as "a lonely spot in the middle of the Atlantic and holding out no temptation for Robinson Crusoeing."[8]

The crews were divided into two watches that alternated on duty for four-hour periods. While the men of the duty watch carried out the various tasks involved in operating the ship and maintaining its gear, the off-duty men had to eat, attend to personal chores such as doing laundry and mending clothes, and get some sleep – but never for more than four consecutive hours. At any time, the off-duty watch might be called upon to help the deck watch take in sail or perform some other urgent task. Two two-hour "dog watches" were inserted between 1600 and 2000 to vary the routine. In these periods, which occurred after most of the day's chores had been completed, men not required for specific tasks would sometimes get together in the forecastle, or on deck if weather permitted. It was as close to relaxation as they could get. Stuart recorded one such occasion, "In the evening the sunshine tempted the men on deck for a general sky lark. Their amusement finished with music & dancing about 8 pm."[9]

In contrast to the sailors, the surgeons Goodsir and Sutherland were men of leisure. They were outside the watch-keeping system and exempt from going aloft or performing hard physical work, and their duties were minimal. They bore responsibility for the health of the crews, to be sure, but during the transatlantic voyage there was little demand for their services – the only complaint on board *Lady Franklin* was toothache. The surgeons were usually called upon to officiate at Sunday services of worship, but at other times they were free agents. As both were keen naturalists, there were many fascinating things to lure them out of their cabins with telescopes, guns, fishing lines, and plankton nets. Greenland whales, fin whales, and white whales were often seen, and amongst the abundance of bird life, Sutherland observed mollymauks (fulmars), looms (murres), dovekies, puffins, kittiwakes, various kinds of gulls, snow buntings, eider ducks, and other species.

Two weeks out from port, the ships entered the colder environment of the East Greenland Current and experienced the first freezing temperatures, snow flurries, and drifting ice. The most spectacular of the ice formations, eagerly anticipated by everyone, were icebergs – immense blocks of freshwater ice calved from glaciers on land, white "ghost ships" moving imperceptibly with the current in their long, slow journey from the east coast of Greenland round Cape Farewell, north into Baffin Bay, and eventually southward to Newfoundland's Grand Banks and beyond, where they would die a slow death in warmer water and air.

Floes of saltwater ice, some miles across and others merely as broad as houses, were less lofty and spectacular than bergs but still of grave concern to the captains. In addition there were relatively small fragments of both fresh and saltwater ice, sometimes called "washing pieces" or "straggling ice" by the whalers. Because they were low in the water, they were hard to see and perfectly capable of smashing a hole in the wooden hull of a ship moving at several knots.

From this point, Penny's ships would be sailing in waters where pack ice and icebergs could be encountered at any time, so he and Captain Stewart posted lookouts aloft and kept well clear of ice whenever the visibility was obscured by fog, precipitation, or darkness. Penny described to Sutherland an incident that had occurred more than a decade earlier during a voyage from Davis Strait back to Scotland on the *Neptune*. They had been sailing confidently through the night when without warning a berg suddenly loomed out of the darkness. The ship passed so close to it that the mizzen boom scraped the ice – a hair-raising brush with disaster.

After rounding Cape Farewell, Greenland's southern tip, the ships altered course into Davis Strait. Having sailed westward for more than a thousand miles, they now had to sail northward an equal distance before crossing to the west side of Baffin Bay. By the beginning of May, they were near Rifkol Bank, where drifting deep-draft icebergs often grounded on shallow offshore banks and turned into stationary islands of freshwater ice. The grounded bergs tended to trap pack ice, building up an extensive barrier astride the whalers' route and forcing them to make lengthy detours away from the coast. Sutherland saw "hundreds of icebergs, with pinnacled tops and overhanging cliffs, streams of ice much broken up, and the impenetrable pack."[10]

May Day passed without the usual initiation of novices, to the disapproval of old-timers on board. Probably Penny was too sharply focused on conditions to let his men indulge in time-consuming charades.

After detouring around the Rifkol ice barrier, the ships made their approach to the Danish-administered colony of Godhavn on the south coast of Disko Island. British whalemen called the place *Lievely*, a corruption of the name Liefde Bai (Bay of Love), used by Dutch whalers in the eighteenth century for reasons one can only imagine. The boats were removed from their lashings on deck and slung from the davits. Men who were inclined to writing put the finishing touches on their letters home, while Penny worked on his official dispatches. At nine in the evening, still more than five miles from the settlement, a whaleboat and several kayaks were seen approaching. Stuart recorded his impressions of the dainty skin boats: "The quick motion of the paddle across and across struck me particularly and I was not a little surprised to see them coming off to such a distance though it is said that they have been found 20 miles off. At last they came alongside and eagerly sprang on deck each one steadying his neighbour's canoe (or kayak) to allow him to rise."[11] The Greenlanders indicated that the winter had been mild – a good sign – and everyone indulged in some casual "troke" or barter. Sutherland remarked on their "filthy" clothes but conceded that "they appeared to be a lot of hearty-looking fellows who could conduct themselves very respectably indeed. They cared little for anything that we could give them, and with the exception of a few tobacco purses, for which they took cotton handkerchiefs, they had nothing to barter."[12]

Late at night, the ships arrived at the settlement – some three or four houses for Danish personnel, a church, storehouses, and a number of crude native huts. The place performed diverse roles for passing whaling and discovery ships, serving as an information centre where captains could learn about ice conditions further north and ship movements on the coast, a post office where they could leave messages for other passing vessels or letters to be carried to Europe by regular steamer, a sanctuary for whaling crews whose ships had been wrecked further north along the coast, and a centre of relaxation where transient sailors could stretch their legs ashore and enjoy music and dancing with the friendly Greenlanders. On this occasion, however, the urgency of their humanitarian mission meant that time could not be wasted. Penny, Stewart, and the ships' doctors went ashore around midnight, visited the sub-governor of the district and his wife, and partook of a meal hurriedly set out by the hospitable Danes. They learned that ice conditions along the coast were unusually open and that several British whaling ships had already passed through on their way north. Penny was unable to persuade any native men to join the expedition, and there was no time to have fur clothing made up for his men, but there would be other opportunities further north. After leaving official and personal letters to be forwarded to England, and dispatches to await the arrival of Austin's expedition, he and his officers returned to the ships and got under way. The stop at Lievely had taken less than three hours.

The first (and easiest) phase of the voyage was now complete. The ships and men now faced five hundred miles of arduous and dangerous ice navigation along the Greenland coast and across Melville Bay to Cape York, before heading across the relatively ice-free North Water to the West Side. Penny had the satisfaction of knowing that no other search expedition had preceded him. He was in the lead, and if any openings presented themselves in the pack ice, he and Stewart would be ready to seize the opportunity to advance further as rapidly as possible.

Among the letters left at Lievely was one to Lady Franklin. After telling her of their rapid passage to Disko and expressing satisfaction with his crew ("as light-hearted men as ever came together under a command"), Penny reported that the region of Disko Bay was free of ice and the prospect for a fast passage to Melville Bay good. "My opinion is my Lady that there has never been such a favourable season for what we are about," he added with characteristic optimism.[13] He also left a letter to await his friend Osborn, commander of HMS *Pioneer*. A few weeks later when Austin's squadron arrived, Osborn would read and forward it (or excerpts from it) on board the returning transport *Emma Eugenia* to Sir Francis Beaufort. On receiving it, Beaufort notified Lady Franklin, who in turn wrote to Margaret Penny with the encouraging news that her husband had reached Greenland safely and was well ahead of everyone else. "He must have made a fine passage indeed," Lady Franklin told Margaret. "The officers of Captn Austin's ships all think so – Sir Francis Beaufort when he writes to tell me of it put a note of admiration, thus! – and I have written to him in reply, 'Our silver Penny is worth his own weight in *gold*.'"[14] The *Times* (28 May 1850) had recently reported that a silver penny from the reign of David I, King of Scotland during the twelfth century, had been found in Northumberland. Perhaps this remarkable archeological discovery influenced the tribute of Lady Franklin and Sir Francis Beaufort.

8

More Hunters on the Track

The more hunters on the track the better.[1]

While making frantic preparations for the voyage, Penny had been well aware that other search expeditions were getting ready to leave for Baffin Bay. In addition to Austin's four-ship Admiralty expedition, Sir John Ross was planning to depart in a small vessel, the United States was sending two ships under Edwin Jesse De Haven, and Lady Franklin was sponsoring a search expedition of her own under command of Charles Codrington Forsyth (the one naval officer with whom Penny had said he would be willing to share a command). The five expeditions would comprise ten ships, all headed for Lancaster Sound where the real searching would commence. Other expeditions were already active in the Western Arctic.

Although no one could know it, the potential area of ship navigation within the Arctic Archipelago would turn out to be unusually small that year, owing to extensive and persistent sea ice. The idea that more hunters on the track would be better was going to face a challenge in those close quarters. Too many cooks in a small galley could spoil the broth. Later Sherard Osborn would look back at the efforts of 1850 and sum up: "Fewer ships, the faster progress."[2]

* * *

Austin's squadron would be a formidable one, comprising 180 officers and men. Officers, some on active service and others on half-pay, had volunteered aplenty, but to find enough qualified men of other ranks at short notice, Austin went on a recruiting trip to northern ports in late February. He was looking for ex-navy personnel and men experienced in whaling or trading voyages, "especially those who have traversed the Arctic seas." But not just *anybody*, he made clear – he wanted obedient men who would support the cause with zeal and refrain from "evil-speaking" and "growling," proclaimed the *Nautical Standard* (2

March). In addition to the forty or so serving officers and men already available, the ships required sixty-four petty officers, forty-six able seamen, and twenty-six Royal Marines, for a total of at least 136 men. Austin urged volunteers to submit names, addresses, and testimonials. Those selected for the expedition would receive six months pay in advance, river pay from the date of their entry, and double pay starting on sailing day and continuing for the entire voyage. These were attractive terms. The *Britannia* (2 March) reported that in Hull alone Austin succeeded in hiring "nearly forty experienced men – mates and able seamen of whalers and other hard-weather craft."

Among his lieutenants were Francis Leopold McClintock, soon to become famous for his contributions to the technique of man-hauled sledging and his later discoveries at the scene of the Franklin disaster, and Sherard Osborn, who would write one of the most interesting accounts of the Franklin search. One of the young midshipmen was Clements Markham, who would become closely associated with the Royal Geographical Society and involved in arctic matters for most of his life.

As naval ships of the desired type and size were not available, the Admiralty had to find suitable merchant ships, purchase or lease them, and bring them to the Thames for refitting. Rumours circulated about the type of vessel to be employed. The *Dover Telegraph* (9 February 1850) reported that the expedition would consist of two steamers and two "dockyard lighters." Employing the latter class of shallow-draft vessels would certainly have delighted William Scoresby, John Ross, Jane Franklin, and William Penny, all of whom favoured small manoeuvrable vessels of shallow draft for the search, but dockyard lighters were definitely not what the Admiralty was looking for. They wanted large ships. A report that the iron screw steamers *Minx* and *Acorn* had been chosen as the tenders was said by the newspaper to be "merely another piece of idle Woolwich gossip," which it considered fortunate as they were "amongst the very worst that could be thought of." The newspaper thought it "more likely that the Archer, new screw steam sloop, now at Devonport, will be one of the vessels selected for the service." But the *Archer* was not chosen either. Four merchant ships were located, bought, and ordered to shipyards on the Thames to be strengthened for ice navigation, modified in other ways, and loaded with stores. The two sailing ships, *Baboo* and *Ptarmigan*, were of the same class, with tonnages of 430. The former was a dozen years old and had been employed in the emigrant service; the latter was a brand-new ship about to make its initial voyage carrying coal to India. The two powered sailing vessels, *Eider* and *Free Trade*, were about 400 tons, with 60 horsepower Boulton and

Watt engines that turned screw propellers. They too were sister ships, so Austin would have two pairs of nearly identical ships, facilitating the exchange of spare parts in case of trouble – a technique first employed by Parry in 1819.

The engined vessels were known as "steam tenders" because their function would be to "tend" the sailing ships – tow them through calms, break channels for them through ice, or speed off under power on important errands. They were rigged as barquentines – square sails on the foremasts but fore and aft sails on mains and mizzens. Tall funnels abaft the foremasts gave them a distinctive appearance. Their ability to utilize the power of either steam or wind, or both together, was a great advantage, but as they could not carry enough coal for prolonged steaming they would have to depend on wind most of the time, reserving the engines for calms and for work among ice.

Naval ships were traditionally given names that would instill courage among their crews and fear among their enemies. The Arctic was a foe like no other, and its "conquest" required not only valour but appropriate ships' names. It would be unreasonable to launch an assault on the formidable arctic ice with ships called *Ptarmigan* and *Eider*: enemies do not tend to cringe in terror when confronted by a brace of birds. The name *Free Trade* might serve well for a merchant vessel, but was it relevant for a naval ship? As for *Baboo*, the Admiralty Lords may have felt that it was too suggestive of a visible majority in a distant part of the empire.

In late February the *Nautical Standard* reported that the names *Baboo, Ptarmigan*, *Eider*, and *Free Trade* would be changed respectively to *Endeavour*, *Perseverance*, *Succour*, and *Refuge*. But apparently the Admiralty still had doubts. Was the name *Refuge* too passive? Would the name *Succour* become the brunt of sailors' crude gibes? In its next issue (2 March) the newspaper announced another revision. The Admiralty's final choice of names would be *Resolute*, *Assistance*, *Pioneer*, and *Intrepid*.

By the third week of February, *Free Trade* was on its way from Leghorn, Italy, and the three other ships had entered docks on the Thames for outfitting. Work on the vessels was allotted top priority and distributed among several facilities downstream from Tower Bridge, including Thompson's yard at Rotherhithe, the Wigram and the Green shipyards at Blackwell, and the Royal Dockyard at Woolwich. Shipwrights, engineers, caulkers, riggers, sailmakers, and other specialized tradesmen were brought in from jobs at Chatham, Deptford, Devonport, and Portsmouth. Men posted to the ships for the expedition were temporarily accommodated in a receiving ship moored in the river and taken each day by boat to work on their respective ships. So great was the demand for labour that convicts were delivered daily, probably from one of the prison hulks moored in the Thames, to assist with unskilled tasks such as scraping, painting, and loading coal. Through March and into April, the work continued apace. From early in the morning, boats shuttled back and forth with tradesmen and labourers, seamen, and convicts, and work sometimes continued by the light of lanterns through the night. Lighters delivered materials and parts from scattered factories and shipyards. Provisions from naval dockyards at Deptford and as far away as Portsmouth were brought in to warehouses at Woolwich.

In addition to the doubling, caulking, strengthening, and insulating carried out on the sailing ships' hulls and various alterations to their masts, yards, and rigging, the steamers required specialized attention to their power systems. Their engines were replaced with ones for which parts would be interchangeable and their boilers were lifted out and overhauled. The most vulnerable parts on the steamers were the propellers – massive, three-bladed metal screws measuring more than eight feet across. These were changed for new two-bladed ones, and to protect them from dangerous ice pressure below the water line, they were made to be removable. Wells into which the detached screws could be lowered were constructed, which made it necessary to completely rebuild the "deadwood" at their sterns. As an additional precaution, each steamer was provided with a spare propeller. Practicality was the foremost consideration. The carved figureheads were replaced by solid unadorned wood, as those of Franklin's ships had been. Spaces below decks were criss-crossed by strengthening beams.

By 8 April, the sailing ships had completed their structural modifications and had thirty tons of coal stowed below. They were then towed downriver by the steam tugs *Advice* and *Jasper* to the naval dockyard at Woolwich to take stores on board and undergo final adjustments to their rigging. Meanwhile, the steam tenders were testing their engines for the first time. Workers arrived on the *Intrepid* at 3 a.m. to get steam up for a few hours. When that was complete, the steamers moved under their own power to Woolwich, averaging an impressive 5.5 knots (6.6 miles per hour). The loading of stores proceeded at full speed, with supplies coming on board simultaneously from the wharf on one side and lighters on the other.

In the days leading up to the departure, the master shipwright and his foremen were busy directing the work, the superintendant of the dockyards was on hand to supervise, officers from the Steam Department were consulting with the chief engineer, Austin and his senior officers were constantly on the prowl, the surveyor general of the navy came to inspect, and members of the Board of Admiralty and several old arctic hands – Back, Beechey, and Parry among them – could not resist a visit or two. There were social occasions as well. At Woolwich, all the officers of the expedition were dined and undoubtedly wined at the messes of the Royal Artillery and the Royal Marines. The festivities were not confined to military establishments. On 25 April, the *Times* remarked, "All the wealthy inhabitants of the town have vied with each other in having the honour of their company previous to their departure."

When the loading was finished, the ships were shifted fifteen miles downriver to Greenhithe. "A large number of the inhabitants of Woolwich assembled on the wharf adjoining the Woolwich Baths to witness the departure of the vessels," reported the *Times* (26 April). At Greenhithe the ships were "swung" to adjust their compasses for deviation caused by iron objects on board. On one of the steamers, the binnacle compass – the critical instrument by which the helmsman would maintain a course – was found to indicate magnetic North 17 degrees away from its proper direction, an error probably created by the engine, the huge iron propeller, and its spare, or some other nearby machinery. The deviation on various headings was corrected with compensating bars of iron.

Not surprisingly, the preparations generated much publicity and London papers kept readers informed of developments. After the ships moved to Greenhithe, they became popular tourist attractions, despite the distance from central London. The *Times* (1 May) reported, "Some idea may be formed of the anxiety to see these vessels previous to departure, when it is made known that upwards of 70 highly respectable persons left the London-bridge station of the North Kent Railway at 7 o'clock a.m. today for the purpose of going on board the vessels at Greenhithe, and numerous parties proceeded with the same object in view by each succeeding train." But when the men of the expedition were paid on the day before departure, Austin ordered a notice be posted at the rail station terminating public visitations. The measure, noted the *Times* (5 May), was "to prevent the inconvenience of crowding when the men were being paid" – in other words, to keep the sailors' money well clear of the sticky fingers of London pickpockets and con men.

At last, as quay-side crowds cheered, HMS *Resolute*, *Assistance*, *Pioneer*, and *Intrepid* started on their mission, assisted by steam-tugs *Advice*, *Comet*, *Dasher*, and *Jasper*. (In many minds, the Northwest Passage and the North Pole were synonymous; was it a coincidence that two of the tugs bore the names of Santa's reindeer?) Another steam-tug, the *African*, helped later in the towing to Stromness, and the steam-tug

Lightning towed the expedition's store ship *Emma Eugenia* (incorrectly referred to in the *Times* on 23 April as the *Maria Eugene*).

Of the five search expeditions intending to leave Britain and the United States for Baffin Bay in 1850, Austin's was the largest in terms of ships, men, and resources, and by far the most complex in its preparations, typifying the Admiralty's approach to polar exploration. The *Nautical Magazine* called it "the most completely fitted expedition that ever sailed from this country for the Arctic Seas."[3] Because the Arctic was a remote and dangerous region, expeditions had to take a capsule of the home environment with them in order to survive and operate effectively. Ships were not only vehicles transporting men to the Arctic; they were also domiciles, shelters from the severe climate during ten frigid winter months, storehouses for provisions and gear, workshops for making and repairing gear, bases for exploration by man-hauling sledge parties, and hospitals for the sick or wounded. Big ships tended to be more comfortable than small ones, and with their large crews they could send out more sledge parties to search, so the official view was that bigger was better. Within the naval establishment, only the hydrographer Sir Francis Beaufort believed that smaller vessels might have some advantages.

The argument in favour of large ships overlooked several important factors. The first was that big ships manned by large crews required a lot of support, which tended to make an expedition cumbersome and dependent. Austin's flotilla required a "freight ship" or transport to carry some of its supplies to Greenland, and as the transport *Emma Eugenia* was slow, it had to be towed from the Thames to beyond Orkney by the steam tender *Lightning*. The sailing ships *Resolute* and *Assistance* were also towed by the tugs *Dasher* and *African* so that their steam consorts *Pioneer* and *Intrepid* could save their coal for the Arctic. Altogether, eleven ships sailed from London in Austin's expedition, but only four of them would enter Lancaster Sound to carry out the search. The other seven were needed simply to get them there.

A second disadvantage of large ships was the length of time needed for preparations. To get the four searching ships ready for sea had required an immense effort affecting a number of shipyards, naval dockyards, engineering works, and factories of one kind or another, along the Thames and elsewhere, drawing skilled tradesmen away from other pressing work. The transport and all the towing vessels also needed to be put in shape for sea, loaded with the necessary stores and coal and adequately manned, putting additional demands on the Thames dockyard facilities and workforce. The complexity of all the operations required to prepare all the ships undoubtedly contributed to their late departure. Austin's ships left on 3

May, almost three weeks after Penny's left Aberdeen. In January, Parry had emphasized the need for *immediate* action. Writing to Beaufort, he said, "Some great men of former days remarked that 'our *expeditions* are generally *tardations*' – and so it is still."[4] Writing to Barrow two months later, as the various search expeditions were still making final preparations, Parry returned to his theme: "I agree with you in the lateness of all our attempts. It is somehow quite true that we are always 6 or 8 weeks too late. But I know not how you and I could help it – for we never ceased to stir the matter up by every means in our power."[5]

A third disadvantage of large ships on arctic expeditions was their sluggishness. Compared to small ships, they were slow to get under way, set sails, change tacks, moor to the ice edge, or come to anchor. Their restricted manoeuvrability could present problems among pack ice, in shoal waters, or in places where sudden changes of course were required.

A fourth drawback was their deep draft, which barred them from relatively shallow waters and increased the risk of grounding on reefs or banks. When Austin's loaded ships started downriver from Greenhithe, the steamer *Intrepid* had a mean draft of twelve feet, six inches, and the others drew from ten to more than thirteen feet. (Five years earlier, Franklin's ship *Erebus* had left Disko drawing seventeen feet, four inches, prompting Commander James Fitzjames to remark, "We are very deep and full as an egg.")[6] W.J.S. Pullen, commander of HMS *North Star*, a few years later championed the use of small vessels in far northern waters: "In all cases of arctic navigation they are much better adapted: no vessel ought to be of greater draft than eight feet; they would then keep off the ground if driving, and rise better in a nip."[7]

Although it was convenient to have a dedicated transport to carry extra coal and food for the expedition as far as Greenland, the transfer of stores to Austin's ships took up valuable time. The ships arrived at the Whalefish Islands on 14 June but did not complete their loading for another ten days.

A major challenge for all arctic navigators was the shortness of the ice-free season. The long, hazardous voyage northward along the Greenland coast to Cape York reduced the potential sailing time on the West Side even further. Whaling captains always departed early from Britain – often in March – in order to maximize their time at the whaling grounds along the coast of Baffin Island, but Admiralty expeditions headed for the same region seldom got under way until mid-May – *tardations* indeed!

* * *

In addition to the government expeditions of Penny and Austin, together comprising six ships and 226 men, a few privately sponsored expeditions were preparing to leave. The old sea dog Sir John Ross had emerged from retirement to join the search. Of all the Royal Navy's arctic officers, he was the most colourful, unpredictable, and contentious, and arguably the most egotistical and least credible. He had joined the navy at the tender age of nine, risen to officer rank, served on HMS *Victory* a few years after Nelson's death at Trafalgar, and was said to have been wounded a dozen times in battle. When Britain renewed the quest for a Northwest Passage in 1818, Ross had commanded one of the first two arctic expeditions. On that venture, his curious decision to retreat from Lancaster Sound, which he claimed to be a cul-de-sac terminating in a range of mountains, had led to scathing criticism and heartless ridicule (including a cruel cartoon in *Punch*) after his former subordinate, William Edward Parry, sailed easily through the non-existent "Croker Mountains" the following year and came very close to completing a Northwest Passage. From this nadir, Ross had miraculously rebounded, embarking on a remarkable private expedition in 1829 on the paddlewheeler *Victory*, the first engined vessel used in the North American Arctic, and returned four years later with almost all of his men alive after abandoning the *Victory* in the Gulf of Boothia. This outstanding achievement had earned him a knighthood.

That had been more than twenty years before. Now he was seventy-three, an improbable age for arctic adventures, yet he was getting ready to depart for Lancaster Sound in a tiny vessel manned by only sixteen men. Many considered him much too old for the job. In 1845 Sir John Franklin's age had been a subject of concern, but Ross was fourteen years older than Franklin had been at that time.

A friendly and supportive letter reached Ross from William Scoresby, the ex-whaling captain and author of the definitive work on whaling and the Arctic, now pursuing his second career as a minister of the church. Like Ross, Scoresby had once been denied command of an Admiralty expedition; and, like Ross, he was in favour of using small ships in the search for Franklin. In 1829 he had offered to serve under Ross and was told that the crew was complete, but held no grudge. In a gentlemanly gesture, he now praised Ross's venture and made a financial contribution to it. According to the *Nautical Standard* (23 February 1850), he told Ross, "In regard to your age, on which remark has been made, my own belief is, that your naturally robust constitution still amply existing, with your anxious zeal to undertake this enterprise, will be a sufficient compensation." A few weeks later, after a newspaper editor suggested that Ross was not fit enough to undertake an arctic voyage, the *Nautical Standard* (9 March) published an angry rebuttal from a man who signed himself "An old sailor who has served under Sir John Ross." The man wrote, "His ability is well known . . . and no one but some rude barbarian or savage of an uncivilized clime, can be so base, so vile, as to put forth such an erroneous assertion . . . when it can be proved that, though [old in] years, he nevertheless possesses a constitution well adapted to the hardships of an Arctic Sea." The letter ended with a parting salvo: "Such diabolical falsehoods are a disgrace to those who uttered them."

Ross never lacked self-confidence and never doubted his own physical fitness. He declared that he possessed a "constitution extremely well adaptd to the climate,"[8] and when addressing a group of notables at Ayr shortly before departure, he told them he had an iron constitution. The claim was not disputed.

He had a tendency to boast, however, and sometimes his statements were questionable, if not downright false. When he stated that during his six years as consul in Sweden he had become well acquainted with "sledging by dogs" a reader who had resided in Sweden wrote to the editor of the *Times* (18 February 1850): "It does not require six winters, or half a winter's, residence in Sweden to ascertain that 'sledging by dogs' is utterly unknown in any part of that country . . . My own belief is, that there is not a single team of dogs employed between the North Cape and Elsinore." Ross's claim, he said, was "all humbug."

Ross regarded steam power as an indispensable tool in the search. He had studied steam engines and their application, had written a book on the subject, and was the only officer in the Royal Navy who had ever taken a steam-powered vessel to the Arctic – albeit an improbable side wheeler that never came back. He now wanted to purchase another paddlewheeler, a packet called *Asp* at nearby Portpatrick, to be accompanied by "a small clipper brig, such as the *Isla* of Aberdeen."[9] He planned to man the *Asp* with twenty-four and the *Isla* with fourteen including officers, making a total of thirty-eight (not thirty-four as he stated in his book). In addition, he intended to tow his own twelve-ton, mahogany-hulled yacht *Mary* unmanned to the Arctic. Three vessels and thirty-eight men: that was what Ross wanted.

Ross had been promised support from the gin magnate Sir Felix Booth, who had financed his earlier expedition, and whose name Ross had bestowed here and there in the Arctic. But in late January his benefactor died in a hotel in Brighton after a fit of coughing. Ross was now forced to scrounge frantically for funds, appealing personally to friends and influential persons. Fortunately, the Hudson's Bay Company agreed to open public subscriptions with a target of £3,000, handle the financial aspects, and donate £500, so he was able to proceed with his plan, but not on the ambitious scale he had envisaged. He would now take one sailing vessel and his little sailboat *Mary*, but no paddlewheeler. This was a cruel blow for the enthusiastic proponent of steam power. One wonders how much more he might have achieved if Felix Booth had lived to sponsor his return to the Arctic in an engined vessel.

After examining ships available at Troon and Ayr in March, Ross selected a new schooner nearing completion at the Sloane and Gemmell shipyard in his home town of Ayr. It could be purchased for £1,350 including strengthening and other modifications necessary for arctic waters. It was 10 April by the time the Hudson's Bay Company accepted the proposal, and then preparations went ahead with the frantic haste common to most arctic expeditions. It would be wrong to say that everything moved like clockwork. In addition to the usual confusion of ordering, receiving, and stowing hundreds of items of food and equipment from a variety of firms, there were overdue accounts, repeated demands for payment, and threats of legal action against Ross and the Hudson's Bay Company. Evidently, Ross had to attest to the receipt of every article, after which the Company was supposed to pay the suppliers, but both parties were very busy, Ross trying to complete a thousand and one things before departure, and the Company trying to raise money and pay bills at the same time. It is hardly surprising that there were some loose ends.

One of the complainants was the manufacturer George Shepherd, who had supplied balloons to the searching expeditions heading to Alaska and Baffin Bay. He had understood Ross to have ordered ten balloons, an apparatus and materials for inflating them, and a press to print messages on satin, and he had sent some of the items to Woolwich to be put on board Ross's yacht *Mary*. Ross, however, denied placing an order, insisting that he had merely been inquiring about prices. Two days before Ross's departure Shepherd offered to pay for the zinc, acid, satin, and apparatus, if Ross would pay for the balloons, and said he could send everything by train from London that night. Ross must have agreed, because the balloons reached Ayr in time, but the dispute about payment continued. Five months later, by which time Ross was in winter quarters in the Arctic, Shepherd asked the Hudson's Bay Company to pay something – even a mere ten or fifteen pounds – as he himself had paid forty to fifty pounds on Ross's account and could not "afford to pay any more."[10]

The Company had sent notices to prominent newspapers in mid-February to advertise the fundraising campaign, and requested the help of customs collectors, magistrates, bank managers, club presidents, and other influential officials in many British cities to solicit donations from prominent persons and the public at large. The response was – to say the least – disappointing. Letters came back from dozens of officials who had failed to raise a single penny. Lloyds of London gave nothing. At Greenwich, the magistrates considered it "inexpedient to subscribe."[11] At Hull, the mercantile interests were "nearly altogether adverse to assist," while bankers showed "a reluctance to contribute anything." At Southampton, a "trifling" amount was obtained. At Goole, the collector of customs had "not been enabled to procure any subscriptions." At Scilly, the merchants and shipowners "decline subscribing to the proposed expedition." Port cities pleaded recent losses of ships, and inland cities pleaded agricultural distress.

Some persons were reluctant to support Ross's private venture because the government was already sending out two expeditions that were expected to do the job. Others considered him too old. The secretary of the Reform Club declared that he was unable to send anything, but he did recommend for the use of the expedition his own invention, the "Magic Stove."

While many bankers, industrialists, and shipowners withheld contributions, many working-class people sent in sums they could barely afford. At Liverpool, "various humble individuals at the Caledonian Hotel in Duke Street" sent a pound and a half, to which the collector of customs added his own two pounds. In Bantry Bay on Ireland's southwest corner, Coast Guard boatmen submitted a total of thirteen shillings. Some contributions were anonymous: from "a friend of the noble hearted Fitzjames" (five pounds); from "a friend to any and every means of search" (ten shillings); from "nil desperandum" (ten shillings); from "an aching heart" (two shillings six pence).

Problems continued long after Ross had escaped into the arctic seas. The wives of his crew members were supposed to receive half of their husband's wages while they were away, beginning on 1 August. In early July, Elizabeth Jones, whose husband was serving as Ross's steward, wrote to the Hudson's Bay Company to say that she had been promised two pounds per month, but on applying had been told that she would get only one pound six shillings. Her husband had trusted Ross, she said, and had left "with full hopes of providing for me." Now she was left "without the means of subsistence." There was rent to pay and food to buy, and times were hard. In late August, the Company received an appeal on behalf of two women in Peterhead, Mrs. Alexander Soutar and Mrs. Jas. Robertson, whose husbands were also serving with Ross. They too had been expecting the half-pay but had received nothing. Both had large families and were "solely dependent on this money." The mother of another man in Ross's crew, Jas. Fraser, was in a similar situation.

As the Hudson's Bay Company was trying to raise money during March and April, Ross was busy with the numerous practical details of the preparations, such as obtaining all the necessary equipment and provisions and hiring the crew. The *Felix* (named for his departed benefactor) would have a total complement of sixteen. Ross's second-in-command was Charles Gerrans Phillips, who had been a lieutenant on HMS *Terror* during James Ross's antarctic voyage. The *Terror*'s commander on that voyage, Francis Rawdon Moira Crozier, was now out with Franklin, commanding HMS *Terror* again, so Phillips had a strong personal motivation to join the search. He had been disappointed in his desire to be given command of HMS *North Star* on its supply voyage for Ross's expedition in 1849, so he had seized the opportunity to sail with Ross's private expedition, withdrawing from a course on steam engines that he was taking in London.

In January, in a letter to Reverend William Scoresby, Phillips had indicated his agreement with views the ex-whaling master had expressed about the search. Phillips added his own opinions, suggesting that Cabinet should discuss the matter, that Russia and the United States should participate, and that public meetings and petitions might help the cause. He also recommended employing some of "our gallant whalers."[12]

Phillips was one of the few officers to express concern about the potential duplication of effort among the several searching expeditions setting out in 1850. He would have preferred to be heading south in Prince Regent Inlet rather than west in Barrow Strait, he told Lady Franklin. "I . . . sincerely wish Sir John [Ross] would carry out *your* plan instead of pushing on in the direction so many other vessels intend to take."[13] His reservations about the destination in no way diminished his enthusiasm, however, and in the days leading up to departure he assisted Ross with various aspects of the preparations.

Ross had hired a doctor for his expedition but the man's use of alcohol went far beyond the preservation of anatomical specimens. Ross discharged him and was fortunate in securing another doctor, David Porteous, at the last moment. The position of sailing master on the *Felix* would be filled by Thomas Abernethy, whose extensive polar experience included six whaling voyages, two discovery expeditions under Edward Parry (one extending over a winter), four years with Ross on the *Victory* in Prince Regent Inlet, four years in antarctic waters, and a subsequent winter in Barrow Strait with Ross.

The two mates, Alexander Sivewright and Alexander Middleton, had sailed many times to the Davis Strait whale fishery. The *Nautical Standard* (16 February) reported that the remainder of the crew consisted of "twelve handsome brawny tars from Peterhead . . . brought up in the whale fishery" – a boatswain, a carpenter and his mate, a cook, a steward, and seven seamen. The crew of the *Felix* was not short on experience.

Another concern for Ross was getting his yacht *Mary* from London to Ayr. She reached the west coast of Scotland by an unusual route. The steamer *Widgeon* towed *Mary* up the east coast to Inverness, then into the Caledonian Canal, which occupies the Great Glen, a ruler-straight fault valley slicing southwestward across Scotland from Moray Firth to the Firth of Lorn. In 1803, work had begun on a bold engineering project, the construction of a navigable waterway from sea to sea through this natural corridor. The waterway would provide a shortcut between the North Sea and the Atlantic Ocean for British warships so that they could cope with a military threat from France on either coast. A sea-level waterway was not feasible because the drainage divide was too high to cut for a canal, but by the relatively new technique of raising and lowering ships by pumping water in and out of locks, the natural obstacle could be overcome. Under the direction of Thomas Telford, the brilliant engineer who later modernized the Scottish road system and the harbour at Peterhead, the work of constructing twenty-nine locks and dredging connecting channels continued for nineteen years. By then, however, the threat of invasion by the French had subsided, and warships had become too large to use the canal. It was deepened in 1847 and is still in service for small craft.

The steamer towed the *Mary* through the locks, over the height of land, and back to sea level on the west coast at Fort William, through Loch Linnhe, southward through the Firth of Lorn and the Sound of Jura, then around Kintyre to Ayr on the Firth of Clyde. To the *Mary* belongs the distinction of being the only sea-going vessel in the Franklin Search to cross Scotland overland, to ascend more than one hundred feet above sea level, and to traverse the Atlantic with no one at the helm. She also had the distinction of being abandoned in the Arctic, but in this she was far from unique, as half a dozen other vessels would meet the same fate during the search period. But whereas the others were either wrecked or abandoned in the ice, the *Mary* would be intentionally left high and dry on the shore, ready for use by Franklin survivors or searching expeditions.

Sir John Ross had a flair for publicity and self-promotion, and his preparations for the search expedition were carried out in front of a supportive home crowd. The launch of the *Felix* on 17 May was a festive occasion at Ayr. The *Aberdeen Journal* (22 May) reported that at a dinner afterwards he reminded merchants and notables that he had promised his "excellent friend" Sir John Franklin to initiate a search if no news came by December 1847, and when he had volunteered to lead an expedition to find him, the Admiralty had refused his offer. He painted himself (not without justification) as a determined underdog who had overcome official barriers and was now ready to fulfill his promise to "the gallant Franklin." He told them he was in favour of using small ships such as the locally built *Felix*, and he praised his mates, each a veteran of dozens of arctic whaling voyages, and the other whalemen who had recently arrived from Peterhead.

When the steamer *Briton* towed the *Felix* out, with *Mary* trailing obediently astern, Ross recorded in his diary that ten thousand spectators lined the harbour, while others sailed on board the *Briton* and the chartered passenger steamer *Scotia*. The cheers, he wrote, were deafening. After dropping the towing hawser, the *Felix* made its first sea voyage, an eighty-mile passage south to Loch Ryan. To prolong his home-town advantage, Ross intended to make a second departure from Stranraer, where he owned a residence called North West Castle. The *Felix* dropped anchor directly in front. Ross's arrival and the expedition's imminent departure for the Arctic had been eagerly anticipated: "Flags floated over various parts of the town, and the shipping in the harbour, during the day, in compliment to Sir John," reported the *Aberdeen Journal* (29 May 1850). On the following day a local steamer towed the *Felix* and the *Mary* away from Stranraer to set off across the Atlantic on the great humanitarian mission.

Unfortunately, the high spirits prevailing at Ross's two departures were offset by serious crew problems. At Ayr the men had asked for a larger pay advance, and when Ross refused to grant it, they got fighting drunk in town. Eight men were still absent when the *Felix* left. At Stranraer the situation was worse. An account from the *Shipping and Mercantile Gazette* reprinted in the *Aberdeen Journal* (5 June 1850) describes the occasion:

> We are sorry to learn, from our Stranraer correspondent, that the Felix, Sir John Ross's vessel, went to sea from Lochryan in a sad state of disorder, from the continued drunkenness of the crew during the whole of the time the vessel was in the Loch. The sailing-master had drunk himself into a state of insanity, *delirium tremens*, and was in his bed; the mate was little better, perhaps worse, for he was furious

> with drink; and the whole of the crew were much in the same state and positively refused to weigh the anchor or make sail on the vessel. This had to be done by the crew of the steamer, and some parties from Stranraer who wished to accompany Sir John out of Lochryan. On reaching the anchorage off Cairnryan, five miles down the Loch, the men insisted upon anchoring, and force had to be used to prevent their doing so. Their gallant old officer, however, is a man of firm nerve, and, with Commander Phillips, showed equal coolness and determination under those difficult circumstances. He had full confidence in his management of the crew, once the vessel was fairly clear of Lochryan.

Speaking earlier at the post-launch dinner in Ayr, Ross had boasted that he had never had to punish a man in his last three commands; he had always acted like a father to his men, treating them as he would his own children. His remarks went over well with the audience, who responded with cheers. What a tolerant and sensible chap, they must have thought. But a modicum of discipline might have kept the unruly crew in order before sailing from Ayr and Stranraer.

There had been a precedent to this sorry episode. In June 1829, when Ross was preparing to depart on his four-year expedition in the *Victory*, he had purchased the Greenock whaling ship *John* (once commanded by William Scoresby Sr) to carry supplies, reasoning that as the two ships advanced northward through Davis Strait and Baffin Bay, the fifty-four men on the whaleship could pursue whales to help to cover the expedition's expenses and provide a share for the crew. At the last minute, the *John*'s crew demanded a guarantee of wages equivalent to those accruing to a full ship, claiming that their late departure would ruin any chances of whaling success. Ross rejected the demands; the crew of the *John* refused duty; and the *Victory* sailed alone. Perhaps it was just as well, for the *John* became an arctic *Bounty* and its captain a polar Bligh. A year later, while on a whaling voyage, some of the crew mutinied in Melville Bay, set the captain and those loyal to him adrift in a boat, and continued under command of the spectioneer, who managed to wreck the ship on the coast of Baffin Island. Most, if not all, of the mutineers perished. The men who had been cast adrift earlier succeeded in reaching safety at a Greenland settlement.

The *Felix* staggered off toward the Atlantic with a badly hung-over crew. Thomas Abernethy, the sailing master, was in the worst shape of all. Despite his extensive polar experience – or possibly because of it – he was totally disabled by drink and was expected to be "laid up" for at least a week at sea. David Porteous, the ship's doctor, severely understated the situation when he wrote to his father hours before departure, "Abernethy has been very poorly today."[14]

Abernethy had turned down the position of ice master on Austin's expedition so that he could go with Ross. In a letter to the governor of the Hudson's Bay Company, later published in the *Nautical Standard* (9 March), he explained that he had seen Austin's ships but preferred Ross's small schooner. "I am of opinion," he wrote, "that his small vessel the Felix, is by far the best suited for the navigation among ice, and will therefore be more likely to succeed in the performance of this service." But John Ross's biographer and descendant, the late Admiral M.J. Ross, noticed that the original letter, now in the archives of the Hudson Bay Company, is not in Abernethy's handwriting, but in that of Ross himself. Like a skilled ventriloquist, Ross had caused his opinions to emerge from the mouth of Abernethy. A testimonial from an experienced ice master, he reckoned, would increase public confidence in the expedition and attract more financial support.

It is likely that Abernethy's real reason for sailing with Ross rather than Austin was not that he preferred small ships but that he disliked serving under strict naval discipline. He knew that on board the *Felix* a few harmless aberrations (such as getting blind drunk before departure) would be tolerated by the commander who treated his men as he would his own children. On a naval ship, however, it would be an offence punishable by flogging. William Parker Snow, second-in-command of Forsyth's *Prince Albert*, heard other whalemen express reluctance to serve on navy ships. They told him they "shouldn't like to be aboard of any ship where there's always a man with a stick in his hand walking by the side of you, or not far off, and where's there's so many officers."[15]

Recognizing that overindulgence in alcohol could seriously undermine the crew's effectiveness, Ross took steps to prevent it. He limited the use of spirits to Saturdays only, and when the ship reached Disko, wrote in his journal, "After sermon on intemperance, I told the people that I was determined to enforce the law against drunkards."[16] His intentions were good, but a sermon was not enough to prevent men who craved alcohol from getting access to it.

* * *

A day before Sir John Ross left Stranraer, two small American brigs departed from New York harbour, bound also on the quest for Franklin. Up to this time the United States had not played an official role in the Arctic. Some of the nation's whaling ships had pushed their blunt bows northward along the eastern flank of the continent, and others had begun to penetrate Bering Strait from the Pacific, but no government expeditions had joined the futile quest for maritime trade routes over the roof of the world. As early as 1753, a private discovery expedition supported by Benjamin Franklin and sponsored by Philadelphia merchants had proceeded up the Labrador coast to the entrance to Hudson Strait, but when a petition to establish a scientific base on Baffin Island was submitted to Congress a third of a century later, it met with governmental apathy. The disappearance of Sir John Franklin's two ships was widely regarded as a British problem whose solution required no assistance from the United States.

This attitude made an about-turn, however, thanks to the initiative taken by Lady Franklin, the spark who ignited so much of the search effort. Aware that American ships routinely sailed to the whaling grounds of Davis Strait and the North Pacific, she sought to involve them and perhaps persuade the government to take some official action. Her emotional letter to the American president in the spring of 1849, which a member of the House of Commons called "the most admirable letter ever addressed *by* man or woman, *to* man or woman,"[17] set in motion the chain of events that led to the first official American expedition into arctic seas.

Lady Franklin hoped to enlist the aid of Russia as well as the United States. "It would be a noble spectacle to the world," she wrote to President Zachary Taylor, "if three great nations, possessed of the widest empires on the face of the globe, were thus to unite their efforts in the truly Christian work of saving their perishing fellow men from destruction."[18] A few weeks later John Clayton, the American secretary of state, replied on behalf of the president, promising that "every proper effort may be made by this government for the rescue of your husband and his companions. To accomplish what we have in view, the attention of American navigators, and especially our whalers, will be immediately invoked."[19] In England this news was received with joy. Summarizing opinions expressed in a London paper, the *Nautical Magazine*, hoped that although it was late in the season to organize an expedition, Captain Charles Wilkes, former commander of the US Exploring Expedition of 1838–42, might manage to reach Baffin Bay "'ere the door be closed for the winter months."[20] This was wishful thinking, however. There would be an American expedition, but not for another year.

The main driving force behind the move to send a search expedition from the United States was Henry Grinnell, the former manager of Grinnell, Minturn and Company, a large New York shipping firm whose ships (including the famous tea clipper *Flying Cloud*) sailed all round the world. Grinnell's concern about the missing British expedition, spurred on in the first months of 1850 by correspondence with the persuasive Lady Franklin, grew by bounds. After first deciding to contribute $5,000 to any expedition she sent out, he raised the sum to $10,000, then said he might go as high as $15,000 if necessary, and finally told her on 4 March, "I have now committed myself for $30,000."[21] In the end he offered to purchase two ships himself, on condition that the US navy outfit and man them. This arrangement, however, was not immediately approved by Congress, which was concerned with more pressing matters such as whether the territory of California, which had been taken by force from Mexico a few years before, should be granted statehood and, if so, whether slavery should be permitted there. Furthermore, many in the government felt that sending American ships to the Arctic to search for some overdue British expedition would not serve the "national interest."

The scepticism and opposition were overcome largely through the efforts of Grinnell himself and another remarkable man, Matthew Fontaine Maury, superintendent of the Navy Department's Depot of Charts and Instruments. While Grinnell enlisted the help of his brother Joseph, a member of the House of Representatives, Maury used his considerable influence with the secretary of the navy.

Earlier, Maury had written a treatise on navigation and had undertaken an ambitious project to compile geographic and oceanographic data from a hitherto overlooked source of information: the logbooks of hundreds of trading and whaling ships that regularly sailed the world's oceans. His research had led to the publication of a wind and current chart of the North Atlantic that greatly reduced sailing times between North America and Europe. This and other scientific achievements had garnered him international recognition and numerous honours. He was now eager to extend American scientific research into the arctic regions, hoping to validate his pet hypothesis that a large ice-free ocean surrounded the North Pole. In July 1849, he had presented a strong case for American involvement in arctic research. Lady Franklin's plea for help provided a perfect opportunity to send an expedition north. A famous portrait of Maury, painted posthumously by an American artist, Ella Sophonisba Hergesheimer, born in 1873, the year of Maury's death, shows him in dress uniform standing beside a large globe. His left hand grasps the hilt of his sword and his right rests upon the model of the earth, palm down over the North Pole – symbolic of his interest in the Arctic and his firm belief in the existence of an Open Polar Sea.

Although Grinnell was deeply moved by the dreadful predicament of Lady Franklin's husband, he was also motivated by the lure of the Northwest Passage. Writing to Maury in mid-February, he suggested that two small, ice-strengthened ships could "start sometime in May direct for Barrow Strait[,] explore the Wellington's channel, no doubt here something would be found from Franklin giving information as to what had been done and what he intended to do – continue on, rescue Franklin – continue on & make the North West passage, go to San Francisco and come home in steamers."[22]

Grinnell's confidence about rescuing Franklin now seems unreasonable, but at this early stage in the search, all the expeditions heading to Lancaster Sound expected to find messages from Franklin indicating his intended course, and they were fairly sure that with this information they could track down the missing navigators. As for the nonchalant way in which Grinnell spoke of the ships continuing to the Pacific, we should remember that three decades earlier, when the British Admiralty sent William Edward Parry in search of a Northwest Passage, they had instructed him to send copies of his journals from Kamchatka across Siberia to Europe after he sailed through Bering Strait. Since that time, a series of maritime and overland explorations had filled in more gaps in the arctic maps and charts, and each cartographic addition and improvement had made a maritime transit more feasible. By 1845, when Sir John Franklin set sail, there was much more justification for optimism than in Parry's time, and Franklin had been ordered to send an officer home with dispatches from the Sandwich Islands or Panama. Confidence was not lacking during the long search for a Northwest Passage.

Like Maury, Grinnell had an intense desire to stimulate research about the arctic region. He would soon become one of the founders of the American Geographical and Statistical Society (now the American Geographical Society), and would be elected as its first president in 1851, although he did not assume office immediately. A historian of the society has called him "the leading spirit in focusing a substantial part of the Society's attention on the Arctic and in initiating its long tradition of interest in polar exploration."[23]

During the first months of 1850, Grinnell and Maury worked feverishly behind the scenes to sway the opinions of legislators. Grinnell's genuine concern for Franklin and his companions, and his passionate interest in arctic geography, combined with Maury's well-founded pleas for oceanographic and meteorological research, convinced many that the United States could play a major role in the remote region between the continent and the Pole. At last Congress empowered the secretary of the navy to go ahead with preparations for the arctic expedition, but by the time the bill passed it was the beginning of May. From that moment on, everything was characterized by extreme haste.

A man on Maury's staff, Lieutenant C.R.P. Rodgers, volunteered to command the expedition, possibly on Maury's suggestion, but he failed to get the job. Grinnell did not see eye to eye with Rodgers's notion that a large vessel of 250 tons or more should be included in the expedition, and that the ships should winter over. He felt that wintering was a waste of time; it would be simpler and safer to send the ships back to the Arctic in the second summer if needed again. He probably did not understand that the most effective method of exploring and searching in the Arctic was not *by* ship but *from* a ship. Sending sledge parties out in late winter from a vessel safely embedded in the arctic ice was the best technique. The season for sledging was approximately as long as the season for ship navigation, and several sledge parties fanning out in various directions could cover far more ground than a ship confined to a linear track. Cost had to be considered too. Rodgers' large-ship plan, probably modelled on British naval practice, would require a capital outlay of much more than Grinnell was prepared to provide.

Before the government agreed that the navy would man the ships, Grinnell thought of recruiting whalemen. He was a part-owner of the whaling ship *McClellan*, which had made several voyages to the Davis Strait fishery from New Bedford, Massachusetts. On making enquiries in that port, however, he learned (as he explained to Lady Franklin in early March) that "all the officers and men fit for such an expedition had gone to California."[24] The lure of gold out West had brought about a serious manpower shortage on the Eastern Seaboard. All he could do was advise the *McClellan*'s captain that during his next whaling voyage he should "look out for Sir John Franklin."[25]

The sequence of letters from Grinnell to Lady Franklin during March, April, and May reveal the rapid march of events leading up to the departure of the American expedition: he intends to send two small ships commanded by Lieutenant Edwin Jesse De Haven, a veteran of the Wilkes antarctic expedition (5 March); he has bought a ninety-one-ton ship (19 March); he has purchased a second ship (1 April); there is still a problem of getting crews (16 April); the bill authorizing an expedition has passed the House and Senate (2 May); the ships will carry relevant books and maps, as English ships do (7 May); portraits of Sir John will hang in each ship (14 May); he intended to call the ships *Lady Franklin* and *Rescue*, but as the former name has been taken (by Penny) he has adopted the name *Advance* (14 May); both ships are at the Navy Yard taking stores on board (14 May); and finally, "I have now the satisfaction to inform you that the Arctic Expedition is positively off" (28 May).[26]

By "positively off," Grinnell meant departed, not cancelled. He and his two sons had just returned from three days at sea on the *Washington*, one of the port's pilot boats, which had been put at their disposal so that they could accompany the *Advance* and *Rescue* at the outset of their voyage. Grinnell had played a vital role in conceiving the venture and had worked strenuously to secure government support. He had bought the ships and they were his. But they were now heading for an unfamiliar environment in which everything was uncertain. Like a proud but anxious father seeing his children leave for college or a distant job, he had fussed over the departure, savoured the last moments, delayed for as long as possible the final adieu, and then with mixed emotions watched them go.

As preparations for the search expedition were getting under way, the *Nautical Standard* (9 March) reprinted an item from the *New York Herald* that had discerned "an intense excitement" in the port. Officers and men, inspired by "a lofty chivalrous sentiment," and keen to "rub off the rust of idleness," would soon be rushing to get a berth on one of the ships. "They love excitement," the newspaper declared, and they were anxious to "revel in the luxuries of an exploring voyage." In one respect, the *Herald* was right: De Haven and his men would get plenty of excitement. Opportunities to revel in luxury, however, would be sadly missing.

The creation of an American expedition was loudly applauded in Britain, with one reservation. Find Franklin? Yes, by all means. But complete the Northwest Passage? Definitely not cricket! The *Aberdeen Journal* (26 June) wrote, "Happy, indeed, should we be if the American expedition succeeded in rescuing Franklin; but it would be too great *magnanimity* on our part to wish them to make the north-west passage, which, year after year, we have endeavoured, for three hundred years and upwards, to effect, but hitherto without success – unless, indeed, Franklin yet comes out through Behring Straits, as we still think he may." The honour of fitting in the last few pieces of the geographical puzzle and of becoming the first to navigate between the Atlantic and Pacific oceans belonged to the country that had sent out so many maritime and overland expeditions, methodically filling in the blank spaces on the charts. It was unthinkable that another country might win the prize, especially one whose government had not previously undertaken a single venture into the Arctic. Similar sentiments had been expressed when James Ross returned in 1849 without finding a trace of Franklin. An anonymous pamphleteer, who identified himself only as "an old officer of rank in the Royal Navy," had written: "The cry is upon England; and England does not stand alone in the world. The eyes of many nations are upon her: their sympathies are aroused: their nerves are braced for the race; and if she be remiss, they will win the prize, and pluck the crown of glory from her brow! Where glory ends, shame begins."[27]

* * *

Although the Admiralty had instructed Franklin to turn southward at or near Cape Walker and find his way to the continental coast, they did not think that he had gone that far. A few months before Austin's expedition departed, Parry – who exercised a powerful influence on Admiralty policy – noted that although John Rae had explored beyond northwestern Hudson Bay in 1846–47 and wintered in the region, he had heard nothing about ships or white men from the Inuit he encountered. As for the region around King William Island and the estuary of the Back River he told Barrow, "I do not think, it is a part *worth exploring*, as, if they had reached that part of the Coast, we should have learned something of them there."[28] Parry made this statement in March 1850. He could not know that bones and belongings of Franklin's men were at that moment lying in the very region he considered *not worth searching*.

The British strategy for the Franklin Search in 1850 is summarized on an anonymous hand-coloured circumpolar map in the archives of the Royal Geographical Society. Although undated, it was evidently compiled in the winter or spring of 1850 before Penny and the others departed. The map vividly conveys the extraordinary magnitude of the area in which Franklin and his men might be located. On the map, this vast region is divided into different zones of searching responsibility. The westernmost zone, stretching from Bering Strait through the Beaufort Sea to Banks Island, was to be covered by the expedition under Thomas Edward Laws Moore, which was already in Alaska, and by the ships of Richard Collinson and Robert John le Mesurier McClure, which left Britain in January to make their way through the Atlantic Ocean, round the tip of South America, and northward through the entire length of the Pacific to Alaska – halfway round the world.

The easternmost zone, including most of Davis Strait, Baffin Bay, and the Parry Channel westward to Melville Island, in which James Ross had already undertaken an unsuccessful search, would be the responsibility of the expeditions under Austin, Penny, and John Ross, now about to depart.

Between the eastern and western flanks of North America, Richardson and Rae would approach the continental coast overland through the Subarctic to examine Banks and Victoria Islands.

For the area extending from Prince Regent Inlet to Victoria Island, south of the Parry Channel, the map caption reads: "Portions of the arctic sea left unprovided for in the search for Sir John Franklin by the Admiralty, but for which the proposed expedition under Cap^{n} Forsyth is intended."[29] Forsyth's venture was known as a "Branch Expedition," because after entering Lancaster Sound with the ships of Austin, Penny, and Ross, his ship would turn southward toward the continental coast, while all the others would continue westward or turn north.

Forsyth would be the only one heading in the right direction.

* * *

Two weeks after Ross and De Haven set off for Baffin Bay from opposite sides of the Atlantic, the little *Prince Albert* got under way from Aberdeen – the last search expedition to leave. It was sponsored in large part by that tightly wound human dynamo, Lady Franklin. She had already worked tirelessly to get Penny his Admiralty command and had succeeded in persuading the United States to send out two ships. Now she intended to dispatch her own expedition as well.

Many volunteers had offered their services, and she had appointed Commander Charles Codrington Forsyth, RN, a friend from the Franklins' Tasmanian days, to lead her expedition. Forsyth obtained leave from the navy and served without pay. He had been determined to take part in the search and had initially hoped to go with Penny, but the Admiralty had said no, it would not be proper for an active naval officer to serve under a whaling captain. (For a whaling captain to serve under a naval officer, on the other hand, was considered to be perfectly acceptable.) Forsyth had evidently applied to Austin as well; in late February he said that if rejected by Austin, he would volunteer for Lady Franklin's expedition.

Like most naval officers of his rank, Forsyth had criss-crossed the world's ocean on a variety of missions – chasing slavers off the west coast of Africa, surveying along the coast of South America, supporting military operations in South Africa, sailing to Burma and Australia. Among the various ships to which he had been assigned since entering the navy in 1826 was HMS *Beagle* under Captain Robert Fitzroy (not at the time of Darwin's voyage, however). Now he was in command of both a ship and an expedition, about to undertake his first venture into arctic seas.

Another man who had initially wanted to sail with Penny but ended up on the *Prince Albert* was William Parker Snow, regarded by a recent writer as "one of the more bizarre characters to step across the polar stage."[30] A restless and adventurous man, a "rolling stone"[31] who knocked about the world through much of his life, he has also been described as "undependable, disputatious, litigious, and possibly paranoid."[32]

Yet he was a man of broad interests, considerable talents, unbounded imagination, and irrepressible energy. After graduating from the Royal Naval School in Greenwich, he had joined the merchant marine, sailed to India, Australia, and Borneo, and worked ashore at various jobs. He had joined the navy, but soon resigned from it, married, immigrated to Australia, and for a time managed a hotel in Melbourne with his wife. His career before 1850 has been described as "up and down . . . high in incident, low in achievement."[33] He told Lady Franklin, "A rough life in many parts of the world had long been mine. Alike acquainted with the snowy Alps and the icy regions of the southern seas, I should not hesitate upon a journey to the arctic shores if ever required to undertake it."[34]

After participating in Forsyth's expedition of 1850–51, Snow would write one of the most colourful narratives produced during the Franklin Search. He was, as the classicist and historian Leslie Neatby has said, "the wielder of a very florid pen."[35] His experience on board the *Prince Albert* whetted his appetite for more arctic adventures, and three years later he would set out on a search expedition of his own, a noteworthy one in several respects. It was the only search to depart from Australia, the only one on which a commander was accompanied by his wife. His ship *Thomas* was the smallest manned vessel ever involved in the search, with only eight people on board. Unfortunately a number of unforeseen circumstances brought the venture to a halt well south of the Equator, with seven thousand miles still to go to Bering Strait.

Snow later commanded a ship belonging to the Patagonian Mission Society in Tierra del Fuego and the Falkland Islands, ran a club in Italy, travelled widely in Europe, and worked at various literary pursuits. A talented writer, he assisted the famous historian Thomas Babington Macauley in England and helped to transcribe and edit the journals of Charles Francis Hall in the United States. (He claimed that he actually *wrote* Hall's book *Life with the Esquimaux* from start to finish.)

Snow had a soft spot for disadvantaged people and a predisposition for getting involved in "good causes"[36] such as promoting an emigration scheme and campaigning for greater safety measures at sea. The plights of oppressed peoples, besieged missionaries, lost explorers, or helpless captives in a foreign country were certain to inflame his mind. When Joseph Wolff, the man whom Lady Franklin considered unsuitable for her private arctic expedition, had been preparing to travel to Bokhara to rescue Stoddart and Conolly in 1842, Snow is said to have volunteered to go with him, but his offer was rejected and Wolff went alone. A few years after the Franklin search expedition, in true Wolff fashion, Snow would propose to travel alone to Ethiopia to secure the release of

English prisoners held by the ruthless King Theodore II. To gain access to the royal stronghold, he intended to masquerade as a deaf and mute lunatic. (What advantage this would have given him is not clear.)

Many of his ideas were unorthodox – original but poorly conceived and utterly impracticable. He proposed a system of floating telegraphic cables crossing the worlds' oceans. (Would they not impede the movements of ships?) He proposed a worldwide maritime safety net consisting of lightships stationed every three hundred miles along major trade routes and kept in position by steam power, to mark the way for passing vessels and store emergency shelter and food for distressed crews. (How would their food and fuel be replenished? How would they be protected against pirates?)

Snow was fascinated by the British search for a Northwest Passage and allegedly applied for a position with Franklin's expedition of 1845 but was not accepted. He followed the progress of the search for the missing expedition very closely. He claimed to possess clairvoyant powers and while sleeping one night in early January 1850 had a powerful vision in which he saw two abandoned ships beset in arctic ice, lifeless bodies nearby, and small groups of men scattered about the landscape, apparently calling for help. The location of the disaster was perfectly clear and remarkably prescient – the region near King William Island and the estuary of the Great Fish (now Back) River. The next morning he wrote to Lady Franklin, proposing an overland expedition from Hudson Bay. A hundred men would leave Moose Factory in southern James Bay "at the opening of summer" and travel northward to Chesterfield Inlet in "small craft." After landing, the expedition would split into three groups and cross the Barren Grounds along separate routes to the search area, presumably on foot. Each group would be divided into three small parties of ten men following the same path at intervals. Snow assured Lady Franklin that he would be honoured to serve in any capacity, without remuneration, and advised her to take up residence in "one of the principal towns in Canada" so that she could supervise the search effort.[37]

Snow had never been to the Arctic, and anyone examining his proposal would wonder about a number of important practical details apparently overlooked in his scheme. How many boats would be needed to carry a hundred men, and where would they be obtained? Would they be large enough to carry all the expedition's food, camping equipment, fuel, guns, and ammunition, as well as the men themselves? Would they be propelled by oars or sails? Would experienced boatmen be available? At what date would the ice cover of Hudson Bay break up enough to allow them a passage? How long would the flotilla take to reach Chesterfield Inlet, a thousand miles to the north? How would the large party subsist while traversing the Barren Grounds to King William Island and then searching in its vicinity? How would they cross intervening lakes and rivers? How would they be able to help Franklin and his men if they happened to find them? And how could Lady Franklin "supervise" from a Canadian town almost two thousand miles from the search area?

Never one to do things in half-measures, Snow outlined an even more ambitious plan in the event that a "public and more extensive Expedition" could be organized.[38] This would involve as many as three hundred men crossing the vast territory between Great Slave Lake and Hudson Bay, and a like number manning ships, presumably sailing between England and Hudson Bay. If it proved difficult to obtain so many men, he suggested, the government should employ convicted criminals. Convicts would be "precisely those required," as they were known to possess "physical powers of endurance" and "almost inexhaustible mental resources." In addition, employing criminals in such a humanitarian undertaking would be far better than "uselessly incarcerating them for years, at an unprofitable expense."[39] (Perhaps his idea of using convicts in the Franklin search had been inspired by the suggestion made almost two decades earlier by George Manby after accompanying William Scoresby on a whaling voyage, that convict settlements should be established in Greenland.)

Snow's search proposals contained a few good points but in most respects bordered on the fabulous. Large parties such as he envisaged could not hope to live off the land, nor could they carry enough food to be entirely independent of game resources, and if they did happen to encounter a large number of men from HMS *Erebus* and *Terror*, they would be unable to provide any real relief because their own resources were so meagre. Yet it is remarkable that Snow's suggested area of search was precisely where Franklin and his men met their end. Whether by reason, intuition, or paranormal vision, he had identified the very region on which search expeditions should converge. Another man, Dr Richard King, who had actually travelled across the Barren Grounds to the arctic coast on a small expedition under George Back, had also pointed to the correct area to search. Both men were ignored, however, largely because they appeared to have exaggerated perceptions of their own judgment and ability, exhibited impatience with anyone who held contrary views, were outspoken in their criticism of the Admiralty's prosecution of the search, and had a great talent for annoying people. Consequently, they were shunned by the naval establishment, and their opinions were never given a just assessment. Lady

Franklin forwarded Snow's letter and search proposal to the Admiralty a month after receiving it, but the matter ended there.

Snow was living in the United States at the time. Intent upon joining a search expedition, he boarded a ship for Britain, but arrived "too late, by a few days only, for Captain Penny's vessels, in which I had a hope held out to me of an appointment."[40] By this time, however, Lady Franklin had decided to send out her own expedition and, impressed by Snow's maritime experience, knowledge of navigation techniques, and passionate devotion to the cause, she appointed him to the *Prince Albert*. Another factor in her decision was that Snow's view on her husband's location concurred perfectly with her own.

Snow may not have revealed to Lady Franklin that his ideas on Sir John's location were based on a vision, but if she had known about his alleged paranormal powers, she might have hired him even earlier, for she herself was an ardent believer in spiritualism. In the spring of 1849, she and Sophia Cracroft had visited the clairvoyant, Ellen Dawson, who in a mesmeric trance " travelled" to the Arctic and "saw" Sir John's two icebound ships and another pair not far off, presumably those of James Ross. Six months later, when Lady Franklin and her niece went back for an update, Dawson revealed that the ships were still beset but were now closer to Bering Strait than to Baffin Bay. In February 1850 the two women consulted another clairvoyant, the son of a naval lieutenant by the name of Morrison. He used a crystal ball to communicate with a pair of friendly spirits, Orion and Gego, who declared that Franklin was alive and well and was now making his way homeward from somewhere north of Melville Island.

Other people also consulted spiritualists in order to discover Franklin's location. One of these was Alexander Maconochie, a naval captain who had served as private secretary to Sir John in Tasmania and who had become famous for his efforts to improve conditions in British prisons (a cause Snow would have applauded). After unsuccessful visits to clairvoyants in Paris and London, he approached Joseph Haddock, the handler of a young woman known as "Emma the Bolton Clairvoyant." On at least six occasions during the year following 21 September 1849, Emma succeeded in "seeing" Franklin. Lady Franklin and Sophia kept abreast of Emma's pronouncements, but were sceptical. They were more favourably impressed with information from the spirit of a dead Irish girl, Louisa ("Weesy") Coppin, whose image often appeared before Captain Coppin and his family, ready to answer questions. When asked about Franklin, Weesy is said to have caused a map to appear on the floor. It showed the *Erebus* and *Terror* and a series of place names scattered from Lancaster Sound through Prince Regent Inlet and across Boothia Peninsula to King William Island, their sequence apparently indicating Franklin's path. Captain Coppin kept Lady Franklin informed of subsequent revelations. He is said to have visited her no fewer than thirty times. Details about Weesy's visions and their influence on Lady Franklin were published forty years later by Reverend J. Henry Skewes, a vicar in Liverpool, in a book about Franklin and "the true secret of the discovery of his fate."[41] Skewes declared that Weesy's vision had caused Lady Franklin to instruct Forsyth to take the *Prince Albert* down Prince Regent Inlet to try to reach King William Island.

* * *

For Forsyth's expedition, Lady Franklin did not try to buy or charter a whaling ship or commission one to search on her behalf; her fingers had been badly singed on an earlier occasion. She learned that Robert White, a shipbuilder and friend, owned a ninety-ton, two-masted sailing vessel built only two years before at his yard in Cowe. It had carried fruit between the Azores and England and may have been employed as a pilot boat. Forsyth and a master shipwright from the Woolwich dockyard inspected the vessel in London's Surrey Docks and gave their approval, whereupon Lady Franklin bought the *Prince Albert* (as it was renamed) and arranged for it to be outfitted in Aberdeen. William Kennedy, who would command the *Prince Albert* on a second search voyage a year later, described it as "a small fairy-looking craft, which might have been taken for a yacht, or pleasure boat, intended for a summer cruize [*sic*] along the picturesque shores of Scotland."[42]

On Thursday, 2 May 1850, the steam tug *Jasper* towed the *Prince Albert* out of the Surrey Docks, down the Thames, past the Woolwich dockyard, past Austin's mighty fleet at Greenhithe, into the North Sea, and up the coast as far as Harwich before setting it loose to spread its sails. The steam tug then returned to Greenhithe to assist Austin's ships, which intended to depart in a few days.

In Aberdeen, fifty shipwrights and carpenters descended upon *Prince Albert* to transform it into a rugged vessel capable of resisting the onslaught of arctic ice. William Hogarth, who had supervised the outfitting of Penny's ships, again directed operations. The *Aberdeen Journal* (29 May) summarized the alterations: "She is doubled fore and aft with extra beams inside at the water-line to resist the pressure of the ice, with plate-iron about the cutwater, bows, and quarters. All the bulkheads have been fitted to resist the cold." According to the newspaper, the *Prince Albert* was almost exactly the size of Ross's *Felix* and was "ketch-rigged with squaresails on her foremast." A week later (5 June), the newspaper announced that the vessel, "having been thoroughly refitted and strengthened, has been taken to the upper end of the dock at the foot of Market Street, where her stores and provisions have been put on board." Forsyth was there, and Lady Franklin had recently arrived from London by steamer "to see the gallant voyagers depart on their hazardous enterprise." She was somewhat miffed to learn that, on seeing the size of the *Prince Albert*, the surgeon hired for the expedition had decided at the last minute not to go. Forsyth told Barrow that the man "came down and took a look at the vessel and then retreated without showing himself to any of us."[43]

Early in the morning of 5 June, spectators and well-wishers began arriving at the harbour, their numbers increasing steadily during the forenoon. Women whose husbands or sons were either missing with Franklin or on their way to the Arctic with one of the other search expeditions came on board with letters, parcels, bits of news from home, hopes for a speedy return, and pledges of devotion and love. They prayed that the *Prince Albert* would encounter their absent menfolk somewhere in the vast arctic spaces and deliver their "silent yet heart-speaking memorials."[44] Commander Forsyth had already said his adieux to his wife and child, and to Lady Franklin, at Hogarth's house. The ship had cleared customs and taken on board bonded stores and inflammable goods. Friends and families of crew members were invited on board to accompany the ship through the harbour; there was hardly room to move on deck. As the ship was shifted into the outer harbour, a huge crowd watched from nearby quays and bridges. The pilot boarded, and the ship's colours were hoisted. It was time to get under way.

In letters to John Barrow Jr, Forsyth described the crew as "a very decent well disposed set of fellows." Not only had they requested prayers by a clergyman before leaving, but when it was time to depart, they were all "perfectly sober."[45] Snow too found this rather unusual: "Not a single instance of drunkenness or insubordination," he recorded with amazement.[46] It was a departure quite in contrast to that of Ross's *Felix*.

As the steam tug *Victory* towed the *Prince Albert* away from the quay, "cheer upon cheer rang through the air; voices shouting out adieus innumerable sounded upon the ear like another Babel; waving of handkerchiefs, hurrahs, hats and caps thrown on high, was the kindly farewell of the honest-hearted Aberdonians."[47] Lady Franklin too watched from quayside, glad to see her little ship off on the search. William Hogarth, who had by now refitted three ships for the Franklin search at record speed, escorted the *Prince Albert* to sea, as he had *Lady Franklin* and

Sophia. Sailing along with them in his yacht for a while, he told Forsyth that his little ship was twice as fast as Penny's. This was welcome news, as Penny was more than seven weeks ahead.

* * *

Forsyth's was the last of the five search expeditions to depart. Ten ships were now heading for the eastern Arctic of North America, and one supply vessel (HMS *North Star*) was presumed to have already wintered there. Other ships were either in the western Arctic or approaching it, and John Rae of the Hudson's Bay Company intended to search part of the central coast and cross to Victoria Island. The search had assumed impressive proportions.

On 9 March 1850 the *Aberdeen Herald* reprinted an article from the *United Services Gazette* which calculated that twelve ships and 615 men were on Admiralty-sponsored arctic expeditions (including that of Franklin, those out searching for him, and those about to depart). Impressive as these figures are, they understate the full dimensions of the far-reaching maritime enterprise that had begun with Franklin's confident departure in search of a Northwest Passage. If one includes the wintering search expedition of James Ross and the three support voyages to Alaska by Henry Kellett on HMS *Herald* (not included in the newspaper's compilation), as well as the private search expeditions of John Ross and Forsyth, and the American expedition commanded by De Haven, the total comes to twenty-one ships and more than 830 men. Of these, Franklin and his two crews comprised only 129 men. All the rest – nineteen ships and about 700 men – were engaged in the maritime search triggered by the disappearance of HMS *Erebus* and *Terror*. This figure does not include the various steam tugs, towing vessels, and transports that assisted at the start of many of the voyages – some going as far as Greenland – or the steamer *Gorgon*, which towed Collinson's sailing ships through the Strait of Magellan to the Pacific. The Franklin search was only in its third year. Considering that maritime expeditions and overland searches would continue until McClintock's discovery of the tell-tale evidence on King William Island in 1859, it is no wonder that it has been called "one of the greatest man-hunts ever staged."[48]

9

North from Disko

In no part of the world are the preconceived schemes of men more likely to be thwarted than in the Arctic Regions.[1]

Having reached Disko ahead of the other search expeditions, Penny held high hopes of maintaining the lead as his ships advanced up the Greenland coast and across the feared Melville Bay to Cape York. That position would put them near the margin of the ice-free polynya of the North Water and enable them to cross rapidly to the West Land. It was not beyond reason for him to think that his expedition might be the first to sight Franklin's vessels or at least find some evidence of their passage. Before leaving Scotland, he had written to Lady Franklin, "I drop this note to say farewell for the present, high in hope of being the means, through Providence, of carrying aid to your noble husband and his brave companions."[2] Now it appeared as if he might actually be able to make that high hope a reality. In a dispatch addressed to the Admiralty on 2 May (which the Danes at Lievely would send on to Europe by the next supply ship), he wrote, "It has been a very mild winter, and very much open water here, extending off 70 miles."[3] In a letter to Lady Franklin, he declared that conditions had never been more favourable. But like all the men who sailed regularly to the Davis Strait whale fishery, he knew enough to restrain the elation generated by an early arrival at Disko. Experience had taught him that pack ice could be a ruthless leveller of skill and good fortune. The ice floes of the Middle Pack, moving in response to variable and unpredictable winds, could easily pin ships against the floe edge and bring the vanguard of the whaling fleet to a halt, negating any advantage earned by bold and capable commanders and enabling the slowpokes and laggards to catch up. (In a modern analogy, cars proceeding on a highway at full speed encounter an obstruction and are reduced to a crawl or brought to a halt, slower cars steadily accumulating behind them.) Maritime traffic jams along the Greenland coast could occur at any time. As Sutherland remarked, "Any one that has been in Davis Strait has learned not to lay plans for tomorrow while there."[4]

After leaving Lievely around 2 a.m. on 3 May, Penny's ships enjoyed good sailing "allong the bould and magestice Island of Disco," as the *Sophia*'s mate Donald Manson recorded in the logbook. They passed Hare Island and by nightfall were off the tip of the peninsula of Nugssuak. Ahead was the large indentation that whalemen referred to as "Northeast Bay," or "Jacob's Bight." In fact it was the broad *embouchure* of Umanak and several other coalescing fiords, roughly eighty miles across. Here they encountered an impenetrable body of pack ice driven in against the landfast ice. Several whaling ships were already there waiting for an opening, among them the *Saint Andrew* from Aberdeen. Seeing the ship he had commanded for more than six years may have stirred up disquieting memories for Penny, such as the time the ice of Melville Bay left the ship with a large part of the keel missing, more than twenty timbers broken, and tons of blubber and oil sloshing in the hold from casks squeezed beyond breaking point. That had been in 1845, the same year in which HMS *Erebus* and *Terror* sailed into oblivion. Franklin and his crews had never come back, and now, five years later, a dozen expeditions comprising hundreds of men were out searching for him. The safety of the missing navigators might depend on whether ice conditions along the Greenland coast retarded the searchers or allowed them to proceed swiftly on the trail. How would Melville Bay treat Penny and his crews this time?

There was no time for musing about bad experiences in past seasons, however. The wind was increasing to gale force, the ice pack was beginning to move, and snow was falling thickly. Penny and Stewart worked their ships eastward to the secure edge of the land ice, set out their ice anchors, unshipped their rudders, and settled down to wait out the ice pressure. The crews were put to work slinging boats from the davits to be ready for towing, getting hawsers from the hold in preparation for warping, and putting up the crows' nests to con the ships through the floes.

The compacted floes filling Northeast Bay made a direct crossing impossible. Penny and Stewart would have to grope their way around the interior of the bay along the long arc of the floe edge. For the present, even that possibility was denied them because there was no space for a ship between the pack and the land ice. Soon the ships were beset about twenty miles northwest of the small settlement of Niaqornaq, and the outlook began to seem less and less favourable.

Day after day the ice barrier persisted and hopes faded. Sutherland later wrote, "The passage across the Atlantic and thus far up the strait had occupied but three weeks. To have advanced so far in so short a time was very satisfactory, and produced pleasant reflections; but owing to the apparently advanced state of the season, from the condition of the ice in this and other deep bays, we had every reason to suppose that our arrival had been at least one week too late."[5] Penny and other whaling captains had observed that early in the season northeasterly winds tended to prevail, pushing the floes away from the land ice and leaving a channel in which the ships could work northward, but the later southwesterly winds tended to hold the Middle Pack in against floe edge, closing off their routeway.

Frustrated in the keen desire to proceed swiftly on their humanitarian mission, the men of *Lady Franklin* and *Sophia* may have found a modicum of solace in not being alone in their predicament. Eight or nine whaling ships were also congregated at the margin of the pack (or, in the case of the *Abram* of Hull, firmly beset within it), all waiting to advance northward to Cape York and cross to the West Side. Penny's ships were up front with the leaders, but positions of the various ships relative to each other were constantly changing. Sometimes subtle current-induced gyrations within the floes, or a general relaxation of ice pressure during calms, or the creation of some open water in brief periods of offshore winds, combined with strong efforts on the part of crews, enabled ships to get close to one another. At other times, despite their earnest endeavours, the inconstant ice carried them apart.

On 11 May, a few men arrived at the *Lady Franklin* after walking about eight miles over the ice from the *Abram* to get medical help. Their captain, John Gravill, the man who had attempted to search Jones Sound on behalf of Lady Franklin, was suffering from a pleuritic condition, and Dr Goodsir sent them back with medicine. On another day, the men of *Lady Franklin* found they had drifted close to the *Advice*, the Dundee ship in which Penny had made efforts to find Franklin a year before, now sailing under Captain Reid, brother to the ice master on HMS *Erebus*. Some of Reid's crew walked over the ice to exchange news. The *Abram*'s men reported that the *Pacific* from Aberdeen and the *Anne* from Hull had each secured a whale earlier (probably in the vicinity of Disko Bay), and the *Horn,* a Dundee ship, had already taken four. As Penny had left Scotland two weeks later than most of the whalers, he was able to perform the duties of fleet mailman, delivering letters from home to the eager crews of the ships he could reach.

On 13 May, after taking advantage of a little sailing room among the floes, Penny's ships spoke the *Joseph Green* of Peterhead, one of the ships carrying coal for the expedition's use in winter. The men transferred fifty bags to the *Sophia*, and on the next day the *Horn* delivered a like amount to the *Lady Franklin*. Another forty bags were on board the *Pacific*, which was yet to been seen. Unlike the ships under Austin, which on their transatlantic voyage to Disko would be accompanied by a large transport, Penny's brigs lacked the luxury of a dedicated freight ship. The whaling fraternity, firmly united behind Penny's search effort, was therefore furnishing vital logistical support.

Progress across Northeast Bay was sporadic, occurring in a series of small increments gained at great effort, separated by periods of suspended motion and anxious waiting. On all sides, ice extended as far as they could see – closely packed floes stretching to seaward and extensive fixed ice extending towards the coast, beyond which was the snow-covered ice sheet crowning Greenland's mountains. For long periods it was a serene and silent seascape, but at times an ominous cacophony of sounds set the nerves on edge. Early one morning, the floes steadily crept closer in, invading and filling every small lead or pool of water. Sutherland described the insidious assault: "The ice all round us seemed as if endowed with life and the power of moving; it emitted deep hollow groans; and, if two or three persons happened to be sitting below in the cabin, a deep grazing sound could be heard against the ship's side, which invariably interrupted the conversation, however interesting, as effectively as if a thunder peal had burst right over-head. Each looked his neighbour in the face, and those looks never failed to bespeak the danger of the situation."[6]

In scarcely any time at all, a stable field of stationary ice floes could be transformed into a chaotic battlefield:

> Floe presses against floe; corners are broken off and overlapped; hummocks and ridges are raised in all directions; blocks of ice of all shapes, sizes, dimensions, and weights, from one to twenty, and even sixty, tons are raised up as it were by an "unseen power." Huge icebergs, of perhaps five hundred million tons weight, plough up floes for miles and miles without being checked in their destructive course. Nothing is to be seen but ice, "living ice," in slow but sure and steady motion, and the surface, which an hour previously presented a level plain of pure white, with pools and meandering lanes of the blue sea, is now one complete wreck; the colour is changed to a mixture of a dirty white and blue, the evenness of the surface is destroyed by masses twenty feet above its former level, and a square foot of open water can not be seen for many miles around. A scene is beheld by the Arctic navigator, which he finds it utterly impossible to admire,

> – a picture to him of wretchedness, misery, and desolation. When such changes take place among the ice, what can the ill-fated ship expect but her share of damage![7]

A large iceberg, responding to deep currents whose direction differed from that of the wind-driven ice, appeared to be driving through the closely packed floes toward the ships. John Stuart wrote in his journal, "From about 11 pm the ice began to set faster around us and the pressure caused by the motion of the berg momentarily increased till at 2:30 am it caused so serious apprehensions for the safety of the ship that all hands were roused out. Both vessels were shoved through the pack ice and were moving as in a heavy swell[,] now riding down a large piece and [anow?] lifted up aloft by the pieces rising from below her."[8] The ships weathered the crisis, but not before ice floes lifted the *Sophia* three or four feet and later dropped her, miraculously intact. The little ships had passed their first difficult test, which gave the crews confidence. A few weeks later, in a dispatch to the Admiralty which he entrusted to the Danes at Upernavik, Penny wrote, "I am sure that their Lordships will be pleased to hear that my little vessels sustained an immense pressure in North East Bay on the 18th ultimo, without the slightest damage, and that they have otherwise shown themselves all that could be wished."[9]

After the crisis subsided and the restless ice became temporarily stable again, Goodsir, Sutherland, and Stuart walked over the ice to the berg that had shortly before seemed so intent on destroying them. Like most medical men who shipped out on arctic voyages, they were intensely curious about the natural world. Not content to merely admire the iceberg from its lower surfaces, they felt obliged to climb to its summit, and not content to merely climb to its summit, felt it necessary to measure its height precisely. According to their aneroid barometer, it was 209.62 feet.

Of the three medical men, Sutherland was the most inquisitive scientifically. As the ships worked slowly north along the Greenland coast, he looked upon every enforced pause as an opportunity to observe, examine, and record every possible component of nature's kingdom, whether animate or inanimate, marine or terrestrial, large or small. Everywhere he was confronted by unanswered questions. How had an enormous block of granite, whose weight he calculated to be 186 tons, come to rest on a small island composed of a different type of bedrock? Why were some icebergs capped by tall pinnacles and spires while others were flat? Why was the air temperature at masthead sometimes higher than that measured on deck? What did ptarmigan eat before willow leaves came out? What was the diameter of a beluga's aorta?

Whenever feasible, he landed on islands to examine their geology, vegetation, and animal life. Through open leads in the pack, he and Goodsir brought up samples of sea water to examine through the microscope, lowered nets to entrap crustaceans, mollusks, and other small organisms, and dredged for bottom life more than a thousand feet below the surface. On their many shooting excursions away from the ships, they collected specimens, mainly of birds, to be sketched, dissected, skinned, or stuffed.

When the ships reached the West Side, the surgeons would be able to enlarge and diversify their collections on land and in lakes and ponds. In this work they would be enthusiastically assisted by certain officers and men. Sutherland would later name new species of fossils for those who found them – five species for officers on Austin's ships (Donnett, Pickthorne, and Ommanney) who collected mainly on Griffith Island. Among his own shipmates he named *Bellerophon Nauterium* in honour of James Knox, "a hardy seaman of the 'Sophia' who, amid all his fatigue, found time to collect it . . . in the extreme northern point from which fossils have been brought."[10] Another fossil he named in honour of "Mr. Manson, mate of the 'Sophia,' long familiar with Arctic service, a zealous observer, and a most experienced navigator."[11] He also named a seaweed *Kallymenia Pennyii* for his commander, and did not overlook himself (*Columnaria Sutherlandii*).

The indefatigable Sutherland also organized a system of daily weather observations, beginning in early May and continuing until the end of the voyage a year and a half later. He recorded data on air and water temperature, wind force and direction, atmospheric pressure, precipitation, cloud cover, and visibility. Air temperature and pressure were recorded every three hours, night and day. Altogether Sutherland was responsible for more than 2,600 temperature observations during the voyage.

* * *

The men used a variety of methods to work their ships through the ice. When tracking the ships along a lead, they sometimes came upon a narrow neck of ice or a mass of small floes frozen together, blocking the channel and perhaps threatening the hull. The men would cut through with ice saws and a variety of smaller tools, and manhandle the fragments out of the ship's path. Sutherland watched a party of men "quarrying among the smaller pieces [of ice], with handspikes and boat-hooks and setting them free."[12] When submerged ice froze to the hulls, the crew resorted to "sallying" or "rolling the ship" – running back and forth in unison from one side of the vessel to the other to rock the ship. In late

May, Stuart wrote, "By means of sallying the ship and prying up with the ice poles a number of pieces of immense size came from under the bilge."[13] Sometimes they resorted to a technique called "centipeding" or "water edging," in which men sat in boats suspended just above water level on each side of the hull and pulled oars on the seaward sides.

Occasionally they entered leads wide enough to allow a brief period of sailing, but even with the captain conning the ship from the crow's nest and the sailors standing by on deck ready to spring instantly to the sheets and braces, it was a risky business. The *Sophia* once got into trouble while beating to windward in a narrowing lead that proved to be a cul-de-sac, and only Captain Stewart's skill in shiphandling avoided serious damage. Sutherland was impressed: "I could have hardly conceived it possible to manage a ship to such advantage with the sails and rudder. On one occasion, when I, and others much more experienced, expected some damage to the rudder by a violent spring astern in the direction of a heavy floe, the slightest pull of the main braces, and hauling the jib sheet close in, changed her position so much, that she not only went clear of the ice astern, but also got into a pool of water ahead, in which she lay for three hours in comparative safety."[14]

A few calm interludes during the slow journey north from Disko provided welcome respite between crises. When the pack ice was closely packed but stable, and no progress was possible, the men were able to catch up on sleep, attend prayers on a Sunday, stroll about on the land ice without having to lean into their tracking belts for hours on end, and even indulge in games. Normal shipboard maintenance and upkeep could not be ignored, however, and the logbooks of Penny's ships record a variety of tasks: pumping ship, setting up rigging (keeping it taut), breaking out the hold (removing provisions and fresh water), cleaning the ship inside and out, greasing masts, tarring backstays, caulking and painting, airing sails and bedding, and rearranging the deck cargoes of lumber they were carrying to house the vessels in for the winter.

On 24 May they were approaching the prominent landmark of Black Hook. It was sobering to realize that it had taken three precious weeks to get there from Lievely, three weeks during which they had averaged only nine miles per day. This meagre rate of advance gave no cause for celebration, but an important national holiday demanded at least a token sign of festivity. It turned out to be an eventful day, as John Stuart recorded on Penny's *Lady Franklin*:

> This being the anniversary of Her Majesty's birthday both vessels were dressed in colours and an additional allowance served out to

> the men who were besides excused from all duty. A sharp breeze of wind which sprang up at noon did not prevent a good deal of sky-larking on the decks & ice but it became necessary to turn to work about 5 o'clock when the ice was fairly adrift. After heaving, and [canting?] and backing & filling for about two hours we were at last driven into close contact with the same berg we had formerly been endangered by. One of our chain boats narrowly escaped being stove between the ships side & the berg but this time the berg rendered us good service allowing us to hold our ground till the ice drove past. Between ten & eleven p.m. the water being pretty clear of ice we cast off making short tacks through openings in the ice. During one of these tacks the ship while in stays made sternway and carried away her rudder but without injuring the sternpost. By the assistance of a hawser sent on board the Sophia we were brought round and subsequently worked ship amongst ice for five hours with no other rudder than the sails and two [steer?] oars down the rudder trunk.[15]

The exposed rudder of a sailing ship was her most vulnerable part. When floes were closing in, a whaling crew could unship the rudder and hoist it out of danger in less than seven minutes. But on this occasion, the *Lady Franklin*, moving too slowly in short tacks to come about completely, had made a stern-board (drifted backwards) and smashed against unyielding ice astern. After some delicate manoeuvres in a lively breeze accompanied by snow squalls, Penny managed to get the ship alongside the land ice. The ship's logbook recorded that the crew "made fast and took the ruder upon deck found it intirely gone[.] Brock [i.e., broke] the between decks out and got the spear [i.e., spare] ruder stock and irons up[.]" Using the spare stock (the post on which a rudder pivots), carpenters, their assistants, and the blacksmith worked feverishly to construct a new rudder. It took fifteen hours to complete the job, the work being "rather tedious from so many bolts and fittings."[16]

Like whaling ships, *Lady Franklin* and *Sophia* were floating workshops. Each ship carried a boatswain, carpenter, carpenter's mate, and sailmaker. *Lady Franklin* also had an armourer (blacksmith) and a cooper. These men had the training, tools, and materials necessary to make emergency repairs on the ship's hull, masts, rigging, and sails. The progress of the ship, and sometimes its very survival, depended on the speed and effectiveness of their work. In addition to dealing with emergencies, they might also be called upon to modify some of the equipment and fittings. As *Lady Franklin* and *Sophia* were on their maiden voyage, facing for the first time the challenging conditions of arctic navigation, it is not surprising that some aspects of sail plans and rigging proved to be less than perfect. A royal mast and yard were needed for *Lady Franklin*, so during a quiet period the crew erected a forge on the land ice and built a snow house over it. Although the mercury dropped below zero Fahrenheit, the blacksmith was able to work comfortably in his igloo; he finished the requisite iron fittings in two days, and the mast was assembled and sent up. On another occasion, a topgallant mast was added to *Sophia*. Sails were sometimes altered to make them draw better, and new ones were made if required. *Lady Franklin*'s carpenter constructed davits for a stern boat. There always seemed to be something in need of improvement or repair.

By the end of May, they had passed Dark Head and were close to the settlement of Prøven. Among the visitors to the ships was a Danish cooper named Christensen who wanted medical advice. Because whaleships usually carried surgeons, they often served as mobile health clinics, dispensing medicines and superficial treatment to the scattered inhabitants of the coast. Despite a "troublesome stutter and a rather imperfect system of English,"[17] Christensen was able to relate how in 1845 he had gone out beyond the islands by boat to intercept some passing ships and had boarded a British naval vessel named *Erebus*. Penny and his officers were an attentive audience: here was a man who had actually seen Franklin and his men shortly before their disappearance! Except for the crews of two whaling ships who had spoken the *Erebus* and *Terror* a week or so later, Christensen was probably the last European to see them alive.

When the Danish governor came on board *Sophia*, he recognized the mate, Donald Manson, who later related in the logbook, "As soon as he saw me he recolected having been on board of a vessel called the Superior I commanded in 1826, – the same vessel I commanded 32 years." The brief meeting had taken place at Upernavik, and in the intervening quarter-century the Danish administrator had probably seen and spoken to hundreds of whaling captains, yet for some reason Manson had remained fixed in his memory.

Unfortunately there was no news at Prøven of HMS *Erebus* and *Terror*, or of HMS *North Star*, which had been sent out in the previous summer to supply the search expedition of James Ross. Nor was Penny able to hire anyone to accompany the expedition; no one could be persuaded to go. Only one more settlement lay ahead. Upernavik, some thirty miles to the north, would afford the last opportunity to find the few extra men that Penny wanted and to get permission from the Danish authorities to employ them for more than a year.

To reach Upernavik required three days of brutal effort, tracking the ships from the land ice, warping them ahead with the capstans, and towing them from the boats. Eating and sleeping were relegated to spare moments crammed in between the bouts of hard work essential to move the ships ahead. Sutherland, who had made the voyage before, remarked, "A very 'ordinary' spell for the whaler is two or three successive days and nights without any rest, and sometimes even a longer period."[18] There were occasional diversions from the intense toil. Once, more than twenty natives from coastal camps, attracted by the marching songs of the tracking crews and the apparent festivity of the occasion, joined the sailors on the ice, took up the lines, and helped them haul the ships along – at record speed. Not for long: the novelty of the operation soon wore off, and the Greenlanders remembered that they had urgent business elsewhere.

At Upernavik, Penny went ashore with John Stuart to make inquiries for volunteers. He needed natives with dog teams for hunting and transport at their winter harbour and for the search parties he would send out in late winter. He also wanted one or more interpreters, so that any Inuit they encountered could be questioned about Franklin. A month earlier, he had written to the Admiralty to inform them that he intended to hire "two Esquimaux and two Danish volunteers,"[19] but so far he had had no success.

Was it realistic to expect to find four recruits on the spur of the moment for an expedition that would take them away from home for more than a year? What inducements could Penny offer a native hunter? Sutherland thought such a man would be attracted by "high wages, a comfortable home on board ship, new acquaintances, the chances of finding wild Esquimaux, and of seeing new lands, and covered with ice too, so genial to his feelings, to say nothing of the high claims of humanity."[20] Like most British visitors, he had seen the crowded and unsanitary huts in which Greenlanders lived and assumed that they would jump at the chance to escape such conditions. He failed to appreciate that the Greenlanders had observed with equal distaste the crowded and sometimes unsanitary living spaces on board the whaling ships they boarded every summer. Would they really look upon a ship's forecastle packed with rough sailors as "a comfortable home?" Would they really want to make "new acquaintances" among men who smelled bad, spoke an unintelligible language, and ate unpalatable foods? Would they be willing to sacrifice their freedom to roam at will over land and sea in order to be confined on a small ship for months on end, following the orders of a captain who could hardly speak their language?

The prospect of earning wages (in the form of a credit at the Danish store) may have interested some Greenlanders, and the opportunity to visit unfamiliar regions and meet other Eskimo people may have had some appeal, but Sutherland's notion that ice-covered land would be particularly "genial" to their "feelings," and that they would be motivated by "high claims of humanity" seems as fanciful as his other ideas on the matter. Such inducements had failed to persuade any natives at Lievely and Prøven to sever the bonds of home and family, and they had no effect at Upernavik either. No one stepped forward to serve.

If Penny could not obtain any native hunters and dog drivers there, could he at least find a volunteer within the small colony of Danes? As these men generally spoke Kalaallisut and sometimes English as well, they would make excellent interpreters, and most were familiar with dog-sledge travel. Penny had already tried to hire one further south: on his way north from Lievely, he had entrusted a visiting native with a letter for Hinrich Johannes Rink, a Danish geologist who had been doing field work in western Greenland for two years and was said to be living at the settlement of Umanak. Several days later, while held up by ice in Northeast Bay, Penny had sent another note to Umanak, by then only about sixty miles away. He asked Rink if he could join the expedition. Both messages reached Rink, and in mid-May, a few days after receiving the second one, he set off for Penny's ships by dog sled. Alas, he was too late – the ships were already out of reach. A few weeks later, Penny wrote to the Admiralty, "I had communication with Professor Rinck (now at Omenak) who would gladly have accompanied me, but I was unable, on account of the distance, to have a personal interview with him, which is the more to be regretted, as I believe he would have been of great service, not only as an interpreter, but also in a scientific capacity."[21] In that assessment Penny was correct: Rink would have been a real asset. A man of diverse interests and extraordinary talents, he investigated the relationship between Greenland's ice sheet and the icebergs of Davis Strait and Baffin Bay, studied the ethnography of the aboriginal people, campaigned on their behalf for political representation, and wrote several books about Greenland. He would have added a new dimension to dinner-table conversations on board HMS *Lady Franklin*.

At Upernavik, Penny had another candidate in mind, a man he had met on some of his whaling voyages. Johan Carl Christian Petersen had come to Upernavik from Denmark to work for the Royal Greenland Trading Department, had married a local woman, sired two children, and become a permanent resident. Now he was assistant governor of the district. In 1845, Penny had told him about his desire to establish a whaling settlement in Cumberland Sound and had tried to persuade him to join the venture, but Petersen had declined. In 1849, when Penny was on his way to Lancaster Sound to make preliminary efforts to find Franklin, he had spoken with Petersen again. Now he was asking him to join an official Admiralty search expedition. Petersen jokingly said "Why not?" before realizing that Penny was serious. Then he decided that it might not be such a bad idea after all.

Having obtained Petersen's consent, Penny and Stuart met with the governor to obtain official sanction. On the next day Stuart again accompanied Penny ashore to discuss with the Danish pastor in the settlement what provisions they could make for Petersen's wife and children during his absence. The pastor could not speak English, and Penny and Stuart could not speak Danish or Kalaallisut. Stuart described the solution: "On this occasion the little Latin I still possessed proved much more serviceable than I had ever expected to find it. The Lutheran pastor . . . was an excellent Latinist and by this means many more things were explained than Mr Petersen could have expressed in his limited English."[22] Through recourse to a moribund European language, negotiations proceeded satisfactorily.

Stuart's remark that Petersen possessed a "limited" command of English is hard to reconcile with Penny's statement to the Admiralty that the man had "a perfect familiarity with the English, Danish and Esquimaux languages."[23] Although he may not have been letter-perfect in English, he appears to have had no difficulty in making himself understood, and he later exhibited skill in translating from Kalaallisut to English. According to Snow, Petersen was "well skilled in the Esquimaux and English languages."[24] By watching him converse with natives (later at Cape York), Snow could probably tell whether the two parties understood one another. Penny had some knowledge of Inuktitut but apparently not enough to make arrangements with the pastor.

It must have been disappointing for Penny to leave Upernavik without any native Greenlanders and only one of the two Danes he had wanted, but he was very fortunate in engaging Carl Petersen, who would serve the expedition well. In addition to translating, he was proficient in hunting and in driving dogs, and he willingly agreed to bring along his own dogs and sledge. Four of his dogs were at Upernavik, and three others, with the sledge, were on a nearby island called Kingatorsoak, being looked after by a Danish friend named Thomas, who had also married a Greenlandic woman and adopted a hunter's way of life. While Petersen packed and got his affairs in order, the men of the expedition wrote letters to friends and families, possibly the last they would be able to send.

Penny put the finishing touches to a progress report for the Admiralty and wrote to John Barrow and Lady Franklin. Many visitors came out to the ships, including the governor, his wife, and their two sons, who dined on board *Sophia* and, according to the logbook, were "well entertained." On 5 June, with Petersen, his gear, and four dogs on board *Lady Franklin*, the ships left Upernavik.

10

Sail versus Steam

Screw not sail shall be the word
To make the Ice obey its lord.[1]

Penny's ships got under way at ten in the morning. The weather was mild by arctic standards, the temperature a comfortable 41°F, the wind light, and the visibility good beneath an overcast sky. Several hours later, they were off Kingatorsoak, where they added to the already crowded ships a sledge and four more dogs, three belonging to Petersen and the fourth purchased from his friend Thomas. There were now eight dogs with the expedition, divided between the ships. As both males and females were present, an increase in canine population could be looked for in the months ahead.

Leaving Upernavik on this most dangerous stage of their journey up the coast of Greenland, *Lady Franklin* and *Sophia* were ahead of all other ships. The whalers that had left Britain two weeks before them had been brought to a halt by the ice barrier at Northeast Bay, where Penny and Stewart had overtaken them. During the month of May, *Lady Franklin* and *Sophia* had steadily moved ahead, becoming the first ships of the season to reach Upernavik – an achievement that suggests something of Penny's skill as a ship-handler among ice floes and bergs. Of the four other expeditions sailing in search of Franklin, Austin's was the closest, but his squadron was still a thousand miles behind and would be delayed for a few days at Disko to transfer stores from his transport before it headed back to England. De Haven's two American ships were abreast of Sable Island, Nova Scotia. Ross was somewhere between Iceland and Cape Farewell in Greenland. Forsyth, the last off the mark, was only now leaving Aberdeen.

Penny had no way of knowing when the other ships had left their home ports or where they now were, but he knew that they must still be in ice-free waters, reeling off up to a hundred miles per day, while his own ships had been slowed to a crawl by pack ice and were unlikely to do better until they crossed Melville Bay and passed Cape York, still three hundred miles away. The gap would continue to narrow until the pursuers met the ice, but by that time the coastal passage might be more open owing to rising temperatures and diminishing ice cover, enabling them to proceed faster than Penny had. If Penny could be the first to reach Cape York and the ice-free region of the North Water, he could then make a rapid voyage to the West Side while the men of the other expeditions were still manhandling their ships along the icebound coast of Greenland.

Before long, *Lady Franklin* and *Sophia* came up against a barrier of ice floes packed in against the landfast ice by southwest winds. From then on, progress was painfully slow. It was only by the most extraordinary efforts that the crews managed to advance thirty miles in six days before mooring the ships to a grounded iceberg near what whalers knew as Berry Island. Six days of back-breaking work to advance an average of only five miles per day! A man on foot could have covered much more.

For men on the ice with tracking belts or in the boats pulling oars, it was as if they were trying to move the rock of Gibraltar ahead. To get a stationary ship in motion required tremendous effort. Once a ship had some way on, the work became easier, but it was still punishing toil and often went on for hours. Determined not to lose momentum, the men were reluctant to ease up on the rope tension, while their officers, eager to advance northward as fast as possible, kept them at the task hour after hour. Penny's crews used a variety of techniques to cope with constantly changing ice conditions. On one day they worked for fourteen hours straight, "'tracking,' 'towing,' 'warping,' 'heaving,' and sailing, up one lane of water, down another, and across a third, round this point, and through that 'nip,' until we got into open water."[2]

The perpetual sunlight was disorienting, disturbing the men's natural diurnal rhythm and deceiving them into working longer, robbing them of their sleep. Sometimes they scarcely knew – or even cared – whether it was night or day. Darkness had been banished, and sleep now took place in the continuous light of the sun and bore little relationship to clock time. The men slept in brief snatches at irregular intervals, often for less than four hours in a twenty-four-hour period. The combined effect of sleep deprivation and physical exhaustion took a toll. When working long hours, the men were issued a quarter to a third more food and spirits than normal rations, but no amount of extra food or drink could replace sleep. Fatigue was their normal state, and in that changeable world of ice, their strenuous efforts were sometimes in vain. "It was exceedingly disheartening work," Sutherland noted, "for many times all that had been gained in a whole day would be lost, in half an hour, by a change of tide or an air of wind."[3]

From their crows' nests, Penny and Stewart studied the ice ahead to plot the best path. Whenever the ships were held up by impassable floes, they walked over the land ice to the nearest island (three miles away on one occasion), climbed to its summit, and scanned the ice fields with telescopes. By ascending a few hundred feet higher than the crow's nest, they could see further ahead and detect significant features not visible from lower elevations, such as narrow leads within the pack ice, distant areas of open water, fractures in the landfast ice, and low-lying rocks. They never failed to sweep the horizon astern as well, to see if any ships were catching up – not only because they wanted to maintain their lead but also because if serious trouble arose, it would be vital to know where help lay. When they were held up for a few days, the summit hikes became regular events, and if a change in ice distribution appeared imminent, a lookout might be left constantly on duty. Later, after several whaling ships joined them, Sutherland noted that one high island nearby usually had two or three ship captains on top watching for a favourable opening ahead. The summit on this particular occasion was almost a thousand feet above sea level, affording close to a map-like view over the ice.

Beyond lofty lookout positions, islands provided opportunities for the surgeon-naturalists Sutherland and Goodsir to record observations on rock formations, plants, and animals. Others could obtain temporary relief from cramped shipboard conditions, get some exercise, and revitalize their senses with the pungent earthy smells. Colonies of cliff-nesting and ground-nesting sea birds attracted shooting parties and egg gatherers, who with relatively little effort could add valuable protein to the ships' provisions. In early July, Petersen recorded that several officers had shot some birds on a nearby island and collected three thousand eggs! Stuart put the take even higher: "So thick were the nests that it was quite difficult to avoid treading upon them and the birds were proportionately numerous. Five thousand eggs, thirty ducks and two sacksfull of down was all that we took . . . but there was sufficient left to [load?] five or six more boats."[4] When they passed boats from HMS *Assistance* containing officers who were shooting looms (guillemots or murres) from the base of an immense cliff colony, they had to pass up the "tempting sport" because their boat was already overloaded.

On some of the islands, they were reminded that others had been there before them. It was not unusual to come upon a silent lichen-encrusted rock tomb from an ancient culture, or a weathered headboard marking the lonely resting place of some long-dead whaleman, its faded lettering giving the merest details of an existence terminated far from home and loved ones. Sutherland remarked, "There is hardly an island on which one lands, from the Arctic Circle to the top of Baffin's Bay but it will be found in a manner consecrated by the remains of some British seaman."[5] On Berry Island, they saw the grave of James Craig, a surgeon on the *Rambler* of Kirkcaldy, who had died a quarter-century before at the age of nineteen, and that of Andrew Turnbull, a lad of only eleven years, serving on the Hull whaler *Dordon*, who had succumbed to some unrecorded accident or illness in 1837. Another headboard marked the grave of a man named McBain who had "died suddenly in the maintop of his ship a few years since."[6] A ship's maintop, forty feet above the deck, was not a good place to die.

Delays at ice barriers also provided opportunities to carry out maintenance and repair. In mid-June, while icebound near Berry Island, the external doubling of *Sophia* was extended higher for greater protection against ice abrasion. With the ship securely moored to the floe edge, the crew ran lines from mastheads to ice anchors, and with blocks and tackle canted the ship over, exposing one side of the hull below the waterline. The carpenters fastened carefully shaped planks to one side and caulked the spaces between them. They finished both sides of the hull in three days and then began making a mast and gaff for a new topsail on *Lady Franklin*.

The inevitable result of the painfully slow progress was made clear on 21 June, when thirteen whaling ships, almost the entire British Davis Strait fleet, were seen to the south, slowly but steadily closing the gap. One of them was *Pacific* from Aberdeen, which, in addition to her normal whaling equipment and stores, was carrying more coal for Penny's expedition. When Captain Paterson caught up, he managed to work his ship alongside, and the unloading began. Two of the other ships had undertaken sealing voyages to the Spitsbergen region, returned to their home ports in Britain, discharged their cargoes, taken on supplies, then departed anew to pursue whales in Davis Strait – and even they had overtaken Penny! Five days later, Austin's ships were sighted. There was now a little sailing room off the ice edge, so Penny sent *Sophia* back to make contact. Stewart returned the next day with letters and a few items that had narrowly missed Penny's departure from Aberdeen but had been sent back to London in time to catch Austin. These included cloth and a small press for printing balloon messages to send aloft from winter harbour.

Like a runner who leads the pack for the entire race but is passed by a rival just before the finish line, Penny felt cheated of a deserving victory. The exciting prospect of beating the other search expeditions to the West Side and becoming the first to discover some evidence of Franklin's whereabouts had seemed within reach, but now it had suddenly become remote. If Austin, with the advantage of steam power, had managed to overtake them despite the difficult ice conditions north of Disko, he would surely be able to forge ahead and reach Lancaster Sound before anyone else. On board *Lady Franklin* and *Sophia*, the men's spirits sagged.

* * *

Austin and his crews had been just as anxious to catch up to Penny as Penny and his men had been to maintain their lead. As Austin had left Britain three weeks later than Penny (from a port almost four hundred miles further away), it had seemed a hopeless task until they learned at Upernavik, to their immense surprise and pleasure, that *Lady Franklin* and *Sophia* were only thirty miles ahead, held up by a barrier of ice. Some may have gloated, but Sherard Osborn, the empathetic commander of the steamer *Pioneer*, exclaimed, "Poor Penny! I felt that fate had been against him! He deserved better than to be overtaken by us, after the energy displayed in the equipment of his squadron."[7] Captain Gravill of the Hull whaling ship *Abram* told them that Penny and his crews had made strenuous efforts and taken extraordinary risks to get where they now were, but could go no further. The ice was so bad, he said, that the whalers would probably have to turn back, as there was now little likelihood of reaching Pond Inlet before the whales started their southward migration.

Osborn was confident that Austin's steamers with their engines and propellers would be able to force their way through pack ice with their consorts in tow, and pull ahead of all the sailing ships. Like boys with a new toy, he and Bertie Cator, commander of the *Intrepid*, had already been making their first attempts to break ice. Osborn wrote, "The 'Intrepid' and 'Pioneer' were now to be seen slyly trying their bows upon every bit of ice we could get near, without getting into a scrape with the commodore; and from the ease with which they cut through the rotten stuff around our position, I already foresaw a fresh era in arctic history."[8]

In the minds of the keen young steamer commanders, the hostile pack ice could only be conquered by the power of engines. Their optimism and their faith in the new technology were reflected in a playful poem penned by Osborn, part of which reads,

The North Water's coming boys,
The North Water's coming!
We are certain now to see the day,
The tenders there will lead the way,
In the North Water coming. –
Steady heave, may ope the Nip,
But steam's a great deal stronger,
We'll win the water by its aid,
Steam a little longer . . .[9]

From the start of the voyage, the steamer commanders had viewed Austin's bluff-bowed sailing ships as slow and unwieldy burdens. Writing from Stromness, Osborn remarked to Barrow that pairing each sailing ship with a steamer was like putting greyhounds in front of hippopotami. The arctic ice that awaited them on the Greenland coast presented a serious challenge, and the superiority of the narrow, sharp-bowed, steam-powered vessels was immediately apparent. After towing the *Assistance* among floes with his steamer *Pioneer*, Osborn wrote, "You would smile could you see us slip through the ice without a shock, whilst our big friend astern ploughs all up before her, and bumps as if to try wood against ice."[10] Although towing in open water was a relatively simple procedure, towing among floes was not. The operations of passing the towing hawser and then of hauling another vessel through narrow, tortuous channels containing drifting ice were much more difficult and dangerous for the crews.

Osborn and Cator were always at the beck and call of Austin and Ommanney, their superior officers on the clumsy sailing ships, a servitude they resented. The greyhounds wanted to cut the lines to the hippopotami and sprint away on their own, but that was not to be. Their mission was to get the sailing ships through the ice. When Austin's four ships reached Barrow Strait much later and prepared to take up winter quarters, Cator would write, "The Ships have been a sad clog to the Steamers. I think if it had not been for us they would not be out of 'Melville Bay' for, with very little exception, we towed them from 'Uppernavik' to where we are now."[11]

More than a year later, after the Admiralty's inquiry into the two government expeditions, the *Athenaeum* would write, "These vessels were, in fact, simply steam tugs – and their power was spent in towing the heavy sailing ships through the ice-encumbered seas. This was like harnessing a spirited war-horse to a cumbrous wagon: – unfettered, what might they not have done!"[12] The two steamers were destined to operate independent of the sailing ships only once – during their homeward voyage in 1851.

* * *

On 1 July, the pack loosened enough to open some navigable leads, but Penny's ships and the whalers were all becalmed and had to resort to manual labour to move their vessels. Austin's steamers now got their first chance to prove their superiority. On the *Pioneer*, Osborn was exhilarated: "At last the hoped-for signal 'take ships in tow' was made; and, with a leaping heart, we entered the lead, having the 'Resolute' fast by the nose with a six-inch hawser . . . Away we went! Past berg, past floe, winding in and out quietly, yet steadily! – and the whalers were soon astern. Penny, indefatigable, was seen struggling along the shore, with his boats ahead, towing, and every stitch of sail set to catch the lightest cat's paw; him too, however, we soon passed."[13]

As Penny's boat crews laboured over their oars, sweating profusely beneath a bright sun that warmed the arctic air to a balmy 54°F, they could see Austin's ships forging steadily ahead, the two steamers towing the sailing ships with apparent ease while the sailors nonchalantly admired the scenery. To the men in the boats, it seemed distinctly unfair that such progress could be achieved without human effort. But when first the *Pioneer*, and then the *Resolute*, ground their way over a hidden reef, the men pulling on the oars felt secret satisfaction. Penny had warned the naval ships by hoisting the flag signal "rocks close ahead," but as Osborn explained later, "We were too busy to see it until afterwards."[14] To hit bottom despite Penny's warning signals, and then tow the flagship over the same obstacle, was unlikely to get the young lieutenant a favourable report from Horatio Austin, the stern commodore of the squadron. For Osborn, conning a steam-powered ship among ice floes for the first time, in an unfamiliar and poorly charted region, the incident was part of the learning process in a school of hard and potentially fatal knocks. Submarine gradients along the Greenland coast were so steep that a moving ship could be on top of a reef before a sounding lead and line could warn of shallowing water; a commander therefore needed to exercise great caution, especially while towing, and should certainly keep his signalmen alert for flag hoists on ships ahead. Fortunately, the ships escaped without serious damage.

Although it was aggravating for Penny's men to see Austin's ships go ahead, they were not ready to accept defeat. Later the log-keeper of the *Sophia* smugly recorded, "At Mide night we passed them[.] They were lying fast to bergs." When Austin's ships halted for the night, Penny's had kept right on going. From the deck of the *Pioneer*, Osborn saw them go by: "And there, sure enough, were Penny's brigs sailing past our squadron, which showed no sign of vitality beyond that of the officer of the watch visiting the ice-anchors to see all was right. 'That fellow Penny is no sluggard!' we muttered, and will yet give the screws a hard tussle to beat him."[15] Stirred to action, four hours later, the naval crews passed Penny's brigs again. Austin regretted that he could not offer to tow them, as the pathway through the floes was narrow and the visibility limited. Belching black smoke into the clean arctic air, HMS *Pioneer* and *Intrepid* drew steadily ahead, towing their sluggish consorts.

Penny was not to be outdone, however. After he and Stewart plotted a different course through a maze of ice floes, the *Lady Franklin* and *Sophia* again drew ahead of the naval ships. Penny had given them "the go-by," admitted Osborn, whose admiration for Penny's shiphandling was growing by bounds. Even the clumsy-looking ships of the whaling fleet were working through the floes more smartly than the naval ships. He could hardly fail to be impressed (and a little embarrassed) when a sailing ship eighty-six years old advanced more rapidly than his modern steam-powered vessel. The barque *Truelove*, the oldest whaler in the British fleet, had been built in a Philadelphia shipyard in 1764 and had been sailing to the arctic whaling grounds for two-thirds of a century. She was now commanded by John Parker, one of Britain's most experienced and capable captains. Parker and the *Truelove* had accompanied Penny into Lancaster Sound the previous summer in search of Franklin, and he had later delivered to the Admiralty Kerr's Inuit sketch showing four ships trapped in the ice.

The friendly rivalries continued, naval crews struggling to beat whalemen, steamers attempting to vanquish sailing ships, Austin intent upon humiliating Penny, Penny doing his best to embarrass Austin. On 5 July, the whaler *Chieftain* managed to precede Austin's ships into the pack, but "steam came to the rescue," Osborn reported, "and after twelve hours' hard struggle we got the pendants again ahead of our enterprising and energetic countrymen."[16] When the steamers and their tows overtook Penny at the entrance to a narrow channel through the floes, Osborn observed, "So keen was he not to lose the post of honour, that as we closed, I smiled to see the Aberdonians move their vessels up into the very 'nip.'"[17] In the British tradition, Penny and Stewart were jealously guarding the head of the queue.

A week into July, the ships were in sight of Devil's Thumb, the tall rock column marking the beginning of the dreaded Melville Bay. Here the ships were confronted with an extensive ice barrier, solid and unmoving.

One after another they arrived and moored to the ice, until "a long line of masts and hulls dotted the floe edge," Osborn related. "The ice was white and hard, affording good exercise for pedestrians, and to novices, of whom there were many amongst us, the idea of walking about on the frozen surface of the sea was not a little charming."[18] While lookouts kept their eyes on ice conditions, men strolled on the ice, visited other ships, played rounders, quoits or skittles, or sometimes simply tumbled about playfully. At times the scene resembled a fair on the ice.

Normally, the naval and merchant services were bitter rivals, each jealous of its own traditions and techniques. Sometimes their objectives and methods came into conflict, especially in wartime when the ruthless predations of naval press gangs on whaling crews generated outrage, but they were all men of the sea with plenty of similar experiences. Austin's crews included many former whalemen, and the whaling ships contained men who had served in Her Majesty's Ships at one time or another. The navy's steam-powered vessels had created a favourable impression throughout the whaling fleet, and the skilful manoeuvres of the whalers had earned the admiration of the naval crews, so there was now some mutual respect. On the ice they had an opportunity to meet and exchange tales of adventures in the arctic whale fishery for exciting descriptions of naval operations in the world's oceans.

Captain Stewart Sr, father of the *Sophia*'s commander and now in charge of the Peterhead whaleship *Joseph Green*, walked over the ice to HMS *Pioneer* and presented a gift of 170 duck eggs. He stayed to dine, and regaled the mess officers with exciting stories of the whaling life. He had spent thirty successive summers on arctic voyages, never able to savour the taste of fresh gooseberries or admire the texture of a standing crop of corn, but always compelled by some strange attraction to return to the Arctic year after year. Osborn was entranced: "I felt a pride in seeing such a man, a brother-seaman, – one who loved the North because it had hardships – one who delighted to battle with a noble foe."[19] To confront a noble foe: this was the bond that seemed to unite the roughshod whaleman and the sophisticated naval officer. Stewart had pursued gigantic whales in icy seas; Osborn had fought enemies of the two-footed sort in China during the Opium War. Both had faced the fury of an angry ocean.

Meeting the *Intrepid*'s commander, Bertie Cator, on the ice, Penny asked if he would like to have a look at the *Lady Franklin*. Cator gladly accepted, and later wrote, "She is a pretty little vessel, but I think too slight almost for this service."[20] For men on board the navy's large, comfortable vessels, this was the usual reaction: the other ships were much too small! But Penny's little brigs had already proved how strong they were, and had shown that among the ice floes they could keep pace with Austin's big ships, and even get ahead of them from time to time, steam engines and propellers notwithstanding. In his journal Cator also recorded his impression of Penny: "He is a nice mild man & at the same time determined fellow & a man of great practical experience & one well calculated to do justice to his government."[21]

As the searching expeditions proceeded northward along the coast of Greenland in fits and starts, invitations to dinner flew from ship to ship. The two young lieutenants commanding the steamers (Cator was thirty and Osborn twenty-eight) were especially hospitable, and Penny returned their kindness, inviting them and others to dine on the *Lady Franklin*. Cator wrote in his journal,

> The previous days dinner with "Penny" was the happiest I expect to spend in these realms of snow & ice, we also drank milk in our coffee which is a great thing to say in 75° N Lat. I was also glad to see the good feeling which shone forth during the afternoon . . . We drank Captn P's wife's health, he returned thanks with great feeling & said he should not fail on his return to tell his "Maggy" the honors he had been done. The spirit of emulation which appears to exalt each individual, as to who should be the successful party to find "Sir John F." without the slightest feeling of jealousy be[ing] exhibited was truly gratifying. It is a day I shall ever remember.[22]

Penny entertained officers from Austin's squadron for tea or dinner on at least six occasions during the month of July, when the ships were brought to a halt by ice. Always keen to educate them about arctic phenomena and the ways of its native inhabitants, he once served narwhal skin to officers from HMS *Resolute* and *Assistance* – not raw as the Greenlanders took it, however, but fried.

* * *

Before the British expeditions left Britain, a few prescient persons had suggested that dog transport could be an effective tool in the search. With dogs rather than men hauling camping gear and food, smaller parties could travel three times as far each day and enlarge the area of search without steadily eroding the men's health by extreme physical effort.

There were, however, cogent arguments against the suggestion. Trained dogs would have to be purchased at Greenland settlements, something that could not easily be arranged in advance and might be impossible to do during a brief stopover. Dogs were not mere commodities to be bought in a store; they belonged to hunters who raised them, trained them, and used them when hunting to obtain food for their families. Even if some dogs could be obtained, British explorers had no experience in their handling and driving, which required specific voice commands and accurate applications of a long sealskin whip. They would have to hire Greenland men, which would also be difficult to do during a short unscheduled visit. On a search expedition, a team of dogs would require a considerable amount of meat, either prepared in advance in the form of pemmican or obtained fresh by hunting, but arctic sledge dogs were not accustomed to pemmican; and the naval explorers – despite their delight in shooting at virtually any animal within range – were not effective hunters in the arctic environment. They were not familiar with the sophisticated methods used by Inughuit and Inuit to hunt seals (the dogs' principal food) on the ice and in the water, and they did not possess the small toggling-head harpoons needed to retrieve them before they sank. They would have to rely on hired hunters to secure enough food for the dogs, and this – if in fact any native hunters could be obtained – would only be feasible while the ship was stationary in winter harbour. As the location of a winter harbour was seldom known in advance, the availability of game could not be predicted: some localities might turn out to be deficient in wildlife resources.

The system of dog transport had too many uncertainties and difficulties to be seriously considered by the Admiralty or even most private expeditions. After trying it in 1830 and 1831 during the *Victory* expedition, John Ross and James Ross had decided not to repeat the experiment. When organizing Austin's expedition, the Admiralty never gave dogs a thought. Dogs were hard to control and required special food, but British seamen obeyed orders and ate almost anything.

Of the several search expeditions in the eastern Arctic in 1850–51, the only one intending to utilize dog transport was Penny's. Petersen's dogs, confined on the ships' decks since leaving Upernavik, needed training and conditioning. On 8 July, after the ships had taken refuge within docks in the land ice, Petersen harnessed up a team to deliver Dr Goodsir to the *Resolute* for dinner with Austin and his officers. The dogs got halfway to the ship but then refused to go any further, and when Petersen continued to urge them onward they mutinied, turned tail, and sped back to the *Lady Franklin*, leaving Goodsir to complete the journey on foot. Austin's officers, watching the episode from the decks of their ships, enjoyed "a good laugh," observed Cator.[23] The team's bad behaviour was not caused by some canine antipathy to Royal Navy ships, or even by dissatisfaction at having been taken away from home, but merely by unfamiliarity with the harnesses and lack of training. One particular dog proved to be especially difficult. Despite Petersen's attempts to break "the unruly 'Sultan'"[24] to the sledge, "he continued refractory," Stuart recorded in his journal.[25] Perhaps the dog was not wilfully disobedient but merely clumsy. Once he tumbled into the water and could not extricate himself. His canine companions gathered round and barked an alarm, and the brouhaha attracted the attention of Stuart, who rushed to investigate and managed to pull Sultan out onto the ice. Clumsy or careless the dog may have been, but he was also smart. Later, when the ships were ensconced at winter harbour, he saved a man from freezing to death on the ice. Somehow Sultan endeared himself to Penny, who appears to have promoted him from the ranks of working sledge dogs to the position of personal pet. Sultan enjoyed this special status during the rest of the voyage and accompanied his master back to Britain at the end of the voyage. Penny gave him to John Barrow, who kept him in London until Sultan grew dissatisfied with the urban scene.

To get the dogs in shape for the forthcoming search, Penny and Petersen hitched them to a sledge whenever feasible and went on training runs over the landfast ice. It was also good practice for Penny, who had never driven a dog team before. On 20 July, he arrived at the *Intrepid* riding on a sledge pulled by five dogs. He joined Cator and other officers for a glass of wine in the cabin, inspected the ship's engine room with great interest, and chatted with the "smoking party" on deck (although a non-smoker himself). He then took some of them for a short spin, one by one. Cator was impressed with the easy motion of the sledge as the dogs trotted over the ice at five miles per hour. Tea and hot toddy were then served on board ship, with the usual Saturday toast to "wives and sweethearts – may they never meet," after which Penny and Goodsir (who had arrived earlier) returned to their ship, a mile away, by dog sledge.

When Penny invited Austin and Ommanney to dinner a few days later, he sent Petersen across to their ships to collect them. The "Commodore" and his first lieutenant arrived regally at the *Lady Franklin* on a sledge pulled by eight dogs. On the next day, Penny used dog transport to travel to the *Resolute* for a meeting. Captain Stewart also had a try at dog-driving, and at the *Intrepid* he offered Cator a ride. The dogs raced off at full speed, and Cator, exhilarated beyond the limit of good sense, attempted to stand on the moving sledge. He managed to insert a foot between two cross bars, almost breaking his leg, and while struggling to recover his balance, took an awkward fall, bruising his ribs. After this experience he took no more sledge rides.

Penny and Petersen undoubtedly enjoyed driving about with the dog team, parading in review before the naval ships and taking some of their officers on joy rides. The jaunts were training exercises for the dogs and their handlers, to be sure, but they were something more. During many voyages to the Davis Strait whale fishery, Penny had come to admire the material culture of the indigenous people of the Arctic, and he was confident that British explorers could do well by adopting some of their equipment and techniques. He and Petersen were demonstrating one potentially useful aspect of native technology.

As the various ships struggled northward along the Greenland coast, the short arctic summer was fading. If Penny's ships managed to reach Lancaster Sound by mid-August, they would have only a month of cruising left before the growth of new ice forced them into winter quarters. Few dared to think that in that short period they could discover Franklin's whereabouts. Almost certainly they would have to freeze their ships into the ice and hope that travelling parties in autumn and late winter could pick up the trail of the missing expedition.

Dog traction could supplement man-hauling but only if the dogs were well trained and in good physical condition. Being in good shape depended also on their being well fed, but the dogs' customary appetite for meat was difficult to satisfy on a moving ship whose stops and starts were governed by the state of the ice. Petersen was anxious to obtain seal meat for his dogs but stalking basking seals on the landfast ice required much time and patience, not to mention a good deal of skill. If the ships were making any headway, or even if they were stopped temporarily, Penny could seldom risk sending Petersen out on a hunt, for a sudden improvement in ice conditions would make it necessary to get under way quickly. To move the ships north as rapidly as possible was the main objective. All other activities were secondary, and as a consequence the dogs suffered.

When the ships were stopped dead, and there was no prospect of imminent change in the ice, Petersen could set off with the dogs and travel over the land floe looking for seals. If narrow leads or small pools of open water existed near the ships, he and others could skirt their margins on foot, or lower a boat and hope to spot seals or other sea mammals when they rose to breathe. But the occasional hunting sorties were not securing enough fresh meat for the dogs, and the ships' pemmican had proved unpopular. "The poor brutes were faring very ill on board ship," observed Sutherland.[26] A well-deserved banquet was in store, however. During the last days of July, the *Sophia*'s logbook recorded "a great number of the unicorn tribe" frolicking about the ships, exhaling noisily, and men from the *Sophia* succeeded in killing one with a gun harpoon. When the narwhal, a female thirteen feet long (exclusive of tusk), was hauled onto the ice, the dogs were alert with anticipation. Before the feast could begin, the ever-inquisitive doctor inserted a thermometer through an incision in the thorax and recorded a temperature of 97°F, some seventy degrees warmer than the environment in which the "beautifully spotted mammal of the deep"[27] had lived. As Sutherland undoubtedly knew, the whaling captain William Scoresby had obtained an identical figure near Spitsbergen more than thirty years earlier.

As the flensing and butchering proceeded, Sutherland carried on with his zoological investigations, noting a rudimentary dorsal fin, identifying in the stomach two species of fish, many cuttlefish beaks, and a number of crustaceans, examining the digestive system, and contemplating the animal's mechanism for thermo-regulation. All this time the famished dogs stood in a circle, patiently watching the narwhal being reduced to anatomical specimens and bite-sized hunks. Finally, they were given their chance. "They ravenously devoured whole masses of the flesh and skin, and lapped up the blood from the floe. Some of them afforded most evident proofs that it suited their tastes, for they ate the same twice or three times in succession." After the blubber was stripped off to render into oil, the dogs were "permitted to exercise their teeth upon the bones."[28] Later they were put to work hauling the rest of the meat back to the ships for future feasts.

* * *

On more than one occasion, Austin generously offered to have his steamers tow ships of the other expeditions. In early July he had apologized to Penny for *not* being able to tow, explaining that the passage ahead was too narrow. A week later he offered to tow Penny's ships but had to change his mind on account of fog. The next day, his steamers helpfully towed the *Lady Franklin* and *Sophia* away from the floe edge.

The accord between the two men, however, was fragile, liable to be shattered at any time by a contrary opinion or an imagined slight. An undated letter (probably a draft) from Penny to Austin reveals a little of the precarious relationship between the commanders of the two government expeditions. Both men were proud, sensitive, and quick to take offence, and as both had difficulty expressing themselves, their words sometimes gave affront when none was intended. Penny wrote,

> I had no idea that in offering to tow the "Felix" & "Advance" off the ice of offending you[.] I never intentionally hurt any one's feelings and I never allow any one to hurt mine.

> My intention on coming on board the Resolute was not to offer suggestions but merely to say (as we had entered into an agreement) that I was to reach over to the West side as far as the ice would allow me to see if this northerly wind was making any alterations.
>
> I am sorry that my good intentions should be misconstrued. Whoever has seen the manner Capt Austin conduct[s] his ship has no reason to doubt his being a blue *jacket*.[29]

Clearly, there had been some misunderstanding while Penny was on board HMS *Resolute*, but Penny had swallowed his pride and apologized, something he normally found very difficult. Holland places this undated episode on the Greenland coast,[30] but as the ships of De Haven and Ross were never together with those of Austin and Penny in that region, it probably occurred later in Wellington Channel.

When the ice was quiet and the ships safely moored to the floe edge, Penny and Austin sometimes exchanged visits and attempted to maintain the appearance of a cordial relationship. From time to time, Austin permitted Penny to survey ice conditions from the loftier crow's nest of one of his large ships. In one of Penny's periodic letters to Jane Franklin (to consign later to one of the whaling ships), he wrote, "I am happy to state my Lady that I have misjudged Capt Austin for I find him a kind hearted sailor and his officers noble fellows . . . The cause we are embarked in purifies the mind from all petty jealiously [*sic*]."[31]

From time to time, mounting ice pressure brought an end to inter-ship visits, shooting excursions, and games. When strong westerly winds drove floes in hard against the floe edge, the crews were quickly recalled to take defensive measures. Once, when Captain Stewart was visiting the *Pioneer*, he suddenly noticed that two whalers were undergoing "nips" and their crews were already assembling survival gear. "The sooner we are all in dock the better," he announced before hastening back to the *Sophia* to take charge.[32] Osborn watched with amazement as the crews of the various whaling ships sprang into action: "Smart things are done in the Navy but I do not think any thing could excel the alacrity with which the floe was suddenly peopled by about 500 men, triangles rigged, and the long saws (called ice-saws) used for cutting the ice, were manned. A hundred songs from hoarse throats resounded through the gale; the sharp chipping of the saws told that the work was flying; and the loud laugh or broad witticisms of the crews mingled with the words of command and encouragement to exertion given by the officers."[33] Across the land ice and over the floes floated the strains of "Rule Britannia," a stirring patriotic song composed a century before by the poet James Thomson to celebrate Admiral Vernon's capture of the Spanish city of Portobello in the Caribbean. Even in the cold and ice-encumbered seas of the Arctic, it was capable of uniting men in common cause.

The men of Austin's steamers collaborated with the crew of the whaler *Alexander* to cut a dock into land ice. The procedure required four hours' work by the three crews, a work force of roughly a hundred men. The dock was 100 feet in length and 80 feet wide, thus requiring 280 linear feet of sawing through ice up to six feet thick. Cator soon realized that the length of the steamers was a severe disadvantage, for the *Intrepid* did not fit into the dock. A few days later, when the two steamers shared a dock by themselves, the men sawed to a depth of 160 feet, with a breadth of 70 feet, requiring a total cut of 390 feet. When the ice pressure slackened, the whalers began shifting out of their docks, and Austin's crews followed their example. Cator observed, "The 'whalers' were very smart in getting their vessels out of dock. In fact they got a long way ahead of us and really to see the way in which they handle their vessels is beautiful to the eye of a seaman."[34]

Ice saws as long as fourteen feet were standard equipment on all arctic whaling ships, so Penny and Stewart were perfectly familiar with their use, but on this occasion their ships had been issued with saws made up at the naval dockyard at Woolwich. These were said to be of an "improved" design with teeth of a different pattern and a heavy ball called a "pitman" added to the bottom. (In a sawmill or shipyard, the pitman stood at the bottom of the saw-pit, hauling down on the saw, while the topman pulled it back up from above). Unfortunately the Admiralty's "improved" saws could not do the job. John Stuart recorded that their teeth "were so badly constructed that after spending two hours in cutting through two fathoms of ice we were obliged to take them on board and hand them over to the blacksmith to be cut."[35] The blacksmith cut the pitmen off the saws and altered the shape and alignment of the teeth. Sutherland observed that their time-consuming alterations had produced ice saws similar to the ones that the whalemen had been using for thirty years. Fortunately, the ice pressure eased off while the work was under way. Had the pressure of the pack intensified while the ships were in an exposed position along the floe edge without effective saws to cut a dock, the situation could have been critical.

On the starboard side, the landfast ice stretched away toward the glacier-capped mountains of Greenland. On the port side, densely packed ice floes extended as far as the eye could see. Between the two zones, the ships were moored head-to-tail along the floe edge, advancing slowly in single file whenever feasible, probing into leads and pools that occasionally opened enticingly among the floes but rarely finding enough room to sail. Through most of July the pack ice looked as if it might endure forever. At the end of the month, the *Sophia*'s mate Donald Manson summarized the outlook in the logbook: "Taking a deliberat [*sic*] look round I must own it has not a Very promising look for us but ice is a changeable body." Three decades of commanding arctic whalers had taught him that however bad ice conditions might appear at one moment, an alteration of wind direction and force could rapidly bring about radical changes.

Manson also knew that the gradual rise in air temperatures during what passed for summer in the Arctic would bring about changes in the structure, density, and geographical extent of the ice, and those alterations could affect the ships' progress. Although the physical changes occurred too slowly to be visible, their cumulative effects were hard to miss. After almost three months, the once-stable platform of landfast ice – "our best friend the land floe,"[36] Stuart called it – was beginning to break up, releasing countless fragments to join the legions of floes already cluttering the sea lanes and impeding progress. Its seaward margin, the convenient natural quay that provided security for the ships, was losing its straight configuration and stability, and thus its usefulness. Even in areas where it was still sharply defined and relatively straight, daytime sunshine and warmth were turning its snow cover into slush, making the operations of warping, tracking, and sawing docks extremely arduous.

In sea ice, the gradual downward migration of salt crystals into the water below makes the ice less saline, lighter, and more buoyant over time. As the season advanced, the ice floes thus tended to float higher and could be more easily dispersed by winds and pushed aside by ships. Icebergs were also undergoing transformation. Summer warmth was weakening their structure, while erosion by meltwater and rain was smoothing their jagged contours, reducing their mass and shifting their centres of gravity. They had ceased to be stable mooring platforms whose lee sides offered protection from wind-driven floes and whose ice or meltwater ponds provided drinking water. Despite their solid appearance, they were now precarious piles of rotting ice, likely to capsize or disintegrate if disturbed. On one occasion Penny's lookouts reported Austin's ships beset and one of them listing badly. Penny sent Goodsir and Stuart to the top of a nearby berg with telescopes to obtain a better view of the situation, but by the time they reached the summit the weather had closed in and they could see nothing of the ships. Science then assumed priority; getting out their Aneroid barometer, they found the height of the berg to be 157 feet. But their researches were suddenly interrupted. Stuart observed, "The barometer had scarcely been registered when a sudden crack & continued trembling of the berg gave us notice to quit and its rather perilous summit was quitted rather more hastily than it was ascended."[37] What the consequences would have been had the berg disintegrated beneath their feet was vividly revealed a few days later when the *Lady Franklin* and *Sophia* were within sight of an immense berg – half a mile on each side and three times as high as the ships' mainmasts – in the process of disintegrating. Over a period of six hours, the men watched with utter fascination as it fell apart, accompanied by "thundering noise and convulsive motions."[38] To see a massive mountain of ice crumble into fragments was a sobering reminder of the vulnerability of small ships in arctic seas. Sutherland later wrote:

> When an immense iceberg begins to tumble to pieces and change its position in the water, the sight is really grand, – perhaps one that can vie with an earthquake. Masses inconceivably great, four times the size of St. Paul's Cathedral or Westminster Abbey, are submerged in the still blue water to appear again at the surface, rolling and heaving gigantically in the swelling waves. Volumes of spray rise like clouds of white vapour into the air all round, and shut out the beholder from a scene too sacred for eyes not immortal. The sound that is emitted is not second to terrific peals of thunder, or the discharge of whole parks of artillery. The sea, smooth and tranquil, is aroused, and oscillations travel ten or twelve miles in every direction; and if ice should cover its surface in one entire sheet, it becomes broken up into detached pieces, in the same manner as if the swell of an extensive sea or ocean had reached it. And, before a quiescent state is assumed, probably two or three large icebergs occupy its place, the tops of some of which may be at an elevation of upwards of two hundred feet, having, in the course of the revolution, turned up blue mud from the bottom at a depth of two to three hundred fathoms.[39]

Late in the season it was prudent to give icebergs a wide berth, but sometimes this could not be done. On one occasion Cator was ordered to clear a passage past a berg with his steamer *Intrepid*. He wrote in his diary, "We passed so close that my *starboard boats* just *cleared the berg*. It was quite awful to see the manner in which it *over hung our decks*. All hands gazed at it with amazement, and if truth were told there were few amongst them but thought and knew what their fate would have been, had it fallen on us. I consider it quite a *temptation of 'Providence'* going so close but when duty calls one must obey . . . I was glad I confess, when we had passed it."[40]

Newcomers to the Arctic were amazed at the seascape of ice. In July 1850, off Devil's Thumb, an officer of HMS *Assistance* later recalled in a letter to the *Times* (4 November 1851), "The immense fields of ice broken up into a thousand shapes, and interspersed here and there with stupendous icebergs, made one fancy that an immense city had been lately overwhelmed by some terrific convulsion of nature, for arches, columns, spires, turrets, towering one over the other, were jumbled together in a confusion resembling chaos."

During an enforced halt in late July, Penny's men tried using kites provided by the Admiralty to drag sledges and boats. On snow-covered ice with a light following wind, a "larger-sized" kite pulled Petersen's Greenland sledge and one occupant at a rate of about two miles per hour, and in smooth water a "single" kite pulled a whaleboat at about the same speed. Whereas the boat could be steered off the wind to some extent, however, the rudderless sledge could merely follow its kite straight downwind, a severe limitation. In addition, getting a kite aloft from the boat proved very difficult because no one could run ahead with the line. Sutherland concluded that kites might be employed to some slight effect in sledging but would not be advantageous in boats.

* * *

As the six ships of the two official Admiralty expeditions – the leaders of the pack – struggled to move northward through Melville Bay, the four ships of the other three expeditions were steadily narrowing the gap. As in the analogy of cars on a highway meeting an obstruction, while the ships in the vanguard continued working slowly through concentrated pack ice, those to the south, benefiting from higher temperatures and more open ice conditions, were able to advance comparatively swiftly.

On 27 July, the lookout in *Sophia*'s crow's nest sighted two unfamiliar vessels coming up from the south. "Supposed to be two discovery ships," Manson recorded in the logbook. Six days later, his observation was confirmed, and the strangers turned out to be Ross's *Felix* and Forsyth's *Prince Albert*. Although Forsyth had departed two weeks later than Ross and had started from Scotland's more distant east coast, he had overtaken Ross on the Greenland coast when the *Felix* was held up by ice. Pressing on northward together, the two little vessels had now caught up with the leaders. By early August, all the search expeditions were within sight of one another in Melville Bay – with one exception. No one had yet seen the two ships of De Haven's American expedition.

Although trailing far behind the expeditions of Penny and Austin, De Haven's *Advance* and *Rescue* were profiting from better sailing conditions. Increasing summer warmth and the resultant ablation of ice floes had acted to their advantage. They met the first pack ice in early July, approximately ninety miles north of where Penny had encountered it two months earlier, and sailed from Disko to Upernavik in a week, whereas Penny had taken an entire month.

About seventy miles north of Upernavik, they encountered a barrier of pack ice extending across their route but an inviting, ice-free channel extended westward. De Haven wrote in his diary, "To the N.W. it was perfectly clear with a dark and clear horizon in that direction. Although I had determined to run to the North and double the pack by the usual route, still the temptation presented of sailing in a straight line for Lancaster Sound could not be resisted."[41] Here was a chance to take a shortcut and beat all the other expeditions to the West Side! Turning away from the Greenland coast, the *Advance* and *Rescue* boldly headed west. After clear sailing for forty miles, however, their progress was again halted by extensive pack ice. What had appeared to be a routeway across Baffin Bay to Lancaster Sound proved to be merely "a remarkably deep bight" in the Middle Pack.[42] What were they to do? With his second-in-command, Samuel Griffin, De Haven considered the alternatives. They could continue to press westward, hoping to get through the Middle Pack as Parry had done in 1824 and John Ross in 1829, or they could turn back for the Greenland coast and resume the traditional coastal course of the whaling ships northward through Melville Bay to Cape York, then westward across the North Water. De Haven decided to enter the pack.

The decision was not popular with his officers. The *Rescue*'s first officer, Robert Randolph Carter, wrote in his journal, "We are sorry that the commo[dore] attempts to cross the pack so low down as all the whalemen and former navigators advise to get more to northward . . . Pity, pity!"[43] Soon the ships were beset and could scarcely move. The sailors feverishly sawed and chopped ice, set out ice anchors, and warped the ships ahead, but with little effect. On 18 July, Carter wrote, "Working hard for nine hours and gained one ships length into thinner ice. The next two hours warped about two hundred yards."[44] A few days later he commented, "This ice business is getting very dull. Work all day like chaineys and make ½ mile."[45] On the next day it was even worse: "Mighty small headway to-day. Worked a good deal didn't go twenty yards."[46]

The hard work, irregular hours, and lack of sleep took a toll on the men, making them irritable and creating tensions. Tempers occasionally flared. On 25 July De Haven wrote in his diary, "I found myself under the disagreeable necessity of punishing one of our men today for mutinous and disrespectful language to an officer."[47] The officer in question appears to have been Midshipman Lovell, who in his own journal recorded on the following day that he had given the offender, Lewis Costa, a dozen strokes with the cat of nine tails.

For three weeks they were confined to an area less than a dozen miles across. Moving the ships forward a few hundred yards each day was not going to get them to Lancaster Sound, still four hundred miles away, so they attempted to return eastward, but backtracking towards the Greenland coast proved to be just as difficult. Advancing through the ice floes without the stable platform of land ice to walk on was dangerous as well as strenuous. Many men took sudden and unwanted plunges into the cold dark water between floes and were lucky to get out. Officer status conferred no special protection. Carter, who had taken a sudden sea-water bath twice, noted, "Every officer but one, Doc, has fallen through the ice and most of the men."[48] Even De Haven was initiated into the Order of the Arctic Bath. Surgeon Kane, later to become a best-selling author, dredged the bottom of his barrel of metaphors to adequately describe the event: "Our commander, who had heretofore miraculously escaped his ducking, while standing on a miniature South America of ice punching with a boat-hook at a little Cape Horn, went down suddenly this morning, leaving a Terra del Fuego of slush and water to mark the place where he had been. He had some trouble in scrambling out."[49]

Their principal method of moving the ships was by warping. They were equipped with two models of ice anchor, the lighter weighing forty pounds and the heavier a hundred. Most of the time they used the former. One or two men would carry it across the surface of the floes or, if there was sufficient open water, it could be taken ahead by boat. After chopping a hole with an ice chisel the size of a crowbar, the men would insert the point of the S-shaped anchor and attach a heavy line leading to the bow of the ship. On deck, the crew would keep the line taut, slowly winching the ship forward and wedging the ice floes apart. But if the floes were jammed tight and the winch imparted too much tension, the line could suddenly part and whip back like an angry snake. A heavy hawser whirling through the air was like a flying scythe.

If their initial heaving succeeded in opening up a channel ahead of one of their vessels, they would set out another ice anchor ahead and warp the ship forward with a lighter line. Kane described the procedure: "Out jumps an unfortunate with a forty-pound 'hook' upon his shoulder, and, after one or two duckings, tumbles over the ice and plants his anchor on a distant cape, in line with our wished-for direction."[50] To reach the desired location often required detours around hummocks, cracks, or rotten ice. "On such parties a ducking is an expected consummation," Kane wrote, "more than once I have seen both man and anchor both disappear together."[51] Being carried underwater by a forty-pound anchor was an experience they tried to avoid.

If the line from ice anchor to ship snagged on icy protuberances, the men had to "crawl, jump, wade, or swim" to get to the trouble spot and lift the line clear.[52] It was hard and extremely dangerous work, and the hours were long. One day the men warped and sawed for eighteen hours, with time out only for meals.

When the ships were in a pool of water or a lead and the men attempted to move the ice anchor to a new position by boat, the boats were sometimes impeded and even stopped by young ice. The sharp jagged edge of new ice less than an inch in thickness could slice like a knife into their wooden hulls. Even the stout ships were affected. De Haven noticed that "on both sides for some feet abaft her iron sheething [*sic*] she [the *Advance*] had been cut into near an inch. I therefore devoted the day to applying a remedy to this serious evil, by covering the vessels; first on one side and then on the other and nailing on a few more strips of iron."[53] Presumably the men hauled on lines from the masthead to cant the vessel over and expose the hull for the work.

The Middle Pack dealt severely with the Americans, providing plenty of hard knocks, severe nips, and narrow escapes from large icebergs. On one occasion, as the *Advance* was endeavouring to get round a berg into an open space, De Haven sent three men onto the shoulder of the berg to get warps ready.

> One took charge of the hawser, and another of the iron crow or chisel which is used to cut the hole; the third, a brawny seaman, named Costa, was in the act of lifting the anchor and driving it by main force into the solid ice, when, with a roar like near thunder, a crack ran across the berg, and almost instantly a segment about twice the size of our ship was severed from the rest. One man remained oscillating on the principal mass, a second escaped by jumping to the back ropes and chain shrouds of the bowsprit; but poor Costa! Anchor and all, disappeared in the chasm! By a merciful Godsend, the sunken fragment had broken off so cleanly that, when it rose, it scraped against the fractured surface, and brought up its living freight along with it. Scared half to death, he was caught by the captain as he passed the jib-boom, and brought safe on board.[54]

The frightening experience may have been further retribution for Costa's "mutinous language" to Midshipman Lovell a few days earlier.

In deciding to leave the Greenland coast and attempt to cut directly across Baffin Bay to the West Side, De Haven may have attached too much importance to the much-publicized transits of Parry and Ross many years before. The whalers, on the other hand, almost always skirted rather than crossed the Middle Ice of Baffin Bay; the longer route was the surest way to reach their destination. British whaling captains had advised Snow of the *Prince Albert*, "Keep to the shore as close as the land ice will permit, and, on no account to be tempted, by any apparently favourable opening to the westward, to proceed in that direction."[55] De Haven had been tempted by just such an opening and was now paying the price. He may have been motivated in part by a desire to avoid risking his ships in the dangerous Melville Bay. A few years later, Kane would write, "Since the year 1819, up from which we may date the opening of Melville Bay, no less than 210 vessels have been destroyed in attempting its passage!"[56] In fact, Kane's statement was badly wrong. A list compiled in 1852 by a British whaling captain, George Harrison, shows that only 84 ships were lost on the entire Davis Strait whaling grounds from 1819 to 1853, the year in which Kane's book was published.[57] Some of those occurred on the Greenland coast, some in the Middle Ice, and some on the West Side. The number of ships wrecked in Melville Bay alone cannot be determined with certainty, owing to incomplete documentation and imprecise descriptions of location, but an examination of the annual lists of British whaling voyages shows that it was somewhere between 46 and 69. Kane's figure of 210 was therefore an exaggeration of from three to five times. If Kane gave this erroneous information to De Haven during the voyage, it may have influenced his decision to depart from the whalers' customary coastal route around the periphery of Baffin Bay and attempt to work through the Middle Pack instead.

11

A Massacre Reported

Two ships . . . destroyed by fire . . . their crews massacred by the Esquimaux.[1]

In the upper reaches of Melville Bay, Penny's small but manoeuvrable brigs drew ahead of Austin's squadron during the first week of August. Close ice conditions often gave the small sailing ships and their more experienced crews an advantage over the naval ships. "We soon discovered that the large ships of Captain Austin's squadron were anything but a match for us among the ice, even with the invaluable assistance of steam,"[2] Sutherland reported. More than once Penny's men saw Austin's steamers get up steam to force a passage but then become firmly embedded in advancing ice floes. By the first of August, Penny's ships were six miles out ahead. While Austin waited out a powerful press of pack ice with all four of his ships stuck fast, Penny's crews worked furiously and increased their lead little by little. On 3 August, the *Sophia*'s logbook recorded, "We kept sailing towing traking and warping till 6 in the morning." Three days later: "Captn Austin is persevering but is a long way behind." The next day: "Captn Austins fleet can be seen from the mast head in a WSW direction and seems much hampered with ice." After that, Austin's ships could no longer be seen. Penny was undisputed leader once again.

On 9 August, the *Lady Franklin* and *Sophia* found a little sailing room and "passed whole cities of icebergs clustered together like formidable fortifications, which the ships dared not approach."[3] A day later they were within several miles of Bushnan Island, whose eight-hundred-foot summit looked like an ideal place from which to survey ice conditions ahead. Goodsir, Sutherland, and Stuart set off confidently over the land ice for shore, but the journey proved more difficult than anticipated. "The ice over which we had to pass was full of holes, which communicated with the water underneath,"[4] Sutherland related. After threading their way carefully across the honeycombed ice for two hours, holding boathooks horizontally like tightrope walkers in case they fell through into the water, they were halted by a broad shore lead between the ice and the land. They managed to climb aboard a small floe which they paddled and poled across to the shore. Climbing a steep slope to the island's summit, they found a large cairn, evidently familiar to the crews of passing whaling ships. Sutherland believed that it had been built in 1818 by men of the Northwest Passage expedition commanded by John Ross, who had named the island after John Bushnan, clerk of HMS *Isabella*. That voyage had initiated the last phase of the long British search for a Northwest Passage, which had led nearly three decades later to the supposedly final expedition commanded by Sir John Franklin. To Sutherland and his companions, the cairn seemed to be a symbol of the centuries-old quest for a maritime passage to the Pacific and an inspiration to them in their humanitarian mission.

From the stone beacon, the three medical men could see in all directions. To the north, part of the second-largest ice sheet on earth extended unbroken from the coast into the interior. To the southeast was the broad expanse of Melville Bay. For thirty years, men had toiled like draft horses to work their whaling ships through its dangerous pack ice and bergs, only to see dozens of their stout wooden hulls crushed to matchwood by the vengeful ice. But the tranquil scene now suggested none of the many dreadful experiences associated with the region. Height and distance disguised its fearsome nature, erased past disasters, muffled the sounds of grinding ice, and reduced everything to a pretty picture.

The *Lady Franklin* and *Sophia* had now been working through Melville Bay for a full month, gaining on the average a mere seven miles per day, a paltry rate of advance achieved only by countless man-hours of mind-numbing, back-breaking effort with tracking belts, capstan bars, and oars. Yet from the summit, the hard work and fatigue could easily be forgotten. The view was magnificent, and nowhere within the scope of their telescopes could the three men see any ships but their own. The whaling captains had all turned back; the crews of Austin, Ross, and Forsyth were far behind, still struggling through Melville Bay; and the Americans had not yet appeared. Penny's expedition might after all be the first to reach the West Land.

Looking westward, the doctors could see a slender glacier-capped peninsula culminating in Cape York, less than twenty-five miles away. The outlook was encouraging: Cape York was regarded as the termination of Melville Bay, and beyond it one could expect to enter the ice-free polynya of the North Water and enjoy comparatively unobstructed sailing to the West Land. Descending from the summit, they boarded their raft of ice, recrossed the shore lead, and made their way gingerly back to the ships, where they learned that Penny and Stewart had also managed to reach the island.

On the following day, the ships made about four miles towards the west, warping along the landfast ice that filled Prince Regent Bay. The weather was getting warmer and the surface of the landfast ice was so soft that the men could not track the ships forward; warping was the only way to advance. Open water could be seen ahead beyond a mass of ice floes. As they forced the ships through the last obstacle, an ocean swell imparted a long-anticipated gentle motion, a sure sign that they had reached the margin of the North Water.

At 5 p.m. on 12 August, a lookout on the *Lady Franklin* discerned a human figure on the shore or the landfast ice. Could it be a shipwrecked sailor? A survivor from Franklin's expedition? Penny brought the ship into the wind and lowered a boat to investigate. The man proved to be a native Greenlander whose wife and baby were nearby, and he readily consented to come out to the ship. An hour later, three more men were sighted ashore and they too were brought out by boat. They all looked down and out – their faces dirty, their skin clothes worn and soiled. Sutherland, who went over from the *Sophia* with Captain Stewart to see the strangers, called them the most wretched creatures he had ever seen. They belonged to an Eskimo group who lived further north than any people in the world, almost completely isolated from culturally similar groups far to the south and west. Encountering them three decades before, John Ross had called them "Arctic Highlanders," and the name had stuck. It was a misnomer, however, conjuring up an image of brawny men wearing kilts, playing bagpipes and eating haggis. They did none of these things (although they did enjoy some meat dishes somewhat reminiscent of haggis). They inhabited a narrow strip of coastal land at the foot of Greenland's continental ice sheet, which Ross referred to as the "Arctic Highlands," but the term "Arctic Lowlanders" would have been more appropriate, as these people lived adjacent to the sea.

Penny offered the men some narwhal skin, which they eagerly consumed. The preliminary gestures of hospitality completed, Carl Petersen questioned them about Franklin, apparently communicating easily in their language. The men were accustomed to seeing whalers pass by, and in the previous year one of them had met Alexander Stewart, Penny's second-in-command, when he had been serving as mate on his father's ship *Joseph Green*. They were aware of the difference between whaling ships and naval vessels, and between whalemen and naval officers, and if Franklin's ships had come this way at any time in the past five years, they had no knowledge of it. They were able to provide one useful piece of information, however. A single ship – not a whaler – had spent the last winter further north, evidently in Wolstenholme Sound, roughly sixty miles to the north. From what they said, Penny and his officers concluded that it must have been HMS *North Star*, carrying provisions for James Ross's search expedition. This was of great interest, because Penny and Stewart were carrying an Admiralty dispatch ordering the *North Star*'s master, James Saunders, to abort the mission since Ross had already returned to England. As Penny would learn later, however, the *North Star* had left Wolstenholme Sound to sail westward on her now-redundant quest for Ross's *Enterprise* and *Investigator*.

The men were taken back to land accompanied by gifts such as iron hoops, empty meat cans, pieces of wood, and knives and needles, all items of slight value to Europeans but immense value to subsistence hunters and their families. Having learned nothing about Franklin from the natives, Penny and his companions turned their eyes and thoughts westward. As soon as the floes parted, they would be on their way across the North Water to the West Land ahead of all the other ships. Before leaving, Penny entrusted to one of the native men a dispatch to be delivered to Austin if he should happen to stop there. In it he reported that they had learned nothing about Franklin's ships but HMS *North Star* had evidently wintered further north.

* * *

A day after Penny and Petersen interviewed the four native hunters in Prince Regent Bay, Austin's squadron arrived at the same spot, now accompanied by Ross's *Felix* and Forsyth's *Prince Albert*. At 10 a.m., as the six ships were working along the ice edge toward Cape York, the officer of the watch on the *Assistance* sighted more natives. Ommanney climbed to the crow's nest to verify the report, then signalled the *Resolute* and requested permission to communicate with them, which Austin granted. With several other officers, Ommanney transferred to the *Intrepid*, which steamed in toward the land. On their way they picked up a boat from the *Prince Albert* containing Forsyth, Snow, and William Wilson, the second mate.

After the *Intrepid* moored to the floe edge about two miles from Cape York, several natives came running down to meet them. When Ommanney shouted for a musket to defend himself, Cator got a gun ready. But the charging "enemy" was only a pair of young unarmed men eager to make friendly contact (and perhaps augment their limited inventory of cast-off metal and wood). Soon a third man arrived, rather older than the others, dragging a small sled. Conscious that he was performing an important service, he delivered – with a dignity appropriate to his mission – a package wrapped in sealskin and addressed to Austin or the British Admiralty. It must have been slightly galling for Ommanney and his fellow officers, after forcing their ships through four hundred miles of pack ice, to arrive finally at this remote place and be presented with a letter written by Penny, who without the advantage of steam power had managed to get there a day earlier, interview the inhabitants, compile a dispatch, and arrange to have it delivered by a native postman.

Ommanney opened Penny's missive on the spot and went on board the ship to write out a dispatch for the Greenlanders, to be forwarded to England on board the next vessel to pass by. The inhabitants of Cape York were being pressed into service as mailmen but as they were being paid for each transaction in useful manufactured goods, it was not a bad arrangement. Ommanney's party donated a selection of articles quite different from those of Penny on the previous day, including beads, an old file, some "wigs" (presumably "Welsh wigs" – winter caps with earflaps, a redundant article of clothing among people whose parkas already had cleverly designed fur-trimmed hoods), and two or three dolls. The dolls, a novel item of commerce in the region, evoked great curiosity among the native men, who immediately turned them upside down to see whether they were male or female. The man who had delivered Penny's letter received more practical items including a knife and a file.

While Ommanney compiled his dispatch, the sailors cavorted on the ice. "Leap-frog, sliding, racing, and all sorts of antics were rife among them for the few moments they were thus idle, and away from their usual occupations,"[5] remarked Snow. The sledge of the postman was pressed into service for a little amusement, but reserved for officers only. "While Dr. Donnet harnessed himself thereto, Lieut. M'Clintock threw himself across in equestrian fashion, and away they both went, rapidly, in a straight line, to try what rate of speed they could accomplish," wrote Snow. "In a short time they returned, and the lieutenant and myself mutually performed similar offices for each other."[6]

Ommanney wanted to see if anything more could be learned about Franklin's ships and the *North Star*, but he had no interpreter. This was an opportunity for the officers to use their copies of the *Eskimaux and English Vocabulary* recently compiled by Captain John Washington, RN, in order "to furnish every officer and leading man in the Arctic Expeditions with a book of ready reference that he can carry in his pocket without inconvenience."[7] It brought together in one handy little volume three vocabularies, one compiled in Alaska during the expedition of HMS *Blossom* under Frederick William Beechey, one from the Igloolik area compiled during the second expedition of William Edward Parry, and a third from Labrador made up by Moravian missionaries. It thus represented an arctic territory some three thousand miles wide. It did not, however, include the dialects of Greenland, where it was now being called into service for the first time. After a guide to pronunciation and grammar, Washington's book presented the regional vocabularies and then listed a number of sentences likely to be useful to men searching for Franklin, with their equivalents in the Labrador dialect of the Eskimo language.

The Admiralty had also supplied Washington's dictionary to the other British search expeditions, including Penny's, and had sent a dozen to New York by Royal Mail steam-packet for the use of De Haven's ships. When Carter, on the *Rescue*, proudly showed a copy to a supposedly illiterate Greenlander near Upernavik in early July, the man examined it with interest, then "pointed out several words spelt wrong for his dialect."[8]

At Cape York, the earnest attempts by Ommanney and his officers to pronounce phrases provided specially for the search mission (but in a dialect spoken more than a thousand miles south on a different continent) may have fallen somewhat short of being intelligible to the native audience. Among the relevant phrases were the following:

Have you seen any large ships lately?
umiak-soarnik tækkolaung-ilasse im-mane?

We are in search of two English ships
umiak-soarnik maggo-inik kenner-pogut

Which have been five years in the ice
okki-onik telli-manik kag-wait akkorng-ane-palai-wut

Have you heard anything of such ships?
umiak-soarnik taimai-tunik tækko-laung-ilasse,
yussa-laung-ilasse-lonet?

Make it known among all the Eskimós or Innuít
kauyi-titsi-niaritse-tók tamak-koning-a innung-nut tamai-nut

That the Queen of England will give a large reward
attan-ersub Victoriab *akkillero-marmago ang-iyowak-soarnik*

To any of the Innuít who will bring news of them
kauyi-titsi-yungnartok taipko-ning-a umiak-soarnik[9]

Even if the natives understood the questions, their replies in Kalaallisut, the Greenlandic dialect, were probably incomprehensible to Ommanney and his colleagues. A few years later, after using Washington's vocabulary and phrase book among natives further north in Whale Sound, Sutherland would write, "We made ourselves tolerably understood to them, but their replies were too hastily made, and far too voluminous, to be comprehended."[10] (The frustrations of attempting to *converse* in a foreign country with the aid of a phrase book may be familiar to modern-day tourists.) Pointing to objects, making gestures, and adopting exaggerated facial expressions all helped to convey meaning, but the conversation – if it may be called such – could not have been very meaningful. Finger signs, body language, and facial expressions often differ from one culture to another. Washington's dictionary and phrase book warned the user, "The Eskimaux make much use of winks and nods in conversing; the former are always intended to convey a negative meaning. A nod, as with us, implies an affirmative."[11] Cator later remarked that the natives had often held up two fingers, but "whether they meant to imply that they had seen two vessels, or one vessel with two masts, I am not prepared to state."[12]

Austin's large expedition, comprising four ships and 180 men, did not carry interpreters, a serious deficiency. Forsyth's small *Prince Albert*, with a crew of only eighteen, had two men who possessed some understanding of Inuktitut from working at fur-trade posts of the Hudson's Bay Company – William Wilson, a mate, and John Smith, the ship's steward. When Wilson went ashore with Forsyth and Snow, he was able to clarify the statements of the natives, or to least to understand their principal points. Several days later, Cator described to Ommanney his impression of the meeting: "Through the interpretation of the chief mate [actually second mate] of the Prince Albert I ascertained that two Vessels had gone to the Northward some time ago . . . and also that one Vessel answering to the description of the North Star, had wintered not far to the Northward of Cape York, and had left this year and gone to the Westward."[13]

A boat from the *Felix* then arrived with Charles Gerrans Phillips, Ross's second-in-command, and Adam Beck, the interpreter they had hired at Holsteinborg. Recognizing the latter as one of their own race, the natives extended a hearty welcome in their language. All the officers crowded around, eager to learn what more they could learn about HMS *Erebus*, *Terror*, and *North Star*, but as Beck knew only a few words of English, he could not translate what the local people were saying. The key to communicating effectively with the Cape York people was Sir

John Ross, because he and Beck were able to communicate in Danish, a second language for both.

Ross claimed to be the only officer in the Royal Navy who could speak Danish. How well he did so is uncertain. Snow said that Ross spoke Danish only "a little" and Beck spoke it "pretty tolerably," but whether Snow was really capable of judging their fluency is open to question.[14] Ross also had some familiarity with Inuktitut, having compiled a vocabulary of more than 1,200 words and a list of useful phrases during his four-year sojourn in the Central Arctic two decades earlier. But Ross was not present – Osborn's steamer *Pioneer* had continued westward with both the *Resolute* and *Felix* in tow, taking Ross away from the place where he was urgently needed. This blunder was to contribute significantly to the confusion that ensued.

Why Ross had not gone ashore with Beck to communicate with the natives is a mystery. Of all the officers in the searching flotilla, he was the only one who had lived among Inuit and made an effort to learn their language. He must have been curious to meet some of the "Arctic Highlanders" he had "discovered" thirty-two years earlier, the primitive and isolated people who had wondered if his sailing ships were living creatures descended from the sun or moon. Furthermore, he knew that it was essential for him to be on hand to translate Beck's translations from Danish into English for the benefit of Austin's officers. What compelled him to turn his back on Cape York and carry on in the *Felix*? His diary reveals nothing; it merely states, "Natives were seen on the land ice . . . Comr. Phillips proceeded with the interpreter to communicate[,] the Intrepid having stood in also for the same purpose."[15] But as Beck could not speak English, how could he communicate with Phillips?

Several days after the meeting with the natives, Ommanney requested the officers who had been present to submit written accounts of the encounter. Their remarks reveal some of the difficulties involved. Beck, from southern Greenland, was not familiar with the dialect of the remote people of the Cape York region. Writing to Ommanney, Cator observed that he "appeared at first, not to be able to converse with the Natives, but after a short time he made them, or appeared to make them understand that we were in search of two ships."[16] Phillips, influenced perhaps by his allegiance to Ross, had a slightly more positive impression of Beck's capabilities: "Our interpreter appeared to understand & be understood by them."[17] McClintock was not favourably impressed by the performance of Beck or of Wilson, who had learned some Inuktitut in what is now northern Canada: "Both these persons experienced considerable difficulty in making themselves understood, and in comprehending the replys [*sic*] of the natives, and frequently had recourse to signs to elucidate their meaning,"[18] he told Ommanney.

Notwithstanding the obstacles, the officers posing questions at the interview came away with a rough idea of the native testimony. What they understood was this: several years earlier two ships whose officers wore caps with gold braid or bands had passed by heading northward. Natives at Cape York may have actually met the Europeans either on shore or on board the ships. More recently, a single ship (presumably HMS *North Star*) had gone northward and wintered, apparently in Wolstenholme Sound; a day or so ago a pair of ships (obviously Penny's) had stopped by and given them the letter for Austin. A severe illness characterized by coughing, choking, and spitting had killed many of their own group somewhere to the north.

Nothing in the demeanor of the native informants suggested that disaster had befallen any of the ships or men they had seen. They gave their statements calmly, with none of the emotion one would expect if they had been aware that a catastrophe had occurred. There seemed to be no vital clues for the Franklin searchers, although the possibility that the *North Star* had passed the recent winter in Wolstenholme Sound was of great interest.

It was time to press on. As the *Felix* had gone ahead, Phillips and Beck were taken on board HMS *Intrepid*, which set off under steam to deliver them to their ship. As the *Intrepid* steamed towards Cape Dudley Digges, Adam Beck seemed disturbed and anxious to communicate something important. He tried to indicate to the sailors that some ships had been lost, but they teased him and "reduced him to tears by declaring that he lied."[19] He wanted to speak to his mentor, Ross, but he was not on board. Cator, commanding the steamer, later wrote, "I observed him writing 1846 on the booms, (he having been taught to read and write by the Danish missionaries)[.] I also saw him endeavouring to make himself understood to the men, but not being able to understand him myself, and having the ship to attend to, I paid but little attention."[20]

Several hours later, Phillips and Beck were transferred to Forsyth's *Prince Albert*, which would deliver them to the *Felix*. On board the *Prince Albert* was John Smith, the second man who had some understanding of Inuktitut. He had not been present at the interview with the natives but now, conversing with Beck, he understood what the Greenlander had been trying so desperately to say, and it was shocking beyond belief. Natives in Wolstenholme Sound had burned two ships and murdered their crews, probably in 1846, the year following Franklin's departure! Beck expanded on the story while Smith translated into English with

Snow at his elbow frantically consulting Washington's pocket dictionary and phrase book. Smith and Snow immediately informed Forsyth and Phillips, who questioned Beck again with Smith translating. Snow took notes of the proceedings, and his record of the interview, dated 13 August 1850, reads as follows:

> At 8.30 P.M. the Esquimaux "Adam" furnished the Steward of the "Prince Albert" with the following information which "Adam" says he received from the Natives communicated with this morning.
> Q. What did the Natives tell you this morning?
> A. Saw two ships with three masts in 1846.
> Q. Where were those two ships
> A. Not far from here (our present position)
> Q. From which direction did they come
> A. From the South. The Officers wore Epaulettes. The Ships were beset in the Ice. The Huskies [Eskimos] came on board of the ships . . . It was in the Winter. A great quantity of Huskies. The Ice set together on the Ships. The men left them. The ships were burnt after they were abandoned (*This last sentence is somewhat doubtful* WPS [Snow's initials])
> Q. Were the ships fishing for whales?
> A. No – Had guns but lost their powder and ball
> Q. What became of the white men
> A. They were all killed with darts & spears by the Huskies. Men were very fatigued and reduced. The place was to the northward of us. (Our position at 8.30 P.M.)
> Q. Did their boats go on shore
> A. They abandoned their ships and went ashore. They made houses in the snow, but did not sleep with the Huskies.
>
> – – –
>
> Captn Phillips & the Esquimaux had been on board of the "Intrepid" during the foregoing part of the day, so that the Esquimaux had had no opportunity of communicating with the Steward before. The Steward was alone capable of conversing with the Esquimaux. WPS
>
> – – –
>
> The Esquimaux had been endeavouring in his broken English to communicate this information to the people of the Intrepid.
>
> – – –

The Esquimaux also states that he was sufficiently long in conversation with the natives to obtain all the information they could impart.

– – –

The foregoing statement is perfectly correct as to the words spoken and delivered in The Cabin of the "Prince Albert" under Captn Forsyth. So far as could be well understood.

William Parker Snow
Clerk in charge
John Smith
Steward Prince Albert[21]

The alarming story related by Adam Beck brought the entire searching fleet to a halt. Apparently, Franklin's expedition in search of a Northwest Passage had met a tragic end in Greenland – both ships destroyed and 129 men murdered! Beck's testimony, remarked Sutherland, "threatened to seal the fate of Sir J. Franklin and his companions, and send us all back to England without advancing any further."[22] For the searchers, there was disappointment that Franklin's mission had failed, sorrow for the brave men who had sacrificed their lives in the cause of geographical discovery, and regret that their own dreams of becoming the means of finding and rescuing Franklin were now utterly shattered.

But Penny's Danish interpreter Petersen was sceptical of Beck's story. How could a small number of primitive hunters armed only with stone-tipped spears and arrows overcome more than a hundred well-armed sailors on two armed and easily defended ships? Why would hunters inhabiting a treeless land, for whom wood was a priceless resource, destroy the ships instead of exploiting the treasure trove to make their lives easier?

Around midnight Austin was informed of the new development, and immediately called a conference on board the *Resolute*. Ross was present at last, and during the ensuing examination of Beck he translated from Danish into English. Opinions as to the reliability of the story were mixed because Penny and his officers had previously met the same natives and been told nothing about two ships heading northward and nothing about a massacre of white men. But as long as there was a chance that Beck's story was true, the search expeditions could not advance further; they must prove or disprove the alarming report. Austin acted quickly and wisely, ordering Ommanney to go on board the *Intrepid*, collect Penny and his interpreter, return to Cape York, and talk to the natives again. (Curiously, he did not send Adam Beck as well.) At 5 a.m., the steamer reached the *Lady Franklin* and *Sophia*, picked up Penny and Petersen, and headed back to Cape York. Around 9 a.m., Ommanney, Cator, Penny, Petersen, and some marines landed. Five native men, including the ones previously interviewed, invited the Europeans to their huts. They had already answered questions from Petersen, a few incomprehensible ones from Austin's dictionary-wielding officers, other queries from Forsyth's mate, Wilson, then similar questions from Beck. Now Petersen was back again, grilling them on the same topic, accompanied by a retinue of officers and marines. They may have wondered if Europeans suffered from some sort of attention deficit disorder, but they listened patiently and replied thoughtfully as the officers posed questions in English and Petersen translated them. Ommanney observed that Petersen "seemed perfectly acquainted with their language and they equally well understood him, we put all the question we could think of relative to two ships[.] Petersen gave the replies in English immediately they came from the mouth of the Esquimaux."[23] The men flatly denied having told Adam Beck that two ships had passed toward the north and had subsequently been burned and their crews killed. Only one ship, a three-masted vessel whose captain they called Sanne, had passed by heading northward. This was clearly a reference to HMS *North Star*, commanded by James Saunders ("Saane"). The ship had wintered peacefully in Omenak Sound, the upper deck closed in with a cover. Natives had often visited on board.

When Petersen left the men and talked alone with some women in the camp, he got the same answers: there had been no massacre. In the evening Penny and Petersen returned to the *Lady Franklin*, "doubtless well pleased with themselves," as Sutherland reported, "for being the chief agents in proving the falsity of the report of Sir John Ross's Esquimaux."[24] They could now set aside the wild story and get on with the search.

After the interview, Ommanney prompted Petersen to ask if any natives wanted to accompany the expedition. A young man called Kalersik (variously rendered as Qalasersuaq, Caloosa, Kallihirua, Kalahierua, or Kalli) volunteered. Certain conditions were laid out, to which he readily agreed. Ommanney took him under his wing, and like Eenoolooapik travelling to Scotland with Penny a decade earlier, Kalersik was subjected to an extreme makeover. Ommanney described the young man's "purification:" he was stripped of his verminous garments, put naked into a tub of water, scrubbed vigorously, given a severe haircut, and then dressed in clothes donated by officers and men. It was harsh treatment, a non-lethal form of ethnic cleansing, but perhaps Kalersik regarded it as a curious ritual of white men, or an initiation into their powerful tribe. "He submitted to it all with patience and cheerfulness and with a conciousness that it was requisite – after this he walked about in his new attire with perfect composure."[25] One officer who observed the young man, now clean, neat, and decked out in a sailor suit, made a comment that seemed to say it all: "He looks very funny."[26] Kalersik's birth name, like his clothes, was discarded. He would henceforth be "Erasmus York," Erasmus being Ommanney's given name and York the British name for the headland near which he joined ship. He would remain on board HMS *Assistance* during the rest of her voyage, fitting in remarkably well.

Parker Snow from the *Prince Albert* was impressed by another native, "a little chubby-face, goodlooking boy of about ten years old" named Aladoonga. "I felt very desirous to make this boy in some measure attached to me," he wrote. How wonderful it would be, he thought, if he could take the boy away from his sterile surroundings, expose him to the outside world, give him an education, make him a "useful member among civilized society," and perhaps render him "serviceable to his own people."[27]

Like Ommanney and many other Europeans, Snow regarded the native people of the Arctic as culturally and technologically impoverished. To lift them up towards decent standards (which is to say, those prevalent among middle and upper classes in Britain) would require taking some individuals away from their homes and families, removing them to Britain, immersing them in an utterly alien cultural milieu, and exposing them to potentially fatal diseases. To their would-be benefactors, the end fully justified the means. Their objectives were not solely altruistic. In taking a young man from Cape York, Ommanney secured a guide for the examination of Wolstenholme Sound (and later a candidate for missionary college), while Snow admitted that the boy Aladoonga "would just suit, not only the views I had for his own and his people's welfare, but also the purpose of an interpreter the following summer at Boothia."[28] Evidently he was still hoping to implement his grandiose plan for reaching King William Island from Hudson Bay. But the "poor little fellow," wearing Snow's colourful cotton handkerchief around his neck, chose to remain at Cape York. "The rude hut, and the state of life, half human, half beast, was to remain his lot,"[29] Snow wrote regretfully.

Of the two young native men from Cape York, which one was better off? Aladoonga probably lived out the span of his natural life – short though it may have been – with his family and friends, in his home environment, whereas Erasmus York died several years later thousands of miles away, alone among strangers in a foreign land.

Transplanting an aborigine to an alien geographical and cultural milieu was a risky affair, and it sometimes backfired. Charles Forsyth, Snow's commander on the *Prince Albert*, had once served on HMS *Beagle* in the South Atlantic under Captain Robert Fitzroy, and he was undoubtedly familiar with the case of "Jemmy Button," one of four primitive Indians captured by Fitzroy in Tierra del Fuego two decades earlier and taken forcibly to England for the supposed benefits of a Christian education. After an absence of three years Jemmy and his companions – save one who had died from smallpox – were taken home to do good among their people, sailing again on the *Beagle* under Fitzroy's command, but this time with the naturalist Charles Darwin on board. During his sojourn in England, Jemmy had become something of a dandy. Darwin noted that "he was vain of his personal appearance: he used to wear gloves, his hair was neatly cut, and he was distressed if his well-polished shoes were dirtied."[30] Arriving at Tierra del Fuego, Jemmy went ashore, immaculately groomed, dressed in a stylish suit and handsome boots, to take up his mission among the nearly nude natives with wild hair and painted faces. He distributed gifts provided by the London Missionary Society, including such incongruous items as "chamber-pots, tea trays and crockery, wineglasses and soup tureens, beaver hats and white linens,"[31] and the *Beagle* sailed away. But Fitzroy's hope that Jemmy and his companions would spread enlightenment among their tribesmen was not fulfilled. When the *Beagle* returned a year later, Jemmy was scarcely recognizable. He had reverted to his former way of life and was now "a thin haggard savage, with long disordered hair, and naked, except a bit of a blanket round his waist," Darwin observed. "We had left him plump, fat, clean, and well dressed; I never saw so complete and grievous a change."[32] Fitzroy's social experiment, although motivated by good intentions, had been a complete failure.

In all likelihood, Forsyth had recounted the story on board the *Prince Albert*, but Fitzroy's failure did not deter Snow, who entertained a similar vision for the youth Aladoonga and the people of Cape York. In a curious twist of fate Snow would meet Jemmy Button face to face five years later while commanding the missionary ship *Allen Gardiner* in the region of Tierra del Fuego. A quarter-century after his abduction on the *Beagle*, Jemmy was a "shaggy-looking man," still living in the native mode, "with long matted hair and eyes impeded by the sting of wood smoke."[33] Snow tried to entice him to join a mission settlement in the Falkland Islands but he declined. One exposure to civilization been quite enough.

* * *

As the several ships rode at anchor off Cape York, Snow met William Penny for the first time. He later described his impressions:

> I stood for a moment gazing at this daring whaler, scanning him from top to toe, to see what sort of man he was that I had originally come from America purposefully to join, and whose energetic spirit I had heard so much talk of, not only on shore, but since I had been among the naval ships that were for a long time in his company. I need not describe him: he is too well known; but suffice it to say that I saw before me, while thus quietly scrutinizing him, the very man after my own heart: the one that would have exactly suited me. I have in my mind's eye now, full before me, his whole picture, – daring, pushing, ardent, enthusiastic – a thorough, frank, hearty seaman – ready and rough, and rough and ready when need be – himself a *working* hand as well as a *directing* hand; – there he stood, bronze-faced, fur-capped, jacketted with spy-glass slung around him, ready for any emergency that might suddenly meet him. Quick and prompt in his action, he denoted the man of firm nerve and inexhaustible resources.[34]

When the ships continued westward, some officers wondered if there could be an element of truth in Beck's massacre story. What if the natives first questioned by Petersen had not been aware of what had happened to Franklin's ships? What if they had known but intentionally held back the information, fearing retribution? What if translation difficulties had corrupted their statements? As long as doubts lingered, the searching expeditions could not turn their backs on the region in which the disaster was alleged to have occurred. Fortunately, all the British search expeditions were now close together, facilitating their discussion of the evidence and implementation of appropriate measures.

As the senior active Royal Navy officer, Austin decided that Wolstenholme Sound, the region in which the Franklin crews had allegedly been killed, must be examined. He dispatched Ommanney in the *Assistance*, with Cator's steamer *Intrepid* to provide a tow. Accompanying the two naval ships, and also receiving a tow from the steamer, was Ross's *Felix*, presumably included in the mission in order to have his interpreter Adam Beck available. The two sailing vessels hove to off Wolstenholme Island while the steamer continued into the sound. Lookouts scanned the shoreline for cairns or anything unusual. Several miles up the sound, the new naval recruit, Erasmus York, pointed out an apparently deserted native settlement, and a party landed by boat to have a look. The settlement had indeed been abandoned, but not by choice. When the officers entered a few of the huts, they were met by grisly scenes. Cator wrote in his journal, "What was our horror to find dead and putrid bodies of the Eskimaux . . . One hut had four poor creatures lying in it, one dressed in his skin dress – all lying closely packed together as if they had been taken ill and died off one by one and not able to move or help each other – these were lying on a blanket – decomposition was very far advanced, in spite of the cold weather, on some of them."[35] The experience was especially poignant for Erasmus York, who came upon the grave of his father (or possibly brother). Scattered among the dead were articles of European manufacture. "Here we found iron hoops and preserved meat tins," Cator wrote. "We were now satisfied beyond a doubt that some vessel or vessels had been here and that, only lately as the paint and Goldners name on the label on the tins were quite fresh."[36]

Leaving the ghost village, the *Intrepid* steamed deeper into the sound. Suddenly someone sighted cairns. When officers went ashore, they found proof that HMS *North Star* had wintered there – messages inside the cairns, graves of four crew members who had died between November 1849 and May 1850, a washtub and three shirts belonging to a marine corporal, miscellaneous articles from the ship – but nothing at all that related to HMS *Erebus* or *Terror*. In pidgin English, an embarrassed Beck admitted to one of Ommanney's officers that he had lied about the massacre.

There may have been a connection between the deaths of the natives and the presence of a naval ship nearby. Historian Ann Savours writes, "These were Greenlanders (Inughuit) struck down by famine or an epidemic, perhaps contracted from the *North Star*."[37] The accidental introduction of harmful pathogens to native people who lacked immunity to them has been an unfortunate concomitant of Euro-American enterprise throughout the world. The natives encountered at Cape York had told of some of their group dying further north after a sickness that involved coughing, choking, and spitting. Such symptoms suggest influenza, which although normally tolerated by Europeans, was deadly among inhabitants of the North American Arctic, even into the twentieth century. Although the number of bodies seen in Wolstenholme Sound appears to have been only four or five, others may have been buried or may have fled the village and died elsewhere.

It is conceivable that other sorts of European introductions played a role in the deaths. Cator's mention of Goldner's preserved meat cans brings to mind two recent hypotheses concerning the deaths of Franklin's men. In the first, the solder used to seal cans after filling them is said to have resulted in lead poisoning, leading to mental and physical deterioration, and eventually death.[38] In the second, unsanitary conditions prevailing in the factories where Goldner's canning took place are said to have permitted the deadly toxin *Clostridium botulinum* to infect the contents before the cans were sealed.[39] To attribute the deaths of the

Greenlanders in Wolstenholme Sound to either of these alleged causes, however, raises the question of why the crew of the *North Star* had not been not affected in the same way. Only four sailors had died during the winter, reportedly from accidents or scurvy.

Famine also seems unlikely. While wintering near the native settlement, the *North Star* had provided medical treatment (well-intentioned but unfortunately unsuccessful) for a hunter suffering from extreme frostbite on his feet. If a paucity of animal food had threatened the survival of the entire community, Saunders would surely have provided some of the ship's provisions to help.

With the massacre story laid to rest, the *Intrepid* left Wolstenholme Sound, cast off the *Felix*, and headed westward across the North Water in company with the *Assistance* to resume the search for Franklin. The investigations at Cape York and Wolstenholme Sound had taken up valuable time. Sherard Osborn remarked that Adam Beck had succeeded in "misleading a large number of Her Majesty's officers . . . and in detaining, for two days, the squadrons in search of Franklin." In an angry outburst, he exclaimed, "May he be branded for a liar!"[40] Cator later called Beck "a great rogue,"[41] and to Clements Markham, a young midshipman on HMS *Assistance*, he was an "outrageous scoundrel."[42]

The only dissenter was Ross. In a final meeting on board the *Assistance*, with Ommanney, Cator, Beck, and York in attendance, Ross maintained his faith in Beck's testimony and suggested that Petersen had invented a completely different translation so that he could remain with Penny's expedition as long as possible while continuing to receive pay. Ross said he would report Beck's story about the massacre in writing to his sponsors, the Hudson's Bay Company, and would send the news to England on one of his four homing pigeons. In view of the fact that no evidence had been found to support Beck's apocryphal report, Cator felt that publicizing it would be wrong and possibly harmful. He was relieved when Ross and the *Felix* went their own way.

There was speculation about Beck's motives in inventing the massacre story. One opinion was that his deception had been "cunningly devised to keep, at any rate, his own ship on a coast whither he could escape to the neighbourhood of his home in South Greenland."[43] But how could he be sure that Ross would keep the *Felix* on the Greenland coast? And even if the ship did remain on the East Side, how did Beck intend to make his way eight hundred miles south to his home at Holsteinborg? If he really wanted to go home so soon, why had he signed up in the first place? A better explanation is that Beck, craving recognition and acceptance among the Europeans, had resorted to the familiar native technique of giving the white men what they so clearly wanted, in this case some important news about Franklin's missing ships and men. If he provided information they desired, even if it was unfavourable, he might be accorded more respect, the search would probably be terminated, and he would get home as desired.

Beck's massacre story remains one of the principal mysteries of the entire search for Franklin, and it has often been re-examined. Some writers have suggested that he did not misrepresent the native information at all but merely misunderstood it because he was unfamiliar with the local dialect. According to others, the Cape York natives did mention two wrecked ships but failed to explain that the events had happened years before and hundreds of miles away; they might have been referring to the abandonment of John Ross's ship *Victory* and his launch *Krusentern* in the Gulf of Boothia in 1832, or to wrecks among ice-trapped whaling ships in Baffin Bay and Davis Strait in the 1830s, or even to the loss of Franklin's ships *Erebus* and *Terror* near King William Island in 1847–48.[44]

None of these explanations is convincing, however. Inuit did not kill any of the *Victory*'s crew, and there is no evidence that they killed any survivors of wrecked whaleships or of HMS *Erebus* and *Terror*. Furthermore, the destruction of the ships of Ross and Franklin occurred from six hundred to eight hundred miles southwest of Cape York and Wolstenholme Sound. Although news, like resources such as soapstone and trade goods such as metal knives, often travelled great distances through successive intergroup exchange, the isolated natives of northwestern Greenland had no contact with inhabitants of the North American Arctic until 1862, when a group of Inuit from Baffin Island, led by the shaman Quitdlarssuaq, reached them.[45]

Most importantly, when Penny's competent interpreter Carl Petersen questioned the Cape York people on two occasions, they said nothing about two ships having gone northward from Cape York, or about ships having been burned and their crews murdered, and they denied having said any such thing to Beck. Erasmus York, a member of the group that inhabited the region, had been present during the exchange with Beck, but he never mentioned a massacre.

This stage of the search had revealed a weakness in the search arrangements. Austin and De Haven, with six vessels between them, had no interpreters. Ross had one interpreter but of doubtful honesty. Forsyth had no official interpreter but was fortunate in having two crew members (ex-Hudson Bay traders) who understood some Inuktitut. Penny was the only commander with a reliable interpreter. As the searching ships met native Greenlanders at various locations from Disko Bay to Cape York, it would have been very useful to have translation capability.

12

To the West Land

On this never to be forgotten day . . . the North Water was gained; its swell, so long a stranger, came to us as a familiar feature of kindlier seas'[1]

Along the Greenland coast, the ten searching ships had undergone a baptism – not of fire but *ice*. After smashing against floes, grazing bergs, getting nipped at the floe edge, being pushed onto reefs, and colliding with other vessels, they emerged as battle-scarred veterans. There are at least twenty recorded incidents in which ships experienced damage of one kind or another to hulls, bulwarks, taffrails, jib booms, bowsprits, standing rigging, boats, davits, rudders, propellers, keels, and even royal masts shaken loose by the force of an abrupt stop against rock-hard ice. None of the ships came through unscathed but, most importantly, none was destroyed.

Penny's brigs now crossed the North Water toward the West Land, enjoying clear sailing for the first time in three and a half months. It was a great relief now to be bowling along under studding sails in an unobstructed sea. Near the Carey Islands, the *Lady Franklin* and *Sophia* reached a latitude of 76°40' N, and Sutherland remarked, "There was nothing to cast the slightest gloom upon our bright prospects of being soon in the spot where our services would be called into requisition."[2] For the crews on the other searching ships, the release from Greenland's icy grasp was equally exhilarating. "The change was wonderful – incredible,"[3] Osborn declared when HMS *Pioneer* entered the North Water.

Penny's first objective on the West Side was Jones Sound, the westward-trending waterway between Devon and Ellesmere islands. His Admiralty orders contained the following: "It being our desire that a certain strait, known as Alderman Jones's Sound, and which would not appear to have been as yet examined, should be searched; you are hereby required and directed to proceed in the first instance to that sound, closely examining the shores for any traces of Sir John Franklin's course, and proceeding, should it offer the means of your doing so, in the direction of Wellington Strait, and on to the Parry Islands and Melville Island."[4] The Admiralty had thus instructed Penny to attempt a westward passage where none was known to exist, along a route merely *thought* to lie parallel to the Parry Channel, but more than a hundred miles to the north. They had assigned him the most challenging task of all. Did they think that he was more likely to succeed than Austin with his four ships, two of them equipped with engines? Did they intend to avoid duplication of effort among the expeditions and cover a broader region in the search? Or did they want to sidetrack Penny while the Royal Navy covered itself with glory in the region through which Franklin had been ordered to sail?

Although Jones Sound lay far north of Franklin's intended course, it had attracted considerable interest since Thomas Lee Sr reported sighting a cairn there in 1848 while commanding the whaling ship *Prince of Wales*. As it was assumed that Franklin would leave progress reports in rock cairns, Lee's news was exciting. Although Lady Franklin had elicited no concrete information from Lee about the cairn's location, and Sophia Cracroft had dismissed his testimony as "a very bungling story,"[5] Lady Franklin thought Lee's report worth following up. Jones Sound had been much in mind when she naively agreed to subsidize the whaling voyage of the Hull whaler *Abram* under Captain Gravill in the summer of 1849. As that voyage had revealed nothing of significance, the sound was still regarded as an important objective in 1850.

Lee's story of a cairn in Jones Sound was one of many unsubstantiated reports, vague rumours, and intentional hoaxes that bedevilled the search for Franklin. Vague and contradictory, it had nevertheless caught the public imagination, inspired Lady Franklin, and excited the Admiralty. Because the sound tended to be blocked with ice until late in the season, it had remained beyond the normal range of whaling activities for three decades and would continue thus after the searching operations of 1850. The cairn story could neither be confirmed or disproved, so it endured, and the longer it endured, the more it was embellished. A seaman named William Millar declared that he had landed from the *Prince of Wales*, seen a hearth and the prints of European boots, and had partly dismantled the cairn before the ship was forced to make a hurried departure. Another man named Moore was said to have been there at the time. Then Lee's son, also a whaling captain, declared that his father had not been in Jones Sound at all. Sophia Cracroft said that Lee Senior, the originator of the story, was "in his dotage" and his son was a "clever fellow of drunken dissipated habits."[6]

For Penny and his men, the prospect of locating Lee's cairn, learning of Franklin's intentions, and following his track into an unexplored region was a powerful incentive. When they arrived off Jones Sound, however, they were unable to enter because it was barricaded by large floes. Sutherland recalled later, "The studding-sails were taken in, and the ships were hauled nearly close to the wind, so as to lay clear of the pack as they proceeded along its edge, expecting an opening to lead westward into the Sound. It was close throughout its entire extent, as far as could be judged from the crow's nest, and it also appeared to be very heavy, and to contain icebergs in abundance. The wind was hourly freshening up."[7] The Admiralty had advised Penny that if he reached Jones Sound "early," found the passage into it "impeded," and had sufficient time to spare, he could go northward to examine Smith Sound. This was now out of the question. Although the passage into Jones Sound was certainly *impeded*, by no stretch of the imagination could 18 August be considered *early*: only a few weeks remained before freeze-up!

After Penny's expedition, Lee's alleged cairn would continue to be a false beacon luring searchers in the wrong direction. The Admiralty considered sending a steamer to look for it; Lady Franklin instructed William Kennedy, commanding her second *Prince Albert* expedition, to make the penetration of the sound his first priority; she later instructed Inglefield to look for the cairn when he commanded the *Isabel*. Robert McCormick attempted to reach it by boat via Wellington Channel during Belcher's expedition of 1852–54. None of these well-intentioned searchers knew where the cairn was located because Lee Sr, its self-proclaimed discoverer, either did not know himself or had invented the story. The search expeditions were seeking the proverbial needle in a haystack, but the needle may have been merely a fiction. The Admiralty contributed to the myth by publishing a chart entitled "Discoveries in the Arctic Seas to MDCCCLII," dated 8 April 1852. The letter X marked the "position of the Prince of Wales, August 13th 1848 according to Captn Lee; having come up Jones Sound & visited a cairn, cooking place & foot prints." The cairn was on the sound's south coast near longitude 89° W, more than 140 miles west of Cone Island near the sound's entrance. Several months later, another official chart, depicting Inglefield's explorations, showed Lee's alleged cairn in almost the same position. Although there was no reliable evidence for its existence, the Admiralty charts gave it an aura of authenticity. (Even today Canadian charts indicate "Thomas Lee Inlet" there.)

Seeing no possibility of gaining access to Jones Sound, Penny kept on for Lancaster Sound in accordance with his orders. At 3 a.m. the next day, the ships sighted two unfamiliar vessels and gradually overtook them.

Colours were hoisted by both parties, and the strangers – to everyone's surprise – proved to be De Haven's brigs *Advance* and *Rescue*. After failing to penetrate the Middle Ice in July, the Americans had regained the Greenland coast and headed north, trying to overtake the British expeditions. On 15 August, a short distance east of Cape York, they had encountered two native seal hunters. A boat was sent ashore and De Haven recorded in his diary, "Mr. Murdaugh who went in the boat could not understand them sufficiently to gain any information further than that a few others of their tribe were living in a hut not far distant."[8] Nothing was learned about HMS *Erebus* and *Terror* or any other naval ships. The lack of an interpreter proved to be a serious disadvantage for the Americans, as it had with Austin's expedition.

The *Advance* and *Rescue* had then followed the coast westward past Cape York, now only a day or two behind the British expeditions. They stopped at the "Crimson Cliffs" seen by John Ross in 1818, where organic material gave the snow a reddish colour. Officers went ashore "to botanize and procure specimens of the snow."[9] Sea birds were plentiful, and the officers secured many for the table. After two hours of shooting from a boat, Kane returned with seventy birds, most of them little auks, or dovekies, which he found "very red in meat, juicy, fat, delicate, and flavorsome . . . the perfection of good eating."[10] On the following day, Henry Brooks, second officer of the *Rescue*, shot a bird that Carter identified as a great auk [*Alca impennis*]. This goose-sized, penguin-like, flightless bird had been regarded as extinct since June 1844. Could Carter really have seen one six years later? He may have been mistaken; neither the journals of De Haven and Lovell nor the book by Kane mention a great auk.

While the British ships were delayed by Beck's massacre story, compelled to return to Cape York to question the natives again and then proceed northward to Wolstenholme Sound to look for evidence of the disaster, and while Austin and Osborn took their ships to Pond Inlet to interview natives, the American ships had forged ahead through the ice-free North Water to the West Side. On 19 August, near the entrance to Lancaster Sound, Kane wrote, "We are among the foremost if not the foremost on the searching ground."[11]

The following day, they came within hailing distance of the *Lady Franklin* and *Sophia*. Penny shouted over the waves, bringing the vessel's officers up to date on developments, reporting the last known positions of the *North Star* and the various searching ships and informing them that Jones Sound was blocked by ice. From the deck of the *Rescue*, Robert Carter, first officer, noted that Penny "didn't speak plain so we lost much that he said." He rightly guessed that "a very genteel man standing by him" was Robert Goodsir. The appearance of the *Lady Franklin* and *Sophia* made a highly favourable impression on the young man of twenty-four and excited not a little envy: "My eyes, what splendid craft the Englishmen had, large, high, roomy, full rigged brigs; every thing looked so clean and comfortable with plenty of room to walk about and a spar deck defying all water. They'll have a nice time."[12] He added, "They had bears meat and more birds than we even." With a "good old English hurra" to encourage the Americans, Penny's brigs drew steadily ahead. The greeting was returned, and after "one cheer more of acknowledgement on each side . . . the sister flags separated, each on its errand of mercy."[13]

Penny's ships, faster sailers than De Haven's, continued to advance and entered Lancaster Sound. The wind was rising and sea was taking on an angry countenance. As the men of the *Lady Franklin* were working aloft, a sudden gust carried away the mast for the main royal and skysail, creating an awkward tangle of rigging and a very dangerous situation. The ship lifted and plunged in heavy seas as the men worked feverishly to deal with the mess. The helmsman struggled to sail the crippled vessel clear of an iceberg "on which the sea was breaking furiously, and sending small showers of spray into the air with a surging noise."[14] Thereafter, they proceeded cautiously under close-reefed topsails in a wicked cross sea that disturbed "the tranquility of the lands men's stomachs."[15]

The nasty weather lashed out viciously at the Americans when they turned the corner into Lancaster Sound. The wind swung into the east and increased rapidly to gale force. De Haven was forced to heave to and Griffin followed suit. After tossing about all night on the tiny *Rescue*, Carter wrote in his diary, "Horrible, wet, uncomfortable, seasick day. May nobody ever see such another unless it be to cure some poor ass of a taste of the sea. Oh wont I be stupid if I ever come to sea again?"[16] Even on the larger and more stable *Advance*, the men were getting a rough ride. Kane wrote, "The sea is short and excessive. Every thing on deck, even anchors and quarter-boats, have [*sic*] 'fetched away,' and the little cabin is half afloat. The Rescue is staggering under heavy sail astern of us. We are making six or seven knots . . . Murdaugh is ahead [at the bow] looking out for ice and rocks."[17]

Ross was not far behind the ships of Penny and De Haven. As the *Felix* approached the mouth of Lancaster Sound, the rising wind kicked up a heavy sea. Ross shortened sail and came about onto a tack that would take them away from the land, "considering it imprudent to run into Lancaster Sd. in thick weather."[18] Conditions continued to worsen and at noon he hove to off the sound's entrance, with plenty of sea room to allow for the vessel's drift. By midnight they were nearing Bylot Island, and Ross wrote in his diary that he could see Cape Hay, "named in 1818 by me."[19] That was a voyage he might have preferred to forget, during which a mirage had fooled him into thinking that Lancaster Sound was blocked by a range of mountains. But it was also a voyage on which he had scattered place names about and landed at various points to claim the land for Great Britain in the usual manner of explorers – a process the late Alan Cooke compared to "dogs pissing on street corners in strange neighbourhoods."[20]

High winds and rough seas make life on a small sailing vessel uncomfortable and dangerous. They also transform dignified men into impromptu comedians, candidates for a prize for the Best Silly Walk. At one moment a man may be moving sedately, at the next prancing foolishly sideways, or staggering drunkenly backward, or making a headlong rush downslope towards a gunwale that is dropping rapidly toward the waves. One can only imagine how hard this must have been for Sir John Ross, a battle-scarred man in his seventies, long before anti-inflammatory medication, arthroscopic knee surgery, or joint replacement. Yet his diary contains no complaint. He was, as Kane observed a few days later when the *Advance* spoke the *Felix* off Admiralty Inlet, "very little stricken in years and well able to bear his part in the toils and hazards of life."[21]

HMS *Assistance* and her steam tender *Intrepid* had entered Lancaster Sound and were off its north coast when the gale hit. On 19 August they passed Cape Warrender and were looking for a suitable place to land and set up a message cairn. Off Powell Inlet, Ommanney, Dr Donnet, and the second master, Krabbé, left the *Assistance* by boat and boarded the steamer *Intrepid* to approach the land under power, but by this time the weather was deteriorating rapidly. The long, narrow *Intrepid* was inclined to roll badly and now took the opportunity to show off its talents for the guests of honour. "The three were scarce on deck until she took one of her rolling fits, and I think I scarcely ever knew anything like it," Cator reported, adding with a touch of smugness, "The little Captain, Doctor, and Krabbé were frightened out of their lives at it."[22] The surf was too high to attempt a landing, so the three officers returned to the more stable *Assistance*, and both ships heaved to. Ommanney signalled Cator to proceed independently in his steamer and rendezvous later further west off Cape Hurd. Shortly afterward, a heavy roll and a brush against ice swept a boat and all its gear off the *Intrepid* into the sea. Cator stayed on watch all night. "It was truly a bitter cold one," he said, "enough to cut one's face in half."[23]

* * *

After sending the *Assistance* and *Intrepid* north on the Greenland coast with the *Felix* to examine Wolstenholme Sound, Austin had headed west across the North Water toward Pond Inlet to look for cairns along coasts in that vicinity and to question any Inuit they happened to encounter. He generously offered a tow to Forsyth, commander of Lady Franklin's private search expedition, so Osborn's steamer *Pioneer* hauled the large *Resolute* and the little *Prince Albert* across the North Water and south to the mouth of Lancaster Sound. Two days before the gale that caught the ships of Penny, Ross, and De Haven, Forsyth cast off the towing hawser and separated from the naval ships. Thanks to the tow from Greenland, the little *Prince Albert* was now among the leaders.

The parting was dramatic, Austin standing at the taffrail of the *Resolute*, waving his hat and shouting over the waves, "Success to you, my fine fellows." One can visualize the excitement among the small crew of the *Prince Albert* as the vessel advanced into Lancaster Sound. Snow wrote, "Here we are, and by the blessing of God, all safe so far; and at this moment the *first*, and most ahead of all the discovery ships, though we left last."[24] Aided by a favourable wind from the east, the *Prince Albert* followed the south coast westward to the mouth of Admiralty Inlet. When lookouts sighted a cairn on top of one of the Wollaston Islands, Forsyth sent a boat party ashore. The men eagerly climbed to the summit and carefully dismantled the boulders. Inside they found no message, no relic of European origin, nothing. Hopes had risen to the clouds; now they tumbled back to earth. The pile of stones was merely an Inuit beacon of some sort.

Such experiences would soon be commonplace among all the expeditions, with a familiar sequence: the thrill of sighting a cairn after hours of eye strain and boredom; the excitement of landing on a strange shore; anticipation and suspense as the men took the cairn apart rock by rock; and finally disappointment when nothing whatsoever was found within. After the landing party returned disheartened to the ship, the breeze began to intensify into a gale and Forsyth decided to heave to. After a tense night in the vicinity of ice floes and bergs, they hoisted sail and continued westward.

Meanwhile, inclement weather was impeding Austin's search operation at Pond Inlet. It was "a longer proceeding than was ever calculated upon," Osborn wrote, "for a succession of thick fogs and strong gales prevented the 'Pioneer' running into the bay, or ascertaining whether cairns or other marks had been erected on the coast."[25] A boat party landed on a point and came upon a number of cairns near an unoccupied winter village. "These we pulled down one after the other, and examined without finding any thing in them,"[26] he related. Later they realized that the cairns had been Inuit markers for caches of seal meat. After leaving some conciliatory gifts in the Inuit huts and depositing messages for passing whaling ships, they returned to the ship, rejoined the *Resolute*, and the two vessels proceeded into Lancaster Sound to catch up with the *Assistance* and *Intrepid*.

By 22 August the entire searching armada had weathered the gale and entered Lancaster Sound. They were now in the waterway that Franklin had been ordered to follow to Cape Walker before turning south to the mainland coast. Ten ships and almost three hundred men – five times as many ships and twice as many men as on Franklin's missing expedition – were now sniffing eagerly to pick up his trail.

* * *

One more vessel was within the confines of Lancaster Sound. Shortly after the gale, Penny encountered HMS *North Star*. It must have a joyful occasion for the sixty-odd men of the transport, who had spent fifteen months frozen into the ice of Wolstenholme Sound – further north than any Royal Navy ship had ever wintered – without seeing another ship or crew. They had endured 116 days without a view of the sun and had experienced temperatures as low as –63°F, which James Saunders, the ship's commander, believed to be the lowest air temperature recorded. After breaking out of winter quarters on 1 August, Saunders told Penny, they had resumed their mission – to deliver supplies and new orders to James Ross at Port Leopold. Failing to encounter Ross's two ships, they had attempted to land the stores nearby, but heavy ice floes barricaded the coasts everywhere. The *North Star* had criss-crossed Prince Regent Inlet, trying unsuccessfully to get into four different harbours, and now Saunders was taking the ship back toward Baffin Bay to examine several alternative depot sites designated by the Admiralty along the north coast of Baffin Island.

Penny told Saunders that the object of his quest, James Ross, had been warming his feet in front of the fire back in England for more than a year, and that five new expeditions had now arrived to take up the search for Franklin. He delivered a letter from the Admiralty informing Saunders that if he had not already landed the stores intended for James Ross, he was to make them available to the expeditions of Austin and Penny to provide them with some security in case of shipwreck or besetment, and to augment their provisions in case any Franklin survivors were discovered. There was a serious flaw in the arrangements, however. The

Admiralty told Saunders to deposit the stores at the Whalefish Islands or Disko for Austin or Penny to collect – but both commanders had passed those places months before. They would certainly not wish to abandon the search just as it was about to commence and retreat across to the Greenland coast, a voyage of several weeks or even months. Freeze-up was less than three weeks off. The Admiralty's orders made no sense.

Penny pointed out to Saunders that if the stores were to be of any use to the searching expeditions, they must be cached in Barrow Strait or Lancaster Sound, where they would be readily accessible. Saunders, however, was no Nelson; he "seemed to dread disobeying his orders,"[27] Sutherland observed, and was reluctant to adopt the suggestion of a whaling captain, even one sailing under Admiralty orders. Although he now knew that several search expeditions were converging on Barrow Strait and might have need of extra supplies, he resolved to follow his instructions to the letter and take the stores back across Baffin Bay. The prospect dismayed Penny. In a dispatch that the *North Star* would carry back to England, he outlined the situation for the Admiralty. If the supplies were taken back to Disko, he wrote, "an otherwise invaluable supply is rendered perfectly useless to all the expeditions at present in this quarter, and we are deprived of what we had reckoned upon, viz. a deposit in the Sound to fall back upon, in case misfortune should compel us to abandon our vessels."[28] When Saunders questioned Penny's interpreter, Carl Petersen, about the feasibility of storing the supplies at Disko, Petersen cited several reasons why that would be unwise: no warehouses were available there; the stores would only be secure from theft if placed in a house; the governor would be obliged to destroy all the spirits and wine.

Penny assured Saunders that if the Admiralty disapproved of his action, he would accept full responsibility. This did not satisfy the *North Star*'s commander, however. Captain Austin could sanction such a change of plan, he said, but he was not there. "In the absence of Captain Austin, he knew not how to act."[29] It must have been an agonizing quandary for the rigid, by-the-rules Saunders, constrained by the strict naval code of behaviour, aware that the arguments put forward by Penny were reasonable yet hesitant to act on them for fear of jeopardizing his career.

Alarmed, but at the same time sympathetic to Saunders's dilemma, Penny wrote a personal letter to his friend Captain W.A.B. Hamilton, permanent secretary of the Admiralty: "You may judge my surprise to learn that he was to proceed to the Island of Disco to land his provisions . . . I have endeavoured to prevail upon him to land some at Navy Board Inlet; however, he has not given me his final answer." Aware that such a course of action might land Saunders in hot water at the Admiralty, he added, "If Mr. Saunders breaks through his instructions by my advice, I trust Captain Hamilton will do his best to see he does not suffer . . . I should be sorry if any blame should be attached to him."[30] This generous plea on behalf of Saunders he sent with his official dispatch, to be carried back to England on board HMS *North Star*.

In the end, Saunders came round to Penny's point of view. A few days after leaving Penny's ships, he landed the stores in one of the Wollaston Islands at the mouth of Navy Board Inlet. The depot was immense. It contained enough food to last a hundred men for a year, packed in almost six hundred barrels, cases, casks, small casks, hogsheads, half-hogsheads, and tierces.

Getting all this material ashore was a major operation for which Saunders deserves credit. He brought the *North Star* as close to shore as he dared and anchored on a poor holding ground off a lee shore in water only slightly deeper than the ship's draft. At 4 a.m. on 24 August, the men began transferring the stores. The containers were slung outboard, lowered into boats, rowed a quarter-mile to the shore, manhandled out of the boats, rolled up the rocky beach on spars and planks, and stacked above high tide mark. The work continued for five full days, the men fortified with extra rations of meat and spirits. At such an unsatisfactory anchorage, bad weather could have threatened the ship's safety and brought the unloading to a halt, but they were lucky. No sooner was the work finished, however, than a gale sprang up. Onshore winds and a heavy groundswell quickly made their situation so dangerous that Saunders ordered the cable slipped, leaving the anchor and sixty fathoms of chain on the sea bed. (They would be recovered accidentally by another ship four years later.) The *North Star* clawed her way off the exposed shore to safety.

The Admiralty had advised Saunders that if he had to leave the stores near Pond Inlet or further south, he should bury them so they would not be disturbed by animals or Inuit. Apparently they considered Pond Inlet to be the northern limit of large animals and Inuit, but in fact the cache was within the range of both. In any case, it was utterly impossible to place six hundred casks and cases *below ground* in a region consisting mainly of permanently frozen rock and soil. By necessity everything was left above ground.

The depot would later become a treasure trove for visiting Inuit. While one doubts that they were thrilled by the taste of lemon juice (there was more than a ton of it), flour (thirteen tons), ship's biscuit (six tons), yeast, vinegar, pickled cabbage, mustard, black pepper, or oatmeal, they might have enjoyed sitting down to campfire dinners of ox-cheek soup and preserved pork, preceded by a tot of rum, accompanied by white wine, followed by a nibble of chocolate (about a ton), and topped off by a glass of port or a snifter of brandy. (The brandy, overproof rum, white wine, and port in the cache amounted to more than two hundred gallons.)[31] These were survival rations deluxe.

Durable non-food items were undoubtedly of much greater interest to the Inuit than the unfamiliar (and to them, largely unpalatable) foods. With a little experimentation, they might have found sperm oil (140 gallons) and candles useful for illumination. Metal and wood were always highly prized, and there were impressive quantities of both. The several hundred casks of various sizes were bound with more than 3,500 iron hoops; the meats, vegetables, and soups were sealed in almost 5,400 metal canisters; and in addition to thousands of wooden barrel staves, there were 3,400 pounds of firewood.

Of no use to Inuit but of great potential value to distressed seamen was coal (40 bags), patent fuel – briquettes of small coal treated with powdered pitch – (38 tons), and carpet boots (124 pairs). Whether distressed seamen would have appreciated the soap (more than half a ton of it) is uncertain.

Naturally, there was tea (663 pounds) and plenty of sugar to sweeten it (5,264 pounds). While it is doubtful that the Inuit knew then how to prepare the infusion, today they rival – and possibly exceed – the English in their consumption of it. Could it be that it all started with some fortuitous "Eureka" discovery involving hot water and spilled tea leaves at Saunders's depot? The cache also contained 1,628 pounds of tobacco. Inuit who had encountered whalemen or explorers had seen some of them smoking pipes, but they themselves had no regular supply of the leaf. Did the depot at Navy Board Inlet get them hooked? (Today the percentage of adults in Nunavut who smoke is three times the national average, and deaths from lung cancer are correspondingly high.)

* * *

As the main function of Saunders's depot was to back up the search expeditions operating in the region and enable them to stay out longer if necessary, it was essential to make its presence known. After speaking with Saunders, Penny continued westward toward Wellington Channel and Barrow Strait. He could easily have notified the other expedition leaders about the location of the depot, if he had known it, but Saunders had not made his intentions clear. Sutherland later wrote, "Although it would have been of signal service to our expedition to have known the steps Mr. Saunders would take, that was denied to us even in the remotest shades";[32] according to Osborn, "Captain Penny left Mr. Saunders under an impression that he was going to Disco, to land his provisions."[33] On the day after parting from the *Lady Franklin* and *Sophia*, HMS *North Star* encountered the *Felix*. Ross sent Phillips, his second-in-command, across by boat to deliver letters and exchange information. Did Saunders tell Phillips about landing the stores at Navy Board Inlet? Evidently not! Ross later told De Haven that the *North Star* was "on her way to England via Disko where she was ordered to land her stores."[34]

Once the stores were landed, the *North Star* proceeded eastward and rounded Bylot Island. Off Pond Inlet the ship spoke the *Prince Albert*, also on its way home. Forsyth consigned some letters and dispatches to the naval ship. Did Saunders now reveal that he had established a depot at the mouth of Navy Board Inlet? There is no evidence that he did. Nor did Forsyth tell Saunders about some relics of the missing expedition that he had picked up at Cape Riley. A few days later, Forsyth sighted the *North Star* again and tried to attract its attention with rockets and blue lights, but the ship sailed on unheeding. The searching expeditions in Barrow Strait, for whom the stores were intended to serve in case of an emergency or an extended search, did not know that the depot existed. They believed the supplies had been taken back to Disko.

After the *North Star* returned to England, it was widely assumed that Saunders had informed the search expeditions of the location of the depot before he left Lancaster Sound. The *Nautical Magazine* remarked, "We have no doubt the supply and its 'whereabouts' left by the *North Star*, is known to Captain Austin."[35] But this was not so – nor was it known to Penny, Ross, and De Haven. Fortunately, their searching expeditions did not have occasion to use Saunders's supplies, but their ignorance of his cache contributed to their decision to sacrifice some of their own limited stores to establish a fallback depot near Beechey Island.

After landing the stores, Saunders could easily have sailed back to Barrow Strait to tell one of the nine searching ships that a depot now existed at the mouth of Navy Board Inlet. The back-tracking would have entailed a voyage of a few days in each direction, allowing him to exit Lancaster Sound around the end of August in plenty of time to get safely through Baffin Bay and Davis Strait before freeze-up. (Whaling ships often stayed into October.) The historian John Brown, who followed the events of the Franklin Search closely at the time, observed, "We have often thought it had been much better had Mr. Saunders endeavoured to communicate with Captain Austin's squadron; he knew, from Captain Penny, they were in Lancaster Sound, and the season was not so far advanced but that he might have spared some little time to accomplish so greatly desirable an object."[36]

On the *North Star*'s return to England, the Admiralty Lords approved of Saunders's decision to land the supplies in Lancaster Sound. In his report to the Admiralty, Saunders neglected to acknowledge that Penny had suggested the change of plans, and he refrained from admitting that he had not bothered to tell any of the other commanders where the depot was located. Saunders's failure to give credit to Penny seems ungrateful and self-serving, in sharp contrast to Penny's magnanimous efforts to clear Saunders of any blame at the Admiralty.

13

Into Lancaster Sound

The great gateway . . .[1]

Lancaster Sound is the only feasible route from the Atlantic sector into the Arctic Archipelago, the labyrinth of islands and waterways lying between the North American continent and the North Pole. In his expedition of 1819–20, Parry had entered the sound and continued westward all the way to Melville Island, some six hundred miles from Baffin Bay, adding to his chart the southern portions of several islands on the northern flank of the convenient east-west corridor. To the south, however, he had only been able to see two small sections of land, so the huge region between his east-west track and the known waterway along the continental coast remained *terra incognita* for the next three decades. It was into this unknown and mostly uninhabited region that the Admiralty had directed Franklin, in the hope of finding a link between the so-called Parry Channel and the coastal route from King William Island to Alaska.

Before the search expeditions of 1850 set off, the Admiralty requested the opinions of what the *Aberdeen Herald* (6 April 1850) called "eminent Arctic officers" – Back, Beaufort, Beechey, Parry, Richardson, James Ross, and Sabine – on the status and location of the missing ships and men. They all concurred in the belief that the men were still alive, and all but one officer thought that the ships were still afloat. As for their location, the seven men pointed to Melville Island and the sea beyond it. They were certain that in the space of five years, Franklin must have succeeded in crossing the Arctic Archipelago to its western limit and possibly even further.

Not one of the officers could imagine that the *Erebus* and *Terror* might have been hopelessly trapped less than halfway across the archipelago, still five hundred miles short of Melville Island. Even James Ross, whose own ships had found it impossible to proceed further west than Barrow Strait in 1848–49, now placed Franklin beyond Banks Island.

Colouring their view, probably, was the apparent ease with which Parry had traversed the Arctic Archipelago in 1819. They may not have realized that he had enjoyed unusually favourable ice conditions. Even so, their unanimous opinion about Franklin's course and location is very curious, because Franklin had never been instructed to continue to Melville Island. He had been ordered to head south after Cape Walker, only halfway along the Parry corridor. If he had succeeded in reaching the coastal route adjacent to the continent, he would never have turned north toward Melville Island.

The expeditions setting off in 1850 to search for Franklin would reveal the true nature of the "unexplored quadrilateral" south of the Parry Channel.

* * *

Lancaster Sound is like the front door of a house containing many corridors and rooms. On entering the vestibule, a visitor may find further progress barred by closed interior doors. Unless these doors open and enable him to pass through, he will be unable to discern the floor plan or determine the dimensions and contents of hallways and rooms. In late August, after the ten ships of the five expeditions had sailed two hundred miles through Lancaster Sound, they found three doors closed. Barrow Strait (continuing west), Peel Sound (leading south), and Wellington Channel (leading north) were all blocked by landfast ice and heavy floes.

A further door was ajar, however. Prince Regent Inlet (leading south) appeared to be ice-free for a considerable distance. This, however, was of interest only to Forsyth, who planned to sail southward in the inlet and cross overland to the vicinity of King William Island. None of the other expeditions intended to proceed on that course, and as the doors leading west and north were closed, they had no choice but to remain in the archipelago's vestibule, hoping that a favourable change in the ice distribution would enable them to advance further before winter locked all the doors for ten months. In this severely restricted ice-free zone, the various crews were to see a lot of each other during the next few weeks as the ships criss-crossed each other's tracks. On every ship, officers and men on deck and aloft scanned the coasts by telescope, constantly on the watch for ships, masts, flags, cairns, tents, or – dared they hope? – men.

After parting from the *North Star*, Penny's ships crossed to the north side of Lancaster Sound to begin searching. The stark beauty of Devon Island's straight, steep, glacier-capped coast, cut at intervals by narrow ravines and fiords, cannot fail to impress any traveller. Vertical cliffs rise more than a thousand feet from the sea, their lower parts mantled by steep talus slopes, their upper portions displaying ancient strata of Paleozoic age stacked in horizontal series so regular and symmetrical that they appear to have been laid down by a stone mason. For Sutherland from the deck of the *Sophia*, the spectacular formations called to mind "endless and perfectly horizontal rows of buttresses," "impregnable fortifications," and "lofty citadels."[2] Later, when HMS *Intrepid* passed by, the horizontal strata would remind Cator of "the two tiers of guns on the 'Rock of Gibraltar.'"[3]

From the practical perspective of a mariner, the height and abruptness of the cliffs made them easy to see and relatively safe to approach. The wind was coming from the west out of Barrow Strait, making it necessary for Penny's brigs to tack back and forth, toward the land, then away from it. "On every alternate tack, the ships were put about so near the land that any arrangement of even a few stones on the beach . . . could be detected," Sutherland observed.[4] After a four-hour detour to the southwest to check ice conditions near Leopold Island, the ships returned to the north coast and resumed their search. Near Cape Hurd and Radstock Bay, the wind rose "and often swept around the bold bluffs with a degree of violence that threatened to carry away our spars, or to upset the ships."[5] Proceeding cautiously under shortened sail, the *Lady Franklin* and *Sophia* approached Cape Riley and Beechey Island. HMS *Assistance* and *Intrepid* could be seen examining that part of the coast, so Penny reasonably assumed that no additional searching effort was required there. That was fine with him; he was eager to get into Wellington Channel.

As Penny's ships turned the corner into Wellington Channel, exciting things began to happen astern. Ommanney decided to erect a cairn and leave a progress report on Cape Riley, at the eastern side of the entrance to what later became known as Erebus and Terror Bay. Leaving the *Assistance* to stand offshore, he went on board the *Intrepid* with several of his officers. The ship steamed in close to the point, and a party landed by boat. Seeing nothing unusual, they erected a beacon, deposited a message, and set off for the top of the tall cliff overlooking the low point of land in order to get a view of ice conditions to the west. The trip turned out to be an ascent worthy of a dedicated alpinist, a severe challenge for sea legs unaccustomed to solid ground. "I never had such a journey in my life," Cator wrote in his journal; "it would have served us all right had we had a good fall for it was really dangerous both getting up & down about 500 feet & nearly perpendicular, some parts at an angle of 70°. Every time one took a step up you were sure to come down two. By the time I got down I was all over bruises."[6]

Safely at beach level again, they were taking a last look around before returning to the ship when someone's gaze was arrested by unnatural elements in the landscape. Closer examination revealed four tent rings, fragments of cloth and canvas, empty Goldner food cans, an iron bar, and bits of rope containing the special marker strand of the Royal Navy. As no Europeans were known to have landed in that locality, it was clear that the site had been occupied temporarily by some of Franklin's men. Here was the first evidence of the missing expedition!

Cator took a pessimistic view: "I'm afraid it tells too true a tale of their sad fate,"[7] he wrote disconsolately. But what fate, exactly? Had the campers been the last survivors of Franklin's expedition? A boat crew escaping from ice-trapped ships? A hunting party sent out from ships wintering safely nearby? There was no way to tell.

Three miles from Cape Riley, across the mouth of Erebus and Terror Bay, lies Beechey Island. On a map or a vertical air photograph, it looks like a frozen drop of water suspended beneath the mainland by a slender icy thread, just about to break free and fall. The thread is a narrow gravel spit almost a mile in length, so low near its fragile connection with the mainland that it is submerged, or almost so, during high tides, providing some justification for calling Beechey an island. At low tide, however, when the sea level is from five to seven feet lower, the spit is exposed, making Beechey into a tombolo or land-tied island – a peninsula with a narrow isthmus. South of its connecting spit, the part-time island broadens to a maximum width of almost two miles and rises in elevation, at first gradually, then steeply, to a level top six hundred feet above sea level.

From Cape Riley, someone sighted a cairn on the summit of Beechey Island. In that man-made monument of rocks must lie a message from Franklin! Resolving to investigate, Ommanney returned to the *Intrepid*, which steamed westward. The best place to land on Beechey Island is on the island's eastern flank facing Erebus and Terror Bay. Here, behind a gently shelving shore, a succession of raised shingle beaches provides the easiest access to the scree slopes that directly abut the flat-topped summit. Owing to impenetrable ice, however, the steamer could not reach this place. Instead, Cator took it across the mouth of the bay through loose floes. McClintock, who had remained on board the *Assistance*, wrote, "We watched the Intrepid with considerable anxiety . . . slowly thread her way thro' the icy labyrinth toward Beechey Island which she succeeded in reaching."[8] Exactly where the officers went ashore on the island is not clear, however; the location is not mentioned in McClintock's journal or in those of Cator or Ommanney (who both landed and climbed to the cairn). Steep talus slopes and near-vertical cliffs extend all round the south and west coasts of the island. At its southwest corner (Cape Riddle), a narrow ravine might have been used to clamber to the top, but as both Ommanney and Cator had remarked on the risk taken in their climb to the summit overlooking Cape Riley, it seems unlikely that they would so soon attempt another dangerous ascent. North of Cape Riddle, along the island's west coast, lies an extensive intertidal zone whose shallows would have discouraged anchoring. It seems probable, therefore, that they landed in Union Bay, on the west side of the spit extending northward from the island.

Buoyed up by the discovery of relics on Cape Riley, and fully confident that they would find in the summit cairn a message from Franklin summarizing the progress of his expedition and outlining his intentions, the men jumped eagerly ashore from the boat. Immediately they spotted more telltale evidence – a fathom of rope bearing the queen's mark, and an empty meat tin. Then they probably clambered up onto the narrow spit and walked south along the spur that leads directly to the summit, the most gradual approach to the high ground on which the cairn was situated. (If this was in fact the route they took, one wonders why they did not notice the three headboards down by the bay.)

When they reached the top of the slope, the six-foot stone beacon was clearly visible. "Our hopes were now raised to the highest pitch,"[9] Ommanney later wrote in his journal. They carefully dismantled the cairn rock by rock, expecting at any moment to uncover a message cylinder or perhaps a sealed bottle or can. "But alas! No record!"[10] The cairn was empty. With pick and shovel, they dug three feet beneath ground level, but again nothing. The purpose of the monument was not revealed. The surface of a broken slab of rock appeared to have been polished and bore an incised straight line, possibly a guide for lettering that had been intended but never carved. What was its unspoken message? On another flat stone lay a few grains of bird-shot, taunting them like clues in a paper chase. If only those tiny spheres of lead could speak!

Beechey Island lies at a crossroads. From its summit, Ommanney and his companions could look east towards Baffin Bay, north into Wellington Channel, west through Barrow Strait, southwest to the mouth of Peel Sound, and southeast into Prince Regent Inlet. Which of these now-frozen paths had Franklin's ships taken?

To Cator, already convinced that some serious disaster had befallen the missing expedition, the lack of a message was "the worst feature we had seen, as it proves I'm afraid that they had no materials for writing a record."[11] Could that be the simple explanation for Franklin's men having gone to all the trouble of erecting a cairn without leaving a note within – no one had thought to bring paper and pencil? Had they attempted to cut some words into a rock but been unable to stay long enough to complete the job? Or had they simply built the cairn to draw the attention of some whaling ship or searching expedition? If so, why? A cairn without a message was like a book with blank pages

Deeply disappointed, the men rebuilt the structure, leaving a record of their visit inside. They were no closer to knowing what had happened to Franklin's expedition. As they started down from the top of the island towards midnight, they sighted a large polar bear on the shore, stealthily stalking the two sailors guarding the boat, unaware of its presence. The animal was acting like a pointer, periodically freezing into immobility as it sniffed the air, savouring the appetizing aroma of this unfamiliar two-legged species. Only moments before the men at the boat became the ursine "special of the day," they saw the bear, jumped into the boat, pushed off frantically, and plied their oars with unaccustomed vigour. The bear dived into the water "and pursued the boat to within 50 yards of the vessel."[12] None of the men in the landing party had carried a gun.

All of the men made it safely back on board and the *Intrepid* steamed back to Cape Riley. At 2 a.m. Ommanney rejoined the *Assistance*. At about the same time, the little American brig *Rescue* arrived at Cape Riley, and her commander, Samuel Griffin, was able to enjoy a bit of one-upmanship, delivering back to Cator the boat that had been swept off the *Intrepid* during the recent gale. He had sighted it "capsized among the ice and picked her up finding oars, sails &c lashed in her."[13] The Americans were now hearing what they called an English accent, and in his journal first officer Carter could not refrain from mimicking their silent "h." Ommanney, he wrote in his journal, had sent his compliments and indicated that he "would be 'appy to see Capt Griffin aboard in the morning."[14] The next day, Griffin visited the *Assistance*. At nearly five times the tonnage of the *Rescue*, the British ship seemed massive, and below she was a "picture of comfort the cabin like a city parlour and warmed by a stove while here we are sleeping with our noses in the open air."[15]

At Cape Riley, Ommanney sent McClintock ashore to expand the search and build a more prominent cairn, conspicuously marked with a pole and a canvas flag, with a new updated message inside. A party from the *Rescue* also landed to inspect the place. They collected bits of leather and cloth and an empty Goldner meat can. Examining the four tent rings, they concluded that the moss-covered stones had lain undisturbed for two or three years. After building their own cairn and depositing a message for their senior ship, *Advance* – not seen since the gale several days earlier – the Americans got under way again and proceeded around Beechey

Island into Wellington Channel, where Griffin was able to meet Penny and pass on the news of Ommanney's exciting discoveries. Once again the "h" disappeared from view in his journal: Penny wanted to know how the Americans had succeeded in getting through Melville Bay without "'aving any experience in ice management."[16] Griffin found Penny to be "a nice fellow" but "not so comfortable as Capt Ommanney."[17]

The *Advance* arrived off Cape Riley on the following day. De Haven spotted the large cairn on the point, took a party ashore, found Ommanney's cairn note, and learned that the first evidence of Franklin had been discovered. Somehow they overlooked the smaller cairn containing the report left for them by Griffin. While ashore on Cape Riley, De Haven suddenly noticed that his ship, which had heaved to about a mile away, seemed to be drifting towards land despite the efforts of her crew to make sail and work further offshore. The *Advance* appeared to be in the grip of a strong current; a crisis was developing. He recorded in his diary:

> I could not embark and get off to her in time to avoid the catastrophe – an effort was made to ware [i.e., wear] her but unfortunately she had got into the influence of a strong tide setting directly on to the cape and her helm became nearly useless. She first struck some grounded masses of ice, passing through them she grounded in 9 feet water at not more than 200 yds from the perpendicular cliff of Cape Riley. Furled sails and took out kedges in hopes of hauling her off but the tide was falling and our efforts were vain. Our hopes now depended upon the rise of the tide[,] good hawsers were attached to the grounded masses of ice outside of us where there was abundance of water to float the vessel. Our anchors were placed and heavy articles about deck were placed in the boats and spare spars launched overboard in order to lighten the vessel as much as possible. As the tide went down she fell over on her bilge. In the evening the Rescue came in sight and making us out in our awkward predicament hove to in the offing and sent a boat in. She was directed to remain near us till further orders should be given . . . At low tide we had but 5 feet water under us, bottom rather hard composed of sand and small stones (the debris of a lime stone bluff 800 feet high which constitutes Cape Riley)[.] The tide rising the ship righted and at 11.30 P.M. (about high water) by bringing a heavy strain upon our hawser she floated off. Warped her to a safe depth and made fast to one of the outer grounded hummocks.[18]

At 2:30 the next morning, they cast off from the grounded ice, picked up their anchors and spars, and sailed around Beechey Island into Wellington Channel.

During the time the *Advance* was aground, the *Prince Albert* had arrived on the scene. Forsyth sent Snow over to offer assistance. As De Haven had the situation well in hand, no help was required, but Snow took a party ashore to look at the site. He read the notes from Ommanney and Griffin and, like previous visitors, collected a number of artifacts. Within a few days, the campsite at Cape Riley had been investigated by parties from no fewer than five ships – *Assistance*, *Intrepid*, *Rescue*, *Advance*, and *Prince Albert* – half the searching fleet. The small hunting camp from Franklin's expedition had become the region's main attraction.

A few years later, in his book *The United States Grinnell Expedition in Search of Sir John Franklin: A Personal Narrative*, Elisha Kent Kane accused Ommanney of having failed to acknowledge that the Americans had been co-discoverers of the first evidence of Franklin on Cape Riley. "Our consort the Rescue, as we afterward learned, had shared in this discovery," he wrote, "though the British commander's inscription in the cairn, as well as his official reports, might lead perhaps to a different conclusion. Griffin, in fact, landed with Captain Ommanney, and the traces were registered while the two officers were in company."[19] In fact, Kane was wrong. Griffin had indeed been there at the same time as Ommanney, but only during Ommanney's *second visit*. He had not been present when Ommanney and his landing party discovered the relics the previous day. The *Rescue* had not "shared in the discovery."

At Cape Riley, Ommanney had told Griffin that he had discovered the relics earlier; Griffin's first officer, Carter, had recorded it in his journal; and it must have been known to everyone on the *Rescue*. Later, when the crews of both vessels were crowded together onto the *Advance* for more than eight months, the finding of relics from Franklin's expedition must have been discussed often. Had Kane not been listening? He persisted in asserting that Griffin and his men had participated in the discovery. Soon after returning home, he wrote to the *New York Tribune* (6 October, reprinted in the *Times* of London two weeks later) that "the discoveries at and near Cape Riley were made at the same time by gentlemen of both nations." This erroneous claim, which denied Ommanney the credit due for being the first to land on Cape Riley and find the evidence of the missing expedition, was not only published in the best-selling book but remained uncorrected through subsequent editions, and has been repeated in subsequent books, including that of Kane's biographer, George W. Corner.

* * *

Forsyth's "branch expedition" was on a mission different from the others. Following the instructions of Lady Franklin, he intended to winter near the southern end of Prince Regent Inlet and send men overland to the vicinity of King William Island, the region to which spiritualists had drawn Lady Franklin's attention and that Snow claimed to have seen in a vision. The essential part of this plan was to locate a secure winter harbour as far south as possible. Passing Cape Clarence, Forsyth decided to examine the harbour in which James Ross's ships had wintered two years earlier. While the *Prince Albert* stood offshore, Snow and a few men manoeuvred their gutta percha boat through scattered floes and brash ice to Whaler Point, the long neck of land that encloses the harbour of Port Leopold.

"As we neared the shore," Snow later wrote, "the whole features of the place came fresh upon me."[20] Although he had never been there before, the landscape was easily recognizable because in London he had visited one of Robert Burford's panoramas on Leicester Square, based on the paintings of Lieutenant William Henry James Browne of HMS *Enterprise* on James Ross's expedition. Commercial panoramas were very much in vogue.[21] Artists' representations of scenes in far-off places, enlarged to impressive size, displayed on moving screens, enhanced by skilful lighting effects, and sometimes accompanied by music, created visual impressions far stronger than the passive word-pictures and small illustrations in published narratives and were more affordable than books. In Burford's panorama, Snow had seen Browne's realistic depiction of Port Leopold and its adjacent heights. Now, there in person, he wrote: "I, almost, fancied that I was again in London, viewing the artistic sketch."[22]

Landing at Whaler Point, the men found two Admiralty cylinders, one attached to a flagpole and the other in a nearby building. "Eagerly did I open them, and take out their contents," Snow related. He found no fewer than four messages inside the copper tubes. "My agitation was so great, that I could hardly see to read, and my hands fairly trembled . . . the excitement was great upon me."[23] Unfortunately, none of the messages was from Franklin. Three had been left by James Ross two years earlier and one deposited only a week earlier by a landing party from HMS *North Star*.

The *Prince Albert* continued southward in Prince Regent Inlet. For fifty miles the ship made good progress, but then impassable ice extending from one side of the waterway to the other brought the little vessel to a halt. To make matters worse, landfast ice barred entry into all potential winter harbours, a remarkable phenomenon for so late in the season. No passage south and no winter harbour: the situation looked grim. After consulting with his ice masters, Forsyth announced that they would turn for home. News of a departure homeward at the end of a long voyage was normally a joyful occasion, but this was different. Although they were almost three months out from Aberdeen, they had been inside the archipelago for a mere ten days. All their optimistic hopes of rescuing Franklin dashed, a "mantle of gloom" descended upon the crew.[24]

Problems with his officers influenced Forsyth's decision to give up the search. In five letters he composed during the course of the expedition to John Barrow Jr., the father confessor for so many officers in arctic service, he complained about his two ice masters. In late August, he wrote, "One is a drunken old fellow that I have never been able to trust to do the most simple thing[.] The other is an ignorant would be mutinous fellow."[25] One of the two was unduly cautious because he had lost a ship under his command years before. Why Forsyth had hired two ice masters is not clear; it proved to be an unworkable arrangement. Problems had arisen as soon as they encountered ice on the Greenland coast. They "never agree with each other about the ice,"[26] he lamented; consequently, "all devolve on me who has never been in the ice before."[27] One can sympathize with Forsyth: never in his life had he conned a ship among ice floes, yet even before reaching Upernavik he had lost confidence in his ice masters and considered it necessary to make all the decisions himself. After encountering Ross's *Felix*, he had wisely decided to continue in company in case of trouble along that dangerous coast.

When the *Prince Albert* came up to the ice barrier in Prince Regent Inlet, Forsyth went through the formality of asking the ice masters their opinions. There was nothing more they could do, they said; it was time to go home. Forsyth concurred – too readily, some would say. Back again in Lancaster Sound, he could see that to the west Barrow Strait was also blocked, but he followed the ice edge northward into the southern part of Wellington Channel. From the masthead, Snow could see Ommanney's *Assistance* beset in heavy floes off Cornwallis Island. Cator's *Intrepid* was probably nearby, obscured by haze. In mid-channel, Penny's *Lady Franklin* and Stewart's *Sophia* were working among scattered floes, and on the eastern side of the Channel Griffin's *Rescue* appeared to be stuck in pack ice. To the north, Forsyth saw an impassable barrier of ice extending across Wellington Channel, and to the south the prospect was equally discouraging. "The only clear water visible," he later wrote, "was that in our immediate vicinity, and in the direction we had come."[28]

As the expedition was not well equipped for wintering, there seemed to be no point in remaining.

Although Forsyth and his men sighted at least four ships during their brief foray into Wellington Channel, they did not communicate with any of them, assist in searching the shore, or offer to carry letters and dispatches back to England. Near Cape Spencer, a boat party went ashore to fill small water casks at a stream, and the men did carry out a brief search for signs of Franklin. After starting for home, the *Prince Albert*, Lady Franklin's great hope, rounded Beechey Island and passed Franklin's summit cairn for the second time without spotting it. If they had not happened to see the grounded American vessel at Cape Riley, they would have reached home totally unaware that any evidence of Franklin had been found and without a small collection of Franklin relics to excite the Admiralty and the public. Even with that information and those artifacts (obtained merely by chance), they headed homeward with a very incomplete picture of the searching operation and its results. Had they not been in such a hurry to leave, they could have learned of the discovery of Franklin's first winter quarters, crucial information that would not reach the Admiralty for almost a year.

Off Bylot Island on the last day of August, the *Prince Albert* encountered HMS *North Star*, also on its way back to Britain. In the exchange of news between Forsyth and Saunders, there was a curious omission. Forsyth said not a word about the discovery of Franklin relics at Cape Riley – certainly the most significant bit of news he was in a position to impart. At that very moment, some of those articles were stored carefully below deck on his vessel, yet he did not show them to Saunders or even mention them. Historian Ian Stone conjectures, "His motives for this secrecy are that, realizing *North Star* would make a swifter passage home than *Prince Albert*, he wanted to retain the news of the only progress in the search made by that time."[29] Forsyth wanted to be the bearer of the first information about Franklin, and his intentional withholding of important information, according to Stone, revealed a certain "deviousness" of character. Saunders may not have been totally frank either, for he does not appear to have told Forsyth about having landed stores at the mouth of Navy Board Inlet.

After seeing all three major waterways beyond Lancaster Sound solidly blocked by ice, Forsyth sailed all the way to the southern tip of Greenland without experiencing any more obstructions. "This was most extraordinary," commented the *Nautical Magazine* later, "and may be said, perhaps, to be hitherto unparalleled in the annals of Arctic Navigation; that a vessel should come the whole way from Wellington

Channel through Baffin Bay and Davis Strait to the Atlantic Ocean without meeting with or seeing any field ice, and even less than usual number of bergs."[30] The openness of Baffin Bay and Davis Strait was in sharp contrast to the closed ice conditions in Barrow Strait and vicinity, where the searching ships were unable to advance.

After a voyage of four months, the *Prince Albert* reached Aberdeen on 1 October. The media and the public waited eagerly to hear what Forsyth could add to the report of Saunders, who had brought HMS *North Star* home a few days before. The artifacts Snow had collected at Cape Riley were quickly delivered to the Admiralty in London and subjected to careful scrutiny by the arctic veterans Parry, Richardson, and Sabine. They in turn consulted Assistant-Surgeon Andrew Clark at Haslar Hospital, who made a microscopic examination of the wood, and Captain Superintendent Peter Richards and Master Ropemaker Burton at Chatham Yard, who traced the provenance of the rope. That four-foot length had a story to tell. As Ommanney and Cator had observed, it contained a strand of yellow worsted identifying it as Royal Navy cordage. But now the master ropemaker revealed that this special marker had not been introduced until 1824, just too late to be supplied to Parry's ships *Fury* and *Hecla* before they left for Lancaster Sound. Furthermore, the rope was made of Hungarian hemp, which had not been used by the navy until 1841. Since that year, only two naval expeditions had sailed to Lancaster Sound, the first commanded by Franklin in search of a Northwest Passage and the second by James Clark Ross in search of Franklin. As Ross had never reached Cape Riley or Beechey Island, the experts concluded that the articles had been left there by a party from Franklin's expedition, probably around 1845 – confirming what the commanders of the searching ships had decided a few months before.

Forsyth sent a summary of his voyage to the *Times* as soon as he arrived. His letter, published on 5 October, boasted, "The Prince Albert is only 89 tons; the last vessel to leave England, the first in Barrow's Strait, and the smallest that has ever been so far in the Polar Seas." He failed to mention that three of the other ships had detoured north from Cape York to search Wolstenholme Sound for evidence of a massacre while he continued west, or that he had been towed more than two hundred miles from Greenland across the North Water to the mouth of Lancaster Sound by HMS *Pioneer*, or that the *Resolute* and *Pioneer* had gone south to search Pond Inlet while he advanced directly into Lancaster Sound. Nor did he acknowledge that Ross's *Felix* and Griffin's *Rescue* were about the same size as *Prince Albert*, or mention that those little vessels had remained in Barrow Strait searching for Franklin while he had hightailed it home. In a voyage that lasted almost four months, he had spent less than two weeks within the arctic islands. One newspaper stated that "the whole of the coasts of Baffin's Bay were also explored without result."[31] In fact, after leaving Pond Inlet, Forsyth is not known to have carried out any exploration at all.

The newspapers knew only as much as Forsyth reported, and they were lavish in their praise. In London, the *Weekly Times* (6 October) declared that "in the whole history of Arctic navigation the voyage of the *Prince Albert* stands unexampled." The *Illustrated London News* (12 October) went even further: "Captain Forsyth's voyage . . . will be handed down to posterity as one of the most remarkable, if not *the* most remarkable that has ever been accomplished in the Arctic Seas." Such extravagant compliments were excessively flattering. Forsyth's only real achievement was to bring back news of the discoveries made by Ommanney at Cape Riley and a few relics left behind by the various landing parties from other expeditions.

Forsyth was quoted as saying, "As we had no harbour to winter in, our returning will save an immense sum to Lady Franklin."[32] A reader might wonder why the other nine ships had not returned for the same reason – to save money – and one might ask whether saving money was really the top priority in a search for more than a hundred men who had been missing for five years. Lady Franklin, sponsor of Forsyth's expedition, found his early return profoundly disappointing. The *Prince Albert*, she confided to Scoresby, "has returned without doing any one thing for which she was equipped and sent – I did not send her to be a messenger & had she been so she shd. have waited till the end of the safe season for farther and fuller information." In her view, the voyage of the *Prince Albert* was a complete failure. "The money spent on her," she told Scoresby, "is utterly wasted."[33]

Within a month of Forsyth's return, she was already contemplating another expedition for the next summer. She was thinking of sending the *Prince Albert* under a different commander toward Boothia again but by a route that no ship had ever traversed – through Hudson Strait, Foxe Basin, and Fury and Hecla Strait. She subsequently abandoned this idea in favour of another attempt by way of Lancaster Sound, which was a fortunate decision. The rapid currents in the constricted channels west of Igloolik Island had defeated Parry's sailing ships in 1822 and 1823, and the *Prince Albert* would never have got through either.

Although Forsyth had contributed little or nothing to the search for Sir John Franklin, his arrival home with the first tangible evidence of the missing expedition gave a boost to fading hopes. The tiresome metaphor of darkness and light was resurrected. If a *mantle of gloom* had fallen upon Forsyth's crew when he announced in late August that he would turn for home, his return to Britain now helped to *illuminate* the situation. As the *Illustrated London News* (12 October) expressed it, "Some faint gleams of hopeful light have at last been thrown upon the gloom of uncertainty which hangs over the fate of Sir John Franklin and his companions."

* * *

Arctic whalemen knew from repeated voyages into Davis Strait and Baffin Bay that the distribution of sea ice could vary greatly from one year to another, yet it was something that the Admiralty and the sponsors of private expeditions often overlooked. When drawing up sailing orders, they tended to be unreasonably optimistic, assuming that each expedition could go further and accomplish more than previous ones. They rarely took into account the variability of ice conditions, and although they always mentioned objectives, they seldom outlined other measures to take if the intended route turned out to be impassable or the final destination unattainable. Although Franklin was given instructions about how to proceed on reaching the Pacific, he was not given instructions for retreat if his ships happened to be trapped or wrecked during the hazardous transit of the island-studded arctic zone. Admittedly, the region's geography was largely unknown, which made it difficult to draw up plans containing precise geographical details, so the Admiralty wisely left Franklin plenty of leeway to make his own decisions on the spot if an emergency arose. It would have been useful, however, if they had laid down some general principles to follow if a retreat became necessary, so that rescuers would know how to proceed. In addition, they should have instructed Franklin to deposit records of his progress and intentions in prominent cairns along the route. They told him to throw bottles or copper cylinders overboard from time to time containing positions and progress reports, but cairns erected on prominent headlands were far more likely to be sighted than small containers drifting at sea. (Only one from Franklin was ever picked up.) Was the Admiralty so confident that Franklin would complete the passage that they considered such a measure unnecessary? Or did they assume he would do this anyway without explicit instructions? His friend Sir John Ross later claimed to have advised him, "Pray put a notice in the cairn where you winter, *if you do proceed*, which of the routes you take."[34] If Franklin really did receive such advice from Ross, he failed to implement it, and he paid the price.

Franklin's disappearance had alerted the Admiralty to the possibility of ships becoming trapped or wrecked in a remote locality and had made them conscious of the importance of communication. In Penny's orders they wrote, "Much of the painful anxiety that now exists respecting the missing ships might possibly have been avoided, if greater care had been taken to leave traces of their progress." Therefore, they told him, "You will consider it rigidly your duty, and a matter of the utmost importance, that every means should be adopted for marking your own track."[35]

His ships would be provided not only with the usual poles and staffs for marking stone cairns but also with red and white lead for making paint. "You will," the Admiralty instructed him, "wherever the colouring of the cliff or shore admits of a mark being made in strong relief, paint a red or white cross, as the case may be, depositing as near to its base as possible, and at right angles with the perpendicular part of such cross, a bottle or other vessel containing a short summary of your proceedings up to the date of the deposit; an account of the state of your supplies, and resources, the health of your party, and your further intended course."[36] The intention was laudable, but if Penny had carried out his instructions to the letter, he would have spent much of his time decorating rock faces with crosses, for the south coast of Devon Island is an endless series of vertical cliffs separated here and there by incised ravines and fiords. His ships would have had to lay off the rugged, exposed coast while men went ashore by boat and clambered up several hundred feet of steep talus slope with paint and brushes to reach vertical faces where a mark could be made – a hazardous operation for both ships and men.

The Admiralty allowed Penny flexibility to cope with unanticipated situations. "We do not deem it advisable to furnish you with minute instructions as to the course you are to pursue," his instructions told him. "We deem it expedient rather, that you . . . should be left to the exercise of your own judgement and discretion."[37] But, like Franklin, he was not given guidelines on how to proceed if it became necessary to abandon his ships.

* * *

Freedom to assess the situation and choose their own courses of action notwithstanding, Penny and the other commanders faced a major disadvantage: their inability to see from above. They could get an idea of the thickness of floes scraping past the hull, especially if some tilted or rafted; they could roughly assess the concentration of ice within a few miles of the ship from the crow's nest; and they could sometimes scan a more extensive area from a coastal hilltop. But they could never discern the overall distribution of ice. Like all seafarers of the time, they lacked the enormous advantage of aerial observations.

Before the search expeditions left England, Charles Richard Weld, assistant secretary of the Royal Society, had suggested that in addition to sending unmanned free balloons aloft to disseminate messages indicating the positions of wintering ships and depots, the searching ships could employ tethered balloons manned by observers. "Supposing an elevation of one mile were attained," Weld wrote, "a very extensive panoramic view would be gained; and, presuming the balloon to be retained captive, I see no reason why this method should not be adopted."[38] Tethered observation balloons had been used to direct artillery fire on battlefields, but they had never been used on ships in the Arctic. Weld may have been inspired by a suggestion made a few months earlier by the manufacturer George Shepherd to provide a captive lookout balloon for Collinson's search expedition to Alaska. The idea had merit, but Shepherd and Weld apparently failed to appreciate the practical problems involved in such an operation.

To lift a basket, an occupant, and a mile of rope, a hydrogen or hot air balloon would have to be large. Shepherd had proposed a diameter of twenty-three feet – almost as broad as the ship that would launch it. It would have to be controlled by a line running to a windlass on the ship. Could a balloon of such a size be inflated on deck and launched without striking masts, yards, or rigging and having its delicate fabric punctured? If the wind shifted or the ship's heading changed while the balloon was aloft, could the crew prevent the line from becoming entangled? Would it be possible to draw the balloon down in a strong wind and enable the observer to climb out safely onto the vessel's deck? (Accidents were possible. Cator saw a small message balloon released on board HMS *Resolute* that "blew against the rigging and went into a thousand pieces."[39])

Yet the idea persisted. A "celebrated aeronaut," Mr Hampton, who had earlier submitted ideas on balloons to the Admiralty and to Lady Franklin, announced a public demonstration in Dublin of "a new and improved fire [i.e., hot air] balloon, called the Arctic Surveying balloon, intended for the purpose of surveying the Arctic Regions, to assist in the noble efforts now being made to discover Sir John Franklin, and his brave companions."[40] Two shows, enlivened by regimental bands, would take place in the Rotunda Gardens, one in the afternoon and another in the evening – admission from one to two shillings (children half-price). Hampton would first launch messenger balloons, "shewing the manner of distributing dispatches, &c.," and this would be followed by "the Grand Experimental trial of the Surveying Balloon."[41] His inflated balloon was no less than forty-five feet in diameter and seventy feet tall, including a basket suspended beneath that could accommodate two observers. Hampton's demonstrations were scheduled for 11 July. On that day, the ships of Penny and Austin, along with several whalers, were struggling through the ice of Melville Bay. A tethered observation balloon – if one had been practicable – would have been enormously useful in conning the ships through the floes.

Showmen in Europe had popularized their ascents in tethered balloons by taking passengers aloft – not only men and women but also pigeons, cats, dogs, ponies, and even bears. A proposal in England to ascend with a tiger had been prohibited by local authorities. In the autumn of 1848, Lieutenant Gale, a noted English balloonist, had launched a balloon in the Cremorne Gardens in Chelsea, carrying a lady popularly known as "The Lion Queen" riding on the back of her pet lion – "a veritable monarch of the forest."[42] Unfortunately only a partial ascent was possible, as the weight proved to be too great for the balloon's lifting power. Gale too had suggested using balloons in the search for Franklin, but on 8 September 1850, a few days before Penny, Austin, and Ross consigned their ships to winter quarters in Barrow Strait, he was killed while landing in France after ascending on a horse slung beneath his balloon.

The search expeditions were equipped neither with captive balloons nor manned free balloons, despite the spectacular popularity of "aerostation" in Europe during the previous two-thirds of a century. Since the pioneering balloon experiments of the Montgolfier brothers and others in France, hundreds of aeronauts had made ascents, the boldest soaring to greater and greater heights and making flights over longer and longer distances. Some had crossed the English Channel; a few had risen over the Alps; a transatlantic voyage had even been proposed. But a balloon voyage across an uncharted, unpopulated, and climatically severe region whose wind systems were totally unknown was a more technically fraught proposition, quite unsuited to the search for Franklin.

Suppose that a few men drifting downwind from their ship in a balloon high above the vast expanses of ice and tundra managed, through some miraculous coincidence, to sight Franklin's ships. What aid could they provide? Even if they managed to land, tether the balloon securely, and reach the ships on foot, they could neither deliver any significant quantity of food nor evacuate anyone to safety. Unless they carried equipment and materials to generate hydrogen gas or to heat air to replace what they had released during their descent, they could not take off again; even if they could, they would be unable to return to their ship unless the wind had reversed its direction precisely. They would sail away to some unknown destination, entirely at the mercy of the wind, without anyone in their search expedition having any idea of what had become of them or being any wiser about the fate of Franklin and his men.

* * *

The most important limiting factor in the movement of the searching ships was sea ice. Without an aerial view, the commanders of searching ships were unable to determine its precise distribution and nature over a wide area. However, by recording sequential observations from their crows' nests as their ships sailed back and forth, and inserting the information on their working charts, they could obtain a rough overall picture.

A published account of the 1850 voyage of the *Prince Albert*[43] was accompanied by a chart showing the distribution of ice in late August in Lancaster Sound and adjacent parts of Barrow Strait, Prince Regent Inlet, Peel Sound, and Wellington Channel. Compiled by Snow from observations from the *Prince Albert*'s crow's nest, it reveals the extent to which the eastern search expeditions were confined to the "vestibule" of the archipelago that summer. The ice edge extended from Prince Leopold Island (off the northern tip of Somerset Island) across Barrow Strait, around Cape Hotham, and across the lower part of Wellington Channel to the coast of Devon Island near Cape Innis. Beyond the open water zone, the pack ice was continuous and impenetrable; every one of the major waterways intersecting near Beechey Island was blocked by landfast ice or consolidated floes. What a striking illustration of the vagaries of arctic sea ice distribution! Three decades before, Parry had sailed west to the very margin of the Beaufort Sea with little obstruction and returned safely in the following summer. A decade after that, John Ross had sailed south through the entire length of Prince Regent Inlet to the Gulf of Boothia. But now the entire searching fleet was confined to Barrow Strait!

The commanders of searching ships had to cope with several types of sea ice. A border of *landfast ice* extended along much of the coastline. Where the ice was flat, like the fast ice along the Greenland coast, it could serve as a quay, a surface on which crews could track their ships or a stable platform into which they could cut docks. But fast ice also prevented ships from reaching the land. This had been of no consequence on the Greenland coast because the captains, making all haste to get through Melville Bay to the North Water, had no reason to get to the land. But now, in the closing days of the navigation season in Wellington Channel when it was essential to find secure harbours for the winter, landfast ice was preventing them from entering bays and inlets that might have provided shelter.

The inlets leading away from Barrow Strait contained *consolidated pack ice*. The closely packed floes often gave an impression of uniformity and stability, but winds, currents, and tides, interacting in complex and unpredictable ways, were constantly altering their distribution. During calms and slack tides, the pack could relax, creating open leads to tempt ambitious commanders, and then, with an alteration of winds or tides, it could close ruthlessly around any ship that had dared to enter.

Because pack ice is an amalgam of ice of different sorts, some of which may have been disrupted by storms, tides, currents, constrictions of the channel, or reefs and shallow banks, it may present an uneven surface. In early September, De Haven's surgeon, Elisha Kane, found out what its topography was like when he and a few others set off from the icebound *Advance* to walk a few hundred yards over the ice to the land near Barlow Inlet on the coast of Cornwallis Island:

> The ice was eight and ten feet thick, studded with broken bergs and hummocks. These fragments were seldom larger than our Rensselaer dining room [his family's summer home in Philadelphia], some twenty feet square, and, owing either to the rise and fall of the tide or the piling action of storms, deep crevices were formed around their edges, partially masked by the snow which had found it is way into them, and by an icy crust over the surface. Alternately jumping these crevices and clambering up the hummocks between them made it a dangerous walk. We had some narrow escapes."[44]

On the *Rescue*, Carter found walking over hummocky ice "tenfold worse than new ploughed ground and quite as bad as loose lava."[45] The zone of disturbed pack ice was virtually impossible for ships to penetrate and extremely hazardous for men to cross on foot. Drifting, jostling, tilting, or rafting ice floes near a ship could evoke awe bordering on terror. In early September, the *Advance* had three ice anchors hooked into the land ice when a flood tide brought heavy floes charging past. The situation was serious, but Kane later described the scene in his usual lighthearted manner: "It was very grand to see up-piled blocks twenty feet and more above our heads, and to wonder whether this fellow would strike our main-yard or clear our stern. Some of the moving hummocks were thirty feet high. They grazed us; but a little projection of the main field to windward shied them off."[46]

On most nights, temperatures dropped below freezing, causing a glistening film of *young ice* to form on calm water. It extended out from the land, spread across leads and pools within the pack, attached itself to the wooden hulls of moored ships, and grew steadily thicker until broken up by waves or melted by daytime warmth. When young ice attained a thickness of several inches and remained snow-free, it was perfect for sliding, skating, or playing skittles; but Penny and the other ship captains saw it for what it really was – an ominous stage in the general congelation of the seas and a serious threat to their safety.

One day in early September, Kane surveyed the frozen seascape surrounding the *Advance*. In one direction there was young ice three inches thick. In another was the main body of Wellington Strait where the ice was "impacted with floes, hummocks, and broken bergs." In a third direction was "a wild aggregation of enormous hummocks . . . totally unlike anything we saw in Baffin's Bay."[47] The outlook was not encouraging.

Although Kane and a few others who wrote journals in 1850 mentioned "bergs" in Barrow Strait and Wellington Channel, it is unlikely that they were referring to authentic icebergs. In the central Canadian Arctic, there is not enough precipitation over the land to form ice caps and outlet glaciers. Most of the icebergs occurring off the coast of Baffin Island are calved from valley glaciers that reach the sea along the Greenland coast. After drifting north, then west across the North Water, they are carried south toward Newfoundland. Although some may be swept into the mouth of Lancaster Sound, and others may be provided by glaciers on the south coast of Ellesmere or Devon Island, they do not travel far before meeting outgoing currents that carry them back into Baffin Bay and southward toward the Atlantic. The navigators of 1850 were probably seeing what today are called *floebergs* – remnants of eroded *pressure ridges* – smaller than true icebergs and made of salt water ice rather than fresh. Pressure ridges could be pushed up high above sea level and deep down below it. When they broke up, the resultant floebergs drifted with winds and currents, sometimes grounding in shallows.

It is possible that Wellington Channel also contained fragments of *ice islands*, thick masses calved from broad ice shelves on the north coasts of Ellesmere Island and Greenland. During their leisurely drifts around the Arctic Ocean, they can break up, and the pieces are occasionally swept into channels within the Arctic Archipelago. Although only revealed as a distinct type of ice by air photography in the 1940s, they must have existed during the search for Franklin.

Experienced arctic mariners could easily recognize ice of different ages: *young ice* that was actively forming in autumn; *one-year ice* that had formed during the previous winter; and *multi-year* or *polar ice* that had endured through several seasons in the Arctic Ocean and then drifted through channels among the islands. The latter kind was extra thick and tough and was avoided whenever possible. When HMS *Intrepid* experienced a nip off Barlow Inlet, Cator described "the awful pressure of this Polar Ice, which is quite different from the 'Baffin Bay Ice' being from 12 to 14 feet thick and Hummocky, instead of smooth." Anyone who volunteered for a Northwest Passage expedition after seeing such ice, he suggested, "deserves to be punished for his temerity."[48] Even the old whaling veterans on board were said to be astonished by the polar ice in Wellington Channel. On the *Advance* Kane observed, "The ice is tremendous, far ahead of anything we have met with. The thickness of the upraised tables is sometimes fourteen feet; and the hummocks are so ground and distorted by the rude attrition of the floes, that they rise up in cones like crushed sugar, some of them forty feet high."[49] The presence of this old ice suggested that Wellington Channel connected with oceanic regions further north, leading some to conjecture that Franklin had followed the routeway into the fabled Open Polar Sea.

14

Wellington Channel

A perfect regatta after "Old Franklin."[1]

Immediately north of the cliffed and shoal-fringed west coast of Beechey Island is a broad bay approximately a mile across its mouth, and a mile and half from front to back. Nestling cosily in a bend of the coast, the long spit from Devon Island marking its eastern limit and a stream delta forming its northern margin, it provided a reasonably secure harbour on a coast that for many miles northward had few safe anchorages. A fringe of landfast ice offered a natural quay where ships could moor and docks could be cut if necessary, and the nearby stream provided a watering place. At first they called it "Beechey Bay," but when HMS *Resolute* and *Pioneer* arrived later from Pond Inlet, Austin would rename it "Union Bay" to mark the fact that seven vessels from four search expeditions had congregated there.

From this bay, the expeditions began to examine the coast to the north, past Cape Spencer and Point Innes toward Bowden Point. From crow's nests, shrouds, and decks, men scanned the land for cairns or other man-made features, and small parties landed whenever feasible to examine the ground more closely. From time to time, ships probed westward toward Cornwallis Island along the southern margin of the pack, looking for openings leading northward, all the while keeping watchful eyes on the ice situation in Barrow Strait, hoping that a break-up of its ice cover would enable them to follow Franklin's intended route to Cape Walker, south to the continental coast, and west to Bering Strait and the Pacific.

The pack ice filling most of Wellington Channel strictly limited ship operations. On its eastern side, the total length of coastline north of Cape Spencer along which ships could sail close enough to survey the land by telescope or put men ashore to search was not more than fifteen miles. On the western side north of Cape Hotham, it was even less.

This short length of accessible coast – say twenty-five miles in all – and the intervening distance across Wellington Channel – approximately twenty miles – could have been covered easily by one ship in a day, given a reasonable wind, favourable ice conditions, and good visibility, but seven ships were employed at the task. Wherever one looked, there were wooden hulls, white sails, fluttering flags, and a plume of dirty black smoke from HMS *Intrepid.* Comparing it to a regatta, as Cator did, was misleading, because a regatta requires central authority and precise organization. Neither existed here: the activities of the various searching vessels were uncoordinated. As commodore of the Admiralty's principal expedition, Horatio Austin was the logical person to direct the search activities of all the other expeditions, but his two ships had not yet arrived from Pond Inlet, and Ommanney, his second-in-command, had not taken the responsibility upon himself.

Ommanney, Penny, Ross, and De Haven directed the movements of the ships under their command, but in deciding how to employ their vessels, the four commanders acted more or less independently, following their own inclinations. Only occasionally, when their ships happened to be moored close together, did the commanders of the three small-ship expeditions consult together and try to synchronize their activities. Forsyth's location was not known, but it was assumed (correctly) that he had taken the *Prince Albert* south into Prince Regent Inlet as planned.

After finding the relics at Cape Riley and examining the summit cairn on Beechey Island, Ommanney sent the *Intrepid* to examine the coast northward towards Bowden Point, the farthest land indicated on Parry's map of three decades before. After about eight hours of steaming, Cator returned, exclaiming that he had "done more work and gone farther than any vessel amongst the different expeditions."[2] It was good work but hardly the occasion for a boast, as none of the other ships had the advantage of engines. How far north Cator went is not clear from the confusing entry in his journal, but it was far enough to see that on both sides of Wellington Channel the coasts extended much further north than Parry had been able to see from Barrow Strait in 1819. Just how far could not be determined, however, because impenetrable pack ice extended from one coast to the other.

At about the same time as Cator's steamer *Intrepid* was probing northward, De Haven ordered his second-in-command Samuel Griffin to take the USS *Rescue* as far north as possible and attempt to reach Bowden Point. After the ship reached Innes Point in a shore lead, Griffin and five men, carrying five days' food, landed and walked northward along the shore, hauling small sledges. They travelled at night and slept in the open during the daytime on the ground cover of lichens and mosses, which Midshipman Lovell called "Irish feathers."[3]

From Bowden Point, or somewhere in that vicinity, Griffin's party looked into a large indentation not indicated on any charts (now named Griffin Inlet). By proceeding along the shore on foot, they had reached a point considerably further north than Cator's steam-powered ship had done. Several years later, Kane would write that Griffin's party had "found at intervals, almost to Cape Bowden, traces of a passing party,"[4] but this statement does not accord with those by De Haven on the *Advance* or Carter on the *Rescue*, both of whom said that no traces had been found.

After Cator's northerly probe with the *Intrepid* failed to reveal a way through the ice barrier, Ommanney ordered him to take the steamer across Wellington Channel and examine the coast of Cornwallis Island south of Barlow Inlet while he followed under sail. When the *Assistance* started west to join her consort, McClintock, her first lieutenant, began making entries in his private journal. "We lead the van," he proudly wrote. "We are pushing across Wellington Straits for Cape Hotham where it is supposed some record of Sir John Franklin may be found."[5] In retrospect, it seems very strange that Ommanney ordered his two ships away from Beechey Island so hurriedly, without making a thorough examination of the vicinity. McClintock's pride in "leading the van" and beating the other expeditions across Wellington Channel to Cornwallis Island was unwarranted, as he would soon learn.

When HMS *Intrepid* reached Cornwallis Island, it moored to a large grounded ice floe off Barlow Inlet. Cator and a small party went ashore and trekked southward along the beach to Cape Hotham, about seven miles away. The soft clay-like mud and sharp stones made the hike as exhausting as his ascent behind Cape Riley had been dangerous. Nine hours later, when they arrived back at the ship, Cator declared, "I think I never walked over such heavy ground."[6] Their strenuous efforts had not uncovered any evidence of Franklin. At Cape Hotham, they had climbed to the top, built a cairn, and left a message (which other searchers would have to climb the mountain to read). They had traversed a landscape devoid of life. Only the crumbling stone walls of a few abandoned native houses revealed an ephemeral human presence in some ancient era – perhaps a seasonal residence of some nomadic hunting band. In sharp contrast to the apparently lifeless land was the richness of marine life in the adjacent waters. Offshore they had seen five polar bears and numerous seals and white whales.

After Cator's party returned, the *Intrepid* cast off from the floe and soon spoke the *Assistance*, which had just arrived from Beechey Island.

A weary Cator retired to bed, instructing the second master, Shellabear, to unship the rudder and propeller if the ice closed in. But Shellabear, whose utter incompetence was a frequent complaint in Cator's journal, failed to carry out orders. The ice came forcibly in, Shellabear dawdled, a severe nip broke the rudder, and ice lifted the ship three feet. The crew raced for their spare clothes in case she went down. Fortunately the pressure eased and Cator went back to bed, but both ships were now firmly beset and remained so for the next nine days. Aside from a hazardous excursion by Cator and a few others over the pack ice to Barlow Inlet to map the harbour (three hours each way, jumping over open leads, one man falling in), the crews stayed on their ships. Ommanney's impetuous rush to Cornwallis Island had resulted in the temporary removal of two ships from the searching force on the eastern shore, including one with steam power. However, Cator had shown that no evidence of Franklin existed along the half-dozen miles of coast between Barlow Inlet and Cape Hotham, eliminating this area from further searching.

Among the various commanders, Penny stands out as the only one conscious of the need for close cooperation among the expeditions in order to achieve their objective. Before HMS *Assistance* and *Intrepid* started across Wellington Channel, he had hoisted signals to them, intending to exchange information about recent developments and discuss future plans. His signal flags had not been noticed (for the second time), so when he later saw that both navy ships were stuck in the ice off Cornwallis Island, he pursued them under sail in the *Lady Franklin*, got within two miles, and walked over the intervening pack ice to the *Assistance*. There he described to Ommanney his earlier encounter with Saunders and HMS *North Star*, and Ommanney told how he and his men had discovered the relics at Cape Riley. After examining the articles they had collected, Penny wondered if the camp could have been made by a small party retreating toward Baffin Bay. As he would relate more than a year later at a meeting of the Royal Geographical Society, "On the 25th of August, 1850, having joined Captain Ommanney on the west side of Wellington Channel, and seen the traces found by him of the missing ships, I considered it my proper course to return to the eastern shore of the channel, with the view of examining those parts more closely than had been done before."[7]

Penny sailed back to Union Bay and on the following morning went ashore near Cape Spencer with Stewart, Goodsir, Petersen, and some seamen to search the ground on foot. Half a dozen miles north of the cape, they came upon the low walls of a circular stone shelter, a hearth, empty soup cans, and many other articles that clearly indicated a European presence. They found

> pieces of oak, such as staves of small casks; the end portions of a small cask with the words "mixed pickles" scratched on them; also larger pieces of oak, such as might have been procured by splitting up the knees or the doubling of a ship, and they were charred at the ends as if they had been in the fire; the bones of birds also a little burned; but there were no beef bones; part of the leaves of a book ... with some markings on them, and part of a newspaper bearing date September 1844; portions of rope, very much chafed, but easily distinguished as belonging to the Royal Navy by the middle yarn; also torn mittens, cotton rags, and blank paper, all of which the wind had driven beneath the stones ... There were a few handfuls of coals in the fireplace, together with birds' wings, tails, heads, feathers, and bones, some of which appeared evidently to have been in the fire. The track of a sledge was discovered, and the marks of the runners, which were very distinct, were found to be about two feet apart.[8]

Among the silent souvenirs of Franklin's expedition was a piece of paper bearing the name "Mr. M'Donald." Penny must have experienced a surge of emotion on seeing this mute reminder of the young Scottish medical student who had sailed with him on whaling voyages to Davis Strait, had written a book about Eenoolooapik's winter in Scotland, and had told Penny a few years later with immense pride that he had been accepted for Franklin expedition in search of a Northwest Passage. In 1845 he had sailed as assistant surgeon on HMS *Terror* under Francis Crozier, fully confident that they would reach the Pacific. The forlorn campsite and its pitiful relics must have been evocative for Robert Goodsir too, for his brother Harry had sailed as assistant surgeon on HMS *Erebus*. This was his second attempt to find him, having accompanied Penny into Lancaster Sound on the whaler *Advice* the previous year.

When Penny met De Haven's ships, he related his discoveries, showed the articles he and his men had retrieved from the campsite, and praised the Americans for their efforts. On board the *Rescue*, Carter wrote in his journal, "Penny seems to be a capital old fellow, sayst [*sic*] that we have done better than all put together to come through the pack without experience and without even a crows nest. Goodsir was with him and seems a fine fellow, little like an author." He added, "I hope we may winter near them if at all up here."[9] Alas, the Americans were not destined to pass the winter in some cosy harbour alongside Penny's ships. An entirely different sort of winter awaited them.

The searchers had now located two campsites, one at Cape Riley facing Lancaster Sound and Barrow Strait and the other north of Cape Spencer facing Wellington Channel. Between the two camps they had found a few relics on the west coast of Beechey Island and a conspicuous cairn on its summit. In the absence of any written documents, they could only speculate about the significance of these sites. Did they represent one or more coastal voyages by boat or on foot southward through Wellington Channel, past Beechey Island, and eastward toward Baffin Bay? Such an interpretation would suggest one or more parties *retreating* from ships trapped or wrecked further north. Or did the evidence represent a wintering site in the vicinity of Beechey Island with two camps established several miles in each direction to monitor ice conditions? The latter interpretation was more optimistic, suggesting they had been *advancing* into the arctic islands on their mission as instructed and might yet be found further inside the archipelago. But if they had wintered at Beechey Island, why was there no evidence of it? Could 129 men have spent a winter there without leaving an account of their experience?

Ommanney's two ships were at Cornwallis Island, and Austin's had not yet arrived from Pond Inlet. Forsyth's *Prince Albert* was assumed to be heading south in Prince Regent Inlet on Lady Franklin's "branch expedition." The five small ships at Union Bay were on their own without a clear plan of action. Penny stepped into the breach and discussed future operations with Ross, Phillips, De Haven, and his surgeon, Kane, who later wrote, "I was very much struck with the gallant disinterestedness of spirit which was shown by all the officers in this discussion. Penny, an energetic, practical fellow, sketched out at once a plan of action for each vessel of the party."[10] Penny proposed that the *Felix* should sail south to Prince Regent Inlet, intercept the *Prince Albert*, and inform Forsyth of the evidence already found; the *Advance* and *Rescue* should attempt to penetrate further north in Wellington Channel; and the *Lady Franklin* and *Sophia* should try to advance westward in Barrow Strait.

One of the aims of Penny's suggested plan was to get in touch with the *Prince Albert*, but Forsyth's little ship had been sailing in Wellington Channel only two days before and had already started for home! Forsyth's failure to communicate with any of the ships in the area was now having serious consequences: the other commanders were thinking of withdrawing one of their vessels from the searching operations to look for him. The choice for this errand was to be Ross in the *Felix*, because his small expedition was regarded as the least capable of making a significant contribution to the main searching operations. But Ross was not willing to be demoted to mere messenger while other warriors sought honour on the field of battle: he insisted on staying in Union Bay. The other commanders still believed his expedition to be under-equipped and somewhat marginal to the search effort. Cator noted later that Austin and Osborn were "doing their best to get Old Sir John Ross to go home" with news of the discovery of traces of Franklin.[11] Their efforts failed too. Ross had no intention of being sidetracked from the search.

Before departing on their various missions, the crews of the small vessels gave Beechey Island and its environs a thorough examination. Penny sent half a dozen men to clamber up the steep side of the long spit and walk south onto the low part of the island. The party, commanded by Stewart, included Sutherland, Goodsir, Stuart, Petersen, Leiper (second mate), and one or two sailors. Penny remained at Union Bay discussing plans with the officers of the other ships. After the group dispersed and Penny was chatting with Kane, someone from Stewart's search party suddenly appeared, shouting that they had discovered graves!

This was electrifying news. Officers from the various ships rushed across the ice to the shore, climbed the spit, and walked south to the main part of the island. On its gently sloping eastern flank adjacent to the large bay that now bears the names of *Erebus* and *Terror*, three painted wooden grave markers could easily be seen at the head of low mounds of stones. In the frost-shattered rock below, three of Franklin's men had been laid to rest, their quest for a Northwest Passage over. The neat inscriptions revealed that two men had died in January 1846 and one in April. John Torrington, a stoker on HMS *Terror*, and John Hartnell, an able seaman on the *Erebus*, had been first to go, only two days apart, and William Braine, a marine on the *Erebus*, had died three months later. The youngest was twenty, the oldest thirty-two. No causes of death were indicated.

Here was conclusive evidence that the missing expedition had passed its first winter at Beechey Island. As Torrington had died *on board* the *Terror* in January, the ship must have been frozen in nearby, and evidently the *Erebus* had been there too. Apparently both ships had passed the winter safely and resumed their voyage when the ice broke up in summer. While it was encouraging to know that Franklin, his officers, and the remainder of the men had been alive when the ships continued on their venture, it was frustrating not to know where they had gone.

Beechey Island now received the close scrutiny it should have had when the officers from HMS *Assistance* and *Intrepid* landed four days earlier to examine the cairn on the island's summit. A more complete picture of Franklin's winter quarters began to emerge. The men found the low walls and wooden beams of a storehouse, wood shavings from a carpenter's work, cinder and iron scraps from a blacksmith's activities, sawed-off beef casks that had served as wash tubs, empty coal bags, scraps of canvas (one with the name *Terror* on it), bits of rope, and signposts to guide sailors toward the ships in fog or blowing snow. Wherever one went on the island, there were signs of the missing expedition. Strong arctic winds, unimpeded by grass, shrubs, or trees, had picked up detritus generated by a winter's residence nearby and scattered it far and wide over the barren mantle of limestone fragments.

The ships had evidently wintered in the large bottle-shaped bay east of the island and its slender isthmus – an anchorage so secure, Osborn thought, that the chances of a ship being driven out of it were "about as probable as any stout gentleman being blown out of a house through the keyhole."[12] Evidently, the crews had continued to sleep and eat on board ship but had used facilities erected on the island for artificers' work and the storage of surplus stores and equipment.

The relics retrieved by the searchers were more than clues; they were emotional links to the men of the missing expedition. Some were disturbingly personal – a blanket-coat, a mitten, a sock, a scrap of paper bearing handwritten calculations for latitude, a key, some bullet-ridden tin cans at a place reserved for target practice, a pathetic little terraced garden of tundra plants, a pair of delicate cashmere gloves set out to dry beneath small stones by some unknown officer and still holding fast against arctic tempests four years later, and of course the sombre grave mounds with their sand-blasted wooden headboards. To Kane, an American aiding in a British enterprise, the diverse articles were merely "the numberless reliquiae of a winter resting-place,"[13] but to Osborn, an officer serving in the Royal Navy, they were "melancholy mementoes of my missing friends."[14]

Sutherland described a puzzling feature, an architectural assemblage of opened cans: "The meat-tins were piled up in heaps in the same regular manner as shot is piled up; each had been filled with loose shingle, and when the tiers of a single layer were completed the interstices were also filled up with shingle. In this way several mounds were raised to a height of nearly two feet, and they varied in breadth from three to four yards. Six or seven hundred tins were counted, and many more besides these were dug up and emptied out in search of documents."[15] This curious feature has never been satisfactorily explained. Kane suggested that the rock-filled tins might have been intended for boat ballast, but ships' boats did not normally require ballast, and if they had done so the cans would have been filled and stacked nearer the shore. Furthermore, the season during which boats could have been used in the icebound bay was short, yet several months must have been required to accumulate so many empty cans, transport them ashore, fill them with rock fragments, and arrange them in rows and layers. The work must have extended over the winter.

Was it simply a garbage dump for empty cans? If so, had the men been so bored that they amused themselves by building a monument, lining up the cans like sailors on parade? Had offenders been assigned the task as punishment? Or was there a more sinister explanation? Could the discovery of some rotten meat have led Franklin to order an inspection of all the cans on board the ships? If so, why had they been filled with gravel afterward? One thing was certain: if the expedition had in fact been supplied with tainted food, the consequences were dreadful to contemplate. Three deaths might have been only the beginning.

Nowhere on the island did the searchers find any messages from Franklin's expedition. Sutherland later wrote, "To us it appeared truly remarkable that winter quarters had been left without leaving records to shew that British ships had been there, if not also to give an idea of the route they intended to pursue."[16]

As the search for evidence on the ground continued, Penny dispatched a boat party under Stewart to examine the coast east of Cape Riley. They left Union Bay at three in the afternoon, rounded Beechey Island, and stopped at Cape Riley, where they read the messages left in cairns a few days earlier by the British and American parties. Continuing eastward along the coast and landing from time to time, they came to the mouth of Gascoyne Inlet where they happened to encounter HMS *Resolute* and *Pioneer*, only now arriving from Pond Inlet. Their crews were totally unaware that any evidence of Franklin had been found. Stewart brought Austin up to date on all the recent discoveries and handed over a copy of the note Ommanney had left at Cape Riley. After warning Austin to give a wide berth to the headlands to avoid reefs, and telling him that De Haven's *Advance* had already run aground off Cape Riley, Stewart rejoined his boat crew, crossed to Cape Ricketts, and explored some distance into Gascoyne Inlet. Finding nothing of significance, he and his companions turned homeward at 1 a.m. and reached the ships at Union Bay four hours later.

When Austin's ships arrived at Union Bay to join the other ships, it was not in the grand style that the commodore of the naval squadron had anticipated. Around the shoulder of Beechey Island they came, the *Pioneer* steaming boldly ahead with the *Resolute* astern at the end of a towing hawser. Suddenly the *Pioneer* hit bottom and stuck fast. Moments before the *Resolute* smashed into her or took the ground herself, her crew managed to slip the tow rope and let go an anchor to arrest her forward momentum. Poor Osborn! On the coast of Greenland he had managed to tow his senior officer's ship over a reef, and now his steamer was perched on a rocky shelf with the tide falling, as Austin, having almost been dragged onto the same reef, looked on again with displeasure. But perhaps Austin himself was more to blame than Osborn; apparently he had not considered it necessary to pass on Alexander Stewart's useful warning to give headlands a wide berth because of shoals. Officers of the Royal Navy did not need to take advice from lowly whaling masters, he may have thought.

The *Resolute* managed to work into Union Bay during the night and moored to the floe edge, while Osborn waited patiently for the flood tide to get the *Pioneer* off the reef. The next morning he had a pleasant surprise:

> I was awoke by a hearty shake, and Captain Penny's warm "Good-morning;" he had come out to me towing the "Mary," a launch belonging to Sir John Ross, in order that I might lighten the "Pioneer," and offered me the "Sophia" brig, to receive a portion of my stores, if I would only say it was necessary. "A friend in need is a friend indeed," and such Penny proved himself; for my position was far from a pleasant one, – on a hard spit of limestone, in which no anchor could find holding ground, and, at low water, five feet less than the draughtt of the "Pioneer," exposed to all the set of the ice of the Wellington Channel and Barrow's Strait, with about another week of the "open season" left.[17]

It is remarkable that in this emergency the crew of Austin's flagship *Resolute* did nothing to help their consort. It was the whaling master Penny who quickly assessed the situation, borrowed Ross's yacht *Mary*, towed her to the stranded steamer behind the *Sophia*, and put her and his own ship and boats at Osborn's disposal. With Penny's help, the *Intrepid* was lightened and floated off at high tide. She steamed into Union Bay and moored at the ice edge near the other ships.

Ross's account of the episode is slightly different. In his "pocket diary" (whose small size and microscopic scrawl make many of the entries impossible to read), he made the following notations. On August 28: "Pioneer run aground under the west side of Beechy [*sic*] Island, sent the Mary yacht and a party to her assistance; " on the 29th: "The Pioneer being lightened by the 'Mary' and boats floated at high water . . . and made fast a little north of the Felix. She recd no damage."[18] Ross's diary thus confirms that his yacht *Mary* was involved in lightening the *Pioneer* and indicates that some men from the *Felix* assisted in the operation, but it says nothing about Penny, and in fact suggests that Ross sent the *Mary* to help on his own initiative and that no one else was involved. That is contrary to the impression given by Osborn's account.

After Beechey Island had been suitably scoured, Penny extended the search in another direction. A small party of men under Stewart, including Osborn and another officer from HMS *Pioneer*, walked from Cape Spencer across the isthmus to the north shore of Erebus and Terror Bay. Here steep cliffs forced them onto the land ice for about a mile and a half, after which they came to an alluvial plain approximately two miles wide extending towards the east and offering a convenient low-level route toward Gascoyne Inlet, which Stewart had reached earlier by boat. Near the eastern shore of the bay, Osborn reported "a perfect congery of sledge-marks showed the spot used for the landing-place, or rendezvous, of Franklin's sledges."[19] The place seemed to be a central hub from which sledge routes had diverged in various directions from Franklin's ships frozen into harbour ice. As Cape Riley to the south had already been closely examined, Stewart and his companions continued eastward on a trail that pointed to a remarkable butte-like sedimentary feature overlooking Gascoyne Inlet, the same "insular-looking piece of land" that Parry had named Caswall's Tower three decades earlier.[20] Near a brackish lake (now called Red Loon Lake), they came upon an old campsite, the litter – a hearth, a couple of bottles, some empty meat cans, several pieces of paper, and scattered feathers and fur – indicating that a hunting party from the *Erebus* or *Terror* had passed that way. "Eagerly did we run from one object to another," wrote Osborn, "in the hope of finding some stray note or record, to say whether all had been well with them, and whither they had gone. No, not a line was to be found."[21] They continued eastward, following the ruts of sledge runners incised as much as four inches into the soggy gravel and shattered limestone, evidence of heavy loads dragged by Franklin's men. Within a few miles of Caswall Tower a sudden gale with blowing snow obscured the visibility and forced them to turn back.

When Penny had travelled across Wellington Channel a few days earlier to exchange news with HMS *Assistance* and *Intrepid*, Ommanney had told him about examining the cairn on top of Beechey Island. Penny now decided to have another look. If Ommanney had failed to see the widespread and clearly visible evidence on the low part of the island, he might have missed something important at the summit as well. Penny's party climbed to the summit to re-examine the cairn. With picks and shovels they dug up the ground all round, but found nothing.

There was some discussion of exhuming the bodies in the three graves to see if any cause of death could be determined. Dr Sutherland, almost certainly the originator of this proposal, was aware that the bodies buried in winter would be "frozen as hard as possible, and in a high state of preservation in their icy casings." The idea was rejected: "There seemed to be a 'feeling' against this really very proper and most important step,"[22] he later wrote. What was to the surgeon a potentially valuable forensic procedure was regarded by others as an unjustified desecration of the dead. Penny himself may have vetoed the idea, either because he personally believed it to be improper or because he deferred to opinions expressed by his officers and men. Exhuming the bodies and performing autopsies on them might not have been favourably viewed in Britain.

Sutherland continued to believe that the dead would yield clues to the Franklin mystery. Two years later he would get his chance to return to the site and attempt the exhumation. In 1852 he would sail as surgeon on Lady Franklin's screw yacht *Isabel* under Edward Augustus Inglefield, who agreed that one of the graves on Beechey Island should be opened. After some of the crew dug down into the permafrost to a coffin, he and Inglefield, working silently in the chilly darkness of a September night, opened the casket and found themselves looking at the well-preserved body of John Hartnell, formerly an able seaman on HMS *Erebus*. "I have had my hand on the arm and face of poor Hartnell,"[23] Inglefield wrote. It must have been an unnerving experience to see the wide-open eyes staring sightlessly from the coffin, a gaping mouth with lips drawn back over the teeth in what looked like agony, unable to utter a word. There was not enough time to thaw the body and perform an autopsy, but Sutherland's inspection of the face and one arm, possibly by the light of a lantern, revealed no signs of scurvy (he probably checked the teeth and gums, and looked for blotchy skin). He thought the man had probably suffered from consumption. Inglefield did not mention the incident in his published narrative, which suggests that he too anticipated an adverse reaction at home. He and Sutherland would later learn that another attempt to uncover the bodies had been made by a party from Belcher's expedition of 1852–54, but they had been defeated by the permafrost and had abandoned the task before reaching the coffins.

Belcher, Inglefield, and Sutherland were unaware that an autopsy had been done by one of Franklin's surgeons soon after the man's death in 1846. When the three graves on Beechey Island were opened a century and a half later for forensic examination, it was discovered that Hartnell's heart was missing and his lungs and liver had been cut open. It was clear that "in the short hours after his death a surgeon on the *Erebus*, probably assistant surgeon Dr Harry D.S. Goodsir, who was an anatomist by training, had attempted to establish the cause of death."[24]

This first autopsy, carried out under the orders of Franklin, appears to have been done by the brother of Robert Goodsir, Penny's surgeon on the *Lady Franklin*.

After Stewart's party discovered the graves on 27 August, a strange story was reported to midshipman Carter by the *Rescue*'s sailmaker. He said that when they had encountered HMS *Assistance* and *Intrepid* near Cape Riley a few days earlier, some of the British sailors had claimed to have found graves on the previous day. This astonishing revelation ignited a brush fire of suspicion about Ommanney's actions and intentions. Had he been withholding information from the other expeditions? Some of his officers were said to have expressed the opinion that the ground evidence at Cape Riley indicated that Franklin and his men had been travelling toward Baffin Bay, and therefore it would be pointless for any of the expeditions to search further west or winter in the region. Yet, after advising the Americans to search the coast *eastward* from Beechey Island, HMS *Assistance* and *Intrepid* had sped off *westward*, towards Cornwallis Island! Carter suspected that in a cairn, either at Cape Riley or on the summit of Beechey Island, Ommanney's men had found a message from Franklin outlining his future course, or possibly his position, but had kept the discovery secret so that they could be first to reach him. He related his suspicions to Griffin, his commanding officer, who in turn notified De Haven, and soon "the officers of the five vessels here were in an uproarious state suspecting themselves of being humpbugged [*sic*] by Ommanney."[25] Carter watched the officers work themselves up into a pretty fair frenzy: "Penny swears to shoot him. De Haven says he'll up helm and go home if it is so. None of us but would eat a small piece of him."[26] However, the flames of distrust burned themselves out, and a few days later Carter admitted, "All our suspicions of him [Ommanney] and Capt Kater [Cator] were as unjust and unfounded as galley yarns generally are."[27]

The readiness with which they all believed the sailors' story and leaped to the conclusion that Ommanney had acted badly reveals that beneath the outward attitudes of friendship and mutual cooperation among the search expeditions there were undercurrents of ambition and rivalry. Any ship's crew that located Franklin would be covered in glory. The ship's commander would be a national hero; its officers and men would be idolized. Financial rewards, promotions, and accolades would fall upon them like rain from clouds. The stakes in the enterprise were very high.

The Royal Navy expedition had by far the best and biggest ships, the largest crews, the most impressive equipment, and the most comfortable quarters, but it also had the most to lose. Its officers and men would look foolish if they were upstaged by a couple of whaling captains, or a veteran of the Napoleonic Wars now in his seventies, or sixteen men sent out by Franklin's wife in a former banana boat, or an expedition from the United States, a country whose government had never sent out an arctic expedition before. Austin's expedition was *expected* to succeed; it *had to* succeed. Before departure, many newspaper articles in Britain had called it "*the* Arctic Expedition" while referring to others as "Lady Franklin's expedition," "Forsyth's expedition," "Captain Penny's Expedition," or "Sir John Ross's Expedition." The two large steam-powered ships of *the* expedition were expected to cleave the ice floes apart and lead the way to Franklin. Austin, Ommanney, and their men wanted desperately to live up to the expectations of the Admiralty and the public. How far would a commander go to accomplish that?

The little ships of the other expeditions were expected do their best too, and even their commanders occasionally merited the adjective *gallant* in the press. But in the public mind, these vessels, small and lacking steam power, were simply not up to the task. What could they possibly achieve in the presence of the powerful Royal Navy armada? In the curious tale about HMS *Assistance* and *Intrepid* breaking away to steal a march on everyone else, all the commanders of small non-engined ships were instantly united in their suspicions of Ommanney, until the tale was judged to be a fabrication.

Nonetheless, Ommanney's abrupt departure for Cornwallis Island with the *Assistance* and *Intrepid* was curious. He and his men had found evidence of Europeans at Cape Riley, and on the western shore of Beechey Island and on the island's summit. As no European explorer had ever landed at these places before 1845, it was clear that Franklin's expedition, or at least a part of it, had been there. Yet Ommanney did not initiate a search of the region east of Cape Riley, or the rest of Beechey Island, or the shore of the large bay adjacent to it. If Penny had not subsequently decided to examine these areas, the most important evidence of all might have been overlooked.

In the note Ommanney left in the cairn at Cape Riley on 23 August, he wrote, "Beechey Island was also examined, where traces were found of the same party" – but in fact he had not examined all of Beechey Island, but merely climbed to its summit to inspect the cairn seen from the sea. His note concluded, "Captain Ommanney proceeds to Cape Hotham and Cape Walker in search for further traces of Sir John Franklin's expedition."[28] Without consulting the other commanders or even notifying them of his intentions, he had decided to push for Cornwallis Island and beyond! But why? Nothing that he and his men found at Cape Riley or in the summit cairn indicated that Franklin had headed west. He might have gone north in Wellington Channel, turned south into Prince Regent Inlet or Peel Sound, returned eastward to Baffin Bay, or even spent the winter in the vicinity of Beechey Island (which, in fact, he did). He headed west without knowing that a second campsite existed north of Cape Spencer and a third between Erebus and Terror Bay and Caswall's Tower, or that graves and innumerable relics of the expedition existed on Beechey Island, proving that Franklin's ships had spent the first winter in the large bay adjacent to it – vital facts discovered soon after by the small-ship expeditions of Penny, Ross, and De Haven. Ommanney had forged ahead impetuously with incomplete intelligence.

His decision to leave Beechey Island did not meet with the approval of all of his officers. Cator felt that he had given in to pressure from McClintock, who did "not care a straw for poor 'Old Franklin'" but thought only of pushing west to Melville Island.[29] Months later, while in winter quarters, Cator would write in his journal, "While on board the 'Assistance' I and Ede [Assistant Surgeon] had a great chat about what the '*People* of *England*' will say on our return, for not finding the 'Wintering Quarters' of the missing expedition, where we first found the 'Traces.' We shall certainly be to blame, we ought not to have given up search, there is no doubt. Had 'Captain O' listened to me he would not have done so. But 'McLintock' [*sic*] who had been on shore with a party, told him it was of no use, he was sure there was nothing to be seen or found again."[30]

Months earlier, while still on the Greenland coast, Austin had decided to split his naval squadron into two parts on the West Side. One pair of ships (his and Osborn's) would examine Pond Inlet and the south side of Lancaster Sound, while the other pair (Ommanney's and Cator's) searched the north side of the sound as far as Wellington Channel. The four ships would meet up again near Prince Leopold Island. In attempting to reach Cape Walker before conferring with Austin at the designated rendezvous, Ommanney appears to have interpreted his instructions rather loosely.

The spirit of competitiveness is basic in human nature, and in 1850 it existed on both individual and collective planes. Rivalries existed between commanders, between ships, between expeditions, between naval men and merchant sailors, and between British and American crews. For the most part the rivalries took the healthy form of trying to outdo one another in sailing speed, shiphandling among ice floes, rapidity of sawing docks, and so on. There were unwritten codes of conduct, of course. Certain actions were unacceptable, such as knowingly putting another ship or person in danger, or neglecting to help others in a crisis. The objective was to locate Franklin, and it was understood that they must work together to achieve this result. To put personal ambition ahead of the team effort would be unacceptable. Viewed in this light, the suspicions levelled at Ommanney are understandable, even though they turned out to be – as far as is now known – false.

Stuart later criticized what he called "the unseemly and undignified race of nine vessels, their different commanders less actuated by any anxiety for the rescue of Franklin, than eager to outstrip his neighbours in a passage to the westward."[31] His bitter broadside, aimed at every commander remaining in the search after Forsyth's departure, seems unnecessarily harsh and does not do justice to the efforts of Penny, De Haven, and Ross, whose crews spent ten days or more searching Beechey Island and environs before crossing to Cornwallis Island. But the prospect of being the first to sight new land or penetrate unknown waters was a powerful incentive for commanders, and a potential motive for attempting to leave other ships behind. "I shall have the honour of being the first vessel ever up in this direction,"[32] Cator wrote in his journal as he took the steamer *Intrepid* north in Wellington Channel. Later, after climbing to the top of Cape Hotham on Cornwallis Island, he would write, "I was the first person known ever to have been on the top of this Cape and certainly the only one who ever wrote a document on it."[33] (Being the only person to write a note on a summit was not generally regarded as a significant achievement; only Cator recognized it as such.)

The Americans, anxious to prove their mettle in their country's first official arctic expedition, were well aware that being the first anywhere in a region full of British ships would count for a lot. When the *Advance* sailed up the east side of Wellington Channel, De Haven wrote in his diary with understandable pride, "It is not known that any civilized being has ever been up this channel so far therefore we may be considered its first explorers."[34] Ambition, rivalry, and competitiveness were powerful and useful forces, as long as they stirred men to greater effort without undermining cooperation and impeding the search for Franklin.

* * *

Ommanney's hasty rush to Cornwallis Island had not conferred any advantage. While exciting discoveries were being made on Beechey Island and adjacent coasts by the crews of the small ships, HMS *Assistance* and *Intrepid* were trapped amid ice floes on the west side of Wellington Channel. Intending to bring Ommanney and Cator up to date, Penny sent a boat party away from the *Lady Franklin* under John Stuart to carry the news to the icebound ships. Stuart summarized the trip in his journal:

> The Captain having promised to communicate to Captn Ommanney with the least possible delay any farther intelligence that might be procured I was dispatched with a boat for the purpose. I was ordered to get to the land floe and haul the boat up to await our return proceeding over the ice to the Assistance. Seven hours walk through the fog according to a course set by compass and kept by the wind brought us to the vessels. They were lying amongst immense broken masses of ice at a short distance from the land ice and within about 2 miles of the land. Both had undergone considerable pressure and the Intrepid had her rudder broken. Captn. Ommanney was of course much astonished at hearing of traces of such a nature having been found within such a trifling distance of the spot at which his people had landed and from which he had made such a precipitate departure.[35]

The news about Franklin's winter quarters did not show Ommanney in a favourable light. Nor did it enhance the reputation of James Ross, whose ships *Enterprise* and *Investigator* had two years earlier spent almost a year frozen into the ice of Port Leopold, only seventy miles south of Beechey Island. Although Ross had sent a sledge party across Barrow Strait to Cape Hurd on Devon Island, the men had returned to the ships within a week, having done little more than leave a message in a cairn. If they had travelled forty miles west to the mouth of Wellington Channel (named by the Admiralty as an alternative routeway for Franklin!), they would almost certainly have seen the abundant evidence of the missing expedition. Cator exclaimed, "*Sir James Ross* will not be able to hold his head up when these reports get to England . . . His old Uncle [Sir John Ross] will not spare him, on his return . . . Anyone ought to be ashamed to own that they ever belonged to such an expeditions as the last. I should for one. "[36]

Having communicated with Ommanney, Stuart started back for his boat:

> After resting for two hours I set out having been charged with letters for Captns Austin & Penny. The fog which had cleared away while we were on board the Assistance again fell before we had walked a mile on our return and our walk back occupied an hour longer than our journey to the Assistance. At the end of eight hours however we succeeded in finding our boat and after waiting for about 1½ hours more to see whether the tide would have any effect in carrying off a loose pack that had set in on the land ice we commenced launching and hauling up the boat among the loose stuff and after an hours

> labour got into open water and by sailing & rowing reached Cape Spencer by ½ past eight in the morning. The American schooner being made fast to floes set in on the Cape I went on board the Advance to inform Captn Duhavn that no further traces had been got on the West side of the Channel.[37]

On the *Advance*, De Haven observed, "Both officers and men were quite exhausted with cold and fatigue."[38] After taking breakfast on board, Stuart and his boat crew resumed their voyage, but with an American passenger. Elisha Kane, De Haven's surgeon, had decided to make a social call on Austin's ships moored in Union Bay a few miles down the coast, so Stuart gave him a lift. After a hard pull against the current, they reached the *Lady Franklin* shortly before noon. Stuart delivered the letters and news from Ommanney while Kane went on board HMS *Resolute*.

Stuart's party had done exceedingly well. Within a twenty-six-hour period, they had reached the ice edge off Cornwallis Island, walked over land ice to the beset ships and back (a total of fifteen hours), traversed a zone in which the boat had to be manhandled out of the water onto ice floes and back again several times, and returned safely across Wellington Channel by sail and oars. They saw eleven polar bears but managed to avoid confrontations. Their venturesome voyage in an open boat among ice floes had achieved Penny's objective – to inform the two icebound British commanders that Franklin's winter quarters had been discovered. It was the second time that someone from the *Lady Franklin* had crossed Wellington Channel with important information for them.

The boat party arrived back at the ship just as the weather took a turn for the worse, heralding the approach of winter. The mercury shrank in its tube; powerful gusts of wind blasted the ships; ice advanced over areas of open water; thick falling snow swirled along the land and ice, obscuring the visibility. The crews cleared snow off the decks and were issued warmer clothing – jackets, trousers, and comforters. Penny and Stewart set out larger ice anchors and hawsers to hold the ships in position. A few days later, the men cut a dock and warped both ships in side by side. HMS *Pioneer* soon squeezed in beside them. After the *Lady Franklin* and *Sophia* left the dock, they again got into difficulty and had to cut a second dock, which subsequently broke apart. On the *Felix*, both hawsers were carried away by ice floes, forcing Ross and Phillips to hoist sail to regain the land floe where they were able to find shelter in the dock just vacated by Penny's ships.

No further searching was required on the east side of Wellington Channel. The coast had been examined as far as possible. The ships could not proceed northward beyond the southern margin of the pack ice. All the commanders were now anxious to proceed to Cornwallis Island, see if Ommanney and Cator had found any evidence of Franklin, and continue westward in Barrow Strait if possible. But the bad weather and unfavourable ice conditions were confining them to Union Bay. Whenever they left their docks and ventured into the moving floes, young ice impeded their progress. "Bay ice forming about the ships,"[39] Ross wrote in his diary, while the *Sophia*'s log-keeper noted, "The ship is quite un manageable amongst the bay or new formed ice." They hoisted the rudder to keep it safe. The signs were clear: the navigation season was coming to an end.

A week of September had passed by the time the ships managed to leave Union Bay and start west. A gale made sailing difficult. Sutherland described a macabre phenomenon: "The ship's bows were encrusted with ice, and the rigging of the bowsprit was fantastically fringed with pendant icicles, which appeared like as many rows of frightful teeth moving furiously over the surface of the water, and ready to seize any unfortunate thing that might come within reach of their horrid grasp."[40] Penny's ships skirted the edge of the pack ice toward what the *Sophia*'s log-keeper (a man experienced in arctic navigation but, like most whalemen and fishermen of the period, lacking in formal education) recorded as "Com Walace Island" (Cornwallis Island). From "Burleys Inlet" (Barlow Inlet), the ships took advantage of a narrow shore lead to work northward along the west side of Wellington Channel. After the crews were mustered aft for Sunday prayers, the ships continued for another dozen miles before the lead petered out. Penny and a few men landed, climbed to a hilltop, and built a cairn. Inside they left a note within a metal cylinder recording their visit and giving news of other ships.

A few days earlier, looking northward from a summit on the opposite side of Wellington Channel, Penny and Stewart had seen open water in the distance beyond the ice barrier. Now they could see it clearly from the west side of the channel as well; it was no illusion. But there was no hope of reaching it; the intervening ice showed no signs of breaking up. "The Wellington Channel seemed to persist in refusing entrance into the sea beyond it,"[41] Sutherland later remarked. As the ice was immobile, the season late, and the navigation becoming more hazardous day by day, Penny had no alternative but to turn the ships around and head southward toward Barrow Strait. He and Stewart had taken the *Lady Franklin* and *Sophia* to a more northly point than any European navigators had ever reached along the coast of Cornwallis Island and had confirmed the existence of open water to the north, but they could go no further.

Penny carried a copy of Franklin's instructions from the Admiralty, in which Cape Walker was identified as the first objective. "But should a permanent obstruction be found to the south-westward of Cape Walker," the orders read, "he is to consider the alternative of passing between Cornwallis Island and North Devon if the strait between them (Wellington Channel) is open."[42] Penny knew from his experience in the whaling trade that the distribution of sea ice could vary greatly from year to year. What if Wellington Channel had become ice-free in the summer of 1846, creating a navigable waterway leading northward from Franklin's first winter quarters at Beechey Island to the beckoning area of open water?

In an undated note on the back of an old letter, he scrawled his thoughts about the northern part of Wellington Channel:

> Sir John Franklin must have been well aware of the presence of this water. Having [positioned?] north of Cape Spencer, from which they would have observed any change in the channel that they could take advantage of . . . also the ruts of deeply loaded sledges passed his first winter at the mouth of the Wellington Chanl. and A watch seat having been [that?] apparently had been sent to examine the channel. These circumstances – with the second clause of SJF [Sir John Franklin's] instructions, + his own + officers well known of references to the Wellins St[rait] is quite conclusive so say my mind thus he has gone in that direction.[43]

To Penny it looked as if Franklin had found Wellington Channel navigable while all other passages were blocked, and had seized the opportunity to advance northward in the summer of 1846 after wintering at Beechey Island – a course of action perfectly in accord with his orders. The seductive image of that expanse of open water beyond the barrier of fixed ice would stay fixed in the minds of Penny and his companions during the months ahead.

In order to travel from Baffin Bay to Bering Strait it was necessary to sail west, but only in a general sense. As the Parry Channel appeared to end in impenetrable polar ice beyond Melville Island, there had to be a detour, and the obvious route was by way of the continental coast, which had already been charted by Franklin and others and was known to be more or less ice-free in summer. Franklin had therefore been ordered to sail southward to the mainland after attaining the longitude of Cape Walker. A course in the opposite direction – *northward* through Wellington Channel – might at first seem to be perfectly futile. Why then had the Admiralty suggested this as an alternative course?

Depending on the type of projection used, maps incorporate distortions of direction, distance, or area, arising from the fact that they are *flat* depictions of the *curved* surface of the earth. But if one looks at a globe, which more accurately represents the true relationship of land and sea – or even a circumpolar map projection – the potential advantage of a northerly route becomes apparent. The distance from Beechey Island to Bering Strait by way of Wellington Channel is at least five hundred miles *less* than the distance along the southern or coastal route, because it is close to a Great Circle – the shortest distance between two points on the earth's curved surface.

Although the Wellington Channel route was significantly shorter than the southern route, it was totally unknown. No one could tell what distribution of land or what extent, thickness, and duration of ice would confront explorers there, or even whether a navigable passage to Bering Strait really existed: Wellington Channel might turn out to be a dead end. If it did in fact lead into an arctic sea, what would that sea be like? Despite the probability that higher latitudes would create lower temperatures and more enduring ice, many believed that the circumpolar ocean (if indeed the region *was* an ocean) was relatively ice-free. However, the existence of an Open Polar Sea could not be proved, or disproved, until someone actually succeeded in getting there.

* * *

Before leaving Union Bay to head west to Cornwallis Island and beyond, the various commanders agreed that before extending the search westward it would be wise to establish a fallback depot at Beechey Island. (Had they known that the stores carried by HMS *North Star* had been deposited by Saunders on one of the islands at the mouth of Navy Board Inlet, they could have dispensed with this measure and saved their own supplies.) Ross agreed to contribute his yacht *Mary* as a secure repository for the stores and as a retreat vessel in case of emergency. The place chosen was three miles southeast of Cape Spencer, near the mouth of a stream whose delta forms the northern margin of Union Bay. The stream was named Mary River, and the seaward face of the delta was called Mary's Beach.

On the last day of August, with the wind howling at gale force, the *Mary* was hauled up above high water. She was left fully rigged, equipped with sails, provided with a stove, and stocked with basic items. Ross donated a cask of flour and one of coal. Austin supplied three casks of flour, some other provisions, and some blankets. Penny gave a cask of beef, one of pork, and one of peas. De Haven added three casks of beef, pork, and flour. In view of the fact that Austin had twice as many men as the other three expeditions combined, his contribution seems a little on the stingy side, even though two of his ships were beset and out of reach at the time. Ross, who could least afford to give away precious stores, made the most generous contribution of all, including the yacht itself (for which he would later receive compensation). Penny sent his two carpenters to help load the stores on board the yacht and later dispatched a party under Stewart to erect a cairn near Cape Spencer with a note advertising the depot's location.

After a long and eventful voyage from London through the North Sea, over the Scottish height-of-land by canal, and across the Atlantic under tow, the little *Mary* would now stand on watch overlooking Wellington Channel in case Franklin or any of the men searching for him should require assistance.

15

West to Cornwallis Island

A racing mania for Melville Island and Cape Walker . . .[1]

By 8 September, the entire searching flotilla, with the exception of Ross's *Felix*, had left Beechey Island and crossed to the west side of Wellington Channel. Ross had fully intended to leave with the others but encountered some last-minute problems. The spectre of departure appears to have thrown his sailing master, Abernethy, into another spell of intoxication, reminiscent of his binges at Ayr and Stranraer a few months before, and the idea of going further away from Greenland had put the Greenlander Adam Beck into a such a funk that he had sneaked ashore unnoticed to mope by himself. After a search party failed to find him, the sails were furled and departure postponed. Beck ambled back later, unapologetic, claiming to have seen a sign bearing a date. He returned carrying the post but had dropped the sign into the snow and failed to find it again. In Ross's diary (8 September 1820) this event is not entirely clear: "Adam Beck found a piece of wood [on/or a sign?] on the North Land on which was a piece of tin with September 1846 on it but did not bring the [illegible]."

The discouraging delay occurred on Phillips's wedding anniversary, so the gloom at not getting away was in some measure offset by a small celebration – a "jour de fate," Ross called it in his diary, taking liberties with French spelling. On the following day, the *Felix* departed at last. She crossed Wellington Channel, stopped briefly at the mouth of Barlow Inlet, which (unknown to Ross) had already been examined by other parties, and then continued south toward Cape Hotham.

As the various expeditions were not in regular communication, their operations involved some wasteful duplication of effort. Cator's party had searched the shore from Barlow Inlet south to Cape Hotham, but this could not be known to the other commanders unless they found his note in the summit cairn or encountered Ommanney's ships. De Haven took his ships to Barlow Inlet, unaware that men from the *Intrepid* had already been there. As the *Advance* sailed into the inlet, young ice slowed the ship just before she ran aground. The bottom was soft and they were able to get safely off on a rising tide. Griffin, exercising more caution, had brought the *Rescue* to anchor off the north point of the inlet among grounded floes and hummocks. De Haven later wrote in his diary, "The inlet itself was a fine protected basin reaching two miles inland, but covered with last winter's ice and no prospect of its breaking up or disappearing this season."[2] The inlet's dimensions and sheltering ring of high ground were impressive, but the extensive ice cover, which appeared to be several years old and prevented them from taking soundings, might impede access to a good wintering position.

Near the entrance to Barlow Inlet, De Haven, Murdaugh, and Carter made their way on foot across the hummocked and crevassed floes to inspect a cairn on the shore. Nearby, as if on guard duty, was a dead seal with a silk handkerchief tied around its neck. The Americans took it to be a bit of British whimsy left to brighten their day, but it must have been the seal that Cator referred to in his journal: "The 1st engineer 'Mr John Ryder' got a 'young seal' and foolishly left it on the 'Ice' with his silk handkerchief tied around it – and what was worse we sailed in a hurry and so he lost all."[3]

The note inside the cairn had been deposited the previous day by Ommanney. It was written, Carter recorded scornfully,

> in the same bombastic style of Capt Ommanney's other document not mention[in]g which way he is going nor where provisions are left &c. but that no traces have been found on Cornwallis Island of which perhaps he has examined one hundredth and that 1/100 poorly if no better than Beechey Island which they thoroughly examined without finding Franklins winter quarters with a conspicuous cairn pointing them out. Perhaps this isn't the place to express such uncharitable sentiments but it does look so ridiculous for that party to be rushing ahead so as to be the first at every place then leaving it unexamined to say that no traces are found there.[4]

Kane too was critical, although a little more discreet. He found Ommanney's statement that no traces existed on Cornwallis Island to be "a somewhat too confident assertion perhaps, seeing that the island, if it be one, is more than fifty miles across, and that observations can hardly have extended beyond the coastline."[5]

Ice came in overnight and barricaded the American ships, so that next morning they barely managed to get out of the inlet. As they prepared to weigh anchor, "The Rescue was first caught, nipped astern, and bodily lifted out of the water; fortunately she withstood the pressure, and rising till she snapped her cable, launched into open water, crushing the young ice before her."[6] On board the little brig, Carter wrote, "the poor little Rescue was . . . popped out like a melon seed."[7]

Meanwhile, HMS *Assistance* and *Intrepid*, after advancing westward to Cornwallis Island on 25 August, had been trapped in pack ice a few miles offshore. The crews built a cairn on shore and made a survey of Barlow Inlet, but most of their time was spent coping with the pressure of the drifting floes. In early September they worked free of the ice and proceeded south toward Cape Hotham, the southeast corner of Cornwallis Island. Cator's steamer towed the sailing ship he commonly referred to (although not within earshot of Captain Ommanney) as the Old Assistance, Old Vessel, Old Craft, Old Mother, Old Matron, Old Lady, and – in moments of extreme frustration – Old Hooker. It was a harmless mannerism, perhaps even a sign of endearment, but it reflected to some degree his perception of the cumbersome sailing ship dawdling passively along astern as he steamed ahead to clear a path through the ice before returning to take her in tow or assist in some other task.

In the private pages of his journal, he referred to his fellow officers in similar vein, even men ("Old Cheyne," "Old Mecham") who were several years younger than he was. Others he named possibly to reflect a perceived deficiency of physical or intellectual stature, such as "Little Ommanney" (or "Little Omm"), "Little Donnett," and "Little Osborn." With the leader of the expedition he took no chances: he was Captain Austin, Captain A., or the Commodore (only once, in a careless moment, the "Old Commodore"). Few persons outside the naval fraternity qualified for this Catorish form of address, although in his journal he once referred to Penny as "Old Penny" (he was forty-one).

Rounding Cape Hotham, the crews of the two vessels established a depot ashore containing twenty days' provisions for ninety men (in case of accident to the ships), built a cairn, and left a message. Soon afterward a couple of careless mistakes brought the *Intrepid* close to catastrophe. Shellabear, the second master, had neglected to lock the rudder in position after it had last been shipped. Ice floes now wedged underneath it and forced it off, leaving the vessel without the means of steering. Then, just as the larger *Assistance* was coming up astern under sail at six or seven knots, the *Intrepid*'s engineer misunderstood an order for slow ahead and instead threw the engine into reverse. "Up came the Assistance," Cator recorded, "just knocked our mizen boom over the 'Quarter.' She was thrown with confusion, had to back her sails and got off with *smashing* her dinghy into a thousand atoms, and nearly smashed two other boats. It is God's mercy we were not cut down to the water's edge – for had she struck us – nothing could have saved us and here we must have ended our expedition."[8] During the night, while moored alongside the ice edge, severe ice pressure put both ships in a very dangerous position. At one time the *Assistance* was pushed over twenty-five degrees from the vertical, but her mooring hawsers held. Along the shoreline only two ship's lengths away was a solid rampart of ice fifty feet high. "If she had gone against it she would have been cracked like a '*Nut Shell*,'" Cator observed, but fortunately the *Intrepid* was protected by a small point of land. "The Floes as they passed coiled up against it 30 to 40 feet high," he wrote. "Although I said Oh! we are all right, I was glad when we got off clear."[9]

For two weeks, the two naval ships had been under relentless attack from ice floes moving south along the coast of Cornwallis Island. Between Barlow Inlet and Cape Hotham, the shoreline is ruler-straight, backed by an escarpment rising directly to an elevation of more than six hundred feet, a slope so steep that on one occasion men on the ships heard thunderous reports and saw huge boulders, loosened by repeated freeze-thaw cycles over many seasons, cascading down its side to make their contribution to the scree slope. The coast, scoured by currents, lacked stable landfast ice against which they could safely moor. The outlook was so discouraging that at one point Ommanney, who bore the ultimate burden of responsibility for the safety of the ships and men, thought of returning to Barlow Inlet to take up winter quarters there. Cator, who privately described his superior officer as "a nervous man . . . very brave but very squeamish when he is not certain,"[10] strongly disapproved of this plan. He wrote, "If we get driven home this year in our attempt to go back to 'Barlow's Inlet,' should we try it – I'll give my views in a proper quarter."[11]

Fortunately a northerly gale brought about a rapid and favourable change. The floes parted, and from the masthead open water could be seen extending westward to the limit of visibility. There was now an opportunity to explore along the south coast of Cornwallis Island and look for a suitable winter harbour. With permission to proceed, Cator ordered steam raised, the rudder shipped, and the screw lowered. Casting off from the ice, the *Intrepid* started along the coast under power, leaving her sailing consort in a reasonably secure anchorage near Cape Dungeness. After steaming several miles west, the *Intrepid* came to a wide, ice-free bay that looked suitable for winter quarters. It would later be named Assistance Bay (why not *Intrepid* Bay, Cator may have wondered, as his ship had discovered it). He steamed back and reported the news to Ommanney and at 3 a.m. took the *Assistance* in tow and headed west again.

The officers scanned the coast by telescope, noting how closely it conformed to Parry's description three decades earlier, but they saw no signs of Franklin's ships. They passed Assistance Bay, continued through what is now called Resolute Passage, crossed another broad indentation (now Resolute Bay), and went as far west as Cape Martyr. The ships then doubled back and moored to the fast ice bordering Griffith Island. On the next day Cator took the steamer westward again and got within a mile and a half of Somerville Island. It was "further than any vessel that we know of to the west and Northward in these latitudes since the days of '*Parry*,'"[12] he proudly wrote. Another record to Cator!

The gale that cleared a path west of Cape Hotham for the *Intrepid* also brought about a drastic disruption of the ice in Wellington Channel. A large part of the pack ice broke apart and drifted rapidly toward the south. Imbedded in the pack and unable to move were HMS *Resolute* and *Pioneer*. Osborn later described what happened: "Imagine, kind reader, our dismay, in finding the vast expanse, over which the eye had in vain strained to see its limits – imagine this field suddenly breaking itself across in all directions . . . and our poor barks, in its cruel embrace, sweeping out of Wellington Channel, and then towards Leopold Island. At one time, the probability of reaching the Atlantic, as Sir James Ross did, stared us disagreeably in the face, and blank indeed did we look at such as prospect."[13] The situation was critical: the ships were drifting a mile an hour; the pack was "fast re-knitting itself";[14] the surrounding floes were too small for tracking or warping; and young ice in the intervening channels was making it impossible to tow with boats.

Osborn now had another opportunity to demonstrate the power of steam. Taking advantage of a fortuitous narrow lead among the floes, the *Pioneer* passed a towing hawser to the *Resolute:*

> Like a prize-fighter, [she] settled to her work . . . The struggle was a hard one, – now through sludge and young ice, which gradually checked her headway, impeded as she was with a huge vessel astern – now in a strip of open water, mending her pace to rush at a bar of broken-up pack, which surged and sailed away as her fine bow forced through it – now cautiously approaching a nip between two heavy floe pieces, which time and the screw wedged slowly apart – and then the subdued cheer with which our crews witnessed all obstacles overcome, and the Naval expedition again in open water.[15]

At one time or another, most of the ships had been temporarily beset in Wellington Channel, but Penny and Stewart, the only experienced ice navigators in the fleet, had managed to keep their ships clear. On the USS *Rescue*, Carter noted, "Old Penny seems to be as cautious as a fox[,] he keeps astern of us all and profits by our experience."[16] If he was inferring that Penny let others go ahead and get into trouble so he could profit by their mistakes, he was wrong. Within the whaling fraternity, Penny was known as a bold captain who usually showed the way to other ships, and on the Greenland coast he had reached Cape York ahead of all the other searching vessels, even those equipped with steam power. If he chose to tread carefully in Wellington Channel, it was because ice conditions demanded it. He did not intend to lead the way into dangerous situations.

The American officers may have regarded Penny as excessively cautious because they tended to be somewhat aggressive in their shiphandling. On the *Prince Albert*, Snow had watched De Haven con the *Advance* north of Leopold Island on 23 August, and he was impressed by "the apparently reckless way in which they dashed through the streams of heavy ice" and "how gallantly they put aside every impediment in their path," with "not a rag of canvas taken in, nor a moment's hesitation."[17] In sailing among ice floes, however, there was a fine line between boldness and foolhardiness. The trick was to find out where that line lay, and this required considerable experience, which De Haven and Griffin lacked.

On most ships there were likely to be a few officers and men who thought their own commanders too timid, and a few who considered them too reckless. Unburdened by responsibility for the safety of the ship and the success of the expedition, it was easy enough to criticize superiors within the covers of one's personal diary. Cator often found cause to grumble about the conservatism of his superior officers Ommanney and Austin. Maybe it had something to do with his younger age, his inferior rank, his impatience, or all three. In addition, he was commanding a steam-powered sailing ship and wanted to show how power could be used to advantage. Time and time again, he bemoaned the fact that Ommanney was curbing his initiative, or that Austin was holding the entire search expedition back. On one occasion he wrote, "I'm disgusted, had I been on a separate command I would have been miles further ahead and not expended so much fuel."[18] Describing Ommanney's tendency to hold back, he wrote: "A worse man on this account in my opinion could not have been selected, for when we consider that we are on a life and death case, we are in duty bound to run risks, at the same time let prudence direct us. But too much prudence or hesitation is as bad as none at all."[19] Yet Ommanney's caution is understandable.

Commanders of searching ships had to assess potential risks and benefits before adopting a particular course of action, and the decision was often a difficult one. Nobody wanted a search expedition to become the subject of another rescue operation.

On 10 September, Penny and Stewart brought HMS *Lady Franklin* and *Sophia* alongside the landfast ice off the southeast corner of Griffith Island and planted their ice anchors. HMS *Assistance* and *Intrepid* were moored about two miles away, and the *Resolute* and *Pioneer* were in the process of making fast. It was the first time in a month that Austin had his four ships together. He had last been in contact with Ommanney's two ships off northwest Greenland. He now learned about the graves of four *North Star* sailors in Wolstenholme Sound, the abandoned native huts containing corpses, and the complete lack of evidence of a massacre of Franklin and his men.

Penny was invited to a cabin council convened by Austin on board the *Resolute*. Cator's discovery of potential winter quarters in Assistance Bay was brought to everyone's attention, and future prospects were discussed. But then Austin and Penny crossed swords. Austin was trying to advise Penny to make a contingency plan in case a retreat became necessary, but his words came out clumsily and Penny thought Austin was telling him not to expect any help from the Navy if he got himself into trouble. Cator, who was present, conceded that Austin was "a difficult man to understand more particularly for a stranger,"[20] but he felt that Penny had been too quick to take offence. The matter was soon resolved, but it would not be the last altercation between the two touchy commanders.

The USS *Advance* and *Rescue* arrived shortly after. There were now eight ships moored to the ice edge. "All the expeditions are together with the exception of 'Old Johnny Ross,'"[21] Cator wrote. Owing to the problems with Abernethy and Beck at Union Bay, Ross was lagging behind the others.

Night-time temperatures were now falling below freezing point, open leads were solidifying, individual floes were being cemented together, and young ice was growing thicker and more extensive. It was not difficult to imagine that during a calm, cold night a spreading film of young ice might stealthily encircle a moored ship and hold it captive in a tightening grip until escape was impossible, or even that rapidly forming ice might bring a moving ship to a halt. Three decades earlier, Parry had observed that when a sheet of new ice "acquired a thickness of about half an inch, and is of considerable extent, a ship is liable to be stopped by it unless favoured by a strong and free wind."[22] Although so thin that one could easily punch a hole in it, and so flexible that a weight on its surface would cause it to bend, young ice possessed an extraordinary lateral strength. A ship under sail "suddenly arrested in her course upon the element upon which she has been accustomed to move without restraint, has often reminded me of Gulliver tied down by the feeble hands of Lilliputians."[23]

It was also possible that floes of older ice driven by a strong wind could surround a ship in motion and claim it for their own. On the first day of September 1849, the ships of James Ross had been beset in Barrow Strait among loose floes that soon "folded over each other, and they were so interlaced, as to form one entire sheet, extending from shore to shore of Barrow's Strait, and as far to the east and west as the eye could discern from the mast-head."[24] After the floes rafted on top of each other and froze together, the ships had been carried eastward into Baffin Bay. Recalling Ross's experience gave the commanders of the nine searching ships something to worry about, for they were in the same region and it was already ten days later in the season.

The searching expeditions, with the possible exception of De Haven's, intended to remain out for the winter. What they needed now were secure harbours in which the ice would hold the ships safely in place for ten months but allow them to get out in the following summer. The question of winter quarters was becoming daily more critical. The Admiralty had advised both Penny and Austin to secure their winter quarters in good time. But as concrete geographical information about specific harbours was not available, they and the other commanders would have to make assessments on the spot. The maps and reports of Parry, John Ross, and James Ross indicated the general delineation of coastlines and perhaps pointed to a few possibilities for wintering, but of these three men, only Parry had sailed further west than Wellington Channel, and he had not had enough time to examine harbours as he went. The approaching winter was now impeding the search for sheltered anchorages. Sheets of new ice were filling bays and creeping seaward from the headlands.

All the good harbours in Prince Regent Inlet were already closed off, as Forsyth had discovered weeks earlier. Union Bay on the western flank of Beechey Island, although regarded as a "snug position"[25] by Osborn, was exposed to westerly winds and ice movement. Barlow Inlet, on the western side of Wellington Channel, had looked promising, but one of the American ships had run aground there, and it appeared to be a refuge for multi-year ice. Thirty years earlier, Parry had outlined the south coast of Cornwallis Island, but his small-scale chart did not indicate any harbours. He had depicted a large island in Barrow Strait and named it after Rear-Admiral Edward Griffith, but it was a fortress-like stack of sedimentary strata with steep cliffs all round, and because he had sailed south of it, he had not noticed its only bay (now called Dyers Cove) on its northwest coast. That cove, however, was not visible to the searching ships because they were all eastward of the island. In any case, it was probably choked with pack ice drifting in from the west.

As for Erebus and Terror Bay, the commodious harbour in which Franklin's ships had spent their first winter, none of the searching ships had actually been inside it. They had found it full of fast ice and floes, forcing them to go round Beechey Island and north to Union Bay. The disadvantages of the place as a winter harbour would be demonstrated two years later when HMS *North Star* arrived in 1952 to serve as a base for Belcher's four searching ships. Her commander, W.J.S. Pullen, learned that getting a ship deep into Erebus and Terror Bay and keeping her in position could be a hazardous undertaking. The crew cut a dock in the fast ice within the bay and began unloading stores and building a storehouse ashore, but strong northwesterly winds sweeping unopposed over the low spit of land between Beechey Island and Devon Island tended to break up the harbour ice, threatening to carry the ship out to sea. Meanwhile, winds from the opposite direction tended to drive ice into the bay and push the ship into shallow water. Currents and tides complicated the situation. The repeated sawing of channels, shifting of ice anchors, adjustment of mooring lines, and creation of new paths through grounded ice to the beach kept the crew from other important tasks, and they had to be constantly vigilant in order to respond quickly to whatever new threat emerged. Pullen kept the boats at davit-head and the furled topsails double-reefed in case they suddenly found themselves at sea within the pack. By mid-September he was in "a continued state of apprehension,"[26] not knowing what to expect next. Two weeks later, the crisis he feared occurred. A severe southeasterly gale drove thick ice floes into the bay, carrying the ship a quarter-mile toward shore with two anchors dragging until she went aground, turned broadside, and heeled "nearly over on her beam ends,"[27] her rudder shattered. Somehow the *North Star* avoided destruction and the men managed to repair the rudder, but the situation remained tense and uncertain for another two months.

Pullen's experience showed that Erebus and Terror Bay, with the low isthmus to the northwest and an open mouth between Cape Riley and Beechey Island, was like a house with open windows on opposite walls through which the winds blew, their velocity increasing as they were compressed in their passage through the gaps. In late November, when the ice at last appeared to be stable, he remarked with relief and a strong dose of British understatement, "It has been a time of great anxiety."[28] He later reported to the Admiralty, "my own opinion as to this place for a winter harbour is very much against it, particularly such a draft as ours is."

The following summer, Inglefield would arrive at the mouth of Erebus and Terror Bay in HMS *Phoenix*, carrying coal and other stores for the use of Belcher's expedition. It was early August and the *North Star* was still firmly frozen into the ice deep in the recesses of the bay. He later wrote, "Erebus and Terror Bay was full of heavy hummocky ice, of great thickness, impervious to the saw or the blasting cartridge, and too rough and too much inundated with deep fresh-water pools to admit the possibility of landing the stores on Beechey Island, or putting them on board the 'North Star' a mile and a half distant."[29] He had no choice but to land everything at Cape Riley, the eastern point of the bay's mouth.

Yet, despite such experiences, it appeared that Franklin had somehow managed to get his ships into the bay in 1845 and out again in the following summer – another example of the vagaries of arctic ice.

The commanders of the searching expeditions in 1850 did not intend to use Erebus and Terror Bay as a winter harbour. Perhaps they were deterred by its enduring ice cover, or sensed that it might be a dangerous place, as Pullen and Inglefield would later find it to be. More likely, they ignored it because they wanted to establish quarters as far west as possible so that their sledge parties would have less ground to cover in late winter. They were still hoping that some sudden opening in the ice of Barrow Strait would enable them to advance a long way in that direction before it became necessary to commit themselves to winter quarters.

To the east of Erebus and Terror Bay there were several fiords (including Gascoyne Inlet, which would later be recommended by Pullen), but they were even further away from the winter searching activities and therefore of no interest. Nobody wanted to retreat toward Baffin Bay. "To the eastward I would not go,"[30] Austin later said, and the other commanders felt the same.

Where then were they to winter? As the days passed, the temperature dropped and the sea congealed ominously around the ships. Cator's discovery of Assistance Bay had added an interesting possibility that the various officers discussed on 10 September when all the ships except the *Felix* were close together at the ice edge south of Griffith Island.

Although it was imperative to find one or more suitable winter harbours immediately, the commanders were reluctant to abandon further searching efforts. Cape Walker, the critical point on the south side of Barrow Strait where Franklin had been ordered to change course and head southward and westward towards Bering Strait, lay only sixty miles to the southwest, and a message might be lying there. After observing a few openings among the floes, Austin decided that Penny's brigs, accompanied by Cator's steamer *Intrepid* with Ommanney on board, should probe into these leads and see how far they could go. It was a decision born of desperation. Late as the season was, no one wanted to give up the search until the last possible moment.

Penny was ready to cast off by 3 a.m., the agreed time of departure, but Cator could not leave until two hours later, owing in part to Ommanney's late arrival on board. Penny was not inclined to waste valuable time and was reluctant to be burdened by a large naval vessel, even one with steam power. His third officer, John Stuart, wrote in his journal, "We cast off & stood over for Cape Burney [on the northern tip of Somerset Island] giving the slip to the Intrepid stmr [steamer] that had been ordered to accompany us[,] her steam & sails combined proving alike useless when matched with us."[31] In Cator's version of events, however, the *Intrepid*, despite her late start, came out the victor: "'*Penny*' got off before us and had three miles good start of us, but thanks to a good breeze, *steam* and *sails* we past [*sic*] him and came to the termination of the open water after a 25 mile run."[32] Then falling snow reduced the visibility to less than half a mile and navigation became hazardous, so Cator turned the *Intrepid* around and rejoined the other ships at the floe edge off Griffith Island. Penny and Stewart, although lacking the advantage of steam power, persevered for a few more hours and closed with the land on the south side of Barrow Strait. Owing to the erratic behaviour of their compasses and the limited visibility, they could not be sure of their position, and young ice was now impeding their progress to a dangerous degree. On the *Sophia*, the mate Donald Manson recorded in the logbook, "At this time we was brought to a stand and knowing that if the Ship got fixed in the ice here we would in all probability drive out of the sound again, we therefore got the Ship on the other tack." On the *Lady Franklin*, Stuart related, "We were unwilling to venture nearer on account of the loose ice lying off it, the Captain considering it too great a risk to venture on account of the danger there was of being froze up in such a situation at that late period. We accordingly put about and by 10 a.m. sighted Captain Austin's vessels, still under Griffith Island."[33]

In view of the falling temperatures, thick snow, and new ice forming, Penny decided to get his ships into winter quarters right away. Instead of stopping to rejoin the ships of Austin and De Haven, he and Stewart kept right on for Cornwallis Island. At 11:30 a.m., they happened to fall in with Ross's *Felix*, only now arriving on the scene. Phillips boarded the *Lady Franklin* and agreed with Penny that they must find shelter immediately, and so the three vessels pressed on together toward the south coast of Cornwallis Island, hoping to find Assistance Bay. Ommanney had described the place and helpfully provided a sketch map, but fog intervened and their repeated risky approaches to the coast revealed nothing corresponding to the sketch, so the ships dodged offshore to await daylight and hopefully better visibility. At 2 a.m., increasing wind velocity forced Penny to double-reef the *Lady Franklin*'s topsail, stow the jib, and set the fore topmast staysail, after which the ship plied off and on the land for the rest of the night. Stewart and Ross took similar measures. Morning brought improving conditions and, to everyone's great relief, the very bay they sought could now be seen ahead. The *Lady Franklin* led the way in toward shore, forging successfully through young ice and making a channel for the *Sophia* and the *Felix*. Late in the morning of 12 September, the anchor cables rattled out of the hawse pipe into six fathoms of cold, dark water. To provide a safe margin of safety, Penny paid out more than fifty fathoms on the bottom. It was the first time in five months that his ships had ridden to their anchors. The crews now had several months of sedentary life ahead and the exciting prospect of setting out in late winter with sledges to find Sir John Franklin and his companions.

16

Beset in the Drifting Pack

A Retrograde Journey Homeward.[1]

The three British commanders all planned to winter, but De Haven was undecided. In his orders he had been instructed to remain out only if they had discovered Franklin's trail and stood a good chance of reaching him. Otherwise he was to bring his ships back to New York before winter set in. On 30 August on board the *Rescue*, Carter wrote in his journal,

> I think that this party ought to endeavour to return home this season as nothing more can be done now[,] all having been examined up to the impassable barrier [in Wellington Channel] and the field for spring excursions too limited for the number of Englishmen to amuse themselves with. There is about enough for Austin to do on this [north] side [of Barrow Strait] and Ommanney on the Southern so that until the next open season Penny, Ross, and ourselves can only play fifth wheel while by starting early next spring [from New York] we may certainly get this far before Austin will be free from his winter quarters. I wish DeHaven [*sic*] would think so, for the prospect of nine months in this little boat while all those John Bulls are in their comfortable cabins is dull.[2]

He added, "If we cant go home lets do something desperate and winter in the pack where if we are not squashed in the spring we'll be the first out of winter quarters." This off-hand remark would prove to be prophetic.

Carter's argument in favour of going home was not unreasonable. The concentration of seven British ships in a small area of open water had already resulted in wasteful duplication of effort. Austin, with two sailing ships, two steamers, 180 men, and all the best equipment, clearly had the means to do the job, while the other expeditions, with fewer resources, would in Carter's opinion merely be "fifth wheels," useless appendages. It was quite possible that the American ships could return from New York to Barrow Strait before the British ships that had over-wintered could manage to extricate themselves from the grip of the ice. They would miss the opportunity to send out sledging parties in winter, but they were not adequately equipped for that anyway, Carter wrote, "not even having a tent, and so few men."[3] He was not over-dramatizing the situation. On board HMS *Pioneer*, Osborn observed of the Americans, "No preparation or equipment for travelling-parties had been made by them, and their fittings for wintering in the Arctic Regions were, compared with ours, very deficient."[4] McClintock called the *Advance* and *Rescue* "wretched little vessels."[5]

If it seems extraordinary that a government search expedition could be sent out without tents for its search parties or crows' nests for its ships, one must remember that the preparations had been carried out in extreme haste owing to the late approval by Congress; the United States Navy did not have dockyards stocked with equipment for polar exploration. In any case, the Americans were not the only ones who lacked resources. Penny's expedition had not been provided with tents either; his sailmakers were busy making some out of spare canvas. They also lacked tracking belts and sledges. As for Ross, he was short of almost everything.

The more generous accommodation on board Austin's large ships had excited envy, and even resentment, among the American officers. After William Murdaugh, first officer on the *Advance*, visited HMS *Resolute*, he described the officers' cabins as

> very fine for winter, each state room connects with a big fire below by flues so that by turning a cock a current of hot air rushes in sufficient to warm and dry a relevée [watch officer] from a four hours rain. I daresay that those fellows came here to spend a pleasant winter as they could scarcely be so comfortable in a house. Quite a contrast with our little "rain come wet me, Sun come dry me" cabin with no stove even, which by the way the English have beside the flues.[6]

It was a strange turnabout. Americans often criticized the English for their cold, damp houses, but now the English crews were enjoying the benefits of central heating while the Americans were sailing the arctic seas in unheated ships.

There was a question of privacy too. On board the *Prince Albert*, Snow observed that whereas he enjoyed a cabin to himself in which he could "quietly read, or write, and study," the American officers shared accommodation "in a truly republican manner."[7]

After visiting the Royal Navy ships at Union Bay, Kane had written, "I had to shake off a feeling almost of despondency when I saw how much better fitted they were to grapple with the grim enemy, Cold. Winter, if we may judge of it by the clothing and warming appliances of the British squadron, must be something beyond our power to cope with; for, in comparison with them, we have nothing, absolutely nothing."[8] When he stayed on HMS *Resolute* for three days during the gale of early September, he slept between clean sheets! The daily routine, he noted, was that of a man of war: the crews were paraded, inspected, led twice daily in prayers, and regimented in every respect. And, as Carter noted, they were rewarded for their efforts: "All the officers and men receive double pay, and the officers [are] certain of promotion after this cruise."[9] Not only were the Americans *not* receiving double pay but they had generously given up in advance any claim to the reward for finding Franklin.

Sherard Osborn, former commander of HMS *Pioneer*, would later salute the American gesture of relinquishing all claims to the reward: "We, I am sorry to say, had acted differently. America had plucked a rose from our brows."[10] The British crews were eagerly looking forward to winning the prize. In late July, when Austin's ships were on the coast of Greenland a hundred miles short of Cape York, Cator had been summoned to the *Assistance* by Ommanney. "I found him preparing a statement of the best manner for the distribution of the *Reward 20,000 pounds*," Cator recorded, "which I perfectly agreed with, should we be so fortunate as to get it. It would give me about *320 pounds*."[11] Ommanney's calculations led to a formal agreement. On 9 August, Cator wrote, "Today we exchange papers made between both vessels – signed by the officers and men of both ships so that we may all share in the prize money should we be lucky enough to get it."[12]

Whether or not they would winter was discussed almost daily on the *Advance* and *Rescue*. When De Haven announced that they would go home if they could not find an opening in Barrow Strait, everyone's spirits rose. When he talked of his desire to winter three hundred miles further west, their spirits sagged. Carter used the analogy of the stock market, writing on one occasion, "Home stock so high a few days ago is not worth shucks now," but two days later, "Home stock is again in the market."[13] If they sailed away while other ships were still endeavouring to find an opening toward the west, they would lose face, so De Haven kept the *Advance* and *Rescu*e probing along the ice edge. Carter confided to his diary the anxiety felt by many: "The young ice is so troublesome now that we apprehend having to hunt for winter quarters soon, lest our dull vessels should be caught in it and drifted S.E."[14] As events would show, their concern was fully justified.

When De Haven's ships joined those of Austin and Penny at the floe edge south of Griffith Island on 10 September, he had still not reached a final decision about wintering. On the following day, he received a memo from Austin announcing that there was no hope for any further advance toward the west and "desiring our co-operation in making journeys over the ice in the spring."[15] At last the dice were cast. There was no chance of advancing further west; the British ships would go into winter quarters. There would be no loss of face now if the American ships left for home in accord with their instructions. De Haven declined Austin's request for assistance in the spring and offered him all his surplus provisions and fuel, keeping one year's emergency supply for his own ships. But Austin was in such a hurry to get into winter harbour that he could not spare time to transfer stores, so De Haven promised to land them near Beechey Island for emergency use, if feasible. He never got the opportunity to do so.

Before the *Advance* and *Rescue* could get under way, the weather took a turn for the worse. De Haven wrote in his diary,

> Soon after midnight the wind veered to the North and increased rapidly to a heavy gale which coming off the ice brought with it clouds of drift snow. The Rescue was blown from her [ice] anchors and went adrift so suddenly that a boat and two of her men were left behind . . . We came near meeting with the same fate for the edge of the floe to which we were fast kept breaking away and it was with great difficulty that we managed to plant other anchors farther in to hold on by.[16]

On board the *Rescue*, Carter wrote, "I have seen days of suffering, hard winds and some cold work but, this was one in which those unpleasant things, combined with danger, made quite an event in the life of every one of us."[17] After blasts of wind drove the ship away from the ice edge, tearing out all four ice anchors, the *Rescue* was unable to regain its former position and had to beat back and forth for the remainder of the night in strong winds, thick snow, and a temperature of 18°F – well below freezing – until Griffin managed to bring her into the lee of the island and drop anchor. But then the anchor dragged, and the ship drifted downwind into an ice floe, smashing the rudder. Manoeuvring with sails alone, Griffin got the ship back into shallow water and let go the port bower anchor for greater security, but it broke before a man could plant an ice anchor. Violent squalls drove the ship away from the land ice, and again they clawed their way back upwind. They retrieved the broken anchor and moored to a grounded floe. During these desperate hours, conditions were atrocious:

> The weather very cold and snowy, while underway constantly looking out for ice which the blinding snow and spray made it difficult to avoid. Our clothes were saturated with the spray in half and hour after starting, but they froze immediately and so were perhaps warmer. But the annoyance was the stiffness of the ropes which were iced and twice their usual size. The head guys, shrouds, and bobstays made a kind of net work which, being dipped as the vessel pitched, soon formed a solid mass of ice out to the bowsprit cap and the sea breaking over the bow rail soon had a solid mass from the knight heads to the windlass and being coated with ice outside she looked like an iceberg under sail.[18]

When they eventually succeeded in making fast to the ice, all the sailors were exhausted, incapable of any further exertion.

Meanwhile, De Haven had managed to keep the *Advance* in position, but the ice to which he had moored was now in motion, setting to the south, and the weather was dreadful. "The thermometer fell to +5 [°F] giving unmistakable evidence of the rapid approach of winter," he wrote, adding premonitorily, "I became uneasy about securing our retreat."[19] His crew cast off the lines and hoisted sail to beat back toward Griffith Island, now twenty miles away, to find the *Rescue*. As the ship threaded its way through ice streams, buffeted by squalls with thick snow, it received some hard knocks, one of which De Haven feared would bring the masts down. Luckily they spotted the *Rescue* lying close inside the island and managed to moor nearby. Griffin reported his rudder broken and stern post split, so De Haven took the unmanageable vessel in tow. He signalled Austin's ships that they were homeward bound, and as Austin made no effort to send letters out with them, the *Advance* and *Rescue* departed towards the east. On board HMS *Intrepid*, Cator was furious that officers and men on other ships were given no opportunity to send mail homeward by way of New York. Although Austin knew of the Americans' imminent departure, he "*would say nothing to any of us, so not a soul was enable[d] to send letters on board,*"[20] he recorded bitterly.

As the two American ships limped along the south coast of Cornwallis Island, De Haven and Kane spotted masts close in to the land. Kane later described the moment: "'Brigs!' said I. 'Undoubtedly,' said De Haven;

and then both of us simultaneously, 'Penny!' On looking with the glass, the masts, yards, gaffs, everything but the bowsprits, were made out distinctly. Lovell was called and saw the same."[21] Murdaugh, rushing on deck half-dressed, took the glass and saw a third vessel as well, a topsail schooner, evidently Ross's *Felix*. Of the total searching force of nine vessels, the ships of Penny, Stewart, and Ross were the only ones safely ensconced in winter quarters.

As De Haven closed with the land for a better look, fog patches intervened, and when they next got a glimpse of the place, only three miles away, there was no sign whatever of the ships previously seen. Kane was dumbfounded. "Four persons saw the same image with the naked eye, and the glass confirmed the details. There was no disagreement. As plainly as I see these letters did I see those brigs . . . Ten minutes of obscurity, we sailing directly toward them, a sudden interval of brightness – and they had passed away."[22] They saw no point in chasing what might have been a mirage, so De Haven turned the *Advance* back onto her eastward track, the *Rescue* still under tow. Unknown to him, Penny and his men in Assistance Bay had been gazing at the ships through the fog patches, wondering which ships they were and where they were headed.

Homeward bound! With favourable conditions, a few weeks should see the *Advance* and *Rescue* sailing into their home port of New York. Men's thoughts turned to girlfriends and families, copious amounts of beer, mountains of decent food, and an eternity of sleep. But De Haven was worried about the "tenacious 'pancake' ice"[23] around the ships. The "pancakes" – circular pans of frozen slush with upraised edges created by contact with each other as they rotate in slowly swirling eddies of surface water – are one of the initial stages of sea ice formation. He observed, "There is nothing to obstruct us in our course except the young ice which increased rapidly in toughness as the wind moderated. Our headway became every moment less and less until 11 P.M. when we were brought to a dead stand with all sails set before a moderate breeze."[24] On the *Rescue*, Carter perceived it as another of those troublesome stock-market-like fluctuations in their prospects: "Home stock was up high but the young ice has caused a great fall."[25]

September 15 was Carter's twenty-fifth birthday, but he was not in a party mood. The ships were off the entrance to Wellington Channel, but despite a six-knot breeze, they would not move an inch. They were held fast. The temperature was still well below freezing point, and the ice, already four inches thick, was growing thicker by the hour. Both he and De Haven now confided privately to their journals the belief that they were facing a winter in the pack ice. When he recalled that Penny had once described the idea of wintering in the pack as perfect madness, his apprehension deepened. "Could ships and men have a more gloomy prospect?"[26] De Haven began to economize on fuel. Even with outside temperatures well below freezing, he did not permit stoves to be lighted below, except for cooking. Wrote Carter, "Our [officers'] apartments are now so cold and damp that the docs call them unwholesome and those of the men are worse."[27]

The consolidating ice had become thick enough to take the offensive against the wooden hulls. The next day, Carter wrote, "The young ice has got so strong now that it plays with our vessels like toys[,] swinging and thumping us about constantly so to keep us constantly at work repairing its damages. To day after almost finishing our rudder it swung the Advance into us nearly crushing our white whale boat and then commenced squeezing us so as to make every timber creak and the poor little brig quake and quiver."[28] To make matters worse, the ice in which the ships were encased appeared to be drifting into Wellington Channel. This was certainly not where De Haven and Griffin wanted to go, but they were helpless to do anything about it. Their ships were passive prisoners of the pack, and a cruel conspiracy of currents, winds and tides was now carrying them steadily northward.

The men followed their usual daily routines, coping from time to time with sudden, dangerous bouts of ice pressure, and watching with apprehension as familiar coastal landmarks appeared ahead, passed by on each side, and receded astern. They passed Barlow Inlet, the potential winter harbour they had previously visited, but saw no chance of reaching it. They passed Cape Bowden on the east coast, which Griffin's shore party had reached. One by one, a succession of nameless and previously unknown headlands slipped slowly behind on both sides of the waterway. On 19 September, less than a week after starting home from Griffith Island, they recorded a latitude of 75°24'30" N. They had been carried fifty miles northward in Wellington Channel into territory never before visited by Europeans. It was ironic. A month earlier, Penny had attempted to press northward but had been stopped by ice. Cator had tried to force his steamer northward but he too had been thwarted. Now the Americans, on their *homeward voyage*, had been carried, entirely by accident, into the very region that the others had been eager to explore.

After describing their prospect of getting free of the ice as "evidently hopeless," De Haven wrote, "If however we could but see the ice *fixed* and the vessels immovably frozen in so that we might venture to remove some of our stores and make arrangements for the winter I would not be at all concerned[,] for our present position is an advanced and favourable one for operating with pedestrian parties in the spring."[29] Despite their hazardous situation he had resolved to continue the search if he could find a secure wintering place somewhere in stable landfast ice. But all round the two ships, the ice seemed perpetually in motion.

For the next two weeks they were shunted back and forth by the restless floes. The ice was unwilling to relinquish its prisoners. De Haven wrote, "All our efforts to cut the Rescue out of the massive floe in which she was imbedded were unsuccessful, gave up all hope of being able to reach any other place than our present, in which to spend the winter."[30] They had attained their farthest north and would get no closer to the newly discovered lands. Still held firmly in the ice, the ships began to move slowly southward again, back toward Barrow Strait.

From aloft the officers could see that the Devon Island coast of Wellington Channel continued northward, while the Cornwallis Island coast was beginning to curve toward the west. Directly to the north, there was the appearance of high land – land that did not exist on the charts, land known only to Inuit if to anyone at all. North of Cornwallis Island, De Haven wrote in his diary, "A wide channel appeared to extend to the west in which direction from the dark and cloudy appearance of the horizon I judged there was a large space of open water."[31]

The journals of De Haven, Carter, and Lovell, and the account published later by Kane, do not mention sighting new land or allocating place names. Perhaps there were too many distractions at the time – such as trying to extricate the ships from the ice, unshipping their rudders during bouts of severe ice pressure, and coping with hefty ice slabs that rafted up dangerously close to the hulls.

The fifty-mile drift northward in the pack ice had taken little more than a week, but the return trip took two months, the two ships moving so imperceptibly that the despondent crews anticipated having to spend the entire winter in Wellington Channel. Occasionally they talked of trying to walk to the wintering British ships, or of taking refuge in some harbour, or of depositing some of their supplies on the land, but none of these courses of action was considered feasible. On at least seven occasions, Kane and other officers made exploratory treks over the ice toward land, hoping to build a cairn and leave a record of their situation, but they were always stopped by a lane of open water adjacent to the shore, or recalled to the ships when the ice began to move about. The outings served mainly to alleviate their restlessness and provide some useful exercise. It became clear that their icebound vessels would be their homes for many months.

De Haven knew that when they reached Barrow Strait and Lancaster Sound again, the current would carry them eastward toward Baffin Bay, further every day from the other expeditions. He and his men were now committed to a winter in the pack. He considered what measures they could take to afford the best chance of survival. What he decided to do was unprecedented. He would gather both crews on board the *Advance*, the larger of the two ships, and use the *Rescue* mainly as a store ship. Heating only one ship would reduce the consumption of coal, and the men would all be together without risk of separation. It was a bold measure because there was no guarantee that the *Rescue* would remain nearby. If the complex movement of floes carried her a great distance away, they would not have access to the provisions they needed for sustenance.

In mid-October the carpenters began the renovations necessary to fit another sixteen men into the already crowded *Advance*. This could only be accomplished by removing most of the ship's stores from the holds and converting those areas into sleeping quarters. After hoisting casks out and depositing them on the ice alongside the ship, the carpenters and helpers enlarged the officers' cabin by extending it into the after hold. They demolished bulkheads, laid planks on beams, shifted the galley down to a former coal space above the keelson, and set up stoves in the cabin and forecastle. On the last day of October, the officers and men of the *Rescue* left their ship and walked across the intervening ice to the *Advance*, taking with them stoves, boats, and personal necessities. Carter, ousted from the home he had occupied for almost half a year, lamented, "The Rescue is a deserted mansion."[32]

Many casks were brought up from the *Advance*'s holds, shifted to the *Rescue*, and stowed on deck or below, but there was not room for everything. Some stores were placed on the ice nearby. Fully conscious of the risk, De Haven wrote in his diary, "Altho' the situation of the vessels may be considered precarious and the stores on the ice not very safe, yet the preservation of the health of our small party rendered it absolutely necessary that this vessel should be broken out [i.e., the stores removed from the hold] so as to accommodate all hands."[33] One threat to the ice depot was from marauding animals. In mid-December "a bear was seen close to the ship, examining with much care the stores which were on the ice. In his efforts to break open some kegs of lard, he rolled them over. This noise attracted the attention of the watch."[34] A musket shot drove the bear off, seriously wounded, and on the next day it was found dead a few hundred yards away.

A more serious threat to the exposed supplies was ice movement. Officers made excursions over the ice almost daily, walking a mile or two from the ship in various directions to look for advance warning of severe pressure. If arctic sea smoke betrayed the presence of open water in the distance, if cracks appeared and spread toward the ships, if leads opened up nearby, if floes began rafting up into hummocks, or if the arctic silence was shattered by the dreadful noise of colliding, grinding, crashing ice, then it would be time to worry. Boats and crudely constructed sledges loaded with provisions and survival gear were kept at the ready, and several times the men came close to abandoning ship. One severe nip lifted the stern more than four feet, sending bilge water downhill toward the bow, where it was out of reach of the pumps and had to be bailed out by a bucket brigade. Miraculously, the stores on the ice escaped destruction while the ice was in movement, but on one occasion the men found their outhouse situated on the other side of a wide lane of water – an inconvenient place for a place of convenience.

After reaching Barrow Strait in the drifting pack ice, the ships began moving slowly eastward in Lancaster Sound. Their progress was slow at first – they seemed to linger within sight of Beechey Island for ages – but in late November they began moving more rapidly, passing in reverse order the headlands and fiords they had seen when they arrived months before. As they were carried along the coast of Devon Island, only a few miles from its steep ice-crowned battlements, De Haven confided to his journal the fervent hope that "the ice will soon become fixed for the winter and not be drifted into Baffins Bay where I am inclined to think that the ice is subject to movement throughout the year, which, with the numerous bergs found there would render our situation a very hazardous one." But on toward Baffin Bay they drifted. By late December, they were getting close to the mobile pack ice of Baffin Bay. He wrote, "Should we get into it and be carried about at the mercy of the elements it would be difficult to say what our fate may be."[35]

17

Into Winter Quarters

Perfectly fixed in the ice. . .[1]

While De Haven's ships were being carried northward and southward in Wellington Channel, those of Penny and Ross were securely in the ice of Assistance Bay. A mere three weeks had passed since arriving in Lancaster Sound, and the navigation season was already over. "It was painful to reflect upon how little had been accomplished by the best endeavours of so many ships and willing crews,"[2] Sutherland remarked. For everyone on the mission, it was enormously distressing to think that somewhere in the treeless, wind-blasted landscape some surviving members of Franklin's missing expedition, with slowly ebbing strength and will, might be looking hopelessly at the approach of their sixth dismal winter, knowing nothing of the efforts being made to save them.

For searchers as well as the sought-after sailors, the harsh, unfamiliar environment imposed strict limits on human activities. The arctic winter was a long interlude separating one brief navigation season from the next. It was a time to retreat from the threatening elements, take shelter on board ice-encased, snow-banked ships, and endure the approaching cold and darkness – in short, a time for human hibernation. The search for Franklin would have to be suspended until sledge parties could set out over the natural highways of ice and hard-packed snow in late winter and spring.

Up to this point, Penny, Stewart, and their crews had been facing challenges familiar to whalemen. Navigating among ice floes and bergs, with all the concomitant threats to the safety of the ship and the survival of the men, was part of their experience. Spending an arctic winter frozen into the ice, on the other hand, was new and therefore intimidating. No whaling crews had ever wintered intentionally on their ships in the Arctic, and on the few occasions that ships had been trapped in drifting pack ice for several winter months, their crews had suffered badly from cold, hunger, and scurvy, and many men had died. Tales of those dreadful events still circulated within the whaling fraternity. But those tragic episodes had been accidental, the ships and crews not adequately prepared for such a contingency. In 1850, on the other hand, wintering was part of the plan. Penny had made careful preparations for it, benefiting from the published narratives of discovery expeditions that had wintered in the Arctic since 1818. His ships had enough food to last three years, a supply of winter clothing, and coal for heating. If necessary, he could seek advice from veterans on other ships: Sir John Ross and his ice master Thomas Abernethy, Penny's next-door neighbours, had passed four consecutive winters together in the Arctic; and Austin's ships, if they could be found, had at least five officers who had previously served on wintering expeditions.

It would much colder than they had ever experienced during the severest winters in Scotland, they knew. This was clear from the narratives of previous commanders. When Parry and his men had wintered at the same latitude in 1819–20, the thermometer above the *Hecla*'s deck had recorded −40°F – cold enough to freeze the mercury – at least once in every month from November to March. In February the temperature had never risen above −17°F, and once it had fallen to −50°F (measured on an alcohol thermometer) while a thermometer located away from the vessel on the ice had recorded −55°F. The temperatures recorded by Parry on his second voyage in 1821–23, and those experienced by John Ross in 1829–33, were of the same order, and Ross had recorded an extreme low of −59.5°F.

Penny's men were also aware that if they were away from the ship when a wind was blowing, they would feel its full force as it swept unimpeded across flat ice and treeless land. Any exposed flesh would freeze more quickly. Although the concept of *wind chill* would not be formulated for more than a century, the phenomenon was obvious to all. "We hade [*sic*] a sharp breeze of wind," Manson recorded in the logbook of the *Sophia* in early November, "which Perhaps made us feel its efects Sooner." The men were familiar with frostbite – there was plenty of that on arctic whaling voyages – but some would get painful reminders when they carelessly exposed hands or faces. Privately, the men wondered how they would react to the extreme cold of midwinter. Would they be able to harness themselves to a sledge in March and April along with half a dozen others and pull a load of more than half a ton for ten hours, day after day, sleeping in blanket bags in small cloth tents erected on the wind-packed snow? They rightly sensed that it would be the most demanding work they had ever undertaken, and the thought was unnerving. No one wanted to let the side down in the great humanitarian enterprise. No one wanted to be found wanting. Each man wondered privately if he would be up to the challenge.

Assistance Bay does not have the neat circular shape of Erebus and Terror Bay or the narrow cleft-like configuration of Barlow Inlet. It is more of an open roadstead, three miles wide across its mouth but only half that distance from front to back. Behind the coast the land rises to a plateau five or six hundred feet above sea level into which several stream systems are incised. Gorge-like in their upper reaches, they become broader as they descend toward the coast, where their sediment loads have built up an extensive floodplain and beach around the bay and a soft bottom offshore affording a good holding ground for the ships' anchors. During the forthcoming winter, the stream valleys would serve as corridors into the interior for hunting parties, and several small lakes in the watershed would prove sources of ice to melt into drinking water, and of fish to enliven the tedious diet of preserved food.

Getting into their final positions for wintering had been a tense, uncertain operation. The ships arrived in Assistance Bay during the brief but violent transition between summer and winter. Floes in Barrow Strait were being driven this way and that, pushing against the hulls, rafting up against the bulwarks, threatening to encompass the ships and carry them out into Barrow Strait. Heavy snow falling on the water among the floes created a thick slurry that Sutherland likened to soft lead. While not firm enough to walk on, it hindered the advance of the ships. To get as close as possible to the land and take up relatively secure positions in the newly forming bay ice, they had to bore through the slush and small floes under full sail, sometimes heeling over so much that boats suspended from davits on the leeward side touched the ice. On deck and aloft, several inches of snow made every shipboard operation particularly hazardous.

After Penny's ships came to anchor on 12 September, there had still been a slight possibility that they would go to sea again before winter, either voluntarily to take advantage of some promising diminution of the ice cover in Barrow Strait, or involuntarily as helpless hostages of wind-driven ice. The crews therefore worked to keep the vessels shipshape, clearing snow off the decks, beating the dangerously heavy coating of ice off hulls, bowsprits, and rigging, overhauling halyards and sheets, sending down the tall royal masts and yards, and replacing the topgallant masts with shorter ones.

Within two days of dropping anchor in the bay, the ships were surrounded by new ice thick enough to run about on, enabling the men to begin shifting some provisions and gear to the land, and make preliminary excursions up the stream valleys onto the plateau. When Penny's crews gathered for Sunday prayers on 15 September, the *Lady Franklin*'s logbook recorded that the ship was "perfectly fixed in the ice." A few hundred yards away on board the *Felix*, Sir John Ross delivered a sermon entitled "Unto us a child is born," inspired by a verse in Isaiah, which he felt "appropriate to the ship being for the first time frozen in."[3] (The worshippers may have been a little puzzled about its relevance.)

Penny was faced with an important question: Where were the other expeditions? It was vital to discover their location so they could keep in touch during the winter. When two ships under sail were sighted outside the bay, there was great excitement. The visibility was poor, so they could not be made out clearly. Some thought they were two of Austin's vessels, perhaps heading back to Beechey Island, but most believed them to be the American ships, perhaps intending to winter in Barlow Inlet. Whatever their identity, there were still four ships unaccounted for. Penny, Stewart, and a few others climbed up onto the high rocky knoll a mile north of the bay, which they subsequently called Prospect Hill. From an elevation of more than 360 feet above sea level, they had an extensive view over Barrow Strait. The place where they had last seen the ships – Austin and De Haven's – at the ice edge was now vacant. There was not a ship in sight. From this vantage point, however, they could not see the western part of Griffith Island or the steep bluff on the coast of Cornwallis Island that Parry had named Cape Martyr.

A few days later, Penny sent a party westward along the coast. Most of the men returned after hiking six miles, but two continued for another four, and as they approached the large indentation now called Resolute Bay, they could see the unmistakable profile of Cape Martyr beyond. To the left of the cliffed headland, only eight miles across an ice-choked strait, the hidden part of Griffith Island now came within their view, and they were rewarded by the sight of Austin's four ships, moored to the ice edge off its north coast, only about ten miles away. But until the narrow strait (now called Resolute Passage) froze over completely, the men at Assistance Bay would be unable to reach them.

Penny must have wondered why the naval squadron had selected such a precarious location. Later he would learn that Austin had initially intended to winter in the bay just eastward of Cape Martyr (Resolute Bay) but had not moved soon enough. Not until a northerly gale began breaking up the landfast ice south of Griffith Island, to which his four ships were moored, had he conceded that it was time to get under way. The weather was ruthless – a temperature of only 5 degrees Fahrenheit, a hard-driving wind, sleet falling. He later called it "as severe a day as ever I experienced in my life."[4] Furthermore, it was Friday the thirteenth (of September).

But it was now or never. Ignoring superstition and the weather, he ordered his two tenders to raise steam. They took their consorts in tow and headed for Cornwallis Island. When Ommanney decided to drop the tow and proceed under sail, Cator's men on the steamer found that the towing hawser, a hundred fathoms (six hundred feet) long, "was so hard the snow and frost on it made it like a bar of iron."[5] With great difficulty, they managed to haul it in, beat the ice from it, and stow it below. The low temperatures were accelerating the growth of young ice, and drifting ice floes were impeding navigation. They could not reach their intended harbour.

After a day of laborious sailing, towing, sawing, and blasting, they succeeded in mooring to the land floe attached to the north coast of Griffith Island, an insecure position but certainly preferable to the open channel. Austin sent Osborn away with the steamer *Pioneer* to look for an opening south of Griffith Island, apparently still determined at this perilously late date to find a means of advancing further westward, but there was no opening there. Resolute Bay was by now unattainable, so Austin ordered Ommanney and Cator to try for Assistance Bay, eighteen miles away (where, unknown to them, the ships of Penny and Ross were already frozen in). The *Intrepid* sallied forth under steam with the *Assistance* in tow, but by this time it was much too late. The sailing ship stuck fast in congealing ice, and in the steamer's engine room, the injection pipes froze solid and had to be thawed with hot coals from the fire box.

The two ships limped back under sail to the floe edge and joined the others. They would get no more chances to find a sheltered harbour. The exposed ice edge in the narrow strait would have to serve as their winter quarters, if they could only manage to hold on. Austin's tardiness in getting his ships into a secure winter harbour had put them at risk.

Successive logbook entries on board HMS *Assistance* during this crucial period reveal a threatening marine environment in a state of violent transition: "young ice forming rapidly"; "so thick as to prevent the ship coming head to wind"; "floe edge breaking away rapidly"; "shifted ice anchors farther in"; floe breaking away"; "small pieces breaking off"; "large pieces breaking off"; "watches ahead on the ice, cutting, breaking and blasting"; "tried mushroom anchor by dropping it from bowsprit, but failed"; "young ice doubling in layers."[6]

* * *

In mid-September on board the *Lady Franklin* in Assistance Bay, John Stuart wrote, "All idea of getting out was now abandoned, and preparations began to be made for wintering."[7] The first thing was to align the ships. No details are given in the logbooks or journals, but they probably oriented them in a north-south direction, as Parry had done in 1819, so that each side of the hull would receive an equal amount of the limited sunlight (and the captain's stern cabin most of all). Aligning a ship required some cutting of ice, then warping the stern to one side and letting her freeze into the new position. Before the ice became too thick, the crews would haul the ship ahead, cut holes in the ice, and raise the anchors. Two weeks later, the *Sophia*'s logbook recorded, "Hove up our anchor and secured it on the gunnel."

With the ships aligned in an optimum direction and held firmly in the ice, the crews could begin housing them in to create wind-protected spaces on the upper deck for work and exercise and to help keep the ships warm. Over a framework of spars and planks, each crew would spread three prefabricated canvas covers from gunwale to gunwale over longitudinal ridge poles amidships. On board the *Lady Franklin*, John Stuart recorded, "The housing cloths were brought up and the midship cloth was placed in its position. It is a very thick & compact material well calculated to shelter a ship from the wind and it will doubtless prove a great comfort."[8] Two days later he added, "The quarterdeck cloth was got up as was also the fore awning but the latter was found not to fit and was accordingly taken down to alter."[9] After the sailmakers finished the necessary modifications, the upper decks of both ships were closed in from stem to stern with "a complete covering through which wind can not penetrate."[10] Their winter domiciles were now secure.

The three ships had anchored far enough from each other to avoid fouling each other's anchors but close enough to facilitate intership visits during the winter. The *Lady Franklin* was about three hundred yards off the beach on which the forge and storehouse were erected; the *Sophia* was eighty yards away, but in which direction it is not clear; and Ross's *Felix* was closer inshore, evidently about halfway between the *Lady Franklin* and the beach. The three crews, amounting to more than sixty men, constituted a small community all engaged on the same mission and facing the same challenges and hardships. With the ships now connected by a highway of ice, the crews could make informal visits to friends on other ships, join together on walks, and participate in games.

On Sunday, 22 September, the crews assembled for divine service. There was much to be thankful for. The ships were safe in harbour, held securely in a stable sheet of ice, and the men were in good health. On the *Felix*, Sir John Ross called his men aft for prayers and treated them to another of his uplifting sermons. One can think of a few biblical passages that might be inspiring to seamen who were safe at last from prolonged perils at sea, and some messages of comfort that would allay their apprehensions about the approaching arctic winter – but to his men, most of whom were in their twenties and thirties and probably worried about the temptations facing their wives and sweethearts three thousand miles away, Ross delivered a sermon on the topic of infidelity!

Like cats in a new neighbourhood, officers and men fanned out from the ships in various directions to become acquainted with the adjacent terrain. From Prospect Hill, Penny, Stewart, and Sutherland saw several small lakes. The largest, almost a mile in length, Penny named Kate Austin Lake in honour of the wife of Horatio Austin. When they reached it and walked on its frozen surface, they could see small fish darting back and forth beneath the clear black ice, suggesting useful additions to shipboard meals. On 18 September, the lake served as an impromptu sports ground; the *Sophia*'s logbook recorded that the crews went "sleding [*sic*] and sliding on the pond." At the vessels the men were put to work dragging yards, booms, boats, and spare provisions ashore to provide an emergency depot and make more space on deck and below. Petersen exercised his dogs along the beach, and hunting parties ventured up the stream valleys in search of small game.

The south-facing anchorage was protected from blustery northerly winds. Several months later, Sir John Ross would describe Assistance Bay – somewhat tongue-in-cheek – as "one of the most delightful localities . . . the most fashionable resort of the most splendid yachts that were ever frozen up within the Arctic regions."[11]

Having discovered the location of Austin's ships, Penny intended to monitor their situation and establish contact as soon as possible. In the last two weeks of September, he dispatched a succession of small parties westward along the coast for seven to ten miles. Two excursions were led by Penny himself, using dogs. On each occasion the naval ships were visible and apparently safe, but open water prevented the men from crossing to them. The short trips not only enabled Penny to keep abreast of the state of the ice in the strait and the status of Austin's ships but also provided exercise for men and dogs. On 3 October, as Penny and Petersen were again sledging westward with their dog teams, they met a party from HMS *Assistance* engaged on a similar mission. Lieutenant George Mecham, the assistant-surgeon Charles Ede, and three seamen had just succeeded in crossing from Griffith Island to Cape Martyr on a new ice bridge and were now heading in the direction of Assistance Bay to see if Penny, and perhaps Ross, had succeeded in getting safely into winter quarters there. Penny now learned that De Haven and Griffin had left for home on 12 September, and concluded that the ships he had sighted through the fog patches must have been the *Advance* and *Rescue*. But whether they had managed to escape from Lancaster Sound before being beset no one knew and many doubted. Mecham intended to look for them in Barlow Inlet if his party could travel that far.

Mecham's party had earlier shot a bear near Cape Martyr, approximately fifteen miles west of Assistance Bay. As they were anxious to keep the skin, a handsome trophy for someone's drawing room (presumably Mecham's), Mecham made a proposal to Penny. If his men could remove the skin and leave it for his party to collect on their way back to Griffith Island, Penny could keep the meat and use it for dog food. They explained precisely where they had left the carcass.

Penny decided to escort the visitors to Assistance Bay. Having already walked for several hours, they were glad of some assistance. Penny seated the two officers with their haversacks and guns on the sledge, along with Petersen and himself, and the four dogs pulled the load at five miles per hour all the way to Assistance Bay. (The seamen had to walk.) "The travellers were right glad to see 'Penny,'" Cator reported later, "for had it not been for him, they would have had severe work of it. For it would have been impossible to have travelled against the wind with 'Ther' [i.e., thermometer] so low."[12] Mecham was impressed by the effectiveness of dog traction and quick to observe that while his own party had experienced great difficulty in hauling their flat toboggan-type sledge, Peterson's dogs easily pulled the Greenland sledge, its platform fixed on top of two vertical runners, Inuit style.

After a night on board the *Lady Franklin*, Mecham and his companions, joined by Captain Stewart and Dr Goodsir, set off eastward with the intention of reaching Barlow Inlet, to see if the American ships had taken up winter quarters there.

Shortly afterward, Petersen, Stuart, Marshall, and eight seamen started in the opposite direction with a dog sledge and a large man-hauled sledge, to retrieve what the *Lady Franklin*'s logbook referred to as the "ded Bair," the meat sorely needed to feed the growing canine population. The two excursions coincided with the onset of a cold snap, which provided all the travellers with a painful but useful foretaste of conditions they would encounter on later sledge excursions. Dr Sutherland's meteorological record shows that during the previous three days the temperature, measured every three hours, had been running between 10°F and 15°F, but on the day the two parties left the ships it was below zero most of the time and never rose above 3°F. Adding greatly to the discomfiture of the men was the wind, which backed from south to north-northeast and increased from force 1 ("light airs") to force 6 ("strong breeze"). Two days later, it rose to force 10 ("storm"). Fortunately both expeditions had returned by this time. After a ten-hour hike during which they suffered from intense thirst and mild frostbite, the eastern party returned to the ships. Reaching Barlow Inlet in such conditions had proved to be out of the question. They had turned back before covering half the distance.

When the western party had not returned by 8 p.m., Penny sent a party of five under Sutherland to meet them and help them back in. By this time the wind was howling, the snow was drifting, and the darkness was only fleetingly illuminated by dim starlight through gaps in scudding clouds. They groped their way like blind men, stumbling over hummocks and breaking through thin ice into pools of water. Fortunately, after covering only a short distance in the dark, they walked straight into the group of returning men. The doctor had brought along some spirits, and prompt application of this efficacious medicine quickened their pace on the final leg of their trek back to the ships, where crews were sending up flares and rockets to guide them in.

The party under Petersen and Stuart had found the dead bear. While they were skinning and butchering it, and stowing the meat on the large sledge, Stuart was "engaged in endeavouring to revive the hands of one of the men that had been very severely frost bitten."[13] The trip back to the ship was harrowing. He later related,

> We set out on our homeward journey but the distance having proved much greater than expected, it was resolved to leave the large sledge at Mr. Mecham's halting place and make our way to the ship as fast as possible with the dog sledge[,] night having by this time fallen and the worst part of the road being still to come. It was lucky that this was done for great difficulty was experienced in sledging among the hummocks we now entered and the violent falls & concussions sustained by the light sledge were such that we momentarily expected to see it break in pieces. More than one of the men had their knees & ankles sprained in consequence of falling through cracks in the ice covered over with snow and Smith the man who had already been

> frostbitten had his hands again frozen. We reached the ship about ½ past 9 at night, not however before being met and assisted by a party sent to meet us.[14]

Mecham and his companions remained stormbound on the ships at Assistance Bay for a few days. It was not time wasted but rather time to get acquainted, exchange news, compare views on Franklin's whereabouts, and discuss searching strategy and methods. When Mecham and Ede made disparaging remarks about the departure of the USS *Advance* and *Rescue* two weeks before, Penny sprang to the Americans' defence, declaring that instead of resorting to bickering and back-biting, the different expeditions should work together in harmony to achieve their common objective. It was not the first time he had stood up for the Americans or attempted to encourage cooperation among the expeditions.

On a still-stormy Sunday morning, the visitors participated in divine service and then joined officers from the three ships for dinner on the *Lady Franklin*. Mecham had the opportunity to meet Sir John Ross and deliver letters from Austin. It must have been with considerable awe, admiration, and deference that the twenty-two-year-old lieutenant sat down to dine with the veteran explorer, his senior by half a century. Ross was a legend in naval circles, having fought in the Napoleonic Wars and reopened the British search for a Northwest Passage before Mecham was born. While Ross had been spending four years in the central Arctic on the *Victory* expedition, Mecham was just learning to walk.

Socializing at Assistance Bay was a relaxing change for Mecham and Ede from the rigid routine on board the naval ships. Still, they were anxious to get back to Griffith Island, supposing (correctly) that their colleagues might be worried about their safety. They set off boldly on Monday, 7 October, during a lull in the storm, escorted partway by Stewart, Goodsir, and Sutherland. To their dismay, their bear skin had disappeared, apparently carried away in shifting ice. Separating from Stewart and the two surgeons, they headed across the ice of Resolute Passage to Austin's ships. On the following day Petersen and several men travelled along the coast and retrieved the abandoned sledge of bear meat and blubber, now nibbled by foxes and frozen solid.

Ten days later, Penny decided to repay the visit by Austin's men. He and Petersen set off along the coast with the dogs, then crossed the narrowest part of the strait on the ice. When they arrived at the naval ships, Penny went directly to HMS *Resolute* to pay his respects to Austin. "Nothing could have been better for all hands," Cator observed on the *Intrepid*, "for had he come to any of us the commodore would have been very jealous, more so as they did not part on the best terms in the world."[15] More than a month had passed since the misunderstanding about fall-back measures, but on this occasion two commanders kept their weapons sheathed. Cator wrote that they "hit it very well this time."[16] They discussed future prospects and agreed on new zones of responsibility for the search by sledge parties in spring, taking into account their winter positions. Austin's ships were best situated to search toward the west and southwest and had the manpower to cover a large area, while the smaller expeditions of Penny and Ross were better placed to search toward the north. Penny's spring travelling parties would therefore examine both sides of Wellington Channel, while Ross's would explore Cornwallis Island. The location of De Haven's American ships was still unknown.

Penny and Petersen accompanied Austin and his officers to a lively concert by the seamen of the *Intrepid*, after which they all sat down to dine. Next morning, Penny and Petersen rounded up their dogs (one had been caught in a live trap set out for foxes), harnessed them to the sledge, and set off. Arriving at Assistance Bay in the early afternoon, Penny handed Ross a letter from Austin describing the spring operations. In addition to outlining the spheres of responsibility discussed at Griffith Island, Austin specified that McClintock would lead a sledge party to Melville Island.

Melville Island had acted on McClintock like a magnet on a piece of iron. The failure of James Ross's expedition to find any evidence of Franklin's passage to the south of Barrow Strait may have suggested to him that Franklin had encountered an impassable barrier and elected to sail westward along Parry's route toward the Beaufort Sea instead of attempting to reach the continental mainland. A search toward Melville Island therefore appeared to stand a good chance of locating the missing expedition. In addition, it seemed likely that new lands could be discovered in that sector. As the organizer of the sledge-hauling system, McClintock was able to convince Austin that he should be the one to search in that direction. This generated some resentment among his fellow officers, who detected in his attitude a whiff of personal ambition. Cator recorded that after he suggested to Austin that two sledges rather than one should undertake the search towards Melville Island, McClintock was "very much annoyed, at my having told the Commodore so, and selfishness shone forth in its worst form, he wants to be the only one that will put his foot on the island."[17] In a meeting of Austin's travelling committee on 11 December, McClintock resisted the idea of having another officer's sledge party with him on the western search, but he was outvoted. A second party under Dr Bradford eventually accompanied him most of the way. Lieutenants Aldrich, Cator, and Osborn all felt that McClintock had behaved badly, and Elliott, when told about the meeting, remarked, "the 'cloven foot' will show out sooner or later."[18]

18

Preparations for Winter

Already had night commenced; next came winter.[1]

In mid-October, as the thickening snow cover on the sea ice of Assistance Bay became firmly compacted by wind, Penny's crews turned to the important work of insulating their domiciles against winter cold. The ships were warm bodies in a cold environment, constantly losing to the surrounding air and water some of the heat generated by their small heating stoves, galley fires, and human bodies. Fortunately, there was an efficient insulating material close at hand. For two weeks the men cut rectangular snow blocks, hauled them to the ships and, like artisans working with gargantuan bricks, laid them in staggered courses to the top of the gunwales, eight feet higher than the ice surface – "ashlar work of great nicety and beauty,"[2] Sutherland called it. The men filled any spaces with loose snow, and to provide access down to the ice, built gangways bordered with sheltering walls of snow, probably using slabs of pond ice for steps. A large volume of wind-packed snow was required for all this work, and as the wind tended to scour the snow cover from the ice, especially around the ships, they had to quarry and transport blocks from ever-greater distances as the work progressed. By early November they were making trips to the land to obtain snow of suitable hardness and thickness. Each day the walls mounted higher, and the ships gradually vanished behind the parapets of snow, assuming the appearance of ghostly white fortresses, their maritime function betrayed only by the stumpy lower masts reaching tentatively toward the darkening sky.

To retard heat loss through the upper deck, the crews laid down a layer of felt, covered it with planking, shovelled on a foot of snow, packed it down, and sprinkled beach sand on top for traction. With housing cloths already installed above the gunwales and overhead, the closed-in spaces above deck added to the insulation.

Despite the felt and wood added to the decks and banking erected around the hulls, the ships inevitably lost some heat to the surrounding atmosphere, and whenever men went through doorways, some warmed interior air escaped and was replaced by cold flowing in. Below the ice, the saline water remained relatively warm throughout the winter – a few degrees below the point at which fresh water freezes – but it still drew heat away from the ships' holds and their contents, some of which were liquid and therefore vulnerable.

On 10 October, the *Sophia*'s logbook recorded, "Our frozen water was all solid in the casks[.] Put them on the ice and will have to take the heads out and use it by melting it." Penny and his crews must have been dismayed by the enormous efforts required to maintain their essential supply of water for drinking, washing, and cooking. Fresh water had been brought on board in Scotland, stored below in wooden casks, and carried several thousand miles. Additional water had been obtained from icebergs along the coast of Greenland. After freeze-up in Assistance Bay, they had supplemented their supply by sawing blocks of pond ice, sledding them out to the ships, melting them in the galley coppers, and running the water through hoses into casks in the hold. But now it was freezing solid down below, so the casks had to be hoisted out and opened, the ice chipped out and carried back to the galley to be melted, and the water run below into the casks where, of course, it might freeze again. A miniature hydrological cycle was in operation on board the ships, and it was consuming precious fuel.

Other liquids stored in the holds were at risk too. When the *Sophia*'s hold was inspected in mid-October, some bottles of pickles were found to have burst. A few days later the ship's log recorded, "The ends of our water casks burst out to day. All our porter and strong ale is frosen solid." Sutherland reported that when the solid beer was melted with eager anticipation, it turned into "a tasteless fluid of a slightly brownish colour, the solid and alcoholic constituents having all escaped."[3] Lime juice, their essential scurvy preventative, also froze, as did vinegar. The mates, who had to make daily entries in the logbooks, were often frustrated by ink that had turned solid. Unlike the beer, however, the ink could be rehabilitated, provided the bottle had not shattered.

When outside air temperatures plunged, the living spaces below deck became uncomfortably cold. For six weeks before the banking was finished in November, the temperature inside Sutherland's cabin hovered near 22°F and never rose above the freezing point of water. It was uncanny to see an uncovered glass of frozen water lying on its side for the entire time without anything spilling out.

Cold air below deck tended to condense the men's exhalations, and their blankets and nightcaps often froze to the hull as they slept. When the temperature rose, their bunks and clothes became damp. Penny ordered the men to air their bedding twice a week, and in mid-November, they covered the bunks with felt and fastened sealskins to the inside of the hulls.

To accommodate a tide gauge and provide access to water in case of fire, a hole was cut through the ice astern of the *Sophia*. From the end of the mizzen boom, which extended horizontally from the stern, a line descended through the hole to a heavy weight on the sea bed. The line led inboard through blocks and ended at a smaller weight suspended above the upper deck. As the tide rose and fell through its four-foot range, the line remained taut, and the vertical movement of the ship – and hence the fluctuation in sea level – was transmitted to the inner end of the line and measured on a graduated scale. The mate, Donald Manson, recorded observations every hour, and every half-hour as times of low or high water neared. Keeping the outside hole clear of ice, he recorded in the logbook on 27 January, was "no small labor as a hand has almost to be constantly taking away the new formed ice, which in a short time has to be remouved again." Sometimes the water froze firmly around the line and caused it to part. The rapidity at which open water could freeze had impressed Parry many years before when he was seeking a Northwest Passage by way of Hudson Bay and Foxe Basin. He had seen from three to five inches of ice form in a cleared channel during a twenty-four hour period, and on one occasion eight inches of ice formed in twenty-six hours.

Victorian reticence restrained the writers of journals from describing sanitary arrangements, but the logbooks reveal that the carpenters built appropriate facilities on the ice in early October. They were usually referred to as "privies" but occasionally (and incorrectly) as "water closets." They were wooden structures, but one can only speculate about their size and appearance. A logbook entry on board the *Lady Franklin* on 11 October mentions the boatswain "lashing up the privy," which suggests that they may have been anchored by ropes to prevent them from being blown away in gales (an unpleasant experience for occupants). Another entry on the following day records the sailmaker "altering awning for privy," which indicates that they may have had canvas windbreaks near their entrances to impede drifting snow. The privies were situated reasonably close to the ships, and the paths leading to them were sprinkled with gravel or sand and lined with snow-block walls. Because it was so difficult to keep open holes from freezing over, they could not be mere latrines built over holes in the ice. They probably resembled the "ash closets" used in British cities, in which excrement fell into ash-filled receptacles that were later emptied. A later entry (23 January) reads, "Cleared out the water closet[.] Carried the refuse a considerable way from the ship." The privies were cleaned out every two weeks or so and periodically shifted to a new site. Galley garbage and other wastes were probably deposited on the ice and may have been removed to the land from time to time.

The sanitary facilities for the dogs were more rudimentary, at least from a human perspective. The eight dogs obtained near Upernavik in early June had increased to thirty by the end of September! They lived on the ice, inside wooden casks at first but later in snow houses built by the sailors, with straw beds to help keep them warm. They did their daily business close to home, and the result was not pretty. "As ther was so maney dogs going about the ship they left a great deal of foulness," Donald Manson recorded in the *Sophia*'s logbook at the end of September. When the situation became intolerable (for both dogs and men), they moved the dogs to a new area. "Shifted the dogs canel [i.e., kennel] a hundred yards from the ship," recorded the *Lady Franklin*'s mate on 8 November.

An able seamen named Alexander Thomson was relieved from other duties and appointed to help Petersen care for the dogs and their numerous pups. The work entailed feeding, cleaning, and helping to train young dogs to pull a sled. Despite his supervision, female dogs sometimes attacked the pups of others, and on one occasion a male dog killed and ate a pup, an incident that may have reflected the dogs' hunger, as fresh meat was scarce and the dogs regarded the pemmican made in England as something less than gourmet fare.

Sutherland remarked on the bond that developed between dogs and men: "Every morning, the moment they discovered that one of us had gone into the open air, they all came out of their kennel and crowded around us, looking for nothing but caresses, which were often lavishly bestowed."[4]

To combat monotony, Penny established a daily routine with plenty of activities. Mornings were the time for work, and afternoons the time for exercise or play, unless certain tasks required attention. Many chores had to be tackled on a regular basis. Slabs of pond ice were cut, sledded to the ships, and stacked alongside ready to be hoisted on board and melted into water when needed. Snow was swept off the housing cloths, cleared from the paths that led to privies and dog kennels, and shovelled away from the doors of the workshops ashore. The ships were cleaned inside and out. Bedding was aired and clothes washed. The dogs were fed and the privies cleaned out. The standing rigging was checked for tautness. From time to time, the holds were broken out for additional food supplies. The carpenter, blacksmith, and sailmaker were constantly busy at countless little tasks. To curb any human tendency toward laziness during the long, dark days, Penny ordered the following statement to be announced and written into the ships' logbooks on 8 December: "It is requested by me that all the crews of both ships or vessels shall be called at 7 AM breakfast at 8 AM and dinner at 12 noon and that no officer shall be in bed after 8 AM and no seaman in bed before 8 PM unless Dr Goodsir shall think otherwise."

Exercise was part of the recommended method of avoiding scurvy and essential physical conditioning for the man-hauling sledge trips that would take place in several months. Many of the chores they regularly carried out required strenuous physical work, but Penny also established compulsory exercise and sports periods in the afternoon. The men marched on deck to the beat of a drum, walked to the headlands that bracketed the bay, played football on a snow-covered "playground" on the ice, held informal foot-races, and sometimes went ice fishing on the ponds. Occasionally they were allowed to indulge in casual frolics such as sliding down hills.

The most popular sport was rounders (or bases), a game in which the ball was bowled in underarm fashion to a batter from the opposing team, who, after a fair hit with a short bat held in one hand, ran around the bases – either all at once or base by base as succeeding batters made hits. Completion of a circuit around the bases placed at each point of the diamond without being sent out by tagging or by a batted ball being caught by a fielder counted as one "rounder." The game was simple and required little in the way of equipment or facilities. Teams were sometimes picked on the spot according to the number of men available, but at other times the men sallied forth to do battle as representatives of rival ships. At no time did they take the game too seriously. If a runner encountered a patch of bare ice (giving new meaning to the expression "sliding into third"), if the ball disappeared into a snowdrift between first and second bases, or if a fielder lost sight of a batted ball against the dark winter sky, such events (serious calamities under normal circumstances) merely contributed to the good-natured fun. Rounders was played on at least sixteen days during the winter, although only five games took place during the two darkest months.

The players could not have imagined that within forty years the sport would become formalized for international play between Ireland, England, and Scotland, or that a Rounders World Cup involving half a dozen nations – Canada included – would be held in the twenty-first century. Nor could they have imagined that the related game of softball would later be played in scores of countries around the world, or that men would be able to become millionaires by playing another variant – baseball – in the United States.

At the end of August, the crews had been issued jackets, trousers, comforters, and sou'westers. In early October, each man received additional articles of clothing: one greatcoat, one waistcoat, one Guernsey frock, one blue flannel shirt, one comforter, one sealskin cap, one hooded cap, one welsh wig (a fur cap with earflaps), one pair of boot hose, one pair of carpet boots, one pair of sealskin mitts, two pair of white mitts, and one pair of drawers. Later in the month, after ice began forming in the men's bunks, each man was issued an extra blanket.

They were thus equipped to venture outside for chores or exercise, but whenever strong winds combined with low temperature, outdoors activities and travel were suspended. Blowing snow, common in the Arctic, can be hazardous. Sweeping across a treeless landscape and an ice-covered sea, an unobstructed wind can easily pick up tiny crystals of cold, newly fallen snow, whirl them into the air, and keep them in suspension. A wind of only ten miles per hour can make it difficult to see things clearly, and a wind of forty miles per hour can reduce visibility to a few yards. It is mainly a surface phenomenon. While a man on foot is struggling through blinding snow, a man thirty feet higher in the crow's nest may have unlimited visibility. Entries in the logbooks during November perfectly describe such a situation: "The drift was flying about very much but over head the sky was clear; (4 November); "Blowing dreadful with the snow so thick that the Sophia no part of her was to [be seen] below the top gallant masts for snow drift" (10 November).

Blowing snow made outside work inadvisable. The *Lady Franklin*'s log (12 November) recorded, "No duty preformed on the ice this day the snow drift being too much to stand." Travel was curtailed as well. Unable to see ships or landmarks, a man could easily become disoriented and wander helplessly in an erratic pattern. Searching for a lost man after the wind had erased his footprints would be almost as hopeless as looking for an missing explorer who had left no cairn messages.

For severe weather, precautions were necessary. There were beaten paths to follow from ship to ship and from ships to shore, partly marked with direction posts, snow blocks, or life lines, and partly surfaced with sand or gravel. In extreme conditions, they were not always enough. On 3 January 1851, the logbook of the *Sophia* recorded an incident under the heading "Sagacity of a Dog":

> On the night of the New Year One of Our Sailors had been calling or paying a Visit to some of his Ship Maits, and on his way to his own Ship laid himself down on the ice, One of the Dogs used all means but to no effect went on board Ship Commenced [pushing?] one of the Sailors unusually and Seemed as if he wanted the man to follow. Which the Sailor [latterly?] did. The Dog led him to his Ship Mait in time to Save him from the [illegible words] which Must have followed.

The log indicates that the hero of this incident was Penny's favourite dog, Sultan. This appears to have been the same incident referred to by Sutherland who mentioned a seaman from the *Felix* who "lay down upon the floe in a state of stupor during night, with the temperature about −30."[5]

Later in the winter, the occasionally tipsy polar veteran Abernethy veered drastically off course in blowing snow after a social evening on the *Lady Franklin*. On the next day Stuart wrote, "Last night poor Abernethy got adrift on his way from this ship to the Felix and from half past five to half past one was wandering about the bay or lying under the lee of hummocks having been unable to find his own ship. He luckily sustained no other injury than a few frostbites about the face."[6] A month later, during a severe gale that lasted for several days, Penny's blacksmith, carpenter, and cooper imprudently embarked on an excursion to the land three hundred yards away, and temporarily lost their way.

In bad weather the men took their exercise on the upper deck within the shelter of the "house," marching and performing drill to the commands of a "corporal" (presumably one of the officers wearing some sort of uniform), the cadence supplied by a fife and drum. The fife player did not perform outside the housing cloths, however, having learned by painful experience that his metal instrument had a tendency to affix itself to his lips and fingers in freezing temperatures.

Mental stimulation was considered as important as physical conditioning, and Penny followed the naval precedent by arranging classes, concerts, and plays. These activities were organized by the medical men, who were well-educated compared to the rest of the crew. Theatrical activities commenced in late October after the ships had been firmly frozen into the harbour ice. The carpenters constructed a makeshift theatre between decks on the *Lady Franklin*, and on the evening of 26 October the crews of all three ships crowded into the "Royal Cornwallis Theatre" to enjoy opening night. It was an evening of amusing songs and skits, and the reviews were favourable. Stuart declared that it "came off in a most satisfactory manner[,] the performers doing full justice to the stylish appearance of the spicy little stage."[7] The venerable Sir John Ross was observed vigorously tapping his toes to the music and making complimentary comments about the performers. Mistakes were made by neophyte actors, but in the spirit of the evening, they merely added to the mirth. "All went to the theatre to laugh and be laughed at," Sutherland remarked, "and those who 'broke down' afforded as much amusement as those who went smoothly and 'charmingly' through their performances."[8] There was a worthwhile bonus for spectators and actors alike: "Extra grog was given the men after the play and they continued to enjoy themselves till midnight in a most social manner."[9]

By the time the second theatrical evening was held a month later, the actors had lost some of their inhibitions and gained confidence, and several other men had overcome their shyness and come forward to make their stage debuts. Others were involved in making scenery, printing playbills, or assisting in various ways "backstage." By necessity, ladies' parts were played by men, but only men without beards needed to apply. Other signs of masculinity sometimes intruded, however. One heroine was quite effective despite muscular arms and calloused hands visible beneath her shawl, a bulge of chewing tobacco in one cheek, and a deep, unladylike voice.

The "Arctic Academy" opened a week later, and from then on classes were held on three or four nights during the week, for three-hour sessions. The three medical officers taught reading, writing, arithmetic, and geography. For some of the seamen, going back to mental work with slates and chalk (paper and pencils were in short supply) was difficult and potentially humiliating. "It was heartless work for the man of thirty-five, who had been married for fifteen years, to sit for hours together, poring over the simplest arithmetical calculations,"[10] Sutherland observed.

Geography was the most popular subject – possibly the only popular one. Many of the seamen had acquired firsthand knowledge of foreign places in their voyages across the world's oceans, and they were able to relate the abstraction of the beautiful maps in Johnston's *Physical Atlas of Natural Phenomena* to the practical reality of their own experience. Trade winds, doldrums, tropical storms, coral reefs, icebergs, the Gulf Stream, the Agulhas Current, and many other natural phenomena were part of their experience, and now they could see their geographical distribution on a global scale. Of particular interest were the maps relating to the region through which they had recently sailed. In their journey from Scotland, they had traversed roughly a quarter of the world's circumference at that latitude. For this northern part of the earth, Johnston's maps depicted geology, glaciers, ocean currents, tides, mean annual air temperatures, the southern limit of sea ice, the distribution of Greenland whales, narwhals, and several other species, the southern limit of polar bears, and the territory inhabited by muskoxen.

On the atlas's title page, Johnston was described as "Geographer at Edinburgh in Ordinary to Her Majesty" and an honorary member of the Geographical Society in Berlin. He recognized that people, like plants and animals, were part of the environment and that the geographical distribution of different human races could be shown cartographically. For convenience he grouped races into three main divisions: "the white and bearded," "the tawny and beardless," and "the black and wooly-haired."[11] Penny's men, who had met some of the tawny and beardless people at various places along the Greenland coast and seen scattered evidence of their former residence along the shores of Lancaster Sound and Wellington Channel, may have examined with interest plate 23, entitled "Ethnographic Map of the World," to see the impressive extent of the Eskimo domain.

In his introductory remarks on physical geography, Johnston included a quotation from Mary Somerville's well-known work *Physical Geography*, a passage that may have inspired the men: "Neither the long gloomy night of a polar winter, nor the dangers of the ice and storm, deter our gallant seamen from seeking a better acquaintance with this ball of earth; even under its most frowning aspect."[12] They had followed Franklin into one of the world's least-known regions, and they might be able to contribute in an important way to European knowledge of its physical geography.

The Admiralty had supplied almost two hundred books for Penny's ships, and Lady Franklin had sent out more for them with Austin's squadron. The Mariners' Chapel in Aberdeen had provided some religious books, and most of the officers had brought their own assortment of historical, technical, and fictional books. Among the non-fiction works in the ships' libraries were the narratives of arctic explorers such as Parry, Franklin, John Ross, and Wrangel and naturalists such as Darwin and Humboldt, and reference books such as *The Admiralty Manual of Scientific Enquiry*. Such was the enthusiasm with which the surgeons taught reading and writing to men deficient in those skills that the libraries, both official and personal, did a brisk business in lending. Some men were illiterate – how many is uncertain – and the reading level of most was undoubtedly low, but the medical men's keen work was creating an appetite for the written word. How strange that this awakening was occurring not in Scotland or England but in a remote and uninhabited part of the North American Arctic.

* * *

In Austin's squadron, twenty miles from Assistance Bay, classes were also held and theatrical productions mounted during the winter months. The "Royal Arctic Theatre" on board HMS *Assistance* was intended to distract the men temporarily from the difficult task at hand. Most of the plays were light farces, and the reviews were always favourable. As on Penny's ships, the officers supplied the creative impulse. It provided an opportunity for them to shed their epaulettes, let fly their imaginations, and indulge in a bit of fun.

At least half of the two dozen officers on the expedition acted in plays or participated in some other way, writing scripts, directing or managing, painting backdrops, constructing scenery, or making up playbills. Involvement in the winter theatrical season was so highly regarded by midshipman Clements Markham (himself one of the actors) that years later, when he compiled short biographies of naval officers who had served in the polar regions, he included details of their contributions to the "Royal Arctic Theatre" and the "Intrepid Saloon." Indeed, in some cases he devoted more space to their thespian accomplishments than to their battle experience, service on foreign stations, and other important details of their professional naval careers. For example, in his entry for John Brooman, the purser of HMS *Resolute*, he spent forty words enumerating his roles in the stage productions of 1850–51 ("acted Mr. Wiffles in '*Done on both sides*;' and King Artaxominous in '*Bombastes Furioso*'" and so on); the only other information he gave about Brooman's naval career was the terse statement, "He died suddenly at Hull, in September, 1858."[13] One would think that Brooman's only professional accomplishment during his entire life in the service had been capering in front of the footlights on an icebound ship in the Arctic. For Charles Ede, Ommanney's assistant surgeon on the *Assistance*, Markham allotted three times as much space to his theatrical contributions as to his 175-mile sledge trip in search of Franklin, a voyage fraught with hardship and danger.

Markham was an immensely talented individual. Only twenty years old at the time, he possessed a knowledge of history that none of his fellow officers could match, and he was not shy about displaying it. After a dinner with Austin on board the *Resolute*, Cator observed, "Young Markham, who also dined with him and who is a very well read fellow in history quite shut 'Tommy' [Horatio Thomas Austin] up – when he advanced any thing and young M did not agree."[14] Austin was seriously outgunned by his precocious subordinate, whose inquisitiveness about the world and its human occupants was extraordinary. "By the age of eleven he [Markham] had written histories of England and all the ancient civilizations of the world, and a year later his interests had expanded to include historical biography, astronomy, geography and natural history."[15] At fourteen, while serving as a naval cadet on HMS *Collingwood*, Markham had made a brief visit to Lima, Peru, and there he had become fascinated by the story of the Spanish conquest of the original inhabitants. The idea of writing a history of the Inca people took root in his young, fertile mind. Now, while wintering amid ice and snow at nearly 75° N, he used his spare time to read about Peru, improve his Spanish, and study the language of the Quechua people of the Andean region from a grammar he had bought in Lima half a dozen years before. It was not time wasted: after returning from the Arctic, he would resign from the navy and depart for Peru to explore the land of the Incas, stopping in Boston for ten days to discuss his project with William H. Prescott, whose great work on the history of Peru he had studied closely during long arctic nights.

Markham's adventurous travels in the Andes were cut short by the death of his father at home, but he returned several years later in an endeavour that would have far-reaching social, political, and economic impacts. While working as a civil servant in 1859, he proposed dispatching three expeditions to the Amazon regions of Peru, Bolivia, and Ecuador to obtain seeds and plants of various species of cinchona trees so that seedlings could be grown at Kew Gardens and established on plantations in India and Ceylon. Cinchona bark was the source of the drug quinine, the only effective preventative and cure for the dreaded disease of malaria, which was endemic throughout the humid tropics and killed millions of people every year. As the bark had to be cut from widely scattered wild trees in South American forests, it was scarce and expensive, and the crude harvesting methods were threatening to eradicate the tree. As leader of one of the three botanical expeditions, Markham succeeded in obtaining young plants and shipping them to London, but the seedlings all died later on their way to India. Fortunately, the efforts of other collectors were more successful, and in time the plantation growth of cinchona trees was securely established. Markham's initiative and vision had paid off; he has been called "the lightning rod and facilitator"[16] of the quest for quinine. Later he helped establish Brazilian rubber trees on plantations in southeast Asia, another biological introduction with enormous economic and social impact. Throughout his later career, he was a strong supporter of exploration in the polar regions. He served as president of the Royal Geographical Society for a dozen years and the

Hakluyt Society for twenty, wrote some fifty books, and was recognized as Britain's foremost geographer.

There is an interesting footnote to the British attempts to cultivate cinchona trees on tropical plantations. It was difficult to keep seeds viable and plants alive during the long ocean voyages under sail from South America to England, and subsequently from England to India. Success was eventually attained by transporting plants in the closed environments of Wardian cases. These terrariums, invented by the London doctor and horticulturalist Nathaniel Bagshaw Ward, had been used to carry live plants from New Zealand to England on HMS *Erebus*, one of the ships for which so many expeditions were now searching in the Arctic.

In June 1850, a newspaper had been started on board the *Assistance*. Called the *Aurora Borealis*, it presented short articles on diverse historical and geographical topics interspersed with bits of silliness. It was intended to amuse as well as instruct, and to contribute to a sense of unity and companionship among the crew. The newspaper once posed the question, "What insect that Noah had with him were these regions named after?" The answer: "The ark tick."[17]

During the winter Austin and his officers hosted several evenings of light entertainment and enjoyed a few masked balls in the *Resolute*'s "Royal Arctic Casino," and Cator played host more than once in the "Royal Intrepid Saloon." Sherard Osborn, who later wrote two books about the Arctic, remarked, "A theatre, a casino, and a saloon, two Arctic newspapers, one of them an illustrated one, evening-schools, and instructive lectures, gave no one an excuse for being idle."[18]

* * *

Soon after the freezing of the ships in Assistance Bay, Penny and Ross initiated efforts to get in touch with Franklin. For all they knew, the missing ships and crews might be somewhere relatively close, conceivably within the range of visual signals from aloft. On 2 October, Penny's ships set off two rockets and one portfire (a signalling device using a chemical compound similar to one incorporated in fuses). They repeated the exercise at least nine time that month and at least half a dozen times in November and December. Sometimes they saw rockets launched from Austin's ships twenty miles away.

In addition to indicating the presence of their ships by rockets and lights, Penny's men periodically launched the hydrogen-filled balloons from which messages printed on silk slips would drop off at intervals as their slow fuses burned away. The latitude and longitude of the sending vessel were indicated, so that any survivors picking up one of the slips could make their way to safety (if they possessed a reliable map, or a sextant, chronometer, and navigation tables, and knew how to use them).

Penny and his men tried other ways to get in touch with the missing explorers. Petersen live-trapped several foxes and equipped them with collars on which the positions of the searching ships and food depots were inscribed. The foxes were then released in the rather futile hope that they would travel to Franklin's men, who would catch them and learn that rescuers were nearby. This innovative method of communication had been used on James Ross's search expedition two years earlier, during which almost a hundred foxes had been trapped and released. The enterprise had a whiff of desperation about it, and care had to be taken that no dogs were loose at the time a collared fox was released. On one occasion Petersen released a message-bearing fox but then saw an unattached dog pursue and kill it.

In mid-October, someone sighted a fox wearing a collar! Could it be a reply from Franklin by return courier? Frantic excitement rippled through the ranks, and a posse of twenty to thirty men eagerly took up the chase. The terrified creature took refuge in the storehouse and, as Sutherland observed, "When it was dragged out, the collar, instead of bearing the names of the 'Erebus' and 'Terror,' as some of us were foolish enough to anticipate, bore the name of HMS 'Assistance,' written but a few days before."[19] After being released from a cage at Austin's ships near Griffith Island, the fox had trotted twenty miles over the ice, hoping perhaps to receive better treatment at the ships of Penny and Ross. The unfortunate animal must have been sadly disappointed.

At the ice edge off Griffith Island, Austin and his men were engaged in the same fruitless activity, turning trapped foxes into "postmen" by attaching message-collars and sending them off to find Franklin. But there was a serious problem; the foxes did not want to leave. They had to be forcibly driven away from the ships. "The departure of a postman was a scene of no small merriment," Osborn reported; "All hands, from the captain to the cook, were out to chase the fox, who, half frightened out of its wits, seemed to doubt which way to run; whilst loud shouts and roars of laughter, breaking the cold, frosty air, were heard from ship to ship, as the fox-hunters swelled in numbers from all sides."[20] Even after being driven off by the noisy two-legged monsters, the foxes tended to sneak back later to resume their scavenging around the ships, perhaps trying to ease some tasty morsel out of another trap, usually with the same unfortunate result as before.

Adding to the difficulty of getting them to search for Franklin was the desire of more than a few officers to obtain a luxurious white fox fur for wives or sweethearts at home. This put them in a clear conflict of interest with the messenger system. To prevent the foxes from becoming trim on elegant feminine garments, Austin declared it an offence to kill any fox found in a trap. Osborn wryly observed, "Of course no fox was after this taken alive; they were all unaccountably dead."[21] Strangely enough, Austin did permit the shooting of free-roaming, un-collared foxes. The official game tally for the *Assistance* and *Intrepid* up to the end of December 1850 was fifteen foxes killed at winter harbour, of which nine had been taken from traps and put to death. Cator was anxious to obtain for his young wife the makings for a white fox fur "muff and cuffs."[22] He succeeded in shooting three, killed one taken out of a trap, and gave another trapped fox to Ward, the ship's doctor. He tried to kill it by injecting a dram of prussic acid but the dosage was too little; the poor fox struggled painfully, sat up, and barked pitifully. The doctor gave it a second dose that mercifully proved fatal.

Some of the officers also wanted to keep the trapped animals as pets – not pets in the true sense, but more like zoo animals, caged or tethered at all times. Even the commander of the expedition had one. When Austin's fox gnawed through its tether and escaped, its hours of freedom were brief. Cator sighted it while rambling over the ice with his gun, and shot it to add to his stockpile of muffs and cuffs. The incident did nothing to improve his reputation in Austin's eyes.

The *Assistance* also had a pet fox. In late December it "broke his chain & seemed so delighted with his liberty that he continued moving along the top of the housing & imprudently approaching the men who were trying to catch him, but the silly little creature was outwitted, a trap was brought & baited with a piece of roasted bear"; the fox found itself back in solitary confinement.[23] Some of the trapped foxes made the mistake of licking the bars of their cages, and were later found with bloodied mouths and sections of their tongues hanging grotesquely from the cold metal. They had to be shot.

With some officers wanting to kill foxes for their fur and others wanting to keep them as pets, the number released with message collars was undoubtedly small, and some liberated ones remained in the vicinity of the ship instead of racing off to Franklin. Osborn wondered whether all their efforts were carried out more "from amusement than from any idea of being useful."[24] The only fox known to have travelled any distance was the unfortunate animal that caused such excitement by turning up at Assistance Bay.

* * *

On board the *Felix*, the Greenland interpreter Adam Beck was an unhappy man. A few months earlier, desirous of returning home, he had concocted a fantastic story at Cape York to mislead the searching expeditions, but his tale had been proven false. Now he was five hundred miles further west, crowded into a small ship with white men who did not speak his language and made fun of him, and he was committed to a winter in the ice in an unfamiliar and uninhabited region. A sense of alienation and isolation had prompted him to leave the ship unobserved at Beechey Island, causing Ross to delay his departure and send out search parties.

At nine a.m. on 5 November, he took his gun and strolled away from the ship. At the end of the day, he had not returned. The wind increased in strength, and all night strong gusts blasted the snow across the ice and swirled it into the lee of the ships. At daybreak two search parties left the *Felix* on foot, one under Phillips and the other under Abernethy, while Petersen headed out from the *Lady Franklin* with a dog sledge. After following Beck's footsteps westward for awhile, the searchers lost his trail in the drift.

They supposed that Beck had felt an urge to visit Austin's ships near Griffith Island, and so he had; but before reaching his destination, he had been caught in the storm and built himself a snow house on the ice, taking shelter inside. From there he could hear loud noises and, on looking out, saw occasional flashes of light and ascending streaks of rockets from the supposed location of Austin's ships.

If he had managed to reach Austin's squadron, he would have witnessed an extraordinary spectacle. From HMS *Intrepid*, a peculiar parade was making its way solemnly across the snow towards the *Resolute*. Eight men dressed as horses, beautifully beribboned and prancing prettily, were hauling a sleigh bearing an effigy of Guy Fawkes, the Catholic revolutionary who had achieved everlasting notoriety 250 years before for attempting to blow up the Parliament buildings in London. An escort of four constables and several great-coated soldiers, armed with muskets and ice axes, marched alongside to prevent his escape. Leading the way was a minimalist band consisting of "a gong, drum, tambourine, and a triangle,"[25] whose military percussion pierced the arctic gloom. Some men carried blue lights; others bore rockets to set off. On the *Assistance*, McClintock found the spectacle eerie: "The effect was very good," he wrote in his journal, "the pale gastly [*sic*] flame of the blue light upon the strange faces & surrounding snow prepared one to receive a shock when the rockets suddenly went off with rushing noise & bright light."[26]

On arriving at the flagship and confronting the commodore, the men sang a few songs and requested some measure of appreciation for their efforts. Austin capitulated without a struggle and promised to send over an appropriate reward later. He chose to supply grog, which the men considered appropriate for all situations, and was given three loud cheers, after which the procession returned to the steamers. Meanwhile, a rival group of celebrants from the *Resolute*, badly outclassed in matters of dress and deportment, were skulking shamefully behind the ship.

In the evening, the *Intrepid* party set out again, toured the fleet, and succeeded in cornering Austin again, this time on the *Pioneer*, where he was now dining with Osborn and other officers. Mounting to the ship's deck, the visitors sang more songs, then retreated onto the ice and proceeded to burn the prisoner Fawkes "alive." The barren, uninhabited, snowy spaces of Barrow Strait witnessed something close to the dreadful public executions of olden days in England, as Guy perished in the flames and spectators cheered. "Guy Fawkes arms were made so as to move by pulling a wire & when he became enveloped in the flames were seen to move about as if in an agony of suffering," wrote McClintock. "Altogether it was a very vivid representation, & one not easily forgotten."[27] The revellers sang more songs, gave more cheers, ignited more rockets, and retired content to their beds.[28] "Never before had such a scene been seen on the ice or in these desolate regions," Cator remarked. On the *Assistance*, Ommanney's protégé Erasmus York was an amused but doubtless puzzled spectator.

York's fellow Greenlander, Adam Beck, making his way over from Assistance Bay on foot, did not arrive until the following day, by which time the show was over. Beck gave several excuses for his absence, none which impressed Cator: "He told us three or four different yarns. The first was that they saw our Rockets and Bon fires, and that he had driven 'Sir John' over, the next was that he had gone out shooting deer and had lost his way, and seeing our lights had come to us, a third was that he had lost his way and spent two nights on the '*Ice*.'"[29] Cator sent him with two men to the *Resolute*, where there was no sign of Sir John Ross, so at least one of his stories was proven false. Cator told Osborn that he thought Beck was "a great rogue."[30] How the officers at Griffith Island managed to understand Beck's statements is unclear. Had he learned enough English by this time to make his meaning clear? Did they use Washington's Esquimaux vocabulary as before?

Beck remained at the naval ships for several days, the purpose of his visit and his intentions a mystery to everyone. Austin wondered if they should escort him back to Assistance Bay, and Osborn and Cator (both wanting to visit Penny and Ross) volunteered to do so. In the end Austin decided to let Beck do as he wished, and after a week he made his own way back to the *Felix*. Whatever explanation he gave for his absence, it was unconvincing. Stuart remarked that his story was "to be received with caution[,] his previous conduct leading us to place little trust in his veracity."[31]

Beck had become a liability, a cause for concern. On two occasions, his unannounced departures from the *Felix* had made it necessary to send out search parties. The first of these absences without leave had delayed the departure of the *Felix* from Union Harbour by a day. He does not appear to have hunted successfully or provided any other useful services to the expedition. Yet his mentor Ross continued to stand by him and steadfastly supported his disproved fable of a Franklin massacre.

* * *

Beck's disappearance was not the only problem in Ross's crew; the dreaded scurvy had made its appearance. Sir Richard Hawkins had called scurvy "the plague of the sea,"[32] and two and a half centuries later, its cause was still unknown. The first case on board the *Felix* was recorded on the last day of October, a few days before Beck absconded, and by the time he returned a second man was showing signs of the disease. Dr Sutherland remarked, "The appearance of scurvy, so early as four months and a half after leaving home, certainly implied that the provisions which had been in use were exceedingly faulty."[33]

Scurvy was common among seamen in all the world's oceans. It was a painful degenerative disease characterized by painful swollen legs and joints, tender gums, loose teeth, extreme sensitivity to touch over the entire body, purple spots on the skin, and extreme physical weakness. It disabled healthy sailors, eroding their capacity to fight the enemy or carry out the physical work essential on a sailing ship; unchecked, it led to death. Although commonly associated with long sea voyages, it was also a common affliction ashore in slums, prisons, military bases, construction camps, and even hospitals – wherever the diet was deficient in vitamin C. It had flourished during the potato famine in Ireland and the gold rush in California, both of which had occurred only a few years before Penny's expedition. It was feared on arctic voyages because most of the proven preventatives and remedies were unavailable. Citrus fruits and fresh vegetables were known to cure the disease (although nobody knew why), but they did not grow in the Arctic. Scurvy grass and mountain sorrel were occasionally used by explorers, but they didn't occur on the tundra in meaningful quantities. Along the coast of Greenland, Sutherland learned that the inhabitants collected a plant they called "bukoblather" and made it into an infusion against scurvy. Crustose rock lichens called *tripe de roche* were occasionally used by voyageurs and explorers in the Barren Grounds, but only when starvation stared them in the face, as it had on Franklin's first overland expedition in 1820.

Although the anti-scorbutic properties of fresh meat were not widely recognized in the scientific community, British whalemen had observed that the Greenlanders lived almost exclusively on a meat diet and seemed free of the disease. They also preserved the skin (*maktak*) of white whales and narwhals to prevent scurvy during the winter. Penny and his men had sometimes eaten fried narwhal skin with their tea, and on one occasion had been able to offer it to Greenlanders. It was more beneficial if eaten raw, but this was usually abhorrent to Europeans.

Imitating the native practice, the whalemen too consumed meat, fish, or fowl whenever they could get it. Through repeated voyages, they had achieved some familiarity with the habits of arctic animals, and the same harpoons they used to capture whales could be used to fasten to smaller sea mammals or retrieve ones they had shot before they sank or drifted out of reach. As their one-season voyages were relatively short, scurvy was rarely a problem, but when ships were occasionally trapped during freeze-up and forced to endure the winter months in the pack ice, scurvy always took a dreadful toll.

Few of the naval officers who took part in arctic expeditions were trained or equipped to hunt arctic animals on land or at sea. Some of the gentlemen officers, accustomed to recreational hunting in Britain, took their own fowling pieces or muskets on the voyages, but the methods to which they were accustomed were not appropriate in arctic conditions. There were no faithful gillies to arise from the tundra and lead the "sportsmen" to a cornered caribou, no packs of dogs to run a fox, wolf, or hare nearly to death so that "hunters" on horseback could gallop up for its execution. From their ships, they often shot polar bears, beluga, narwhal, and seals in the water, but mainly for amusement (which they dignified by the name of "sport"). Most sank or were swept under the ice by currents. During wintering voyages, the treeless arctic landscape provided little cover for hunters and therefore put a premium on stealth and knowledge of the terrain, but European explorers in unfamiliar regions were likely to be deficient in both. They did have considerable firepower, however – certainly enough to kill muskoxen and caribou if they could get close enough.

Birds were a different matter. Standing in a row at Lord So-and-So's estate to blast away at game flushed out at their convenience was no longer an option, but shooting from a boat at the base of the Arctic's many colonies of cliff-nesting sea birds provided huge tallies of feathered flesh. Sometimes a single shotgun blast killed dozens.

In contrast to Ross's crew, half of whom came down with scurvy, Penny's men were unaffected owing to the fresh vegetables and meat carried on board the ships and the "country food" secured in the Arctic. More than three years earlier, Sutherland had suggested that scurvy was caused by a lack of antiscorbutic foods, potatoes in particular. In 1849 Penny had kept potatoes on board the *Advice* for months, packed in compost near the stove and watered daily. By extending this same technique, he now made potatoes part of the shipboard diet through most of the voyage. Potatoes planted in Scotland in the spring of 1849, harvested in the autumn, stored there over the winter, and brought on board the ships before departure in 1850 had served them well over the summer months; they would last until mid-March 1851, by which time the last of them would be almost two years old. A popular dish with Penny's men was sliced raw potatoes topped with vinegar and pepper, a concoction used a year or two earlier in California to combat scurvy among gold miners. Raw carrots were often on the menu too. Beef, kept more or less frozen since entering the Arctic by hanging in the rigging, was reserved for special occasions.

In addition, Penny's men had been making efforts to obtain fresh animal food as they went along, harpooning seals and narwhals, catching fish, shooting birds, collecting eggs, and trading with natives on the Greenland coast for caribou meat. After taking up winter quarters in Assistance Bay, *Lady Franklin*'s crew had enjoyed considerable success at the ponds. Men could often be seen scattered over the ice, squatting at their holes, periodically breaking away the skim of newly formed ice as they jigged patiently with lines and hooks. In the last nine days of October, they caught more than two hundred arctic char. While most of these were small, less than a foot in length, flavoursome salmonid fish nonetheless constituted an important addition to the food supply and helped prevent scurvy. When Ross, Phillips, and Porteous came over to the *Lady Franklin* on Trafalgar Day, Penny's cook served up a hearty dinner of fresh arctic char for the guests. The *Felix* crew, on the other hand, appear to have been comparatively unsuccessful at fishing. Ross recorded in his diary that they caught two dozen fish in late October but mentioned no others.

With the long winter still ahead, the early appearance of scurvy on the *Felix* was alarming. Penny was confronted with a dilemma. Should he share his antiscorbutics with Ross at the risk of jeopardizing the health of his own men and undermining their ability to haul sledges in the crucial searching operations scheduled for the spring? Should he attempt to nurse back to health the men of an expedition that had so far proved ineffective and showed no signs of becoming otherwise? He and his surgeons decided that Ross's immediate needs could not be ignored, and several days later, Ross wrote in his diary, "During last week Mr Penny made me a daily visit and sent me some potatoes and carrots for the 2 who are improving."[34] Two days later, Penny delivered more carrots, and thereafter he and Stewart visited often, usually bringing raw vegetables, lime juice, sugar, or beer. Dr Porteous of the *Felix* was doing his best to cure his patients, but as he lacked effective antiscorbutics, his efforts were of limited use. "The eagerness with which our preserved and fresh vegetables were sought after," Sutherland observed, "was a proof that they excelled in nutritious properties the preserved potatoes, so abundantly in use on board the 'Felix.'"[35] By the end of December, Ross's crew appeared to be scurvy-free.

19

The Festive Season

Monotony was our enemy, and to kill time our endeavour.[1]

As the ice grew steadily thicker around the three little ships in Assistance Bay, the sun made its appearance a little later each morning, reached a little less high in the sky each noon, and sank below the horizon a little earlier each evening. Although the change from one day to the next was imperceptible, the men were aware that the balance between night and day was slowly tilting from the endless sunlight of mid-summer days to the continuous darkness of mid-winter. Days without sun would make all outside activities much more difficult and hazardous, and there was a psychological consideration as well. The sun was the giver of life, the supplier of warmth, the provider of hope after the gloom of night, and it was *meant to be visible* during the day. At Aberdeen and Peterhead, it never failed to rise. It might be temporarily hidden above clouds, or low in the sky in mid-winter, but it was *always there*. In Barrow Strait, however, some twelve hundred miles closer to the Pole than those Scottish cities, the cheering golden disc took a three-month sabbatical every year. No one on Penny's ships had experienced such extended darkness before. As the sun's celestial arc diminished day by day, a feeling of apprehension – even foreboding – spread among the men.

Then came the day on which the sun never rose completely, the day on which only its upper edge could be seen above the horizon, and only for a short time around midday. Now a glimpse of a mere crescent became a precious moment, after which the men on deck or out on the ice silently watched the disc sink beneath the horizon. When it was no longer visible from the deck, they climbed aloft on the ratlines for a longer look. Early in November it vanished altogether.

The disappearance of the sun was not immediately followed by total darkness, but by a twilight glow produced by rays refracted from below the horizon, a partial illumination whose intensity diminished day by day. The men's eyes adjusted to the lengthening semi-darkness outside their wooden walls and adapted to the low light provided by candles and lanterns inside. Their entire world now appeared somber and cheerless, but it could be brightened by activities. On the very day that the sun dropped below the horizon, teams from the two ships played at rounders, doing their best to follow the flight of fly balls between bat and mittened hand and to track the erratic bounces of balls over the hard-packed, bumpy snow (rounder grounders?).

About the time of the winter solstice, a few men went out onto the ice to see if they could discern written words. John Marshall, the *Lady Franklin*'s mate, recorded in the ship's logbook on 21 December, "This being the shortest day could read small print with tolerable ease at 30 minutes past noon." Two days later, on the *Sophia*, Donald Manson wrote, "At noon, made an atemp at reading a book of middle sised print on the flow, but it [was] rather difficult and that with good young eys." (His boast of good young eyes may have been a humorous touch; Manson had been sailing on arctic whaleships for about forty years.)

Was 22 December really the "shortest day" in Barrow Strait? The calendar day, like all the others in the year, was twenty-four hours long, and the length of time in which the sun blessed the earth and its creatures in those latitudes had been zero for over a month, so in neither sense was it the shortest day. It was, however, the day on which the sun at noon stood directly over the Tropic of Capricorn, twenty-three degrees south of the equator, its illuminatory effect in the northern hemisphere at its annual minimum.

Air temperatures would probably not reach their lowest level for another month or two. The men were not greatly worried about extreme cold because it could be kept at bay by warm clothing and stoves, but they could not insulate themselves against darkness, so it was a relief to have the solar regime on the upswing after the solstice. From that moment on, all the Franklin searchers in the region – in Assistance Bay, off Griffith Island, and wherever De Haven's crews happened to be – could look forward to a slowly increasing level of illumination, culminating in the reappearance of the sun in a month and a half. And of course they could look forward to celebrating Christmas and welcoming the New Year, despite the darkness.

Penny and his officers intended to use of the festive season as an antidote to winter depression. At noon on the day before Christmas, the crews of all three ships sat down below deck on the *Lady Franklin* for a fancy dinner. Toasts were proposed to Queen Victoria, King Frederick VII of Denmark (much appreciated by Petersen), Franklin and his men,

Lady Franklin, the commanders of the other search expeditions, and the crews at Griffith Island. "Soon after," Manson recorded in the *Sophia*'s logbook, "dancing commenced and was Keept up to a leat hour when then it became Visable that the sailors hade been drinking Some thing Stronger than Water." But the men (with a few exceptions) behaved well, and the evening passed in a spirit of harmony and good will.

Penny declared Christmas a holiday from work. Men strolled about on the ice, made forays to the land, and visited friends on other ships. On board the *Lady Franklin*, hectic preparations for a gala evening were under way. Carpenters were at work on a stage, backdrop, and seats; performers were rehearsing; Goodsir and Stuart were rushing about, correcting the timing of acts, fussing over costumes, and attending to a multitude of last-minute details. The play, eagerly awaited, was presented on 28 December. Manson walked across the ice from the *Sophia* at 7 p.m. to attend, and later employed his unorthodox spelling to record his impressions in the logbook:

> On my entry I found the Seats almost filled up but on Threeding My way to what May be called the pit then I beheld Sir John Ross Suported on his left by Captn Penny and on his right in the senter frunt Seats wer Commander Phillips Mr Stewuart of Sophia in one of the Frunt Sides Seats and all the other officers in Succession. The play consisted of S^{t}. Patricks Day and concluded with a number of songs. Great Credet is due to the Surgen & Asistant Surgen who took the management & acted a part in the Performance.

Invitations had been sent out for a grand masquerade ball on the eve of the New Year. At the appropriate time, the partygoers, suitably masked and garbed, assembled for merry-making, singing, and dancing. Regrettably, the costumes worn on this occasion are not a matter of record, but one assumes that there were a few "ladies" at the ball to provide dancing partners.

The year 1851 began with a strong reminder that although the sunless twilight was getting brighter, winter was far from over. At noon on 1 January, the outside temperature was −35°F, with a moderate breeze from north-northwest and occasional snow squalls. These conditions did not deter Sir John Ross and his officers from walking across the harbour ice from the *Felix* (a distance of 144 yards) to join Penny, Stewart, and others at a special midday dinner on board the *Lady Franklin*. The menu featured roast beef and potatoes from Scotland.

* * *

Twenty miles to the west, Austin's four ships huddled at the floe edge near Griffith Island. On the day before Christmas, the commander ordered an issue to each man of a pound of beef, a pound of rice, a half-pound of flour, a half-pound of treacle, a quarter-pound of currants, a "preparation of chillies," and a half-gill of brandy. McClintock declared, "I never knew of any man who has less idea of stinting his people than Captain Austin; this gratuitous supply in addition to the daily allowance is sufficient to feast the men for a week."[2] Later, officers visited back and forth between the ships. Cator enjoyed a dinner of mutton and plum pudding on board Osborn's *Pioneer*, writing in his diary later that night, "I think I never enjoyed a better dinner, since I have been out here."[3] Afterwards, officers retired to the *Resolute* for a magic lantern show put on by John Krabbé, the ship's second master, then joined in a singsong in the gunroom. Towards midnight, after more than one brandy and water, they drifted away to their respective ships. Returning to the *Intrepid*, Cator found his second master Shellabear and his first engineer Ryder "both as drunk as fiddlers." Moments later, Ommanney and some others from the *Assistance* came on board, "all as drunk as they well could be," and they all talked with unusual animation till 2 a.m., after which the visitors headed back to their ship. From the deck of the *Intrepid*, Cator watched them "making tacks and stern boards and heaving to, tumbling about the snow hummocks" until they at last reached their home port around 3 a.m.[4]

Divisions were held on Christmas morning, followed by divine service. Casual intership visits occurred in the afternoon, to exchange good wishes. The subtle glow cast over the spectacular scenery by a sun still well below the horizon was a source of wonderment and pleasure to the strollers. "The day was really as fine as one could have wished," Cator wrote; "to the South was one of the most lovely skies imaginable, such as only these regions can produce. It consisted of a pale yellow, blended together with a light red and bluish grey. The 'moon,' which was shining beautifully, threw her pale light over the whole."[5] In the evening the four ships held festive dinners. The lower deck of the *Resolute* was tastefully decorated; the cakes and puddings bore inspirational slogans as well as small flags of Britain and the United States accompanied by the motto '"the heart that can feel for another."[6] Sadly, the Americans were not present to appreciate the friendly gesture. Where they were, nobody knew.

On the *Intrepid*, the second master and first engineer were in their cups again and somewhat quarrelsome, but the general mood was pleasant. Cator went quietly off to bed at ten, "thinking of her I love and hoping that next Xmas would be spent with her."[7] After divisions on the next day, the men were sent out onto the ice to disperse the lingering fumes of Christmas cheer, while officers indulged in more intership visiting and dining.

On New Year's Eve, the crew of the *Intrepid* serenaded their officers, but high winds cut short plans to extend the tour to the other ships. Cator, who had taken such pride in being the first European to reach a high latitude in Wellington Channel, and the first man ever to write a message on top of Cape Hotham, now reached eagerly for another record. As midnight struck, he recorded in his diary, "I wrote January 1, 1851, and I believe I was the first to write 1851 in the whole squadron."[8]

New Year's Day started slowly. Cator, who had not retired until 3 a.m., slept in till 11:30, confessing that he was "somewhat heavy from sundry glasses of '*brandy and water*'" and did not risk consuming as much as a bowl of soup until afternoon.[9] Later he walked over to HMS *Assistance* for a grand dinner put on by Ommanney, and quite unexpectedly was able to play a part in shattering another record. At Ommanney's dinner there were nineteen officers – "the largest number of officers that ever sat down to dinner together in these regions,"[10] he wrote in his diary.

Two weeks later, Cator reached for yet another record. On a very cold day after low temperatures had turned the mercury in thermometers from a liquid to a solid state, he filled a bullet mould with mercury and left it on deck to freeze. Half an hour later, he loaded his pistol with a quicksilver bullet, erected a wooden barrel stave an inch thick, and blasted a hole through it. He then cut out part of the wood to keep as a souvenir of the event, recording in his journal that he had been "the first to think of doing it and putting it in practice."[11]

* * *

While more than 240 men at Assistance Bay and Griffith Island were passing the winter solstice and celebrating their rituals of Christmas and New Year, they wondered what had happened to De Haven's ships and crews. Those who concluded that the Americans had reached their home port of New York and were now spending the holiday with their families or friends were envious. Others, supposing that they had been caught in drifting pack ice for the winter, were apprehensive.

On the day of the solstice, De Haven and his officers had been within sight of Cape Warrander, near the threshold of Baffin Bay. Since drifting out of Wellington Channel, they had already had their full share of crises. The *Advance*, the larger of the two ships, on board which everyone was now living, had been in serious jeopardy on several occasions. Severe pressure in early December had raised the ship's stern five feet higher than the bow, a position it would retain during the next six months. More than once the men had been ready to abandon ship and make a desperate attempt to reach land hauling sledges loaded with tents and provisions. But what would they have done on that barren coast in midwinter with only a few weeks of food on hand? Reaching the land would merely mean that they had avoided being lost with their ship, only to perish more slowly ashore from starvation and exposure. And even if they could somehow manage to survive for a time, how could they escape from the Arctic without a ship? The dismal outlook depressed the men's spirits, and the absolute hopelessness of the situation may have contributed to a few cases of antisocial behaviour that occurred before Christmas.

Although the American ships had been bought for the search expedition by Henry Grinnell, their crews consisted of naval men who were governed by its rules and code of discipline. Every Sunday morning on board the USS *Advance*, divine service was accompanied by a reading of the Articles of War to remind seamen of the regulations of the naval service and the scale of punishments laid down for specific infractions. For several offences, an American commander was allowed to order up to twelve lashes with a cat-of-nine-tails, the iniquitous weapon of nine braided and knotted cords fastened to a stiff handle. Only a *plain* cat-of-nine-tails was permitted, however, not one made more hurtful by the addition of wire or other material. After hearing the articles read out once a week with the sombre warnings of lashes, courts martial, and death by execution, a man who committed any of the actions deemed to be unacceptable could not plead ignorance of the law; he simply had to take his punishment.

The law was no deterrent for Edward Wilson, a seaman on the *Advance*. On 27 November, De Haven wrote in his diary, "I was under the painful necessity of punishing Ed. Wilson today for insubordination."[12] He ordered the man to receive a dozen lashes. In fact, Wilson got thirteen strokes because the boatswain's mate somehow failed to stop at twelve. (Was he absent-minded, or did he enjoy the work too much?) Despite the extra stroke, De Haven felt that the boatswain's blows had been too gentle, and he briefly considered punishing him as well. Perhaps he had been inspired by the example of Charles Wilkes, his former commander on the US Exploring Expedition of 1838–42, who is said to have ordered double the legal number of lashes on "no less than twenty-five instances" during the four-year voyage, and who once had a man flogged round the fleet.[13]

To many, the situation of the ships looked hopeless. To deaden their panic and pain, a few of De Haven's men set their sights on the unguarded spirits stored on board the now-unmanned *Rescue* encased in ice nearby. Two weeks after his flogging, Wilson and two other men went unnoticed to the abandoned vessel, broke into the hold, knocked the head out of a rum cask, and proceeded to get what De Haven described as "outrageously drunk and riotous."[14] Wilson was given another twelve lashes, the others nine. De Haven feared that it was not the last occasion on which he would have to punish Wilson and one of his companions, both of whom had established a solid reputation for insolence. He was right; on the next day he had to book the incorrigible pair for "mutinous and insubordinate conduct" – a dozen additional lashes each.[15] In little more than two weeks, Wilson had received thirty-six strokes of the cat. These events do not appear in the pages of Kane's best-selling book about the Grinnell expedition. He may have thought it would serve no useful purpose to wash the navy's dirty laundry in public.

The floggings did nothing to improve morale, but the arrival of the winter solstice, and close on its heels Christmas and the New Year, lightened the gloom. On 22 December, spirits rose with the knowledge that from this day forward the days would grow steadily brighter and the ships' drift path would soon be curving south towards the returning sun and open water of the Atlantic. Kane wrote with exhilaration, "The solstice! – the midnight of the year!" He was less successful at the reading test than the British at Assistance Bay and Griffith Island, however, probably owing to an overcast sky: "We could not read print, not even large newspaper type, at noonday."[16]

The next few days continued depressingly dark; on 23 December there was "perfect darkness!" and on the next day "utter darkness." On the following day, however, the joyousness of Christmas brightened the scene. De Haven summarized the occasion in his diary:

> This being Christmas and all the hands inclined to be cheerful I afforded them every facility during the few hours of light that we had. Both the crew and officers engaged in foot races. In the evening a glass of punch was served to each man in the ship[.] This together with a dinner made up of all the good things that were at my disposal put them all in the best of good spirits. They asked for permission to toast and cheer me and the officers generally, which was granted, and a hearty cheer they gave us. At 7 PM the colours of both vessels were hoisted and a number of rockets and blue lights were burned which displayed several transparencies of a national

> and a patriotic character. At 8 P.M our theatre opened with all the embellishments in our power in the way of scenery, which to be sure were very limited.[17]

Was there ever a theatre to rival in primitiveness the one on board the USS *Advance*? On the upper deck beneath a felt housing cloth at temperatures that rose no higher than –4°F, with ensigns draped over boxes and casks to demarcate what passed for a stage, thirty-three men assembled for an evening of entertainment. The actors presented a comedy entitled "The Blue Devils." Highly popular with the audience was the heroine Annette, played with a heavy brogue by a six-foot Irishman named Daly, whose sole qualification for the role appears to have been an absence of beard. The dainty Annette tripped about the stage "with mastodonic grace," wrote Kane, who confessed to be no lover of plays but nonetheless enjoyed the "tawdry quackery" of the amateurish farce, and declared that the performance had been "one continual frolic from beginning to end."[18] At the conclusion of the play, the audience and actors joined in the singing of the "Star Spangled Banner," and the French cook Henri de Roque was allowed to have a go at the "Marseillaise." A sailor named Bruce danced a hornpipe to the accompaniment of the "orchestra" (one Jew's harp), presents such as jackknives were distributed, and Kane's medicinal supply of alcoholic spirits was reduced by several bottles of brandy.

Christmas on board the *Advance* was as jovial as they could make it under the circumstances. Two crews were crowded onto one small icebound ship, drifting helplessly through the arctic cold and darkness toward uncertain conditions in Baffin Bay, the outcome very much in doubt. There was no cause for excessive celebration, but some manifestation of thanks that they were all still alive and in reasonably good health seemed appropriate, and De Haven recognized the importance of keeping up the men's morale. He allowed the main brace to be officially spliced at least twice on Christmas (and who knows what the recalcitrant Wilson managed to organize in the way of an illicit supply of spirits?). New Year's Day featured sledge races, a generous issue of rum, and a splendid dinner ending with a cake baked half a year earlier in the United States.

Of the four search expeditions wintering in the eastern Arctic, only De Haven's was in a truly perilous situation, yet they managed a Christmas celebration. Considering their inadequate equipment for wintering and the crowding of almost three dozen men into a small ship that had previously held only half that number, it is a wonder that they could go beyond mere survival and enjoy the occasion.

20

The Sledging Season: A False Start

God aiding us we will do our duty.[1]

At Assistance Bay, the year 1851 came in with a reminder that the coldest period of the year was still ahead. On the second day of the new year, the mercury plunged to −37°F, and for almost four days it rose no higher than −30°F. On 13 January, Sutherland's thermometer was "as solid as a leaden bullet,"[2] the first time during the winter that its mercury had frozen. Astern of the ships, the fire-holes through the four-foot ice froze over repeatedly, requiring frequent clearing. Ice closed sphincter-like around the line leading to the *Sophia*'s tide gauge, and the men had to use saws to free it. Water stored in casks in the holds froze, and the crews were obliged to chop out ice to melt into drinking water. A cask containing frozen beef was given similar treatment. Snow drifted in tightly around the vessels during a gale, filling the carefully groomed paths that connected them and led to privies and dog kennels. Outside work and play were suspended, and exercise was limited to marching around the closed-in decks.

Temperatures of −30°F or below occurred on half the days in January and February, and on seven of those days the thermometers registered below −40°F, cold enough to freeze the mercury again. On 7 February, with a strong gale driving snow furiously across the ice and burying all paths, Penny cancelled outdoor activities, and the men remained inside their ships. When Dr Sutherland inspected the medical chest, he discovered that "everything fluid in it was frozen quite hard; castor oil became as hard as horn, and contracted about one-twelfth of its volume; all the syrups and the various preparations were one mass of ice, and in some cases the bottles were found broken; the only loss that was sustained by this was that of the glass."[3]

A few joyful events helped to dispel the gloom of darkness and the discomfort of bone-chilling cold. On 4 February, the anniversary of Penny's marriage to Margaret was celebrated at a special dinner on the *Lady Franklin*, attended by officers from all three ships and featuring "some prime fresh Aberdeenshire beef and some honest Morayshire potatoes."[4] Congratulatory speeches were made, and the men put on a concert afterward. A week later, the seamen, who always kept a sharp lookout for any plausible excuse for a celebration (and of course an extra tot or two of rum), asked for a holiday in honour of the anniversary of Queen Victoria's marriage to Prince Albert in 1840. Penny, who may have seen the royal couple pass by his house during their visit to Aberdeen in 1848, granted their request, and a good time was had by all. The ships, festooned with fluttering flags, provided a blaze of cheerful colour in the austere winter landscape.

At noon on 7 February, the upper limb of the long-unseen sun rose above the horizon, casting its rays tentatively around the top of the world. The men, wrote Sutherland, "beholding the glorious sun bursting forth into view, and illuminating the southern sky, after an absence of ninety-five days, could not deny gratitude to Providence."[5] A few days earlier, John Leiper, *Lady Franklin*'s second mate, and Thomas Abernethy from Ross's *Felix*, had enjoyed an advance showing from the top of "North West Hill," several hundred feet above sea level, but now the top of the golden orb was visible to everyone on board the ships. John Stuart noted that it made little practical difference because the sun could only be partly seen for a short time each day, and candles would be still be needed below decks for a long while yet. For everyone else, however, the occasion was a joyous reunion with an old friend.

The reliance on candles and oil lamps in the closed-in upper deck spaces and the living quarters below had an unfortunate disadvantage. It produced soot, which was deposited on cabin bulkheads and ceilings, clothes, books, and everything else that came in its way. The men were constantly coughing it up from their lungs. Ineffective evacuation of air through the chimneys exacerbated the situation.

* * *

Near the end of February, the mercury in the thermometer remained frozen for four days, marking the time of maximum cold. Exercise periods were held for two hours a day on the housed-in deck of the *Lady Franklin*. Concerts, theatricals, classes, informal get-togethers, and card games kept the men occupied, while the medical officers carried on their scientific enquiries. Sutherland's insatiable curiosity led him to measure the conductivity of sound in snow, the rate at which ice formed in the fire hole, the vertical range of temperatures inside the cabins, the water content of mustard and cress grown below deck, the formation of salt crystals on the surface of the ice, the rise of temperature when a man lay inside a snow cave, and the effectiveness of goggles (wood or gutta percha) and veils (some green, some black, of various thicknesses) in preventing snow blindness. When a ringed seal was killed, he examined its stomach to see which species of fish it had consumed.

As air temperature rose, more outdoor activities were permitted. Parties of men strolled out on the hard-packed snow to the points bordering the bay. Games were held on the ice. Some men went to one of the lakes to try for salmon and cut through more than six inches of ice to lower their lines, but despite a few nibbles they caught nothing.

* * *

Maritime expeditions engaged on the search for Franklin did not spend ten months of the year frozen into the ice for the sole benefit of a second navigation season. They did so also in order to send search parties away from the ships in late winter and spring. Small parties, each consisting of an officer and from six to nine men, could travel on foot over snow-covered land and frozen sea in almost any direction, hauling their food, fuel, camping equipment, guns and ammunition, behind them on sledges, while their ships remained secure in the grip of the ice. Virtually all the searches by ship-borne expeditions were carried out on foot in April, May, and June. In 1857, McClintock would reckon that since the expedition of James Ross in 1848–49, a hundred sledging parties had covered more than 44,000 miles looking for Franklin;[6] his own expedition in 1857–59 would add significantly to that total.

The summer navigation season was short and dangerously variable. Landfast ice barred many passages; drifting floes threatened the survival of ships; charts were incomplete and often misleading; harbours of refuge were scarce. As ships cruised, officers and masthead lookouts could scan adjacent coasts by telescope, but closer examination required landing parties, and it was unsafe to put men ashore on an exposed coast if floes were in motion nearby. Travel on the land after the snow cover melted was difficult for a number of reasons: streams and lakes imposed barriers to men on foot; the sparse vegetative cover of lichens, mosses, and sedges were spongy and very tiring to walk over; bare ground (other than bedrock) was often a waterlogged morass. William Edward Parry, the first to winter in the Arctic Archipelago, observed at Melville Island on 9 June 1820, "We pursued our journey at half-past five P.M., and found the ground extremely wet and swampy, which made the walking very laborious; but we remarked that our feet always came to the frozen ground at the depth of eight or ten inches, even in those parts which were the most soft."[7] The loads they carried on their backs increased the bearing pressure underfoot, pushing them deeper into the top "active layer" of the permafrost.

Over most of the arctic land area, millennia of low air temperatures have created permanently frozen ground, or *permafrost*, which in some areas extends hundreds of feet below the surface. Rock perpetually below the freezing point of water is little changed, but silt, sand, gravel, and other sediments whose interstices contain water become solid. During the brief summer, the slanting rays and limited duration of sunlight may succeed in melting the very top of the permafrost – but the permanently frozen ground underneath prevents melting snow and rainfall from percolating into the subsoil, leaving the surface saturated. Furthermore, as the landscape has only emerged from beneath continental ice within the last several thousand years, it is not yet effectively drained by a mature network of streams and rivers, so excess water cannot flow quickly away to the sea.

Another factor making extended trips impracticable in summer was the lack of a suitable baggage vehicle in which to carry tents, sleeping bags, spare clothes, food, stoves, and fuel. Parry and a dozen men travelled over the tundra of Melville Island with a light two-wheeled cart, but pulling it over soggy ground was exhausting, and when an axle-tree broke, they abandoned it – with relief, no doubt – and trudged back to the ship without it. During the 1851 sledging season, a party under McClintock would find the wheels from Parry's cart and use parts of them for fuel. A few years later, men from Belcher's western division came across them and took fragments as souvenirs. A hundred and thirty years after Parry, when a Canadian geologist tramped across the tundra of Melville Island, the ruts made by the wheels were still visible.

After ships established winter quarters in the autumn, water bodies and the surface of the ground froze, facilitating travel on foot. Straits, ponds, and streams that in summer had constituted barriers became solid natural highways, connecting rather than separating. Sodden ground became firm and easier to walk over. As snow accumulated, the prospects for sledge travel improved. However, excursions in October and November were risky because temperatures and daylight hours were declining, and in mid-winter, travel away from the ships was out of the question. Although the durable mantle of snow, firmly compacted by cold and wind, was suitable for walking and dragging sledges, the searchers could not utilize it until the days became brighter and warmer.

By mid-April, conditions were more permissive. Parties could now set out and benefit from rising temperatures and longer days. In May, travel conditions were optimum; in June, the sun would begin to melt the hard snow, turning it into mush and making the work of dragging heavy sledges much harder. On the sea ice, melted snow could accumulate several inches deep, soaking the men's boots and the lower parts of their sledge loads, and preventing them from camping. Leads would begin to crisscross the ice and extend along the shore, imposing barriers to movement, while on land the snow melted into torrents of meltwater. Sledge travel rarely extended beyond the end of June.

For sledge parties, there was therefore a rather narrow window of opportunity, open for less than three months, not even a quarter of the year. It was to take advantage of the favourable conditions in this brief period that expedition commanders were willing to freeze their ships into the ice for most of the year. Search parties could fan out radially from each ship to cover a vast territory, in stark contrast to the narrow linear pattern of a ship's track during the short navigation season. As only a few men would be required as caretakers of icebound ships, most of the crews could participate directly in searching activities.

There were limitations, to be sure. Overland travel was restricted by steep slopes, high elevations, fields of frost-shattered rock (felsenmeer), glacial erratics, and areas devoid of snow cover. Contrary to the popular perception, the central North American Arctic does not receive much precipitation, and the winds sweeping unobstructed across the treeless tundra redistribute the fallen snow, denuding high exposed areas and forming drifts in the lee of rocks and ridges. A large part of the landscape ends up bare.

Stream valleys and ravines in which snow tended to accumulate could serve as travel corridors if their gradients and valley slopes were suitable, but adjacent windswept ridges with exposed felsenmeer or dormant ground vegetation could be difficult to cross with loaded sledges. Most search parties therefore kept to the coasts unless taking shortcuts over headlands or peninsulas. HMS *Erebus* and *Terror*, after all, were not amphibious, so there was no use looking for them inland. Although snow-covered sea ice provided a natural highway for sledge parties, it was not flat and smooth everywhere. It had its own topography, which could include some formidable obstacles.

Another reason for travelling on the sea ice adjacent to the coast was that sledge parties often mapped as they went along, especially in uncharted territory, and the fundamental geographic feature they wanted to delineate was the boundary between land and sea. The rough charts they compiled would show coastlines, indicate prominent headlands and bays, and include some place names for noteworthy geographic features

– not only for the navigational convenience of future travellers but also to support claims of sovereignty for regions that might later be considered up for grabs. Most sledge parties had neither the time nor the equipment to measure water depths or make extensive observations of land heights, so usually neither soundings nor contours would appear on their maps, although a few "spot heights" – the elevations of particular summits measured by portable aneroid barometer – might be indicated.

The system of sledge travel used by the various expeditions of 1850–51 was based largely on experiences gained on the *Victory* expedition of John Ross in 1829–33 and the search expedition under James Ross in 1848–49. The officer who had organized the sledging activity on James Ross's expedition, and who had travelled more than five hundred miles in forty days, was Lieutenant Francis McClintock, a man of great endurance with an inventive mind that was continually devising improvements in equipment and methods. He was now with Austin's squadron at Griffith Island, serving as Ommanney's first lieutenant on HMS *Assistance* and directing sledging operations. Another veteran sledger from James Ross's expedition was Lieutenant W.H.J. Browne on HMS *Resolute*.

McClintock had devised a multi-tiered strategy in which one or more "supporting sledges" would accompany the "long-party sledges" at the outset, carrying enough food to feed them for the first several days so that they could keep their own supplies in reserve. Later, the supporting sledges would set out depots to enable the long-party sledges to reach home along the same path even if their provisions had nearly run out. Thus the long-party sledgers, benefiting from extra food and fuel on both outgoing and incoming voyages, could stay out longer and cover greater distances. Another feature of his search plan was the establishment of advance caches of provisions in the autumn along the routes to be taken by the sledging parties several months later. Austin had sent out depot-laying parties in October. These autumn man-hauling trips gave officers and men a valuable introduction to camping and travelling over ice and snow, but it was a rough initiation. McClintock later wrote an article for the ship's newsletter, *Aurora Borealis*, which, tongue firmly in cheek, he entitled "How to Prepare Oneself to Enjoy the Winter":

> Join an autumn travelling party, walk thirteen or fourteen miles through deep snow, in as many hours, you are unable to get through the luncheon of frozen pork; your feet becoming too cold to stand still, and fingers so numbed as to be unable to hold a knife; endure sundry frost-bites throughout the day, sufficient to keep you in a

> constant state of alarm about your nose; just before encamping, tumble into a crack in the ice, and wet both legs; wait two hours for your pemmican supper, by which time you will be cool all over; devour it in two minutes, and lay down in a blanket-bag, to shiver all night – at least, as much as the confined space allotted to each person in the tent will permit. In the morning, when you protrude your head without the bag, like a tortoise from beneath its shell, a shower of frost, dislodged from the covering, falls thickly over your face and down your neck. With feet cold, almost without sensation, you hurry through your breakfast of luke-warm watery chocolate, and get your fingers benumbed in vainly endeavouring to draw on your frozen boots. At length, the united efforts of the party get them *nearly* on; and you are able to limp along dispirited, unrefreshed, hungry, and miserably cold, and with a thorough conviction that the skin will be chafed off both ancles [*sic*] long before night again requires you to exchange your boots for a blanket-bag. Repeat this process, until it is scarcely possible to continue it any longer, then in all probability you will enjoy the remainder of even an Arctic winter.[8]

* * *

With the final plays and concerts behind them, the crews of the *Lady Franklin* and *Sophia* began to focus on the approaching sledge trips. The prospect of getting away from the stuffy atmosphere below deck on the cramped ships and possibly discovering the missing expedition was exhilarating, but there was plenty of work to do before departure. Penny's ships had not been supplied with the necessary equipment for the search parties, not so much as a sledge or a tent. Almost everything they needed had to be made on board. Carpenters, sailmakers, blacksmiths, and cooks worked feverishly to produce the necessary items, with other men assisting where feasible. The number and variety of articles required for their trips was impressive. Dr Sutherland compiled a detailed account of the equipment to be used, and its preparation.[9]

The fuel and food for forty days weighed 830 pounds per sledge. This would decrease as the men used fuel and food during the trip. The camping gear and extra clothing weighed 592 pounds per sledge but this would increase as a result of condensation inside the tent at night. The total weight at the start would thus be 1,422 pounds, drawn by seven men – an average drag weight of about two hundred pounds per man. Considering the men's cumbersome homemade footgear, the soles made of leather, gutta percha, or even cork, crudely fastened to canvas uppers, with multiple socks and blanket wrappers stuffed inside, the man-hauling excursions over the variable surfaces of snow, ice, and beach shingle were likely to be arduous in the extreme. On 29 March the men set up the tents on the ice and made note of alterations that would make the task easier.

* * *

In late March, Cator and another officer from Austin's expedition arrived for a brief visit. The twenty-mile crossing from Griffith Island, mainly over firm ice and hard snow, was well within their capabilities, but on this occasion a strong headwind introduced a new hazard. Cator later wrote, "I got a very important part of my body frost bitten, walking against the wind. It might have been serious, then I do not know what my poor little wife would say on my return."[10]

In July, on the Greenland coast, Cator had extolled Penny's virtues, but now he came up with a slightly different assessment. "He is a terrible arbitrary fellow to those under him," he wrote, "although to us, he was as kind as it was possible for a man to be. His officers and men nearly all hate him."[11] Cator did not hesitate to record derogatory comments about individuals, whether colleagues on his own expedition or men on others.

* * *

By this time the sledges of Penny and Stewart were all packed. Stuart noted that the visitors were "not a little surprised at the perfect state of our equipment considering our small means."[12] A few days later, after the guests had set off to return to their ships, assisted for some distance by Petersen and his dogs, the ice of Assistance Bay presented a lively and colourful scene as Penny's crews carried out trials with fully laden sledges. Flying small flags with inspiring inscriptions, they were hauled by the various teams around a circuit of a mile or so. "The day was hot," Stuart observed, "the road was soft, the sledges were heavy and last but not least we ourselves had for five months been very lazy and all these circumstances concurring firmly impressed the beleif [*sic*] on all the draggers that the travelling would be very hard work. After dragging past the Felix and cheering and being cheered we pitched tents, unpacked sledges and got into the bags then up tents[,] repacked and dragged on board."[13]

The exercise revealed a striking difference in performance. One sledge hauled by only seven men went as fast as another dragged by a team of eleven. When the crews were switched, the same sledge forged ahead. Close examination showed a very slight difference in the curvature of the runners. The faster sledge had a rise of 1/16th inch per foot both forward and backward from the centre line of the runners. Carpenters got quickly to work, carefully reshaping the runners on all the other sledges.

Penny was alone among the expedition commanders in planning to use dog teams as well as man-hauling parties. Since arriving in winter quarters seven months before, he and Petersen had been taking the dogs out often. In September and October, they had repeatedly travelled west to monitor the status of Austin's ships. On several occasions, they had driven over the ice to Griffith Island to bring visitors back to Assistance Bay or escort some of Austin's officers home. More than once, they had gone out to assist incoming parties back to the ships. Penny intended that two dog teams, one driven by Petersen and the other by himself, would support the man-hauling sledge parties by delivering and collecting messages, coordinating activities, depositing supplies, going ahead to locate camping areas, evacuating injured or frostbitten men if necessary, and providing a rapid communication link with the ships.

By this time Penny had become a reasonably capable dog driver, and other officers had given it a try, although no one could rival Petersen in directing the dogs with voice commands and the long whip. In April, Sherard Osborn, commander of the steam-powered *Pioneer*, came over from Griffith Island to visit Penny. After a few days, Petersen took him home by dog sledge and returned immediately to Assistance Bay, having covered forty-five or fifty miles in one day, a convincing demonstration of the potential of dog traction. A man-hauling party with a loaded sledge would require intense physical effort by the men for ten or more hours and would cover a mere quarter of that distance.

On the forthcoming searching trips, the dogs would be fed a mixture of "oatmeal mixed with melted fat, and a very small proportion, perhaps one-sixth or one-eighth, of 'soup and bouilli.'"[14] This was packed in bags of about 150 pounds, pressed flat so that they would freeze into easily stowable packages. In the absence of fresh meat, this was the best they could provide, but a few days before departure Petersen had the good fortune to shoot a bear, which the cooks promptly butchered. The animal was lean and his stomach empty, but the meat was enough for several good dog dinners before departure, with some left over to carry on each sledge.

One evening, Sir John Ross came to dinner on the *Lady Franklin*. The conversation turned inevitably to the forthcoming searches, and Ross, an engaging raconteur whose own accomplishments tended to figure prominently in many of his stories, talked confidently about what the *Felix* men would achieve. Stuart, for one, was sceptical. "He speaks of their travelling," he recalled later. "I fear it will not be much and I think it would be much better if he left it alone with his small means."[15]

As the departure time crept closer, Penny wrote a farewell note to Ommanney at Griffith Island, saying that he and his sledging parties would do their best. His irrepressible optimism was undisguised: "There is no doubt in my mind but we will be able by two months to say what has been their tale."[16]

He also composed a letter to Margaret (which he would not be able to deliver for another half-year). "How happy we should be if the Almighty should spare me to return in safety to you," he wrote. "Kiss all the children for me and tell them how their Father devoted his life if it should please God to exact it of him in the noblest cause man could face."[17] The sentiments, although awkwardly expressed, were heartfelt and sincere.

On 14 April, and again two days later, the crews hauled their much-improved sledges around the bay for practice. Excitement was rising as the departure day neared. On 16 April they shifted the six loaded sledges to the smooth ice on which they would be travelling, and parked them overnight. Early the next morning, Penny read a prayer to both crews assembled on board the *Lady Franklin*:

> O Lord God, our heavenly Father who teachest man knowledge, and givest him skill and power to accomplish his designs we desire continually to wait and call and depend upon Thee. Thy way is the sea and thy path in the great waters . . . We therefore draw night unto Thee for help in the great work which we have now to do . . . We most humbly beseech thee of Thy goodness O Lord to comfort and succour all those who are in trouble[,] sorrow[,] need[,] sickness or any other adversity[,] especially such as may now be exposed to the danger of the deep or affected with cold or hunger.[18]

After Penny made a few uplifting remarks, the travelling parties, comprising forty-one men, made ready to set off. "There was not a drooping spirit among us," Sutherland noted, "for every face beamed with delight, and each seemed to feel proud of the laborious duty upon which he was entering."[19] The eager crews walked to their sledges, took up the drag ropes, and got under way, cheerfully assisted for a mile or two by officers and men from the *Felix*. Ross, less able to assist physically, remained on his ship. Sutherland later recalled, "I shall never forget the venerable appearance of his grey locks, as he waved his hat, or fur cap, with its light blue veil, and joined heartily in the cheers of those we left on board."[20] Ross had asked to address the search parties before they left, but Penny refused. Writing to Margaret, he said, "Such is the man that I would not let him make a speech to my crews before starting[.] I told [him] I had commended them to the care of their maker and wanted no speeches[.] Poor old man it was a great blow to him but I could not let him interfere in any way." Relations with Ross at the harbour had been less than perfect during the winter, he told Margaret. Ross's expedition had been "very ill provided"; scurvy had affected seven of the *Felix* crew as well as Ross himself, and Penny had been the means of saving their lives. He confessed he had no admiration for Ross: "He is an utterly selfish man and his name is proverbal [*sic*] for false statements."[21]

The flag-flying flotilla of sledges proceeded eastward toward Wellington Channel, keeping slightly offshore on the snow-covered sea ice. From the start, it was tough going: sledge runners sank into snow made soft by sunlight. Physical exertion led to perspiration and thirst, and the drinking-water supply was rapidly exhausted, making it necessary to stop and light stoves to melt snow, placing an unexpected burden on their fuel supply. Two of the homemade kettles were ruined by adding snow while they were hot. Falling snow accumulated on the men's clothing and later melted. Some of the nails securing shoeing to the runners of a sledge began to pull out, and another sledge was found to be too low to clear deep snow. At the end of the day they pitched their tents on the ice, ate supper, and wiggled their weary bodies into sleeping bags. Warmth from bodies and stoves melted the snow beneath the floor cloth and soon their clothing, blankets, and wolf skins were damp, and of course heavier.

Penny and Petersen, who had remained at the ships for a while, caught up with them on the second day. With their dogs, they now performed the useful service of scouting ahead to look for the best route through rough ice and select the next campsite. When there was no easy path through ridges and hummocks, the draggers were forced to use three crews to manhandle each sledge in turn, which required time-consuming back-tracking. Slippery soles on their clumsily made boots made hauling more difficult and caused many falls.

During the second night, the dogs launched a stealthy raid on Stuart's sledge and consumed seventy pounds of bear meat before any of the exhausted men took notice. Sutherland declared that no harm would come to the animals because they had each eaten only four pounds of meat! Thereafter, the men alternated on watch during the night.

After rounding Cape Hotham, the parties trudged northward with Wellington Channel on their right. To their left was the remarkable escarpment of Cornwallis Island's east coast, rugged and steep. Most of the time they kept to the natural highway of the sea ice, but when rough ice placed impassable barriers in their path, they shifted to the narrow snow-covered beach between the shoreline and the foot of the steep talus slopes. All this coast between Barlow Inlet and Cape Hotham had been examined on foot the previous August by Cator's party from

HMS *Intrepid* and *Assistance*, without finding the slightest evidence of Franklin. Penny's objective was simply to get past it as quickly as possible and strike northward into regions that had not been searched.

Penny had intended that the crews would start off hauling their sledges by day and then gradually shift to the nighttime in late April. In general, travelling during the day was characterized by good visibility, tolerably warm air, and relatively low risk from frostbite, but these advantages were offset by softer snow, greater effort, slower progress, and a serious risk of snow blindness. At night, hauling the loaded sledges over the harder surface was easier and faster, but in the lower temperatures the incidence of frostbite tended to be higher, and affected areas on the face were more difficult to see in the dim light.

The contrast between daytime and nighttime conditions could be blurred or even reversed by changes in the weather. One day might be unusually cold and windy, but a night might be overcast and relatively warm, so it was difficult to adopt a rigid schedule for travelling. They had to remain flexible, adjusting their itinerary in response to weather and snow conditions.

As air temperatures dropped to −30 degrees F outside – and even to −25 inside the tents! – exposed skin became extremely vulnerable. Frostbite could lead to sloughing off of skin, open sores, and in extreme cases, gangrene and death. The only preventive was adequate footwear and clothing, and perhaps some sort of facial covering to guard against wind. On the march, the sledge haulers regularly inspected each other's faces for the telltale grey patches of frozen skin, and in the tents each man examined himself thoroughly before turning into his blanket bag to sleep.

The dazzling reflection of sun's rays from surfaces of snow and ice could cause serious eye irritation, intense pain, and sleeplessness and render a man sightless for several days. Captain Stewart was severely affected: he wore glasses with horn frames and clear lenses – of no use in filtering sun's rays. Others tried veils of various materials and colours, but as Sutherland had ascertained, "Neither a neutral tint, nor black or green veils, afford even a slight compensation for undue exposure."[22] Even inside the tents in daytime, the brightness of the white walls could exacerbate the affliction. (He recommended brown tents in future.) To men who were snow blind, the doctors applied an eye salve of wine of opium with water and made them rest with their eyes covered. This, however, sometimes overheated their eyes and made their condition worse.

On 21, April several men complained of not being able to see, and on the next day they were worse. Sutherland, described their symptoms: "violent pain and swelling in the eyes. The conjunctiva was [*sic*] deeply congested, and the eyelids were in a highly tumefied state. The most partial admission of light into the eye by separation of the eyelids, was accompanied by excruciating pain, as streams of water gushed from them, and became frozen upon the faces of the poor sufferers."[23]

Man-hauling the heavy sledges was very hard work. For the men it was a curious sensation to be sweating profusely inside their bulky clothes while their exposed faces froze into crusty grey patches and their beards were decorated with "barber." To sooth chapped lips, they applied hogs' lard. Some wrapped mufflers around their mouths.

Despite the long hours, strenuous physical demands, and the discomfort during sleeping hours, the men maintained good spirits. Snuggled in their blanket bags after hours on the march, socks, boots, mittens, and water bottles tucked alongside their bodies to keep from freezing, they enjoyed a tot of rum, a pipe, and a hot meal (or at least a warm one) before retiring. With the tents close together in their small temporary village, jokes, jests, and good-humoured taunts were tossed back and forth; problems and setbacks were related with wit and irony. The silence of the arctic landscape was broken from time to time by a doleful melody on a flute or a sprightly jig on an accordion.

North of Cape Hotham, the sledge parties were moving slowly against a cold northerly wind, and cases of frostbite were common. Fuel was being depleted faster than expected to provide water for the men's persistent thirst. On the next day, after unloading two sledges and caching most of their supplies, John Stuart and James Reid started back for Assistance Bay, taking with them a few men who were suffering so badly from snow-blindness that they were unfit to haul. The rest pressed on past Barlow Inlet.

A week after leaving Assistance Bay, Penny and his men reached the cairn they had set up eight months earlier after landing from the *Lady Franklin* and *Sophia*. When they camped five miles further on, Penny called the officers together to discuss the situation. He pointed out that their stoves and kettles were not good enough, that certain of the sledges still required alterations, and that their allowance of fuel was insufficient to cook and provide drinking water for another month. Without water and warm food, their survival would be at risk, and they might not be able to get back to the ships. Should they go on, or should they return temporarily to Assistance Bay?

All agreed that the best course of action would be to turn back, make the necessary improvements at the ships, and come out again better prepared. To have to retreat from the advanced position they had attained in Wellington Channel, at the very threshold of unknown territory, was a cruel blow, but they had little choice. In calculating fuel requirements, Sutherland had underestimated the magnitude of the thirst problem. Water was essential to avoid dehydration while working hard, but melting snow into drinking water had consumed as much fuel as cooking.

After leaving two sledges and most of their provisions and stores, Penny's men started back with minimal loads on the remaining sledges, each hauled by thirteen men. They covered the forty-five-mile route to Assistance Bay in two days, and the work of modifying the equipment began.

* * *

With the sledging season, the spirit of competitiveness between different expeditions had come once more to the fore. In early April, Ross had sensed that the rivalry between Austin and Penny was pushing them to initiate sledging earlier than was reasonable. Over at Griffith Island, Osborn confessed that "great fears were entertained lest our long parties should not beat those of the 'Lady Franklin' and 'Sophia' in time and distance; a piece of *esprit de corps* highly commendable, no doubt, but which, I blush to say, I took no interest in, having gone to the Arctic regions for other motives and purposes than to run races for a Newmarket cup, or to be backed against the field like a Whitechapel gamecock."[24] Not all of his fellow officers were as reasonable. Cator, unofficial Grand Master of the Gloat, sneered when he heard that Penny's parties had returned after a mere nine days, "'Penny' and his party had returned on Saturday, finding it so cold, that they could not get on or stand it. Now he was the man, that stated at the 'Admiralty' that *whalers* were the men to be employed in these seas, what did we know about it. So here it is proved . . . which [expedition] can do most so far, half our work is completed . . . and he obliged to turn back."[25]

Cator might have been less critical of Penny's parties if he had been out with a man-hauling team himself; but, like Austin, he had stayed with his ship. In fact, Penny's sledgers had not been defeated by cold but by critical problems with equipment and by a shortage of fuel. To go on with these serious deficiencies would have been irresponsible. They had experienced very cold weather, but constant vigilance by officers and visual inspections by the men themselves had detected cases of frostbite before they became serious. Feet, especially those encased in restrictive leather boots rather than canvas ones roomy enough for three pairs of socks, had often required warmth and gentle massage at day's end to restore circulation. But all the men had returned to the ships without having suffered serious injury. "Not so much as a finger nail was lost by one of them," remarked Sutherland proudly, "although the temperature for three successive days was generally 60°[F] below the freezing point of water."[26] He and his medical colleagues Goodsir and Stuart had been conscientious in treating any frostbite promptly.

In a letter to Margaret, Penny described the aborted trip. "I have no doubt when you read this you will think what sort of a man is my Wm turned out to be[,] out sleeping in a sleeping bag in a cotton tent in a snow storm with thick drifts and a low temperature ie 30°[F] below zero for eight days." Every time they camped, he told her, he had gone from tent to tent to check the men's physical condition. "How thankful should I be to the Almighty for allowing us to come of[f] without a frost bit[e] while Capt. Austin['s men] who have returned are all frostbitten more or less."[27]

Although only a few of Austin's auxiliary parties had returned to Griffith Island by the time Penny's men reached Assistance Bay, it was clear that frostbite had been a serious problem. More details would emerge when the long-distance parties returned. The Western Division, comprising the parties of McClintock and Bradford, accompanied by three supporting sledges, were badly affected. In a summary of his journey, McClintock later wrote, "Frost bites were constantly seizing upon our hands and faces, but to these we soon became indifferent as they were easily remedied." But the frostbites may not have been so *easily remedied* or the men so *indifferent* to them, for he went on to say, "Yet many suffered very severely – eight men were disabled, and of these two were delirious [when] they reached their ships & [all] but one of them recovered."[28]

The man who failed to recover was George Malcolm, a thirty-three-year-old Dundee man on HMS *Resolute*, captain of the sledge *Excellent*. After his feet were badly frozen, he was exempted from pulling and was later carried on a sledge until he could be sent back with one of the auxiliary crews. During his long ordeal, gangrene spread up his legs to his abdomen, and when he arrived at the ships he was "all but *dead* from *Mortification* of the leg,"[29] Cator observed. He died two days later and was buried on Griffith Island.

Cator thought that Malcolm's death could have been prevented. "There is no doubt great blame attached to '*Old Bradford*,' and also to McLintock [*sic*] for allowing him to go on five days after he was [frost-]bit, and to walk alongside of the sledge instead of on it. He should have been sent back immediately . . . . But 'Mr. McLintock' is a hard hearted man."[30] After noting that Dr Bradford had told men with frostbite not to worry as it was only skin deep, Cator wrote, "He might doctor a 'Bear' very well, but he should never come near me."[31]

Several years earlier, McClintock had undertaken the longest man-hauling search up to that time, travelling with his leader, James Ross, and a dozen men for five hundred miles in forty days, from Port Leopold southward through Peel Sound and back. The party had suffered badly from foot blisters and frostbite. Five of the men were disabled, two had to be hauled on the sledges, and all of them later required medical attention.[32] Had McClintock not learned from this experience?

In view of the frostbite injuries, amputations, and death among Austin's sledging parties, Penny's decision to fall back to the ships in late April had been, as he told Margaret in his ongoing letter, "a prudent step."[33]

* * *

Austin's men had established some advance depots in the autumn to enable the spring parties to travel further, but Penny had decided against it. Had he attempted similar trips, weaknesses in the sledge equipment might have been identified and rectified in time for the spring journeys. But in late September, everyone had been hurriedly making essential preparations for wintering. Whereas the sledges for Austin's ships had been made at the Woolwich dockyard and were ready to go into operation as soon as the sea ice was suitable, the Admiralty had not provided Penny with any, so his carpenters and blacksmith had to construct their own, as well as most of their camping equipment, on board ship. It is highly doubtful that they could have prepared all the necessary gear in time for sledging in early October, a mere three weeks after taking up winter quarters, with so many other things to do to make the ships secure for winter. Consequently, the spring searching parties set out without previous sledging experience. After their return in April, Alexander Stewart observed, "No one of the whole party, except Mr. Petersen, having ever travelled before, it need not be wondered that we found a few things deficient."[34]

Although Penny lacked an experienced man-hauler such as McClintock among his officers, he had one advantage over Austin. As his ships were smaller and securely frozen into a stable sheet of ice within a bay, only a few men were required to stay on board as ship-keepers, and almost all his men could be directly involved in the sledging. On the first attempt, only three men out of the total complement of forty-five remained at the harbour, and at one time or another every man in the expedition – with the sole exception of the steward – had participated in sledging. That physical effort contributed to their fitness.

Austin, on the other hand, had larger ships in a less secure position, so it was necessary to keep more of his men on board in case drifting ice carried them out to sea while the sledge parties were absent. In addition, a much higher proportion of his men were regarded as unfit as on Penny's vessels. In February, the crews of HMS *Assistance* and *Intrepid* had been rated according to their physical fitness. Only a third of the men were considered to be Class 1 – *fit for long journeys*. Another third were Class 2 or Class 3 – *fit for short journeys*, or *able to lay out depots*. A full quarter or more were regarded as *unfit*, and were not to be involved in sledge trips.[35] Out of a total complement of 180, Austin kept about seventy-five men with the ships – 40 per cent of his total complement – whereas only one of Penny's men (2 per cent of his crews) did not take part in the sledging.

* * *

As the sledge parties from Austin's ships were fanning out across the vast territory lying to the west and south of Griffith Island, Penny and his men at Assistance Bay frantically tried to rectify the shortcomings of their travelling equipment. To solve the problem of nails pulling out of the shoeing, the carpenters attached strips of hardwood between the wood and the iron, fastened them with two-inch screws, and made them secure with long vertical bolts. Some of the sledges had proved to be too short, but unfortunately there was no wood suitable for making new runners. The sailmakers modified the lacings on the tent doors to make them more weather-proof. The blacksmiths manufactured larger stoves by surrounding tin containers of a gallon and a half capacity with sheets of galvanized iron, and their fuel supply was doubled. The men's daily allowance of sugar was increased by half. Their daily rum ration was also increased, from half a gill (2.5 ounces) per man daily on the first outing to one gill (5 ounces) for the forthcoming attempt.

In the eighteenth century, Royal Navy sailors had been issued enough rum per day to result in widespread intoxication within the fleet. Although commanding officers, including such noteworthy leaders as Nelson, Hardy, and Cochrane, resorted to severe disciplinary measures, the plague of drunkenness continued, so in 1824 the daily issue per man was reduced to a quarter-pint (five ounces, or one gill) per man each day, diluted with water to form "grog." Five ounces of the strong rum issued on ships, writes a historian of rum in the Royal Navy, "was equal in strength to at least four double whiskies today,"[36] while another authority compares its effect to that of six whiskies.[37] These comparisons are imprecise because there is no specific amount of whisky in a social drink – either single or double – and the proof (or strength) is not specified; but the amount was unquestionably capable of reducing a man's judgment and reaction time.

Two years later, wine measure was replaced by imperial measure, and this had the effect of increasing the amount of rum issued by one-fifth, counteracting the changes made in 1824. Concern mounted again, and in January 1850, a few months before the Admiralty search expeditions of Austin and Penny set out for Lancaster Sound, a Grog Committee of three admirals and seven captains was struck to consider a further reduction of the rum issue. The problem was not merely that some seamen might become intoxicated by consuming their daily ration, but also that determined drinkers could manage to scrounge a good deal more than their share and proceed to get downright blotto. As the rum did not have to be drunk "at the tub" under an officer's supervision, teetotallers and seamen of modest drinking habits could save their portions for thirstier crewmates in return for various favours or articles. Bullying undoubtedly played a role in such exchanges, and young sailors were often victimized.

When the committee issued its report in March, it drew attention to the consequences of drunkenness: men under the influence of rum sometimes fell overboard from the deck or from aloft, making it necessary for others to risk their lives rescuing them. The prospect of drunken men working below at night during battle, near lighted candles and gunpowder, was especially alarming.

The committee recognized that as punishment had not been an effective deterrent, additional preventive measures were needed. They recommended that the amount of rum be again cut in half, that it be diluted before serving with three times as much water, and that it be issued only once a day at noon rather than twice as before. Midshipmen, first-class boys, and certain other ranks were not to receive the rum issue unless their commanding officer specifically allowed it, while cadets and second-class boys were not to receive it at all. Those deprived of the privilege of drinking rum, or who elected not to receive it, were to be given monetary compensation. However, these measures did not go into effect until October 1850. As the crews of Austin and Penny were already in the Arctic, a daily issue of five ounces of rum per man was still in accord with regulations.

Sailors regarded the daily issue of spirits on ships as an inviolable right, and they understood that it extended to arctic sledging trips as well, for what were sledges but miniature ships, each with a name, a flag, a commander and crew, a daily routine, a code of discipline, and enough supplies and equipment for a voyage? Like Royal Navy ships, the sledges were acknowledged to be the property of the queen. Setting out from Her Majesty's Ship *Assistance*, for example, was "Her Majesty's Sledge *Reliance*," commanded by Erasmus Ommanney.

Each of Penny's sledges had started off carrying rum weighing 48 pounds, and on their second outing they would be taking double that – almost a hundred pounds. Was the pleasurable buzz of a stiff drink at the end of a hard pull really worth hauling the added weight? And did rum really warm them up and help them to withstand cold? Although it wasn't generally known at the time, the ingestion of alcohol dilates blood vessels and contributes to loss of the body's core heat, inadvisable in a cold environment. The issue of rum on sledge trips was never questioned, however. Perhaps it was worth the additional labour and loss of body heat. It gave the men something to look forward to as they slogged along hour after hour, straining into their harnesses and perspiring into their heavy winter clothes while their fingers, toes, and faces became numb with cold. Rum contributed to a pleasant ambiance before turning in, a bit of normality in a long day that was mainly physical drudgery, and to a sound sleep in the blanket bags after eating.

21

Second Attempt

Pas à pas on va bien loin (motto of sledge *Perseverance*).[1]

After ten days of rest and recovery, with the travelling gear modified and extra fuel and rations packed, everything was ready for a second assault on Wellington Channel. By 6 May the weather was warm enough to travel at night and benefit from a harder snow surface while avoiding the sun's harmful rays. At 8 p.m., the man-hauling teams left the ships and started again for Wellington Channel. As much of their stores had been left at depots, the sledges were light and the men travelled fast, covering thirteen miles in eight hours before camping for the day near the depot laid down by the men of HMS *Assistance* the previous August. The next night they rounded Cape Hotham, turned northward, crossed Barlow Inlet, and reached the southernmost of the two advance depots. The food cache there had been disturbed by bears and foxes, but fortunately the damage was slight – only a few bags of bread and pork had been damaged or consumed.

After repacking the supplies in their gutta percha barrels and making the cache as bear-proof as they could, the men retired to their tents to sleep, pulling their blanket bags over their heads to block the light of day. At four that afternoon they resumed the voyage. With the wind at their backs, Stuart and his crew rigged the floor cloth for their tent as a sail and sped along effortlessly – until a sudden gust capsized the sledge, to the amusement of the other crews. Reaching the "upper" or "fifty-mile" depot in less than five hours, they all made camp, had a meal, and awaited Penny and Petersen, who had lingered at the ships again. Their dog teams trotted up after a rapid journey of eight hours during which they had stopped only once for a meal. Penny and Petersen were accompanied by Alexander Thompson, Petersen's assistant in caring for the dogs.

Penny's entire sledging squadron was now assembled and ready to carry the search for Franklin into unknown territory. According to his plan, they would now diverge onto different paths. The two dog teams (Penny and Petersen) and two man-hauling parties (Goodsir and Leiper) would continue north on Cornwallis Island. Three other sledges would cross the ice of Wellington Channel to the coast of Devon Island, where they would establish a depot to use on their return voyage and then split again: two sledges (Stewart and Sutherland) turning north along the coast, and one (Stuart) heading south to re-examine the area around Beechey Island.

From the beach at "Separation Point," the parties destined for the northern part of Cornwallis Island watched as the other teams hauled their sledges onto the ice of Wellington Channel and set crude sails to use the power of the wind in their crossing to the east side. Penny, Petersen, and Thompson then started north with their dog teams, leaving the northward-bound man-hauling parties to follow at their own plodding pace.

On 12 May, from the summit of a hill roughly four hundred feet above sea level, Penny looked eastward across Wellington Channel to Devon Island, only twenty miles away, where the parties under Stewart and Sutherland must by now be advancing along a parallel course. The coast on which he was travelling now began to curve westward, and an expanse of unfrozen water was visible. Wellington Channel seemed to be opening out into what Penny called a vast, "newly-discovered sea."[2] It would later be named Queen Victoria Channel. Two days later, having advanced in a northwesterly direction, he could see a large island and another segment of land beyond it. None of this territory had ever been visited by Europeans.

In two directions, northeast and northwest, no horizon was visible, which suggested the existence of channels leading out of the enclosed sea. In his initial instructions from the Admiralty, Penny had been ordered to sail westward through Jones Sound to reach the north end of Wellington Channel. Had he now located one or more connecting straits between the two bodies of water? The sledging parties under Stewart and Sutherland would soon discover that the northeast sector is not a sea-level passage but merely low-lying land separating Jones Sound from Wellington Channel. The gap toward the northwest, on the other hand, would turn out to be a navigable waterway, one that would soon exert a powerful influence on the search for Franklin. It is now called Penny Strait.

* * *

There was great excitement when Penny's favourite dog, Sultan, spotted a bear and gave chase, trailed by a howling posse of eager dogs and – just as eagerly but not as rapidly – by Penny, Petersen, and Thompson. The bear's camouflage, speed, and familiarity with the terrain proved decisive, and five hours later the weary men and dogs returned exhausted and unsuccessful to their camp. It was a huge disappointment, as only a four-days' supply of dog food remained.

At the northern tip of Cornwallis Island, they left the land and started across the ice toward a large island approximately ten miles offshore. Penny named it Hamilton Island in honour of the Admiralty's secretary, W.A.B. Hamilton, who had done so much in connection with his expedition. (It is now officially called Baillie-Hamilton Island, a hyphen inexplicably inserted before his surname.) "The moment we landed," Penny related, "I set out to a bold headland, or I should say rather the south-east point of the island; but I found no traces of the missing expedition."[3] This point he named Cape Washington to commemorate the man whose Esquimaux dictionary and phrase book had helped the officers of the search expeditions communicate with native hunters at Cape York in Greenland nine months earlier. Travelling along the island's east coast to its northern termination, they found no cairns or relics of Franklin or anyone else. "But lo and behold! to our surprise a strait, and nothing but clear water, opened out before us."[4] He called the place Point Surprise. To the west of Hamilton Island, several small islands were visible. Across the open strait lying to the north were two islands, and beyond them a mountainous coast trending east and west.

In August of the previous year, Penny and his men had found a campsite made by some of Franklin's men near Cape Spencer on the eastern coast of Wellington Channel. Now that the northern part of the channel was seen to lead into a large expanse of water and possibly into the unknown polar sea, it was easy to believe that the missing ships – or at least some boats from the expedition – had come this way. But had they been advancing or retreating?

Penny tried to imagine himself in Franklin's shoes. If he had seen from the summit of Beechey Island that Barrow Strait was blocked by ice in the summer of 1846, but his watch camp had reported that Wellington Channel was ice-free, he would probably have sailed north, crossed this large open area beyond Cornwallis Island, and continued into the uncharted polar sea, which many believed to be free of ice. The lack of evidence of Franklin's passage was easy to understand: if his ships had been making progress in ice-free waters, he would not have wasted valuable time to send parties ashore to erect cairns. And as he had not deposited any messages during an entire winter at Beechey Island (for some inexplicable reason), he would have been even less inclined to do so while his ships were under way, advancing at last toward Bering Strait

– the end of the Northwest Passage. The absence of messages, therefore, did not necessarily mean that Franklin's ships had not come this way. It might mean only that they had been proceeding rapidly on their mission and could not spare the time to stop.

Penny wanted to extend his search, but there was only a two days' supply of dog food left. Walrus floated fleetingly past on ice pans; seals peeked provocatively from open leads; sea birds bobbed buoyantly offshore; but as the men had no means of retrieving animals shot in the water, they sensibly held their fire.

Eskimo dogs have a marvellous capacity to survive on little, but their efficiency declines when food is scarce, and of course prolonged starvation leads to death. Penny and his two companions did not want to be deprived of dog traction more than a hundred miles from the ships, so further exploration was out of the question. They had to start for home. Before heading back, Penny looked at the inviting expanse of open water extending toward the northwest and came to an extraordinary decision. He would return to the ships, get a boat, haul it all the way back on a sledge, and launch it into the beckoning sea. With a boat he would be able to extend the search further and perhaps discover traces of Franklin's passage. His search by dog sledge had extended only as far as Hamilton Island. Several other nearby islands remained unexamined, and by the time he returned with the boat, melting might have exposed some evidence previously hidden beneath the snow.

On 17 May they started south, travelling at night. One of the dogs was no longer strong enough to pull, and even when unfastened from her trace, was unable to keep up with the others. Unwilling to leave her behind to starve, the men decided to shoot her. Penny hoped that the meat could sustain the other dogs, but they refused to touch it. To lighten the loads the men walked and ran for hours beside the sledges, bathed in sweat, until they were near collapse. The following day was Sunday, and after Penny read prayers they set off again. The dogs – despite not having eaten in twenty-four hours – pressed on valiantly, sensing that the path led homeward. They badly needed food, so Petersen stripped off his sealskin parka and trousers, cut the tattered sweat-soaked garments into bite-sized pieces, and scattered them over the snow. (One presumes he had other clothes beneath.) The famished dogs devoured it eagerly, hair, perspiration, and all. It was nourishment enough to get them to the depot at Separation Point, where the animals were rewarded with pemmican. At midnight on 20 May, the two teams arrived back at the ships.

The harbour was quiet, the other sledging parties still out on their searches. Until Penny's carpenters, sailmaker, and blacksmith returned, he could not construct a sledge large enough to transport a boat north to the open water. While awaiting their return, he decided to inform Austin of his discoveries and intentions, so he and Petersen sledded with their dogs along the coast and across the ice to Griffith Island. Boarding HMS *Resolute*, Penny described his experiences to Austin. Other officers gathered round to hear his news. Among the listeners was Cator, who asked Penny what the area beyond his furthest point was like. Penny replied that "he imagined it to be a quantity of islands or archipelago and . . . that he [Franklin] had got into the polar basin and . . . he thought no traces would be found."[5]

Penny was on hand for the celebration of Queen Victoria's birthday on 24 May. The four naval ships were decorated with flags, and Austin entertained those of his own officers who were not away on sledging trips and his two visitors from Assistance Bay. That night Penny left for Assistance Bay under what Cator thought were mysterious circumstances.

> At 8 am the Quartermaster reported to me that "Penny" had gone about 12.30 last night. I could scarcely believe such to be the case, but on further enquirey [*sic*] it proved to be quite true. The cause no one can explain. The "Commodore" declares he has had no row with him. About midnight, he [Penny] walked down to the "Assistance" and said he was going, it being a fine night, and had come to say good-by [*sic*]. I can not help believing that he and the "Commodore" have had a row, however much the latter may say no, for he has such a nasty dirty way with him, however we must hope that he has not.[6]

Cator's suspicions were correct. Penny had departed in a huff following a disagreement with Austin. In the afternoon, as the two men were strolling together on the ice, Penny had talked enthusiastically about the open water north of Cornwallis Island. Austin questioned him closely and was openly sceptical of his report. He could not see how there could possibly be a large expanse of open water beyond Cornwallis Island when there was solid ice far to the south where his ships and those of Penny and Ross were frozen in. Months later he declared, "I could not well understand that he had been in an open sea, circumstanced as we were."[7] Although he himself had sailed from Greenland to Lancaster Sound by way of the ice-free polynya of the North Water, he seemed to be unaware that polynyas could exist elsewhere.

To Penny, Austin's incessant interrogation and incredulous comments indicated disbelief and lack of trust. This raised his ire, and his ire was never subtle. Austin later said, "I found that Captain Penny was angry, and I felt that he would not allow me to make such inquiries."[8] That evening, when Austin, Penny, and several officers sat down to a special dinner on board HMS *Resolute*, the atmosphere was tense. Austin remained aloof as Penny talked to others, probably describing the expanse of open water he had discovered. Afterward, Penny asked Austin if he could lend some men to help drag the boat up to the open water, but Austin refused, saying that he could not spare any men. In a somewhat bewildering statement much later, Austin recalled the conversation: "At this time I really had no assistance to give to Captain Penny, but I am by no means prepared to admit that that Captain Penny asked me for it. My impression is, that I observed to Captain Penny that I regretted that I had not the means, but that we must have a boat in that water. What I meant by that was, that Captain Penny and myself, that is with my assistance, should have a boat in that water."[9]

To Penny, Austin's protestation that he could not spare anybody to assist was a smoke screen. If he had sent a hundred men out on sledge trips, as he told Penny, there must still be eighty at the ships. Surely *some* could be spared to help! At least two officers – Cator of the *Intrepid* and Allen of the *Resolute* – volunteered to go north with Penny to help man his boat, but Austin would not allow it. The two commanders said goodnight coolly and retired to their quarters. At midnight Penny and Petersen roused the dogs and left Griffith Island for home.

Arriving back at Assistance Bay, Penny found the *Sophia*'s carpenter hard at work modifying a large sledge so that it could carry a six-oared whaleboat, and the sailmaker preparing a long canvas cover so that the boat could be converted into a tent for cooking and sleeping. Manson and the dog-handler, Thompson, were packing food and fuel. It was almost the end of May, already past the optimum season for travelling over ice and snow, and conditions were deteriorating day by day, but there were still not enough men at the harbour to haul the boat a hundred miles north to open water. After the first of the sledge-hauling parties arrived back and rested for a few days, they were pressed into service, along with every available man at the ships save Penny and Petersen, who would catch up later with the dogs, and a ship-keeper or two to remain at the harbour. On 4 June, fifteen men started off, dragging the long sledge with the boat perched in a cradle on top, and a smaller sledge packed with food and gear. Two days later, Stuart's party came in from their arduous month-long trip to the region beyond Beechey Island. A mere thirty-four hours after returning to the comparative comfort of the ships, they inserted their worn and weary bodies into the loathsome tracking belts again and set off to catch up to the boat haulers and give them a hand.

A few days later, two men returned to Penny with bad news. The new sledge had proved unequal to the task of carrying the boat. The party had reached Cape Hotham but could go no further. This setback, so late in the season, would have induced some commanders to cancel the project, but Penny was not ready to admit defeat. He decided to have another sledge made from whatever lumber was available. Time was short – the haulers were waiting at Cape Hotham – but the carpenter and blacksmith, with the generous assistance of two carpenters from the *Felix*, speedily crafted a stronger sledge. At midnight on the next day Penny and a few others set off hauling it. They reached the encampment at four in the morning, and at five in the afternoon everyone got under way again. The boat cavalcade consisted of the new sledge with the boat on top, two other man-hauled sledges, and two dog sledges. Altogether, there were twenty or twenty-one men, including Penny.

At the lower depot, they encountered Sutherland and his men on their way back to the ships from the eastern side of Wellington Channel. They had turned for home at a latitude of 76°20' N, leaving Captain Stewart's party to carry on for a few more days along the coast of Devon Island. Almost all of them volunteered to assist immediately in getting the heavy boat north to open water, but Penny told them to continue to the ships, rest for a few days, then transport some food to the coast of Wellington Channel for use during his return journey. After exchanging "three British cheers," the two groups separated. "How pleasant it is to see with what enthusiasm they perform their duty,"[10] Penny remarked in his travel report.

Travel on the land was now very difficult. Sledge runners sank deep into the melting snow, and the once-solid streams were in flood, imposing barriers across their path. On the sea ice, the men were forced to the make lengthy detours around zones of hummocky ice, often wading through meltwater. The physical exertion was telling on Petersen, so Penny sent him back to the harbour accompanied by Stuart while the boat haulers and support personnel continued northward. Soon they encountered Goodsir's party returning from their journey around the top of Cornwallis Island. Their news, like that of Sutherland, was discouraging: not a trace of Franklin had been found.

Toiling at their drag ropes for as much as twelve hours a day and sleeping fitfully in damp blanket bags, Penny's men pressed on without complaint. When they sighted open water from a hill, everyone's spirits soared. The following day, 18 June, they arrived at the water, launched the boat, and loaded it with provisions and gear. The boat party consisted of eight men: Penny, his second mate John Leiper, his carpenter

Daniel Henry, the *Sophia*'s cook James Knox, and four seamen from the *Lady Franklin* – James Davidson, William Bruce, Alexander Thompson, and James Hodgston. They climbed aboard, pushed off, waved goodbye to the men of the support party, hoisted a close-reefed sail, and set off into the newly discovered sea.

22

Penny's Boat Voyage

A feature in Arctic discovery which is truly anomalous . . .[1]

After nine months on board ships frozen into the ice, followed by many back-breaking days hauling sledges, Penny and his seven companions felt a sense of freedom as the whaleboat heeled to the wind and lifted to the waves. They were on the sailor's natural element again, experiencing the familiar sounds of wind and water, the creak of sheet blocks, the rhythmic slap of waves against the hull, the occasional crash as the bow plunged into a deep trough. To men who had had little to do all winter other than housekeeping chores, it was exhilarating to be under way again, actively engaged on the mission for which they had volunteered a year before. They were eager to pick up the trail of Franklin and his men somewhere in the open sea that Penny had discovered.

When the wind abated, they took to the oars, continuing along the coast of Cornwallis Island as it gradually trended more and more westward. Later in the day, the wind rose to a dangerous pitch and they were forced to land. After carrying the provisions up the beach, they hauled the boat up, propped it upright, and rigged the housing cloth, transforming the craft into a narrow tent in which they cooked, ate, and slept. Whether they slept on the narrow floorboards or on planks laid along the thwarts is not clear, but eight men crammed into a boat with an overall length of under thirty feet, a beam of less than six feet, and a depth of a mere three feet could not have enjoyed much comfort.

Their first night was very difficult. The weather deteriorated until a gale was blowing and thick snow falling. The euphoria of embarking on the search faded slightly when Penny looked outside the next morning and found that "what was open water yesterday, was now covered with ice to an extent of twenty miles."[2] It was as if fate was teasing them, for on the other side of the packed floes they could see open water near Hamilton and Baring Islands and a water sky extending toward the horizon to the northwest, but they were unable to go further. Their frustration mounted day by day as the ice, driven by persistent northwesterly winds, continued to blockade the coast. They occupied their time walking the beach, investigating distant rocks that might be cairns, climbing nearby hills to check ice conditions, and strolling inland in search of game. In leads and pools among the pack ice, they saw numerous white whales, narwhals, walruses, and seals. An entire week passed and they were still confined to the place they named "Disappointment Bay."

When a wind shift began to disperse the floes and create open spaces, Penny decided to get under way. They hauled the boat across a mile and half of ice to reach the water, then returned to transport the supplies in a small sled brought for the purpose. When the boat was loaded, they set off in a lead, reached open water, and sailed to Hamilton Island in three hours. Grounded and rafted ice had built a barrier along the beach, but they managed to bring the boat in near the island's southwest corner and haul it up for the night. The next morning they spent a few hours searching the shore, then sailed east to Washington Point, where they were stopped again by extensive ice. Unable to reach the coast, they hauled out on land ice for the night. Before turning in, Penny stepped out from the covered boat for a last look around, and found that westerly winds had driven the ice a hundred yards away from the land. They were adrift! All hands emerged swiftly from their sleeping bags, and by midnight they had managed to shift the boat and all their gear to the land.

Ice kept the coast blockaded, and the men could do nothing useful except explore and hunt, neither of which brought results. But on Sunday, 29 June, things began to change. A northeast wind began pushing ice floes away from land, creating enough room to set sail. They hurriedly carried provisions to the shore and prepared to depart, but first they gathered for worship. Penny never failed to hold a service on Sunday, whether travelling by ship, boat, or sledge. He also kept the day free of work unless absolutely necessary. At Disappointment Bay, he had allowed the men to hunt on Sunday but afterward had been assailed by guilt. "A small still voice whispered to me," he confessed later. "'Is it a work of necessity?' I feared it was not."[3]

Most expedition commanders insisted upon strict observance of the Sabbath. Austin's men started every day with a prayer session. On the *Felix*, Sir John Ross regularly brought his men together for worship and usually delivered a sermon. Sunday was a day of rest, a day for pure thoughts, he believed. On one occasion, he noticed that Phillips, his second-in-command, had been *reading a novel* on Sunday! Ross quietly slipped a note between the pages: "Remember that thou keep holy the Sabbath day."[4]

Penny and his companions sailed back west, rounded the southwest point of the island and proceeded up the west coast to its northwest corner where they were once more stopped by ice. From the top of the "bold, almost perpendicular, headland"[5] Penny assigned names to features on a few islands he could see to the west. Only half a dozen miles to the north were Margaret Island (named for his wife) and Dundas Island (named for one of the Admiralty lords). Suddenly he saw what appeared to be two cairns on the latter. The discovery lifted the spirits of the men, who were "quite beside themselves with ardent hopes that they should prove to have been built by those for whom we had hitherto sought in vain."[6] The next day they attempted to cross the strait, but ice floes in rapid motion stopped them halfway across and they retreated to Hamilton Island. Here, however, the shore was mantled with rafted floes, and they had great difficulty in finding a suitable place to land and haul the boat up. Their position, Penny later wrote, was

> anything but a pleasant situation. But it was the best that we could get; indeed, I may say, the only place where a boat could be hauled up from the mercy of the drifting ice. On the one hand we had a perpendicular snowy cliff, on the other the ice pressing up and squeezing in two fathoms water, with a loud grinding noise, and tumbling over in huge blocks upon the terrace, on which we had taken shelter. The tide was very rapid, and the ice always continued in motion; no one slept, owing to the terrifying and groaning noise which it kept up all night.[7]

The next day – they were now into July – the situation was unchanged. The men explored and hunted, bagging a few hares for the pot. They could still see the cairns across the ice-filled strait and with the telescope some were confident they could see signal poles mounted above them. How frustrating to be held back by the moving ice! That night Penny was awakened by a slight sound near the boat. Thinking it might be a bear nosing about, he grabbed his pistol and cautiously peered out. What he saw was not a polar bear but just as alarming:

> Immense hummocks were tumbling over and over, with the pressure within a few yards of us. No one waited to put on his clothes. For each flew to the provisions, and conveyed them up to the face of the precipice, and then to the boat to attend to its safety. The ice on which it rested was broken into several pieces, and thrown very much from its level, by the pressure among the hummocks around

> it. In the middle of the channel it was truly fearful, and could be compared to nothing but an earthquake. Some pieces were rising to a height of twenty feet, and tumbling down with tremendous crashing and rendering. We again turned in beneath our covering; but little sleep was obtained, for every one was peeping from beneath the housing-cloth.[8]

Later they succeeded in shifting a few hundred yards to a better position but the rampaging ice kept them fenced in for another four days, during which they continued to roam about the island, searching in vain for evidence of Franklin and erecting cairns containing records of their visit. "I do not think there is a spot of Hamilton Island but has been gone over,"[9] Penny wrote. Along the shore the men found several pieces of driftwood, alien visitors to a treeless environment. Most appeared to be in their natural state, unaltered by human hands, carried north from distant forests by rivers and ocean currents. One, however, was distinctly different: a piece of English elm showing clear signs of having been worked with tools. It was a splintered part of a plank about a foot and a half long and an inch in thickness, bearing traces of a substance resembling pitch or tar. It did not appear to be scarred by contact with ice, which suggested that it had been transported by human agency. Had it been cast overboard from one of Franklin's ships, dropped by a landing party, or left by a sledging team sent out from the wintering ships at Beechey Island? Or could it have originated from a whaling ship in Lancaster Sound? (An oar from the whaler *Friendship* had been found at Cape Hotham in the previous summer.)

Although Penny drew no hasty conclusions, the elm did appear to be one more piece of evidence strengthening the case for Franklin's passage: a campsite on Cape Riley, the winter quarters at Beechey Island, a watch camp near Cape Spencer overlooking Wellington Channel, sledge ruts incised into the shore, open water in Queen Victoria Channel – and now a sawn piece of wood of a species foreign to North America.

After being confined to Hamilton Island for more than two weeks, they succeeded at last in reaching Margaret Island. From its summit, Penny could see across another narrow channel to Dundas Island where one of the cairns stood. They crossed the strait and landed. Approaching the cairn with one companion, Penny felt elation at the great discovery he was about to make. In that man-made monument of stones, he would find the first message from Franklin, which would surely reveal his route and destination

It was nothing but a pile of rocks.

They walked to the island's north coast and continued westward around the coast, walking along the beach where feasible and clambering awkwardly over ice hummocks piled up around the base of steep bluffs. At the island's southwest point they collected two more pieces of driftwood.

Back at the boat with nothing but disappointing news to report, Penny and his men resumed their boat voyage, heading north in a shore lead. Before they were brought to a halt by ice, they sighted two large cairns on the part of Devon Island lying to the north, which Penny referred to in his travelling report as Prince Albert Land. Would these prove to contain the long-sought dispatch from Franklin, or would they turn out to be empty like those they had already encountered? Were they purpose-built postboxes left by the men of HMS *Erebus* and *Terror*, or beacons built by ancient people of the Arctic for some mysterious reason, or simply some natural geologic features masquerading as clues to the great mystery?

For the moment, however, they could not reach the cairns. The ice was closing in and there was no suitable camping place handy, so they retreated toward Margaret Island. After several narrow escapes among moving floes and overturning ice blocks, they hauled out safely at its southeast corner. The next day they crossed to the coast of Hamilton Island, and after camping they woke to much-improved conditions and started for Baring Island, about thirty miles to the west. They reached it in nine hours, sometimes sailing and at other times pulling on the oars. There the men followed their usual procedure, searching round the coast on foot and examining any prominent rocks or suspicious objects. They found several pieces of driftwood but no evidence of Franklin's presence. From the summit Penny took bearings, surveyed ice conditions, and built another message cairn. There were hundreds of birds' nests but only a few contained eggs; a dozen was all the men could collect for the larder. Their provisions were getting dangerously low.

After a fruitless fourteen hours on the island, they set off for Prince Albert Land. They soon came upon a pair of walrus swimming. Penny was aware that these fierce animals, up to a ton in weight and armed with foot-long tusks, were capable of capsizing a boat and should therefore be given a wide berth, but he and his men were not only short of food but also out of fuel for the stove. They decided to risk an attack. From close range, they fired a ball into one of the animals, with no noticeable effect. They then fastened with a hand harpoon and line, while shooting repeatedly, and finally delivered a killing thrust with a lance. Somehow they managed to strip off part of the tough hide and remove blubber and meat without proper flensing tools while standing in the open boat.

The wind had died, so the men took to the oars. After labouring for twelve hours, they landed on a point of land extending out from Prince Albert Land. On top of the bluff stood one of the cairns they had sighted from Dundas Island. "The moment we landed," Penny reported, "I ascended to the top of the hill to examine the cairn."[10] It was obviously a structure made by human hands, and he excitedly took it apart rock by rock. This time he was overjoyed to see a message within.

His hopes quickly tumbled to the ground. It was a note left two months earlier by Alexander Stewart, whose sledging party had arrived after a lengthy voyage around the eastern and northern sides of Wellington Channel. They had reached their turn-back point and started on the long journey back to the ships, but not before looking over the open water that stretched enticingly westward beyond Cape Becher. Stewart wrote in his travel report, "Here, then, there was an opportunity for going ahead, if only we had a boat."[11]

Penny walked several miles north until he was overlooking a sizable indentation (now called Inglis Bay). "From the point where I stood, at an elevation of from six to eight hundred feet, nothing could be seen to the N.W. but open water and a watery sky."[12] It beckoned to him as strongly as it had to Stewart: "I erected a cairn, and took another view of the expanse of water that was before my eyes – oh to have been here only with my two little vessels, what could we not have done in the way of search?" Then he added, "But I greatly fear, if we had, the missing ships are beyond our reach."[13]

He could see that the open water stretched far to the northwest through the gap between Devon and Bathurst Islands (today called Penny Strait), and there was no sign of land beyond: "That there is a large arctic sea beyond this channel, in which the ice is constantly in motion, there can be no doubt: for where could all the ice have gone to? Where does the comparatively fresh driftwood come from? It must be from America or Siberia, and that through a body of drifting ice."[14] Penny felt that ships penetrating into this unexplored arctic sea would probably not be able to overtake Franklin, but they might discover valuable clues about the direction he had taken and would certainly acquire useful information about the distribution of land, sea, and ice.

This was the terminal point of Penny's exploration, his farthest north. With food and fuel running low, he and his men started back for Assistance Bay like horses returning to the stable. Using sails and oars, dodging ice floes, and landing from time to time on Dundas and Hamilton islands to climb to hilltops and plot a safe course ahead, they reached the place on Cornwallis Island where they had launched the boat in mid-June and found the sledge that had carried it. They had taken a full month to reach Prince Albert Land by boat but a mere sixteen hours to return, owing to the diminution of ice. July had by now nearly run its course. All the travelling parties sent out by Penny had returned to the ships weeks before.

Penny and his boat crew were now caught in a difficult transition period between the optimum times for different methods of travel. It was too late to haul the sledge along coastal beaches because the snow cover had melted, and too early to proceed along the coast by boat because Wellington Channel was still largely icebound. Hauling the sledge on the sea ice would be feasible but only if they could find a way to cross the open shore lead to gain access to a relatively flat ice surface. If they could manage to cross the lead, how far would the highway of ice take them? Somewhere to the south they would inevitably encounter open water, after which the sledge would be useless and they would need a boat. In the narrow shore lead between the land and the ice, boat travel would be feasible but discontinuous; wherever they encountered a barrier, they would have to manhandle the boat out of the water onto the ice. Then, in order to transport the boat to the next segment of open water, they would again need the sledge.

Penny decided to continue with both boat and sledge. Unfortunately this proved impracticable. It took the men four hours to drag the loaded sledge a mere two miles over ice covered with meltwater. Although neither conveyance would be useful without the other, together they were too heavy to drag. Reluctantly, Penny decided to discard both. They would *walk* back to Assistance Bay.

The men set to work mending footwear and clothing in preparation for the eighty-mile slog to the ships. As they would be limited to what they could carry on their backs, they would have to leave behind most of their gear, including the tent. Penny described the day of departure: "At 6 A.M. I called the people out, and got breakfast; packed up six sleeping bags to be taken with us, stowed away all our spare clothing, felt blanket, tent, &c., in the boat; thirty pounds of bread and a broken case of pemmican were also placed in the boat, and the whole was covered over with the housing-cloth of the boat."[15] They took only six sleeping bags, as two men would remain on watch for bears while the others slept. Most of the driftwood they had collected on the islands was left behind. After using a tent pole as a balance to ensure that the men's knapsacks were of equal weight, they left the place they called Abandon Bay at noon and started south, along the shingle beach. Four hours later, they came to a stream in full flow. "We had to strip and ford it," said Penny, who would long remember the shock of the frigid water.[16] On the next day they arrived at their upper depot, where they enjoyed some food and coffee, "followed by a glass of spirits, the first we had had for nearly three weeks,"[17] and then some well-deserved rest.

They set off again at six in the evening, travelling at night so that they could sleep more comfortably in the warmth of day. After a few hours of trudging, they came to another stream and again had to shed their clothes to wade across. With the air temperature was only slightly above freezing, it could not have been pleasant. Penny observed that at the height of the spring melt, the stream had carried large boulders out of a narrow ravine and scattered them over the sea ice far out from shore. What havoc that would have caused if a sledge party had been passing at the time, he thought.

Rain poured down during the night, drenching the weary travellers. They stopped to rest but could not sleep. Then they were off again: another stream to ford, a wide bay to cross on dangerously rotten ice. After twenty-five miles, they "lay down upon the beach, but slept none."[18] At four in the morning of 25 July they came to Separation Point, where two and a half months before, Penny's man-hauling sledge teams, eager and full of hope, had diverged onto their respective routes, only to return weeks later having found no evidence whatsoever of Franklin and his companions. Now Penny was bringing the last of all the search parties home.

Detouring out onto the rotten sea ice to get round a stream mouth, two men fell through and got thoroughly soaked. A gale-force wind and drenching rain added to their discomfort. They were still fifty miles from the ships. Penny later summed up the situation: "Some of the people would be laid up by such another night; and all of them were very much fatigued. There was nothing for us but push on at our best."[19] And so they did, but a very welcome surprise awaited them: they came upon a whaleboat that Stewart and his men had dragged from Assistance Bay for their use. Launching it into the shore lead, they pulled for several miles as far as Barlow Inlet. Beyond this point, however, the ice was fast to the shore and they could see no open water, so they hauled the boat up above the high tide mark, turned it over, and began walking again. They still had more than twenty miles to go. Seven hours later, the eight men – exhausted, wet, and cold, with clothes and boots in tatters – rounded the east point of Assistance Bay and made their weary way across the ice toward the ships.

* * *

While Penny and his boat crew were away, signs of spring had begun to appear in Assistance Bay. Each day until the summer solstice, the sun rose higher in the sky, remained visible longer, and nudged the mercury up in the glass. In July, the temperature dipped slightly below freezing on only five nights, and the mean monthly temperature was a full 67°F higher than it had been in February. The rapid melting of snow and ice was transforming the landscape in explosive bursts of violent activity. Sherard Osborn, who had travelled across from Griffith Island to await news of Penny, was a fascinated witness to the sudden disgorging of pent-up meltwater from the narrow valleys incised into the edge of the plateau. "Three large ravines broke open whilst I was in Assistance Harbour," he wrote, "and the thundering sound of the ice, water, and shingle, which swept down, and soon cut a broad channel for many yards through the floe in the bay was a cheering tune to the gallant fellows who were looking forward to being released from their winter imprisonment."[20]

As midsummer days slid by with no sign of Penny's party, the crews of the icebound ships at Assistance Bay made preparations for sea. No one knew when the ice holding them in place might suddenly break up and drift away. Relatively warm water flowing off the land was already melting the ice around stream mouths; the harbour ice was thinning and fracturing; open water in Barrow Strait was becoming more extensive. After the long arctic winter, the ships needed a good deal of attention before they would be fit for an ocean voyage to Britain, thousands of miles away. The snow banking that had insulated the ships all winter now had to be removed so that the ice surrounding them could be cut with saws. For almost ten months the harbour ice had held them in its grip. Although the ships had been growing lighter with the steady consumption of food and coal, the weight of the snow walls and drifts around the hulls had been depressing the ice attached to the hulls, forcing the vessels deeper into the water in opposition to their buoyancy. When the ice was sawed around them, they suddenly "rose out of their beds" like corks released under water, and floated free.

The ships' housing cloths were removed; the planking protecting their upper decks was lifted and the felt insulation beneath taken up. All the equipment and stores the men had shifted ashore in September and October would have to be transported back and stowed below – but first the ships had to be ballasted. The casks of fresh water that kept the ship stable during her Atlantic crossing had been expended, and Sutherland estimated that from thirty to forty tons of ballast would be needed to make the ships seaworthy again. Empty casks were checked, tightened up, repaired if necessary, taken ashore, filled with stream water, and transported back to the ships and hoisted on board. They may have been lowered by block and tackle through hatchways into the bottom of the ships, or perhaps placed on the upper deck so that water could be siphoned by hose into empty casks below. Beach gravel and shingle were also brought out for ballast. While the men worked ashore filling the casks, the dog teams made trip after trip, hauling the casks out to the ships over the deteriorating ice. Once the ballast was securely packed deep in the holds, the blacksmith's forge and anvil, along with sails, spars, cordage, and spare provisions, were removed from the collapsing snow-block storehouse near the beach, sledged out to the ships, and stowed. Standing rigging was inspected, tightened, and tarred. Topgallant and royal masts and yards were sent up, sails bent, halyards, braces, and sheets reeved. Bowsprits, jib booms, and hulls were scraped and painted. Cables were bent to the anchors. When Sutherland returned from a visit to Austin's ships in late July, he found the two ships "as neat and trim as if they had been newly launched."[21]

Despite widespread relief at having come safely through the harsh winter, a cheerful optimism generated by the exuberance of spring, and the pleasant anticipation of returning home, the men at Assistance Bay were worried about Penny's boat party. A month had passed since they had set off into uncharted waters a hundred miles from the ships. The risks associated with sailing a small open boat among ice floes in that remote region were high. Had they survived the dangers? If so, did they have enough food and fuel to make it home?

On 25 July, the weather was overcast and rainy. The temperature rose to the upper thirties (F), but strong squalls made it feel cooler. Around ten that evening, before retiring for the night, Sutherland and Petersen were chatting on deck when, looking eastward toward Cape Dungeness, they saw a group of men on foot. The visibility was poor, and they wondered if some men from the ships were still out enjoying a late ramble on the ice. A telescope revealed that it was Penny and his companions. The glad news spread quickly, and by the time the weary voyagers reached the ships, flags of welcome were flying and half a hundred men were wildly cheering them in.

They were in pitiful condition. "The men and their commander were drenched to the skin, and most of them had either their toes or heels exposed through their boots to the ice, or the rough and sharp-pointed fragments, of which the surface of the land is composed, over which they often had to perform their weary march for whole days."[22] Food, warmth, dryness, rest, and undoubtedly a good wash and a change of clothes were what they craved.

* * *

A few days after the arrival of Penny's boat party, when the wind was favourable, Sir John Ross sent off his two remaining carrier pigeons to report the news to England. The birds were not set free directly because they might return to the ship. Instead, they were enclosed within a cage suspended beneath two hydrogen balloons that would drift (Ross hoped) several hundred miles downwind until a slow match released the winged messengers for their long flight back to Ayr, Scotland. Ross sent a letter to his wife and a report to his sponsors, the Hudson's Bay Company. He had generously offered free air-mail service to others, and in his diary a "list of notes sent by pigeon" contains a total of eleven messages from Penny, Austin, Ommanney, Cator, and others whose names are not legible.[23] If all these notes were sent off on this occasion, they must have been extremely brief, presumably written on tiny pieces of very thin paper inserted into capsules attached to the birds' legs.

Unfortunately the balloon contraption had insufficient buoyancy to lift off cleanly. Over the ice the basket rose and fell, dragged and bounced, dipping into meltwater ponds and colliding with hummocks, all the while pursued by a pack of dogs, followed in turn by a posse of men anxious to protect the pigeons from the dogs in the event of a crash landing. At the edge of open water, the balloon, with the baskets and birds, lifted off at last and soared away, struggling to gain altitude, while dogs and men came to an abrupt halt. With apparent satisfaction, Ross recorded in his diary the successful departure of the pigeons: "They went rapidly out of sight S.S.E. true."[24] But the logbook of the *Lady Franklin* recorded another account of the fate of the balloon and its caged avian aeronauts: "At 4" 20 minits it fell in the water and drowned the pidgeons about 3 miles distance from the ships." Did Ross fail to see the tragic splashdown, or did he just refrain from recording it in his diary?

* * *

The navy men over at Griffith Island had also been concerned about Penny's boat party. Austin's search parties had criss-crossed the region to the south and west without finding the slightest indication that HMS *Erebus* and *Terror* had ever been there, and as no further searching was required in that quadrant, Austin saw no reason to keep his expedition out for a second winter. He had already decided to take his ships back to England as soon as they were extricated from the ice, but he could not depart until Penny returned and confirmed that no evidence of Franklin had been found in the sea north of Cornwallis Island.

When Sutherland visited the ships at Griffith Island during Penny's absence, he noticed that Austin, Ommanney, "and indeed all the officers under them, seemed to have as much anxiety about our commander's return as we had ourselves."[25] But their anxiety, he perceived, "related to themselves only, and not to its source": if Penny did not return soon, they might have to help search for him, and if they failed to find him in the next two months, they would have to remain out for a second winter. Even if Penny did return safely, he might reveal that he had located conclusive evidence of Franklin's passage north of Cornwallis Island and request Austin's help to extend the search into the polar sea. The prospect of another winter in some remote and unexplored region further north was an unattractive one, yet Austin could not leave the scene if his men, ships, and steam power were needed to find Penny or to follow Franklin. Deserting an ally would not go over well with the Admiralty or the British public.

Austin's men seemed to be demoralized by the negative outcome of their extensive searches to the west and south. One of their shipmates had died from severe frostbite, several had undergone amputation of parts of the feet, and some still bore open sores. They had their fill of daytime darkness, swirling snow, and piercing cold. Their welfare was ultimately Austin's responsibility, and Sutherland observed that he "looked upon those under his command, one hundred and seventy-nine men, like a fond parent."[26] Yet among his crews, there was "a general depression of tone, both in health and spirits, which no one sought to conceal."[27] They were not in control of their own destiny: everything now depended on Penny. As Sutherland observed, "Everyone looked in the direction of home, if Mr. Penny did not on his return take upon him to keep them all out a second winter."[28] The uncertainty weighed heavily on Austin's shoulders, and he sent "messenger after messenger" across to Assistance Bay to obtain news of Penny's boat party.[29] John Stuart, writing up his journal on board the *Lady Franklin*, recorded visits in July by Robert Dawes Aldrich and Walter Waller May from HMS *Resolute*, Erasmus Ommanney, Frederick G. Mecham, and William Dean from HMS *Assistance*, Sherard Osborn and T.R. Pickthorne from HMS *Pioneer*, and John Ward from HMS *Intrepid*.

On the day after Penny's return, an officer from the *Pioneer* and a seaman from the *Intrepid*, who had spent a week at Assistance Bay, made their way back to Griffith Island and gave Austin the welcome news that Penny and his companions had at last returned. They had searched an extensive body of open water north of Cornwallis Island to the best of their abilities. But had they found any traces? Austin desperately wanted to hear more.

Penny intended to get together with Austin, display the rough chart that he and Stewart had compiled from observations taken by the various search parties, and describe his boat journey. Together the two commanders could review all the search efforts, compare views on Franklin's probable course, and decide how to proceed. But the physical strain of the journeys by sledge and boat had so sapped Penny's energy that for ten days he remained at Assistance Bay, incapable of starting for Griffith Island. It was, as he now knew from harsh experience, an awkward period for travel, whether on foot or in boats. Assistance Bay and Griffith Island, although connected by a highway of stable ice through most of the winter, were now separated by a hazardous, shifting network of rough and rotting ice floes, meltwater pools, and leads of open water.

On 4 August he composed two letters to Austin. The first was a private note replying to a letter from Austin. After Penny's impetuous midnight departure from HMS *Resolute* more than two months before, Austin had regretted having denied him assistance to haul a boat north to the expanse of open water. To make an apology in person – "to effect a reconciliation," Ommanney would later say – he had crossed to Assistance Bay with some companions, but by the time they arrived, Penny was on his way north with his boat party.[30] Austin had stayed on the *Lady Franklin* almost a week, waiting in vain. Before returning to Griffith Island, he left a contrite letter for Penny, in which he said, "If you are offended, the desire of an honourable mind must be to make an apology. If I have offended you I never intended it."[31] Now Penny, in an uncharacteristically personal note, acknowledged Austin's gesture and declared it a "sufficient apology to any generous-minded man," adding rather pointedly, "however much his feeling had been injured by expressions and marked neglect."[32] He then commented briefly on his search operations. He had found no traces of Franklin; rapid tidal currents between islands made it "dangerous even for a boat, much more so a ship, unless clear of ice"; and "the conclusion I have arrived at after what has passed under my observation is to return to England after I have had an interview with you."[33]

The second letter was an official message in which he stated, "I forward an outline of search by the expedition under my charge. There are no traces of Her Majesty's ships 'Erebus' and 'Terror' further than a bit of English elm. As soon as I am at liberty I will come to Griffith Island and wait an interview."[34]

In the first letter to Austin, he did not mention the piece of elm, and in the second he did not appear to attribute much importance to it. In neither letter did he suggest that the search should be pressed northward through Wellington Channel. On the contrary, he clearly stated his intention to return home.

On 6 August, he set off by boat, accompanied by Lieutenant Aldrich of HMS *Resolute*, who had been visiting at Assistance Bay for two weeks, "unable to get back on account of the breaking up of the ice, and the thaw having filled the gullies with torrents of water."[35] Again, drifting floes in the strait forced them to turn back well short of their objective. The Greenlander Adam Beck managed to reach Austin's ships by paddling his kayak across lanes of water and carrying it over intervening ice floes, but although he was able to deliver dispatches from Ross, he had not been entrusted with the letters Penny had written. Austin would have to wait a little longer before seeing them.

Three months earlier at Cape Becher, Penny had regretted not having his two ships there to follow Franklin's presumed track toward the northwest. But he had admitted, "I greatly fear, if we had, the missing ships are beyond our reach."[36] When he returned to Assistance Bay on 25 July, he had told his officers that "he had seen nothing of the missing ships except a bit of elm, which might have belonged to them, and that his opinion was they had gone away to the north-westward far beyond his reach."[37] On the day he attempted to reach Griffith Island with the letters for Austin, he wrote in his journal, "We have no hope now of being of any use to our missing countrymen, whose fate will for ever remain in obscurity."[38]

Despite any solid evidence, Penny believed that Franklin's ships, after wintering at Beechey Island, had gone through Wellington Channel and toward the northwest in 1846. During the intervening five years, they had probably gone so far into the polar sea as to put them beyond help. And yet, if they had been trapped or wrecked soon after gaining the polar sea, they might still be within reach. Would it be wrong to abandon the search now? When he discussed the alternatives with his second-in-command, Alexander Stewart, Sutherland observed that Penny was "in a sad dilemma" about what to do.[39]

* * *

The *Lady Franklin* and *Sophia* were ready for sea, but one major task remained – to saw a channel out through harbour ice to the open water of Barrow Strait. During the winter months, Sutherland had been recording the growth of ice and the thickness of its snow cover at sixty different locations in the bay. Before the travelling parties set off in early April, the rock-hard ice had averaged almost seven feet thick, with three inches of snow on top. By the end of July, when they began the sawing operation, ablation had destroyed the snow cover, reduced the ice thickness to four feet in places, and made it softer. Nonetheless, to cut a channel for a distance of more than a mile was still a formidable task.

The work teams took up positions, one on each side of the proposed cut. They erected tall tripods, attached the long heavy ice saws to suspended blocks and tackle, and began hauling them up and letting their weight carry them downward through the ice. Time after time they repeated the operation, shifting the tripods and advancing slowly toward the open water. The logbook of the *Lady Franklin* recorded their progress: on the first day they sawed 210 yards, then on successive days 249, 235, 204, 195, 295, 230, 225, 270, 250, for a total length of almost a mile and a half.

The work was brutally hard and often frustrating. As approximately 85 per cent of the ice was below water level, the water would rise up into the crack after they made a cut. On cold nights it refroze, forcing them to repeat the operation the next morning. Standing sometimes in pools of meltwater on the surface of the ice was uncomfortable; they had been issued with new calf-length leather boots for the job, but they were neither warm nor waterproof.

Some thought that the men of Ross's ship were not contributing wholeheartedly to the enterprise. "The Felix's people assist in a small way," Stuart wrote in his journal, "but they seem very indifferent about the matter apparently thinking that we must cut our way out and consequently they can easily follow us."[40] In the end, Penny's work crews sensed that their efforts had been in vain, for as they had been toiling day after day, the ice had been deteriorating and breaking up on its own. After a sudden and extensive disruption of the ice within the bay on 8 August, Stuart remarked, "All our fortnight's sawing has gone for nothing and nature has done in an hour what we should probably have taken another fortnight to accomplish."[41] Yet the sawing had been necessary, as there had been no way to predict the rate of natural breakup.

On 11 August, most of Assistance Bay was ice free. The three ships were riding to their anchors when round the western point of the bay came Austin's smoke-belching steamers, towing the sailing vessels. Early in the afternoon, the four naval ships dropped anchor nearby. As soon as they were securely moored, Stewart went on board HMS *Resolute* bearing the two letters Penny had written a week earlier and the outline chart they had compiled of the region north of Wellington Channel and Cornwallis Island. Penny arrived soon afterward and in Austin's cabin gave further details about their explorations. At some point in the discussion, Ommanney joined them.

The question now facing the two commanders was whether or not to extend the search through the open water discovered by Penny. If they did so, some or all of the ships would almost certainly have to stay out for a second winter, as only a month remained in the navigation season. Penny's orders, however, warned against any "needless, reckless, and hazardous continuance in the arctic regions."[42] His three years' supply of provisions, already reduced by half, would have to support his crews for another year and a half – more if the ships were trapped in the ice – and he might have to provide for Franklin and his men, who on leaving Beechey Island had numbered 126 – three times as many as on Penny's two brigs. His orders told him to "secure the return of your ships to this country" in the summer of 1851 unless there was "a certain prospect of coming up with any part of the missing expedition" or some "unforeseen impediments" had arisen.[43]

On board the seven ships in Assistance Bay, nearly 240 men waited anxiously to see what the final decision would be. While the commanders were meeting in Austin's cabin, officers travelled from one ship to another, exchanging ideas and opinions. John Stuart reported "whole bevies" of visitors on board the *Lady Franklin*. He learned that in Austin's squadron there had been many discussions about where Franklin had gone and where he should be sought, but nothing close to agreement: "Wellington Channel, Jones Sound, Smith Sound[,] Whale Sound &c. &c. have all their advocates and supporters . . . and Captain Austin appears in a pickle as to which he will look at. I suppose in the end . . . he will come to British ground as directly and almost as quickly as ourselves. Our course I believe is home as hard as possible indeed there is nothing else for us to do."[44]

What transpired between Penny and Austin on board HMS *Resolute* is uncertain; contradictory statements would emerge later. As Austin did not keep a daily narrative and Penny's journal later vanished under suspicious circumstances, there is only one surviving account written at the time by a participant or witness – the journal of Erasmus Ommanney, who was present part of the time. He wrote:

> Proceeded to Capt. A.[,] found Penny with him giving an account of his search and opinion regarding the necessity of prosecuting it in the direction of his discovery – His firm conviction was that there was nothing to justify the risk of incurring another winter on his part, [he] expressed an anxiety to act in concert requesting one of his brigs should accompany a steamer if considered requisite. P. does not consider the straits safe navigation for any vessel[,] a

> strong current of 5 1/2 miles per hour setting between the islands which narrow the straits to small passages – the islands abound in wild fowl, [he] examined the shores of them most thoroughly without meeting a trace – [he] can hardly suppose it probable that Franklins expds. could have gone by without [Penny] meeting some trace or record of their progress – [he] picked up a small piece of elm[,] one side had been painted lead color[,] the other bleached by exposure, impossible to say how it came there or from what vessel as so many pieces of wood have been picked up along these shores – [he] would not state decidedly to A. whether it was his opinion that further examination was unnecessary, but even if there was clear water he would not extend his search up the Wellington Channel beyond the 20th Aug. – [he] was of opinion from the accumulation and heavy barrier of ice extending across the [Wellington] straits the expanded portion of the channel southd. that it could not clear out this year.[45]

Penny appears to have argued against further searching by way of Wellington Channel, and his remarks (as reported by Ommanney) agree generally with the sentiments he had expressed in his two letters, written a week earlier, which Austin had just now received and read. The two men were in agreement: no certain traces of Franklin had been found, and embarking on a thrust up through Wellington Channel would be risky and contrary to their orders. They should therefore depart immediately for Britain. But before doing so, Austin wanted Penny to state unequivocally *in writing* that no further searching by way of Wellington Channel was necessary. Such a document would clear him of any responsibility for abandoning the search in that direction.

In the relationship between these two men, there were strong currents of suspicion and distrust. Penny was reluctant to give the written statement Austin desired. If he did so and Austin then decided unilaterally to take his steamers up through Wellington Channel, Penny might be excluded from the region that his expedition had discovered and explored so thoroughly; furthermore his negative judgment about the necessity of further searching might be perceived as unduly cautious, even cowardly. At the same time, Austin could not depart for home if Penny felt that further searching *was* necessary. Neither commander wanted to continue the search but neither wanted to be seen as the cause of any curtailment of it. The two stubborn commanders were engaged in a game of bluff and counter-bluff, each worried about his own reputation and what the other might do to impugn it.

Linguistic subtleties were not within the scope of Penny's talents, and the verbal sparring with Austin put him in a sour mood. Suspecting that he had been backed into a corner, he became frustrated and irritable. His rising ire was easily apparent to Ommanney: "Penny struck me as labouring under very strong excitement and spoke in a wild manner, making pointed remarks quite uncalled for as A. treated him with marked courtesy and diffidence. On leaving the ship P.['s] conduct was such as could not be passed unnoticed, hardly deigning to reply as he went over the side."[46] Once again the fragile accord between Austin and Penny had been shattered, and Penny had fled the scene wearing his anger on his sleeve.

Late in the afternoon Austin, accompanied by Ommanney, went over to the *Lady Franklin*. "Now, Penny," he said, "let us set aside all feeling . . . We are going home non-successful. Depend upon it we shall be called upon closely . . . It is now necessary that we should take upon ourselves a certain amount of responsibility, and there must be a little correspondence pass between you and me."[47] Penny gave him a letter he had already prepared, which read, in part, "I beg to state that you already have my chart and expressed opinions, which I again send you in writing; i.e. that I have not found any trace otherwise than a bit of elm, which you have seen, and I can only judge when I see Wellington Channel what course I will then pursue. My orders are such as would not justify me in passing another winter in the Arctic seas."[48]

Austin read it but was not satisfied, because Penny was intimating that if he saw that Wellington Channel was open as he passed he might return to the search. That would put Austin in a difficult position for he would be obliged to cooperate. After insisting on a more definitive response, he returned with Ommanney to the *Resolute* to await Penny's answer. When no reply had arrived by midnight, the officers retired for the night on their respective ships. Finally, at 0200, a letter was delivered to Austin, and with considerable relief he read Penny's words, "Your question is easily answered. My opinion is, Wellington Channel requires no further search; all has been done in the power of man to accomplish, and no trace has been found. What else can be done?"[49] This careless statement would return later to haunt and hurt him.

But at last the matter was decided – they were all free to return home! By 0230 the anchor cables of the *Lady Franklin* and *Sophia* were coming in through the hawse pipes, sails were being unfurled, and the decks were being secured for sea. The little brigs made their way out of Assistance Bay, their arctic residence for the past eleven months. A few hours later, Austin's four ships got under way, and his steamers gave Sir John Ross's *Felix* a helpful tow for the beginning of the long journey back to Britain.

23

A Thousand Mile Drift among Ice Floes

> The traveller, however well qualified, is wholly at the mercy of the mighty ice-power which reigns in the arctic regions.[1]

When the British expeditions started for home on 12 August, they still had no idea what had happened to De Haven's ships, last seen ten months earlier. Austin, Penny, Ross, and their crews would have been relieved to learn that the Americans were safe, and astonished to know that they were located off the coast of Greenland – at almost the same latitude as themselves but eight hundred miles to the east.

After being blown away from the ice edge in Barrow Strait during a September gale in the previous year, they had been trapped within newly forming ice, carried northward in Wellington Channel and back again, then out through Lancaster Sound and southeastward along the mountainous coast of Baffin Island toward Davis Strait and the open Atlantic. Thirty-three men huddled together on the larger ship, *Advance*, while the unoccupied *Rescue*, embedded in the floe a hundred yards away, kept company like a faithful dog. Drifting helplessly within a desolate white world of ice and snow, the men had amused themselves with amateur entertainments and celebrated Christmas as best they could. A special dinner on the first day of the New Year featured extra helpings of food, plum cake, and a few bottles of wine. But Kane confessed, "There was no joy in our merriment: we were weary of the night, as those who watch for morning."[2]

In the United States, families and friends of the icebound officers and seamen, the expedition's sponsor Henry Grinnell, the oceanographer Matthew Fontaine Maury, congressmen who had supported the cause, and members of the public who had been inspired by the dream of a glorious American solution to the Franklin mystery were all ignorant of the location and state of health of De Haven and his crews – as people in Britain were of the status of the men serving under Austin, Penny, and Ross.

De Haven and his companions settled in for a long winter's drift. The consolidated pack ice of Baffin Bay, the ships embedded within it, drifted slowly southward. The vast moving seascape was far from featureless, however, and far from quiescent. Pressure ridges, hummocks, and rafted floes pushed up in level expanses of ice. Narrow leads of water appeared from time to time among the floes, introducing variety and danger into forays away from the vessel until they froze over or closed. Each day changed the appearance of the white neighbourhood within their vision, creating a new jumble of topographic forms. During a particularly violent disruption in mid-January, floes rotated and collided, chasms opened and closed among them, and vibrations coursed ominously through the timbers of the ship. The contortions of the ice were accompanied by a discordant symphony of sounds that set the men's nerves on edge. In his diary De Haven described "the horrid music of a line of hummocks heaving up at 250 yards on our Star[b] side."[3] The men gathered on deck carrying blanket bags and wearing knapsacks filled with survival gear, their small emergency sledges loaded. The crisis passed in a few days, but others would follow in the weeks and months ahead.

Despite the cold, the gloomy twilight of winter days, and the constant apprehension regarding ice movements, the crews enjoyed a splicing of the main brace on 8 January, the anniversary of Andrew Jackson's defeat of British troops near New Orleans in 1815 (an unnecessary battle that had occurred after peace had already been declared in Europe). A repeat performance of the popular play "Blue Devils" that evening ended with a spirited singing of "The Star-Spangled Banner," followed by applause and cheers. In late February, on one of the coldest days they had yet experienced, they celebrated George Washington's birthday with another play. With an outside temperature of –46°F, the enclosed theatre on deck could not be warmed above –30°F, "probably the lowest atmospheric record of a theatrical representation,"[4] Kane noted.

It was with a specially intense joy that the men saluted the sun's return after an absence of eighty-six days. As the moment grew near, they lined up on the ice alongside the ship to glimpse the golden globe as it peeked above the horizon, a unique "conjunction of sunrise, noon-day, and sunset,"[5] observed Kane. In his diary De Haven noted the pleasing sight of their "old friend" the sun,[6] and Lovell remarked, "it is high time the old fellow ['Sol'] came back again for the long absence of light has made us all have pretty white faces."[7]

A case of scurvy had appeared on board the *Rescue* surprisingly early in the voyage – only two months after leaving New York, while the ships were still off the Greenland coast. A month and a half later, when the two ships commenced their ice drift in Barrow Strait, Surgeon Kane had submitted to De Haven a list of recommendations to combat the scurvy: vacate the forecastle every day for two hours in order to shake and fold bedding into small piles; pack spare clothes away in bags; scrape lockers and the deck; use a drying lamp to lower humidity; keep the galley fire burning all night to dry wet clothes; subject the men to "periodical ablution of the entire person"; and distribute extra preserved meat and vegetables to the sick.[8] A few days later William Benson, the man ill with scurvy, was transferred to the *Advance* so that Kane could make him "scrub and exercise."[9]

In late September, Kane had reported three more cases on the *Rescue*, and added with evident frustration, "It becomes my imperative duty, under these circumstances, to urge immediate and active measures in accordance with the recommendations contained in my previous communication – many of which have not yet been acted upon. Otherwise, serious consequences will result to the efficiency of the expedition under your command."[10] After the crew of the *Rescue* crowded onto the *Advance* in late October, Kane urged the elimination of salt provisions and a more rigid enforcement of diet and cleanliness, "both of which owing to the peculiar habits of the sailor require a constant supervision."[11]

Every month Kane gave his commanding officer a report recommending specific measures against the disease, yet the men's health continued to deteriorate. In late December, he wrote, "The symptoms of scurvy among the crew are still increasing and becoming more general. Faces are growing pale; strong men pant for breath upon ascending a ladder; and an indolence akin to apathy seems to be creeping over us."[12] By the end of January, the situation was worse: "Pains in the joints, rheumatisms, coughs, loss of appetite, and general debility, extend over the whole company."[13] The men's appetites had declined. Whereas in October the thirty-three men had eagerly consumed a total of thirty-two pounds of food per day, now, despite the greater cold, they could manage only half as much, and that with reluctance.

In mid-February he suggested increasing the ration of preserved meat and potato, and allowing unrestricted access to dried fruit. He insisted again on a daily cleaning and airing of the forecastle, and emphasized the need for "renewed systems of cheerful exercise" for all hands.[14] De Haven acknowledged Kane's report about the scurvy situation in his diary, and wrote, "In order to check its rampages amongst our small community the rations have been changed. Fresh provisions are issued four times a week and stewed apples supplied daily together with lime juice."[15] By this time most of the officers, including De Haven and Kane, were experiencing pains in joints and bones, diarrhea, and loss of appetite. Of the two dozen seamen, only five were free of symptoms. (Not surprisingly, four of the five men *unaffected* were cooks or stewards.) In addition to administering quinine and salts of iron, Kane applied external rubs of muriatic acid, a treatment commonly used for the treatment of syphilis.

In March, Kane told De Haven that eighteen of the thirty-three men were unable to perform duty and only three men remained free of scurvy. In mid-June, he pointed out that temperatures in the forecastle were only in the mid-thirties (F) and sometimes dropped below freezing, which made it difficult to dry clothes; stoves should therefore be permanently erected. A week later, he submitted a written reminder: "I would respectfully call your attention to my communication of the 15th upon which you have declined acting favorably."[16] It was imperative to conquer the disease, he said, lest they were forced to spend a second winter in the ice.

In Kane's opinion cleanliness, good ventilation, wholesome food, and regular exercise were the keys to preventing and curing scurvy. This was an accepted model of treatment at the time and would remain so until the discovery of vitamin C early in the twentieth century and the recognition of its vital role in preserving health. The measures advocated by Kane were not easy to implement on a small ship drifting within pack ice through an arctic winter, however. Space was limited, potable water was in short supply, fresh vegetables were unavailable, and outdoor exercise were often inadvisable owing to cold. The crowding of both crews onto one vessel exacerbated the problems of cleanliness and ventilation; the forecastle in which the seamen lived was stuffy, full of exhalations, body odours, reeking clothes, and dense smoke from pipes and from lamps burning bear fat. Kane monitored the situation religiously. His constant attention to the health of the men and his frequent badgering of De Haven to implement his recommendations were in large part responsible for the men's survival without a single death from scurvy.

When a few recalcitrant seamen objected to Kane's code of cleanliness or rejected his antiscorbutics in favour of the usual staples of hard tack and salt meat, he resorted to cunning and stagecraft, pretending that the coffee-coloured water he prescribed for sponge baths was really a powerful medicinal fluid, or feeding them what he touted as a highly efficacious experimental medicine (actually a mixture of lime juice and olive oil), followed by a paste of mashed-up raw potato, sauerkraut, and molasses.

As inactivity was said to promote scurvy, the men exercised, when weather permitted, for three hours or more a day, despite the pain that accrued to knees, elbows, and backs and the absolute exhaustion that followed. "The exercise," Kane wrote, "consisted of foot-ball and sliding, followed by regular games of romps, leap-frog, and tumbling in the snow. By shoveling away near the vessel, we obtained a fine bare surface of fresh ice, extremely glib and durable. On this we constructed a skating-ground and admirable slides."[17] Few medical men would have disagreed with the treatment in mid-nineteenth century, but in retrospect, exercise may have been counter-productive. Those affected by the disease might have benefited more by rest.

Routine chores also served to get people outside and active: collecting snow to be melted into water, keeping the fire hole clear of ice, breaking out provisions from time to time, clearing snow from the housing. After the severe disruption of ice in January scattered and buried most of the stores deposited on the ice, the men made repeated outings with picks and shovels to dig into snow-covered hummocks and pressure ridges to attempt to recover some of the lost provisions and gear. In December, a wide crack opened in the ice between the ship and an outhouse, and a few weeks later the shifting of floes carried away both of their outhouses. Replacements were constructed.

Carpenters removed the bowsprit of the *Rescue*, broken in the violent January episode, and replaced it with a new one made out of spars, after the requisite iron fittings had been created by the blacksmith in a forge erected on the ice. Many trips were made across the ice to the *Advance*'s vacant consort to mend her shattered bulwarks, excavate snow and ice around her stern, and repair her damaged stern post.

Month after month, the men of the *Advance* and *Rescue* drifted in a glacial world beyond sight of land most of the time, surrounded by a seemingly limitless icescape in which few creatures were observed. But imperceptibly they were moving southward, the season was advancing, temperatures were rising, and the ice was showing signs of slackening. In late March, leads began opening and closing among the floes, and soon after this "breach in our prison-walls,"[18] the men sighted frost smoke far to the south – a certain indication of open water ahead. Since mid-February they had seen no land, but on 19 May they sighted a cliffed headland on the coast of Baffin Island that they took to be Cape Searle. They were now entering Davis Strait, where the distance between Greenland and Baffin Island is only half the width of Baffin Bay. The south-flowing current, squeezed into a narrower compass, picked up speed until they were drifting eight miles per day, about 60 per cent faster than before. In early March, De Haven had remarked in his diary that their drift was "so slow that I am fearful that we shall not be liberated from the pack until the season is so far advanced as to render it so inadvisable to renew the search for the missing navigators."[19] Now the prospect of returning to Barrow Strait to resume the search was looking brighter.

A week after sighting Cape Searle, the ships and their enveloping ice floes turned the corner past Cape Dyer, the easternmost extremity of Baffin Island, and headed toward the open Atlantic. Cape Walsingham was sighted on 27 May, and Mount Raleigh, discovered by John Davis more than two and a half centuries earlier, was visible on the last day of the month.

The widening leads were soon occupied by seals, narwhals, and even a bowhead whale, as well as a variety of water birds pausing on their flights northward – ducks, geese, guillemots, skuas, gulls, dovekies, little auks. Land birds were less common but at least one raven was seen, and for a while in June a flock of snow buntings, probably in need of rest and social interaction, landed boldly on deck and paraded about; one even perched jauntily on Carter's outstretched hand. Polar bears and foxes, frequent visitors throughout the winter, still wandered around the ships, attracted by cooking smells.

The increasing availability of wildlife meant more health-restoring fresh food for the scurvy-ridden men, although walking over the snow-covered ice with firearms was becoming more hazardous, and tasks on board ship were demanding attention. The hunters' catches included many small birds, especially little auks: more than fifty were killed on some days. Seals and narwhals were occasionally shot but seldom secured before they dived or sank. Marauding bears were shot at whenever they came close, and were often wounded and sometimes killed. One bear, struck in the thigh, led the men a bloody chase until they came upon the ball on the snow; he had bitten it out of his own flesh. Lovell declared bear flesh "very good eating, for those who like their meat flavored, with fish and seal."[20] Kane and some others ate bear liver, which is often toxic, but luckily they suffered no adverse effects. Carter noted, "Foxes are eaten as a great delicacy but I dont like them,"[21] but he did find little auks "most excellent eating."[22] De Haven considered dovekies to be "tough and fishy."[23] Every animal killed, large or small, contributed to the recovery of the scurvy patients, who were allotted extra portions.

* * *

Now began the work of preparing for a sea voyage. Thousands of miles of ocean sailing lay ahead, so everything had to be made shipshape on the two little brigs. After the interior of the *Rescue* was tidied up and dried out with stoves, her crew moved back on board with their belongings. The makeshift sleeping places for the extra men on the *Advance* were torn down and the normal quarters reconstructed. After that, although neither crew could be said to be enjoying the "luxuries of an exploring voyage" that the *New York Herald* had predicted before the departure of the expedition, they were far more comfortable than they had been while crammed into one small vessel during the past half-year.

One loss to the *Advance* by the shift, however, was the *Rescue*'s first officer, Robert Randolph Carter, a fervently religious man who objected strongly to performing work on Sunday, a practice not in accord with the *Sabbath Manual* published by the American Tract Society. When the two crews crowded together on board the *Advance*, Carter had made his views known to De Haven, after which work was banished on Sunday and church services were held regularly, with Carter officiating and serving as chaplain. As soon as he returned to his own ship, however, De Haven cancelled the next service, and no others are mentioned in his diary.

On board the *Advance*, the housing cloths were dismantled, folded, and stowed. Sails were unbent, dried, mended where necessary, and refurled. Rigging was set up. Halyards, braces, falls, or footropes exhibiting excessive wear were replaced. Boats were painted. The leather on boom jaws was renewed, and the masts were re-coppered at points of friction. The men shovelled a trench between the two ships and spread salt on the ice to speed the melting of a channel to enable them to come together again. On both vessels the men cleared snow and began cutting away the ice in which the ships were held. The *Advance* was in a particularly bad position, the stern still standing five feet higher than the bow, with a massive agglomeration of thick ice clinging tenaciously to the hull.

Gales in March and April helped to break up the consolidated floes and expand the open leads. Another occurred early in May but did not deter the officers of the *Rescue* from crossing the ice to attend a celebration of De Haven's thirty-fourth birthday. The special dinner was accompanied by the last remaining bottle of wine and featured a cake baked at home more than a year before.

As flocks of birds continued to fly overhead on their northward migration, movements of the ice intensified, and the breakup was eagerly awaited. It occurred in dramatic fashion early in June. De Haven wrote in his diary, "The whole of the immense field in which we had been imbedded for so many months was sent in all directions leaving not a piece exceeding 100 yards in diameter. Both vessels were liberated[,] ourselves partially so for the ice in which our stern is imbedded still adheres to us from the main chains aft keeping our stern elevated in its uncomfortable position."[24] Breaking free of this tenacious encumbrance of ice – referred to in the various journals as an "inveterate companion," "pertinacious friend," and "fast-sticking friend" – proved to be difficult. They tried saws; they tried wedges; they warped with strong hawsers; they considered gunpowder but could not spare enough. In desperation, they hoisted sails, but all without effect. The ship "stands motionless amid the ice," wrote Kane "although her wings are spread and tense."[25] Through one entire night, they kept full sail on and advanced "about 100 yards dragging our piece of ice with us," breaking their two best six-inch manila hawsers in the process.[26] At last their efforts succeeded in dislodging the ice mass, and the ship was afloat again. For several months, the *Advance* had been nose-down in the ice, sloping one foot in sixteen, an annoying challenge for weary muscles and aching joints. Now Kane happily exclaimed, "Even keel again!"[27] More than ten feet of the false keel had been ripped off by the ice, but no leak was detected. On 8 June, the two brigs shipped rudders and set sail together for the coast of Greenland. De Haven – incredibly – had resolved to resume the search.

Yet there was still ice to contend with. The *Advance* experienced several hard knocks from small floes, and one of her boats was damaged when they grazed a tall "hummock" (probably a floeberg, or remnant of a pressure ridge). After manoeuvring through scattered pack ice, they found themselves in the open sea, but the wind and swell were increasing. De Haven wrote in his diary, "This sudden pitching as it were into a gale with rough sea from so long a confinement to perfect smoothness caused many both officers and men to become sea sick – altogether we had a most uncomfortable time of it."[28]

Crossing Davis Strait, the ships became separated. After intersecting the coast near Holsteinborg, they sailed north, stopped separately at the Whalefish Islands, and reunited at Lievely. Here they lingered for five days, the men eating as much fresh food and drinking as much beer as they could obtain, dancing the polka and waltzing with native women, and gradually recovering their health and motivation. From a Danish supply ship, they obtained potatoes, a pig, some beer, and woollen clothes to replace their worn-out rags. Greenlanders sold them skin clothing, ptarmigans, and birds' eggs.

From Lievely, the two American ships started up the Greenland coast toward Melville Bay and the North Water, retracing the path they had taken a year before. They reached Upernavik on 7 July, the same date as in 1850. Beyond this point, as they worked through heavy pack ice, they encountered the British whalers *Jane* and *Pacific*, both returning south after an unsuccessful attempt to break through a formidable ice barrier in Melville Bay. More than a dozen other retreating whalers appeared during the next few days, including the American vessel *McClellan* from

New London, Connecticut. Most of the captains boarded De Haven's little brigs to exchange news, and after hearing the harrowing story of their winter drift, generously donated fresh food such as beef, venison, turnips, potatoes, eggs, and vegetables, as well as brandy, malt, and some recent newspapers. But whereas the British captains "would not hear of taking pay," Captain Quayle of the *McClellan* charged them "Yankee profit prices" for some potatoes and a small stove.[29] There was a certain irony here: Henry Grinnell, the benefactor who had purchased the *Advance* and *Rescue* for the American expedition, was a part-owner of the *McClellan*.

The Americans told the whaling captains about the exciting discovery of Franklin's first winter quarters at Beechey Island more than a year before. This news had not yet reached any place outside the Arctic because Forsyth had left before learning about the discovery of the graves, and the ships of Austin, Penny, and Ross were still icebound in Barrow Strait. At Lievely, De Haven had left dispatches containing this vital information to be forwarded to the British Admiralty and to Lady Franklin at the earliest opportunity, and he had undoubtedly informed the Danish authorities there. As his ships were heading into the ice again and might become trapped for a second winter, he wanted to spread the information as widely as possible so that it would be sure to reach the Admiralty.

Surgeon Kane provided captains Walker and Patterson of the *Jane* and *Pacific* (and probably other captains) with a memo summarizing the main events of the previous summer and the winter drift of the American ships. A few days later, the *Prince Albert* arrived, returning to the Arctic on another search expedition sponsored by Lady Franklin, this one commanded by William Kennedy, a Canadian fur trader. The two whaling captains went on board the *Prince Albert* for breakfast and showed Kane's memo to Kennedy, who copied the pertinent details. Soon after parting from the whaling ships, Kennedy overtook the Americans, who supplied more details about the events of the preceding summer and their harrowing winter. The news about Beechey Island "set our little society in a turmoil," Kennedy later recalled.[30] Had it not been for De Haven's courageous decision to return to the Greenland coast after the perilous ice drift, Kennedy would have reached Lancaster Sound and begun his search without knowing anything about Franklin's first winter quarters.

Like De Haven, Kennedy was conscious of the need to inform the authorities in Britain. As he was not planning to return to England for a year and a half, he decided to send passenger pigeons home with the news. Before the departure of the *Prince Albert*, Lady Franklin had given him detailed instructions about sending messages by pigeon. One was to be sent off "only in case of particular events, or of decided intelligence" about the missing expedition. The birds were to be marked with numbers from one to ten, with red or black bars painted beneath the wings, each number and bar having a specific meaning. Their tails would be coloured red to make them easily visible. A cross on the breast would indicate good news; no other mark was to be placed on the breast unless Franklin and his men were safe. She listed fifteen abbreviations to be used in the messages that would be attached to the pigeon's legs, nine representing individuals (including Penny), five for places, and one for the month. For Lady Ross, the pigeon alleged to have flown all the way home from Ross's ship in the Arctic in October 1850, there were special instructions. She was "to be sent off *only* under the joyful circumstance of Sir John's individual rescue & safety" and was to be marked with a cross on the breast, and bear the number seven.[31]

One pigeon was released from its cage on the *Prince Albert* with messages attached to each leg. The winged messenger circled the ship several times and then returned. A second attempt had the same result. A few days later, two birds were sent off. On their wing feathers was marked "Authentic trace of Sir John Franklin, C. Riley," and to each leg was sewn a note addressed to Lady Franklin, 21 Bedford Place, London, reporting the discovery of Sir John's first winter quarters at Beechey Island, the drift of the American ships, and De Haven's intention of resuming the search.

But, reported the young French naval officer Joseph René Bellot, who had volunteered to serve without pay on Lady Franklin's search expedition, "The pigeons wheeled round the ship all day, and came for their food; at night one of them was eaten by the dogs."[32] The men tried to frighten the birds away from the ship with gunfire, but the lure of food and comfort was too strong and the sound of gunshots no longer alarmed them. Another pigeon bearing a message and ready for takeoff was killed by dogs before it could be released.

The ships of De Haven and Kennedy kept company for a while but made little progress against the ice barrier in Melville Bay. Kennedy then decided to follow the example of the whaling masters. He turned south again, intending to attempt a crossing to the West Land at a lower latitude. The Americans carried on – against impossible odds. Kane described the situation in mid-August as hopeless. The deteriorating landfast ice was calving floes into the already ice-choked waters north of the Devil's Thumb: "We are surrounded by floes, and the channel outside is a compacted surface of floating rubbish . . . As far as the eye can reach, the sea, and, by refraction, the air, is studded with bergs, apparently concentrated about our anchorage." Hundreds of icebergs could be counted, all of them unstable and liable to disintegrate or capsize at any moment, and many were perilously close.[33]

De Haven's persistence was admirable but unwise. As in the previous season, he rejected the advice of whaling captains who had spent year after year sailing, tracking, towing, and warping their ships up the Greenland coast and across Melville Bay. A dozen of them had told him that the ice of Melville Bay was impenetrable in this season, and they had themselves returned southward. Kennedy on the *Prince Albert* had done likewise, on the advice of his sailing master John Leask, a veteran whaling master who had served as ice master on HMS *North Star* in the previous season, and who warned De Haven that to push further into Melville Bay could only lead to disaster. But instead of acting on their information, De Haven kept struggling against impossible ice conditions, putting his ships and scurvy-weakened crews in grave danger of being beset for a second winter.

On board the *Prince Albert*, Bellot noted that De Haven's decision to press forward into ice-choked Melville Bay was not popular with his crews. "The sailors tell us they will desert at the first land,"[34] he reported in mid-July. Two weeks later, he wrote, "Our men are not so well disposed to help the Americans as at first, because they do nothing themselves, and have themselves told our sailors that, come what may, unless the ships should get under sail [for home], they will not haul, so afraid are they of having to pass another winter yonder."[35] After the *Prince Albert* turned south, Bellot wrote, "I much fear there is a mutiny among them, our renunciation of the northern course having probably excited the apprehensions of their men. What a life is that of officers who are obliged to be armed against their own men, and to keep weapons out of their reach! For they have guns only for the officers, whilst each of our men has his own."[36]

It was not only De Haven's seamen who were discontent; the reality of the situation was apparent to his officers as well. They were trapped in the ice, unable to advance, and were still hundreds of miles from Lancaster Sound. Only four weeks remained in the navigation season. Health and morale were low after their long winter drift. "The spirit of the hunt had left us,"[37] Kane admitted. On 19 August, to everyone's relief, De Haven decided to turn for home while there was still a chance to get clear. Fortunately, he and Griffin succeeded in extricating their ships from the pack ice and made good time down the Greenland coast. They cleared Holsteinborg on 6 September, and reached New York separately, the *Advance* arriving on 30 September and the *Rescue* several days later.

Meanwhile, the *Prince Albert* had crossed successfully to the West Land at a latitude of 72° N and passed Pond Inlet by the end of August. By the time the American ships reached home port, Kennedy and his companions were sailing through Lancaster Sound, bound for Prince Regent Inlet to take up the search.

24

Homeward Voyage

We were given to understand that little remained to be done but proceed to England.[1]

In mid-August, as the American ships were about to abandon their quest on the Greenland coast and return to New York, the seven British ships that had wintered at Assistance Bay and Griffith Island also got under way for home. The three expedition commanders were well aware that if they returned to Britain after only one year in the Arctic, they would face severe criticism. Shortly after taking up winter quarters ten months earlier, Austin had outlined to Ross his plans for searches in late winter. The extensive sledging operations, he believed, would "satisfy the Public mind."[2] Ross had replied that "nothing could be better arranged than your plans to Satisfy the Public mind, which I fear is *now all* that can be expected."[3] Almost half a year before the sledging season began, these two officers, commanding a total of five ships, had already decided that there was little prospect of finding Franklin! The important thing, in their minds, was to *satisfy the public mind*. Now, several months later, having failed to find any signs of Franklin's passage during the sledging season, they felt that it was of paramount importance to present an image of having done their best. Each attempted to do this in a different way.

* * *

Austin had resolved to make a detour northward on his way home to examine Jones Sound and, if possible, Smith Sound. Even before his ships got free of the land ice at Griffith Island and headed toward Assistance Bay for the final meeting with Penny, his officers had known of this plan. Osborn later wrote, "We had been given to understand some days previously, Jones Sound was to be our destination."[4] In a memo to Ommanney, Austin outlined his reasoning: further searching in the region they had already scoured by sledge would be unnecessary and not in accord with his orders. "But looking to the instructions of the Lords Commissioners of the Admiralty to Captain Penny, relative to the search of Jones Sound, which he did not succeed in carrying out . . . I have determined to attempt a search in that direction."[5] In this memo there is a strong hint of disapproval, even blame: Penny had *not succeeded in carrying out his orders*! Austin did not bother to mention that Jones Sound had been completely blocked by ice when Penny's ships arrived off its entrance a year before, or that Penny's decision to reach Wellington Channel by way of Lancaster Sound had been perfectly in accord with his orders. Yet, despite the implied criticism of Penny, Austin's intention to look into Jones Sound was not entirely without merit. A cairn was said to have been sighted there a few years earlier; a month remained in the navigation season; unlike Penny, he had steam power at his command; and it may even have occurred to him (as it had to Osborn) that Jones Sound might offer a better avenue of approach to the open water Penny had discovered north of Cornwallis Island than icebound Wellington Channel. It is curious, nonetheless, that he did not discuss his intentions with Penny.

As the four naval ships headed out through Lancaster Sound, Austin and a few of his officers transferred to Osborn's *Pioneer*, leaving Ommanney in charge of the two sailing ships. The two steamers would examine Jones Sound while the two sailing vessels continued across to northwest Greenland to wait off Cape Dudley Digges until the steamers caught up, after which all four vessels would return to Britain.

Jones Sound is the westward-trending waterway separating Devon and Ellesmere Islands. Its funnel-shaped entrance opens eastward into Baffin Bay, and in the mouth of the funnel, Coburg Island stands like a sentinel, splitting the entrance into two channels – Glacier Strait to its north and Lady Ann Strait to its south. Thirty miles west of Coburg Island, a large peninsula extends southward from the north coast, pinching Jones Sound into its narrowest point, only about twenty-five miles wide. Further west, the sound opens up to a width of forty to fifty miles and extends for another 140 miles. At its extremity, it slants off toward the northwest through the narrow channel now called Cardigan Strait – a potential routeway to Wellington Channel and eventually Bering Strait, and thus a passage that Franklin could conceivably have taken. However, the size and shape of Jones Sound, and its outlet at Cardigan Strait were not known at the time. No ships had ever entered it.

Austin's two steamers proceeded parallel to the east coast of Devon Island as it gradually curved northwestward into Jones Sound. The scenery was magnificent, the mountainous land crowned with snow-covered ice caps and outlet glaciers. According to the report Austin would send to the Admiralty after reaching England, they coasted along the "southern shore of an extensive sound in a north-west direction, about forty-five miles" and arrived at a "fixed barrier of ice" that stretched from shore to shore. As the ice made it impossible to advance any further towards the west, they deposited a message in a cairn on "a remarkable conical island" (now called Cone Island) and, "having closely examined both sides without discovering traces of the missing expedition,"[6] made their way out of Jones Sound through Glacier Strait. Austin later stated that on looking westward from the conical island, there had been "every appearance of a well-defined outline of land stretching across and terminating it [the sound] to the westward,"[7] although he could not be absolutely certain that no outlet existed.

No observer, even in clear weather, could possibly have made out the true configuration of the coasts or the full extent of the waterway from a distance of 140 or more miles, owing to the curvature of the earth's surface and the overlapping of successive headlands and mountain ranges at different distances. Furthermore, no position in the mouth of Jones Sound provides a line of sight through Cardigan Strait, the "back door" of the sound. So, even if Austin's officers had been able to see the end of the sound at such a great distance, they would not have been able to see its outlet. The *Nautical Magazine* concluded that Austin's researches had "crushed our hopes of any trace of Franklin's ships being found in Jones Sound."[8] Apparently the magazine's editors did not wonder whether the sound might in fact have an unseen exit at its western end, or whether the hundreds of miles of unsearched coastline within the body of water might contain Franklin's ships and men or cairn messages indicating their intended course.

Like Captain Gravill, who had attempted to take the whaling ship *Abram* into Jones Sound on Lady Franklin's behalf two years before to look for the cairn reported by Captain Lee of the whaler *Prince of Wales*, Austin had gone no further west than Cone Island – barely within the entrance to the sound. It must have been especially galling for Osborn, whose "ice quartermaster" on the *Pioneer*, Robert Moore, claimed that he had been with Lee on the *Prince of Wales* during her alleged deep penetration in 1848, and told Osborn that they had sailed westward within the sound for fourteen hours on one day and continued on the next, covering a total distance of 150 to 160 miles, and they had then seen "open water as far as we could see from the masthead."[9]

But for Osborn, the failure of the two steamers to penetrate Jones Sound was not altogether a bad thing. He noted "the unpleasant fact that we had parted from the ships containing our main stock of provisions, without the means of following up any traces, should we be happy enough to discover them, of the poor missing expedition."[10] The steamers *Pioneer* and *Intrepid* were carrying food for only three months, the rest being on board the *Resolute* and *Assistance*, which had been sent to the coast of Greenland. Without enough provisions to last over the winter, the steamers could not risk being beset inside Jones Sound. What had Austin hoped to accomplish other than creating the impression that he had succeeded in searching Jones Sound whereas Penny had failed to do so?

The steamers passed out of Jones Sound and through Glacier Strait, then headed eastward across the North Water. Before parting from the sailing ships, Austin had told Ommanney that he would examine Smith Sound as well as Jones Sound, but instead he headed directly across the North Water to Wolstenholme Sound. A party landed to inspect the abandoned native village near the site of HMS *North Star*'s wintering in 1849–50 (even though it had been investigated a year earlier). A few days later, several native men came out to the ships in dog sledges made entirely of whalebone, delighted to receive some pieces of wood and metal. Ice conditions then deteriorated badly, and the two ships were separated more than once among jostling floes. With the aid of the engines, the crews worked feverishly to move their ships forward, eager to reach Cape Dudley Digges and regain contact with the two sailing ships. Things were getting tense: they were almost into September, birds were flying south, and temperatures were dropping. Pack ice was consolidating around them, and icebergs were on the move. Sherard Osborn, commander of the *Pioneer*, summed up the desperate situation: "To stop was to be beset for the winter, and be starved and drifted Heaven knows where."[11]

After the ships had been pushed several miles apart by the movement of floes, the *Intrepid* was subjected to severe pressure, and before steam could be raised, the vessel was driven violently onto the underwater tongue of a berg. Masses of ice towered above the bulwarks, threatening to crash onto the deck, and the entire ship was pushed upward until the taffrail was forty feet higher than the level of the adjacent floes. The ship's stern was high enough for Cator to walk beneath it on the ice to look for damage, and the ship listed so precariously that he ordered the hatchways battened down in case she fell over onto her side and took in water through the deck. Emergency provisions were brought up, and boats were placed on the ice. For two days, disaster threatened, the floes pressing the ship against the berg with varying force, sometimes rafting up violently against the lower side of the hull and at other times breaking asunder and allowing the ship to slip down the berg. Three times the *Intrepid* was elevated to a dangerous position and then mercifully lowered again nearly to her normal level. The last period of pressure and uplift, accompanied by the dreadful sound of cracking timbers, demolished the rudder. Yet somehow the ship survived. In Cator's opinion, "No human power could possibly have saved the vessel from utter destruction, nothing but the intervention of Divine Providence could have done it."[12] Had the ship gone down, the men would have had almost no chance of surviving. The iceberg was too steep and slippery to provide a refuge, the temperature was near freezing, and open boats would have had little chance among the densely packed, wind-driven floes.

On 1 September, HMS *Pioneer* spoke the *Resolute* and *Assistance*, which had been waiting outside the zone of pack ice. Ommanney had been worrying what would happen if one or both steamers failed to make the rendezvous, a contingency not covered in the orders he had received from Austin. Now the contingency he had feared had occurred: one steamer had turned up, but the other was missing. Austin, now on board the *Resolute* again, decided to send the *Pioneer* to search for the *Intrepid*. To Osborn, the assignment was daunting: "The single steamer was, alone, unaided, to penetrate the pack and seek her missing mate. Find her, if she could; if not, winter, and seek with foot parties, both this autumn and next spring."[13]

The *Pioneer* loaded enough provisions for two years, discharged several ill or weak men to the sailing ships, and took on healthy replacements. After prayers and an emotional farewell speech by Austin, Osborn's steamer embarked on its mission. Two hours later, Osborn and his crew heard the sound of gunfire, alerting them to a recall signal flying from the masthead of the *Resolute*. The *Intrepid* had been sighted!

"The evil was now averted," wrote Osborn; "the whole squadron was united."[14] On 5 September, the four ships set off homeward.

* * *

After departing from Assistance Bay, Sir John Ross, whose small, under-equipped expedition had accomplished little during the winter, resolved to pay another visit to Beechey Island, even though it had been thoroughly searched by men from several ships the previous autumn. After getting a tow across the mouth of Wellington Channel, the *Felix* separated from Austin's squadron, a party went ashore, and a cursory and fruitless examination ensued. Ross then took the *Felix* across to Greenland north of the Middle Pack and sailed south to Disko, where he took Adam Beck before a Danish magistrate to make a deposition under oath regarding the controversial interview with the natives of Cape York more than a year before. Ross was still insisting that Beck had made an accurate translation of their remarks and that Franklin's ships, after wintering at Beechey Island, had returned across Baffin Bay to northwest Greenland where they were burned and their crews murdered. A month earlier, at Assistance Bay, he had composed a letter to Archibald Barclay of the Hudson's Bay Company, his principal sponsor. As reported by the *Times* (4 October 1851), he stated, with his customary measure of melodrama, "It is my intention, if possible, to make a thorough inquiry into the truth of Adam Beck's report of the ships having been wrecked in the space between Whale Sound and Cape York. This, indeed, may detain me another winter, and with 16 men I am not very able to cope with a numerous tribe of hostile savages; but I *must* and *will* try." To call the natives of the Cape York region *numerous* and *hostile* was mere theatre. When Ross had first encountered them in 1818, they had been very few in number and altogether peaceful, and in the summer of 1850 he had not even gone ashore to meet them.

As Ross did not have enough food on board for a second winter, he intended (he said) to take possession of the provisions supposedly left at Disko by HMS *North Star* a year earlier and then sail northward past Cape York to search for the wreckage of Franklin's ships. But Saunders had not taken the supplies to Disko, having instead followed Penny's suggestion and landed them at the Wollaston Islands in Lancaster Sound without telling anyone. Ross was therefore obliged to "defer" his search and shape a course for Scotland.

Ross's diary, however, contains no reference to the alleged plan to return north, or even to using the supplies supposedly left by Saunders. According to the diary, the *Felix* dropped anchor at Disko on 30 August, and Beck, whose services as interpreter would have been essential if Ross did return to Cape York and Wolstenholme Sound, was discharged the same day. On 1 September, the crow's nest was taken down, and the ship was ballasted and watered. The following day, after Beck's wages were settled with the authorities, the *Felix* left for Scotland.

Two weeks short of Scotland, doubtless recalling some of the predeparture frolics sixteen months before, Ross braced up his crew to withstand the emotional stress and unsavoury temptations associated with homecoming by delivering another sermon against drunkenness. He arrived in Ayr to a hero's welcome, beginning with a thirteen-gun salute and "hearty cheers from several hundreds of anxious spectators who crowded the piers, while the vessels in harbour displayed their colours." After landing, "the gallant Admiral"[15] was proclaimed a freeman of the burgh and entertained at a special public dinner in the King's Arms public house. More than a hundred guests listened as he gave an exciting account of the expedition, richly embellished with false statements, including the following: the *Felix* was the only ship that had not suffered damage; Ross had discovered the graves on Beechey Island; Ross had sailed further north in Wellington Channel than any other ship; Ross had discovered land beyond Wellington Channel; Ross's interpreter had revealed the place where Franklin's ships had been wrecked; Ross's second-in-command, Phillips, had carried the flag donated by ladies in Ayr to "the utmost extremity of Cornwallis Island."[16] After telling his audience that whoever commanded the next search expedition must be able to speak Danish, Ross modestly mentioned that he was almost the only officer in the Navy who did so and declared that he was ready to serve if invited (great applause).

Ross complimented local firms that had provided supplies for the expedition. "A comparison of the stores among the different discovery ships in the Arctic regions was made, and those of the Felix were declared to be far superior to any of the others," he announced. (In fact, no such survey took place. Indeed, Ross had regularly received antiscorbutics from Penny to combat scurvy among his crew and had been forced to beg provisions from HMS *Resolute* near Griffith Island and obtain more from Ommanney's depot near Cape Hotham. According to his own diary, a boat sent twice to Ommanney's depot in July 1851 had returned with a total of 1,360 pounds of preserved meats, as well as 300 pounds of chocolate, tea, and bread.)

Ross was well known (although perhaps not at Ayr) for taking liberties with the facts. Lady Franklin once said, "I think that Sir John Ross is deranged in one point, viz. That he *cannot* speak the truth."[17]

* * *

While the ships of Austin and Ross lingered briefly in Baffin Bay, those of Penny hurried homeward, pausing briefly to send a party ashore at Cape Hay, Bylot Island, in case any new orders had been left there by whalers (they found nothing), and to land briefly near the entrance to Pond Inlet to question Inuit (none were present).

As they proceeded southward, the Middle Pack of Baffin Bay kept them hugging the Baffin Island coast, frustrating Penny's intention of crossing to Greenland to return Petersen and his dogs to Upernavik. Halfway to Davis Strait, they managed to work away from the coast through ice streams and bergs until they struck soundings near the submarine bank off Rifkol, to which British fishing vessels often came for cod. The men set out fishing lines, without success. "We are likely to go into Lievly [*sic*] tomorrow if the wind holds," Stuart wrote, "as it is thought a likely place for orders being sent to in consequence of the North Star having been ordered to land her provisions there last year."[18] But ice now became more troublesome, made more dangerous by the occurrence of thick fog, and after several close brushes with disaster, Penny decided to bear away for Scotland. The *Lady Franklin* was a faster sailer than her consort, and Penny was anxious to get home as quickly as possible, so he crowded sail onto the little brig and drove her hard across the Atlantic toward Scotland, leaving Stewart and the *Sophia* to follow at their own speed.

As the ships had sailed southward through Davis Strait, the weather had been growing warmer, and a succession of sunny days made the dogs on deck uncomfortably hot, irritable, and inclined to fight. On board the *Sophia*, Stewart gave an order that his men found very difficult to carry out – to put most of the dogs to death. Sutherland wrote,

> This was a great trial to the seamen, who were all exceedingly attached to them; and to the man who had charge of them, and who had watched over them during the long winter, it was a source of great grief and not a few tears. Although four were kept alive . . . there was not a person on board but felt ashamed of the deed of extermination that had just been perpetrated. I must confess that we might justly recoil at the idea that we had destroyed in wholesale manner the poor brutes that had served us so faithfully, and had been our companions for such a length of time, the moment that they began to be troublesome, and their services were no longer required.[19]

On the following day a powerful impact suddenly jarred the *Sophia* and set her masts shaking. While sailing at six knots or more, she had collided with a fin whale. It was almost as if the whale had intentionally rammed the ship as punishment for the heartless execution of the dogs.

There is no indication that a similar annihilation was carried out on board Penny's *Lady Franklin* – nothing in the log, and nothing in Sutherland's book. In his journal, John Stuart mentioned that they had "got clear of some of the dogs" on 11 August, before leaving Assistance Bay, but he probably meant that they traded or given them to officers on board Austin's ships. Certainly, Penny arrived home with his favourite dog, Sultan, which he gave to John Barrow, and another dog, Tommy, which is said to have later been taken to New Zealand.

Four weeks after departing from Assistance Bay, the *Lady Franklin* was approaching the northern tip of Scotland. In his cabin Penny wrote a long dispatch for the Admiralty, summarizing events since the departure of his sledge parties from the winter harbour in mid-April. He described how he and his men had sighted a large expanse of open water during their sledge trip north of Cornwallis Island and recounted their exploration of the region later by boat. As for the tense meeting with Austin at Assistance Bay, he wrote as follows, as quoted in the *Times* (12 September 1851): "On the 11th of August Captain Austin's ship[s] entered our harbour in their progress to the eastward. His parties had penetrated as far as ships could hope to go, yet, like our own, [were] unsuccessful in finding the least trace of the missing expedition. In fact, none had been found such as would warrant the risk of a second winter, and, my orders being such as left no alternative, I determined on immediately returning to England, if no instructions to the contrary should be met with."

As the *Lady Franklin* drew near to the Shetland Islands, a steamer was sighted, and Penny ordered rockets fired to attract her attention. The *Lady Franklin* turned into the wind, a boat was lowered, and Penny boarded the vessel, which turned out to be HMS *Tartarus*, engaged on a fisheries survey. After he explained to her commanding officer, Lieutenant-Commander Risk, that he had been engaged on the search for Sir John Franklin and had an urgent need to report to the Admiralty, Risk took Penny on board and headed immediately for Aberdeen, leaving the *Lady Franklin* to follow under the command of the first mate, John Marshall. The steamer arrived in Aberdeen at 4 a.m. on 10 September, and two hours later Penny was on the express train for London. As he sped southward over the rails, the following message was delivered to the Admiralty from the Strand station of the Electric Telegraph Company: "Penny, Arctic Search landed Aberdeen H.M.S. Tartarus. Express train with Despatches."[20] Owing to the invention of Morse code and the extension of the electrical grid in Britain, news could now travel even faster than trains.

25

Penny's Arrival

They were Captain Austin of Scarborough town
Brave Gravel and Penny of much renown
And Captain Ross and so many more
Has long been searching the Arctic shore.[1]

The news of Penny's arrival spread quickly. At the royal retreat at Balmoral, Queen Victoria and Prince Albert received relevant dispatches and a copy of Penny's outline chart within a few days, sent by W.A.B. Hamilton at the Admiralty. After examining them with great interest, they sent the documents back to London. Victoria and Albert were very interested in the progress of the search expeditions. Albert had donated an organ for Penny's expedition, and another for Forsyth's little vessel, which bore his own name. The queen had an intense interest in the native inhabitants of British North America and assumed them to be her subjects. In 1843 she had met some Ojibwa people, and soon she would have an opportunity to meet a family of three Inuit from Baffin Island at Windsor Castle.

What the Admiralty and the general public knew about the searching activities in the eastern Arctic was very limited. A year had passed since Forsyth returned from Cape Riley with some relics left by men from Franklin's ships, but as he had departed too soon to learn about the discovery of Franklin's winter quarters, nothing was known yet in Britain about that very exciting development. What further progress had been made by the other nine ships during the recent winter was also unknown.

Several days before Penny reached Britain, a whaling ship from Davis Strait had arrived at the port of Berwick-on-Tweed at England's northeast corner. Captain Ord of the *Tyne* brought a dispatch for the Admiralty written by Captain Parker of the *Truelove*, enclosing a letter entrusted to him two months earlier by Surgeon Kane of the American expedition. Parker was one of a dozen whaling captains who had encountered De

Haven's ships near Upernavik after their long winter drift in the pack ice. Like the other captains, he had obtained a copy of Kane's memo, and somewhere on the whaling grounds of the West Side he had consigned it, along with his own covering letter, to Captain Ord, who was about to leave for home. The documents were published in the *Times* on the day before Penny arrived at Aberdeen.

Parker had been among the captains who accompanied Penny into Cumberland Sound in 1840, explored its shores, and introduced commercial whaling. Nine years later, the two men had sailed into Lancaster Sound together looking for signs of Franklin, following which Parker had returned to England bearing Captain Kerr's account of the Inuit story about ships frozen into the ice and the Inuit sketch allegedly depicting their situation. Now Parker and Penny were linked in the search again. They were friends but also, to a degree, rivals. Penny was about to land at Aberdeen and reveal the exciting news that Franklin's first winter harbour had been discovered, but Parker had scooped him by sending Kane's memo to England on another whaling ship, with a covering letter.

Kane's letter provided the Admiralty and the British public with the first news of the sensational discoveries on Beechey Island a year before – the headstones and their inscriptions, the storehouse and blacksmith's shop, the gravel-filled meat cans, the countless artifacts littering the landscape. "There can be no doubt," wrote Kane in the *Times* (9 September), "that the cove between Cape Kiley [Riley] and Beechy [Beechey] Island, facing Lancaster Sound, was the first winter station of the missing vessels." (The newspaper's editors and typesetters, hurriedly reading Kane's handwritten memo, did not do justice to the unfamiliar arctic place names, and apparently did not have recourse to adequate maps. According to the newspaper, Parker had written from "Stolstinburg" – their version of Holsteinborg.)

Kane's letter had this to say about the ice drift of the *Advance* and *Rescue*: "The American vessels, while proceeding homeward, were frozen in opposite Wellington Channel, drifting during the ensuing winter from a latitude of 75.25 throughout the channel and sound into Baffin's Bay." His wording did not make it clear that they had drifted to 75°25' N *in* Wellington Channel, but Parker's covering letter clarified the matter. The American ships, he explained, had been heading homeward when "a gale sprang up and drove them up Wellington Channel 50 miles, and afterwards they were frozen in." Neither message revealed that the Americans had sighted and named land beyond the limit of European experience. Parker may not have known about this, and Kane, who was probably in a hurry to complete the letter before the ships separated, may have given priority to the crucial news about Franklin's winter quarters rather than the discovery of new land. Or perhaps De Haven had urged his officers to say nothing about their geographical discoveries until an official announcement could be made at a later date. Whatever the reason for this omission, it was later to lead to confusion.

The news about Franklin's winter quarters, communicated by Kane via Captain Parker and relayed to the Admiralty by Captain Ord, was vitally important and certain to inspire hope. But it was woefully incomplete: an entire year was still missing from the record. Nothing whatsoever was known yet about the wintering of the expeditions under Penny, Austin, and Ross, or their sledging operations in spring. Had they ascertained Franklin's course from Beechey Island? Had they found him? Or were they still persevering in the search? Penny's arrival promised to bring the record up to date.

In London, Penny delivered to the Admiralty a dispatch he had compiled at the winter harbour just before the sledge trips, summarizing proceedings since the previous fall, including the discovery of the graves, storehouses, and watch camps on and near Beechey Island. He also presented a dispatch from Austin, dated 12 August, and the report that he himself had written at sea shortly before reaching Scotland. The dispatches summarized the searches carried out by sledge parties and revealed that the three British expeditions were now on their way home. When these documents were published in the major newspapers, the reading public had a general outline of what the government expeditions had accomplished – and not accomplished – in their voyages of almost a year and a half duration. Their discovery of Franklin's first winter quarters was tremendously exciting news, but there was widespread disappointment that they were coming home after passing only one winter in the Arctic, having learned nothing further about the missing ships and crews. The ships of Austin and Penny were all intact, their crews were healthy, and they had food enough to last another winter. So why had they abandoned the search?

Some readers in England may have thought it curious that while the seven British ships were heading homeward, the relatively inexperienced Americans, after spending a dreadful winter in drifting pack ice, had bravely resumed the search! And those who followed developments closely might have been conscious of a certain irony in the fact that the discovery of Franklin's first winter quarters – which could have been revealed a year earlier by Forsyth if he had only stayed a few days longer in Barrow Strait – had been announced by an expedition from the United States of America.

Penny also wrote a long letter to the Admiralty, one that must have put the lords commissioners on full alert. Writing from the Ship Hotel on the Strand, he explained his early return by saying he had been "fettered" by his instructions, which had left him no alternative but to return home. He pointed out that none of the commanders had been aware that HMS *North Star* had established a fallback depot at the Wollaston Islands in August 1850 (on Penny's suggestion). They had all assumed that Saunders had followed his orders to the letter and had taken the supplies across to Disko, where they would be useless if Franklin or any of the search expeditions needed them. It was "beyond a doubt," Penny now declared, that Franklin's ships had gone up Wellington Channel and northwestward through Queen Victoria Channel, and he urged that they should be "followed with the utmost energy, determination, and dispatch." Although an ice barrier had blocked Wellington Channel in 1850 and 1851, he was confident that a steam-powered vessel "would do with safety what nothing else could accomplish." In the open water of Queen Victoria Channel, he had seen walrus, seals, migratory ducks (a month earlier than they were observed further south!), and an abundance of driftwood, including a "piece of drift wood of English elm, evidently to me a bit thrown overboard from the Erebus and Terror . . . And from all these circumstances of improved food and fuel, it is impossible to doubt that Sir John Franklin's party may be still living in this most northern region."[2] Penny then made an astounding request:

> In conclusion I venture to put before your lordships the importance of immediate steps for the prosecution of the search; and I entreat you to send up a powerful steamer without delay to follow up the work, which we have failed to accomplish for want of fitting means to do so . . . If the promptest measures are adopted, and a steamer prepared with all possible dispatch, she may easily reach Lancaster Sound before it is closed by ice for the season; and for the command of such a vessel, to be accompanied by the Lady Franklin and Sophia, whose officers and crews are to a man willing and eager to return, I beg to offer my services.[3]

How can one explain this extraordinary letter? In it Penny advances opinions contrary to those he had expressed before. Not only are the sentiments in stark contrast to all his previous statements, but the letter has a clarity – an eloquence, one might almost say – that is uncharacteristic of his usual writing. Suddenly all is certainty rather than speculation. It is now *beyond a doubt* that Franklin had gone up Wellington Channel and away to the northwest, where he *must* be sought with a steamer, and sought *immediately*! A month before, at Assistance Bay, he had assured Austin that no further searching up Wellington Channel was necessary. "All has been done in the power of man to accomplish and no trace has been found," he had written. And in his dispatch to the Admiralty written at sea on 8 September – a mere four days earlier – he had stated that not the "least trace of the missing expedition" had been recovered in the waters north of Wellington Channel. Yet he was now calling the piece of elm evidence that Franklin had gone that way and pressing the Admiralty to search that very region without delay!

It is not unreasonable to think that Lady Franklin, that magnificent manipulator of men, was partly or even wholly responsible for Penny's radical change of heart. She may even have composed his letter to the Admiralty herself. On that same day, at her home in Bedford Place, she penned her own letter of support on his behalf to the lords commissioners of the Admiralty: "I implore you then not to abandon your brave countrymen and Her Majesty's loyal and faithful subjects to their fate at the very moment when the pathway to them has been found." The way to locate her husband and his companions, she insisted, was to send Penny out again, equipped with a steam-powered ship. Never one to overlook details, she even had a specific vessel in mind. She was generous in her praise for Penny: "As a skilful, experienced & brave commander, in the Arctic seas, & as a humane, kindhearted man, Captain Penny is I believe inferior to none," and she assured the Admiralty Lords that his "premature return home" was a result of his having misunderstood his orders, "which assuredly were not meant to make his prolonged search dependent on the discovery of successive documents of the course taken by the missing ships."[4]

The two letters written on the same day, one by Lady Franklin and the other ostensibly by Penny, outline the same plan of operation, even to the point of recommending that a supply vessel be sent north in the following summer. If they were not in fact both written by Lady Franklin, they were certainly composed by two people acting in concert, Penny providing information and opinions about his recent search, and Jane Franklin putting a concrete action plan into acceptable wording. Although Penny was probably being criticized for returning after only one winter, his reputation might be safeguarded if it could be shown that he had followed his official orders conscientiously and could not have penetrated the ice barrier in Wellington Channel without steam power.

Unwilling to leave any stone unturned, Lady Franklin also wrote to Henry Grinnell, the American initiator and patron of De Haven's expedition, on 12 September 1851. She said, as reported in *The Leader* (1 November):

> Captain Penny's letter is to-day before the Board of Admiralty, urging them in the strongest terms to despatch instantly a powerful steamer to Wellington Channel, in which quarter, to the north-west, he has discovered the passage which there can be scarcely a doubt the ships have taken, since it is the only opening they have found anywhere . . .
>
> . . . I have written a strong letter also, and I know that Sir F. Beaufort, Captain Hamilton, and Mr. Barrow, who are all at their posts, will allege the necessity of prompt measures with all their powers. Should we fail we must look to America alone as our resource.

The letter was published in various newspapers. Its thinly veiled appeal for American help was intended to put pressure on the Admiralty. The alarming prospect of another nation solving the mystery of Franklin's whereabouts, and conceivably rescuing him and his men, was seldom out of mind in Britain. According to the *Illustrated London News* (20 September), there was already talk in the United States of "fitting out the steamer wanted, and of robbing England of the glory of the enterprise."

When Penny's proposal to take a steamer back to Wellington Channel was made public, it brought forth highly favourable comments in the press. The *Illustrated London News* (20 September) reopened its file of light and dark metaphors, declaring, "A ray of hope has now broken in upon the darkness." The *Athenaeum* (20 September), after expressing regret that Penny was not still in Wellington Channel, stated that he still had enough time to search in that region "if the Admiralty will for once act without being trammelled by board-room ceremonies and official etiquette, and place a powerful steamer instantly at his disposal." The *Aberdeen Herald* (20 September) wrote, "There can be little doubt . . . that the voice of the country will support Captain Penny in the urgent entreaties he is now making with the Lords of the Admiralty to give him a powerful steamer, for the purpose of pushing up Wellington Channel, into the hope-inspiring field of investigation he has been so fortunate as to discover."

"The pathway to them has been found,"[5] Lady Franklin had written. "The trail has certainly been struck," echoed the *Athenaeum* (20 September). "The door seems to have been opened to us," proclaimed the *Dublin Evening Packet* (4 October). Some newspapers went even further. The *Illustrated London News* (20 September) declared, "It has been proved almost beyond the possibility of a doubt that the Great Arctic Ocean has been discovered, that Sir John Franklin has sailed into it, that, in confirmation of a supposition entertained long since, this vast Polar Sea enjoys a milder temperature than the Arctic regions previously known, and that it abounds with animal life." In fact, none of these things had been proven but it did not take much to bring the centuries-old dream of an Open Polar Sea out of the closet. Thanks to Penny's discovery of a large expanse of open water at high latitude, allegedly with a milder climate and more abundant wildlife, the hypothesis would enjoy a revival in the weeks to come, not only among newspaper editors but also within senior ranks of the Royal Navy and among the members of the Royal Geographical Society.

Despite many favourable comments in the press, it is hard to believe that Penny's proposal to return immediately to the Arctic was sincere, for the region was already undergoing its autumn freeze-up. In the previous year, his ships and those of Austin and Ross had barely made it into winter quarters by mid-September, while De Haven's vessels had been ensnared within drifting pack ice and held fast for nine months. A year before that, the ships of James Ross had been trapped in the ice as early as 1 September. Yet Penny was now saying – on 12 September – that he could reach Lancaster Sound before it froze!

How long would it take for the Admiralty to refit a steamer and Penny's two brigs for an arctic voyage, stock the ships with all the provisions, gear, and equipment needed for a winter in the ice, and hire crews? If the ships could get away within two weeks – highly unlikely – they would not reach Disko until late October, by which time the expanding Middle Pack might seriously impede or prevent altogether their crossing of Baffin Bay. Even if they could somehow manage to reach Lancaster Sound before its drifting pack ice congealed into an impenetrable barrier, and get as far west as Barrow Strait, they would find all the potential winter harbours in the region sealed off by landfast ice.

It was an unrealistic proposal, and Penny more than anyone else would have known the reality of the situation. His request for a steamer seems to have been more of a gesture than a serious proposal – a gesture probably conceived in the active mind of Lady Franklin in her relentless attempts to extend and intensify the search for her missing husband.

The premature return to Britain of her "Silver Penny," the bold arctic navigator who had promised to bring her husband back, had been a dreadful blow. As Sophia Cracroft confided to Scoresby, "This return of Captn Penny has told visibly upon her frame, both mentally and physically."[6] Jane Franklin herself acknowledged it. In a letter to Grinnell, published later in the *Leader* (1 November), she wrote, "My heart is growing weak, and my health is sinking." The situation was terribly depressing. Her husband had been gone more than six years. Could the missing men – if indeed any of them were still alive – survive yet another winter until the search could be renewed the following summer? This dismal outlook may have caused her to put her support firmly behind the idea of Penny's returning immediately to Wellington Channel with a steamer. It seems likely she suggested the desperate measure herself.

Penny may have found the proposal, however impractical, somewhat appealing. If he was placed in command of an expedition strengthened by the inclusion of a steamer, there was a slim chance he could succeed in reaching Franklin's ships and rescuing the surviving members of his crews. Even if he failed to find Franklin, he might be able to add significantly to the charts. In his whaling voyages, he had always demonstrated a keen desire to extend the limits of geographical knowledge. A decade earlier he had brought Cumberland Sound back into the scope of European awareness. Now the prospect of blazing a trail into the fabled Open Polar Sea may have inspired him to propose – or at least agree to – a rash adventure.

In the months before Penny and the others had set out on their mission, Parry and several other prominent arctic officers had expressed the opinion that Franklin would be found near Melville Island, or possibly in the sea beyond it. When Penny returned a year and a half later, they had to reassess this conclusion. Neither Penny nor any of the other commanders had succeeded in getting much past Barrow Strait, 350 miles short of Cape Dundas, the southern extremity of Melville Island, which was still 200 miles from the ocean beyond it. On 12 September – only a day or two after Penny reported to the Admiralty – Parry sat by himself surrounded by charts and reports spread all over the floor and desk. Writing to Hamilton, he said "The plot does indeed thicken . . . I do not seem to have arrived at one definite conclusion." He was, he said, "wading through the evidence of Penny and his people, and licking it into shape under the several heads of inquiry,"[7] attempting to reconcile their discoveries with previous knowledge and hypotheses.

When Parry, Beaufort, Beechey, and James Ross had the opportunity to interview Penny, they agreed with his conclusion about Franklin's probable course. The *Times* (26 September) noted "their unanimous conviction that Sir John Franklin had taken the passage to the northwest out of Wellington Channel, and that he must be sought by taking the same route." The idea of sending out another expedition in 1852 to follow Penny's track was already being discussed. Although the lords commissioners of the Admiralty acknowledged Penny's "zeal" and the "generous offer of his services," they "declined to accept" his proposal to go back in a steamer. It was simply too late in the season.[8]

Penny's bold offer to return immediately to Wellington Channel indicated to the public that he had not abandoned the search in 1851 after all but had merely lacked the means – steam propulsion. It justified his own early return and made Austin out to be unnecessarily cautious, possibly even negligent. With four large ships – two equipped with steam engines – the naval officer had failed to "follow up the clues discovered by his subordinate," declared the *Illustrated London News* (20 September). The *Literary Gazette* (27 September) posed the question that was on many people's minds: "Why, in the name of humanity, did not Captain Austin detach one of his steamers at the first opportunity, – both of them being, we may observe, in good order, – to explore Victoria Channel!" (Queen Victoria Channel was also referred to as Victoria Channel and the Queen's Channel. Its official name is now Queens Channel.)

At this point, the verbal sparring between the two commanders in Assistance Bay on 11 and 12 August was unknown in Britain. Penny had not mentioned it and Austin had not yet arrived back. But Penny had delivered to the Admiralty the dispatch that Austin had written at Assistance Bay, in which he had stated that after receiving Penny's "official reply" to his letters, he had not deemed it necessary to extend the search up Wellington Strait. The Admiralty lords now wanted to know what exactly Penny had said to Austin in his *official reply*. They requested a copy, and when Penny responded, he gave the wording of the careless note he had finally delivered to Austin, in which he had stated that in his opinion no further searching was needed in Wellington Channel. This seemed to absolve Austin of any wrongdoing in the matter and shift the blame onto Penny.

On 9 December, the *London Chronicle* printed the letter Penny had written to the Admiralty in September. Referring to the meeting with Austin on 11 August, he said, "I begged him to give me one of his steamers, and I would take the Sophia, and would act as pilot, and go up the [Wellington] channel, and wait to see if the ice would clear away; that he would thus have the advantage of my practical experience." Now it was Penny who appeared to have had a legitimate reason for returning early. Without steam power, his little ships had been unable to do anything more, and when he had offered to pilot one of Austin's steamers up through Queen Victoria Channel, Austin had refused to make a steamer available. Again it was Austin's turn to appear culpable. As successive pieces of information reached Britain, public opinion swept back and forth like flotsam on a beach.

* * *

A week after Penny had boarded HMS *Tartarus* north of the Scottish mainland, the *Lady Franklin* arrived at the mouth of the Thames, the first of the overwintering search vessels to reach England. Taking on a pilot, she made her way upriver to Woolwich and made fast alongside a coal hulk. The brig was said to be "in excellent condition, and remarkably clean, and with as healthy and robust a crew as ever sailed in any region, and without a single complaint among the men on board."[9]

When the *Sophia* arrived three days later, Captain Stewart hired a steam tug for a tow up-river and tied up to another hulk. The dockyard requested the ship's top gallant masts and yards to be sent down and the sails unbent, but as the *Sophia*'s logbook noted, "Our people declined and seemed wishful to have a little rest." They surely deserved it, so dockyard hands were sent over to perform the work. The stores and equipment remaining on board the ships were inventoried and taken by lighter to warehouses at Deptford.

Three officers appointed by the Admiralty (Beechey, Parry, and Richardson) travelled to Woolwich to obtain precise information on the Beechey Island discoveries, the various relics recovered, the occurrence of cairns, and fundamental geographical matters such as terrain, ice, tides, currents, and wildlife resources. They interviewed captains Penny and Stewart, surgeons Goodsir and Sutherland, and mates Marshall, Manson, and Stuart, on various points.

Accompanying the report they submitted to the Admiralty was a list of articles found in the vicinity of Beechey Island by Penny's men. Altogether there were more than a hundred items, including many pieces of canvas, cloth, rope, wood, iron, tin and copper, charcoal, coal sacks, and bread bags – precisely the sort of litter one would expect to see near wintering ships and at the campsites of small hunting or lookout parties. There were also a number of distressingly personal items: "1 old brace for trowsers, 1 duck waist belt (flannel lined), part of a white shirt (collar &c.), 1 pair old marine black cloth trowsers, 3 old stockings, part of a sheepskin glove, part of a worsted mitten, 2 pieces of a neck comforter, 2 small pieces of cotton handkerchiefs, 1 small towel, part of an old blue cloth waistcoat."[10] These were sad reminders that there had been men as well as ships at Beechey Island in 1845–46 – men whose whereabouts and status were still an utter mystery.

In October, Penny and Stewart played host to another distinguished group of visitors, including four lords of the Admiralty, the surveyor of the navy, and the director of Engineering and Architectural Works. This was more of a courtesy call than an inspection, but the dignitaries agreed that the ships appeared to be in excellent shape despite their winter in the ice. (Penny had named islands in the newly discovered Queen Victoria Channel after three of the Admiralty lords – Baring, Stewart, and Milne – which may have inclined them toward a favourable view.)

The *Lady Franklin* and *Sophia* had served the purpose for which they had been purchased in 1850. Emptied of their leftover provisions and stripped of much of their gear, they were ready to be towed to Chatham dockyard by the steam tug *Myrtle* and laid up there for the winter. It would not be the end of their association with William Penny and the arctic regions, however.

* * *

At the end of September, Austin's squadron arrived home. Before reaching the Thames, HMS *Resolute* and *Pioneer* hove to off the Norfolk coast so that Austin could send official dispatches to the Admiralty. Up to this point, Penny's early return had appeared to be a result of his inability to extend the search beyond Wellington Channel because Austin had refused to provide a steamer. Now it became clear that Austin himself had abandoned the search without taking advantage of his own steam power. "Nothing can justify Capt. Austin's presence just now in England," was the opinion of the *Athenaeum* (11 October 1851).

A few days later, his ships reached Woolwich, and on 6 October copies of the crucial notes that had passed between the two commanders at Assistance Bay two months earlier appeared in the *Times*. Penny was back at home in Aberdeen at the time, reunited with Margaret and the children, when an item in the newspaper caught his eye. The article, whose short length failed to suggest the impact it would have, was headed in the usual manner "The arctic searching expeditions," and it began with the familiar words, "We have been favoured by the Admiralty with the following." It contained four short notes, three by Austin and one by Penny, all of them written at Assistance Bay on 11 and 12 August when the three expeditions were together, preparing to return home. There in black and white for all to see were Penny's exasperated and self-incriminatory words, "Wellington Channel requires no further search."

It now looked as if Penny, rather than Austin, had caused the early abandonment of the search. Two hours later, he was on his way back to London, furious that the Admiralty had released to the press the notes exchanged on 11 August between Austin and himself – notes that Austin had promised not to use to Penny's disadvantage. In a letter to the Admiralty on 10 October, he said, "I think this is not fair to myself as I had no opportunity of making any explanation of the circumstance, which is quite necessary to a right understanding of it."[11]

He then listed a number of grievances, very personal in nature: Austin had refused to help him take the boat north to the open water; Austin had rejected his offer to pilot one of the steamers to that region; Austin had shifted onto his shoulders the responsibility of saying whether the two expeditions should prolong the search or not; Austin had kept badgering him for a written declaration that further search was unnecessary, even though he had already given a verbal reply. Penny declared that he had always treated Austin with respect (Austin and Ommanney might have disagreed), but his goodwill had not been reciprocated: "I do not hesitate to say that I have received from him very unkind and, as I think, very unfair treatment treatment."[12] In these injudicious remarks, Penny's bitterness leaps off the page. The cumulation of perceived slights dating all the way back to their first awkward meeting at the Admiralty before the expeditions departed had been discharged like a man o' war's broadside.

Not surprisingly, the Admiralty took affront at his criticism of the naval officer they had appointed to command the four-ship expedition. Five days later, the secretary wrote to Penny, telling him that he would soon be given an opportunity to express his views.

26

The Inquiry

You will have an opportunity of proving any of the allegations contained in your letter.[1]

In the face of widespread disappointment about the early return of the two government expeditions and their failure to trace Franklin's progress further than Beechey Island, as well as Penny's complaint of unfair treatment from Austin, the Admiralty decided to establish an "Arctic Committee"[2] to examine the two commanders and their officers and ascertain whether or not the decision to give up the search after only one winter had been justified. It was partly a means of saving face, for criticism had been directed not only at Austin and Penny but also at the Admiralty. Since the spring of 1848, the lords commissioners had dispatched eight ships and more than three hundred men to Lancaster Sound and Barrow Strait; all of them had scurried home after only one winter without getting much beyond Griffith Island. There was no disputing that their sledging parties had covered a vast area and mapped long stretches of hitherto unknown coastlines, contributing greatly to the evolving chart of the complex archipelago; but the primary objective had been to find Franklin, not to make new discoveries, and all they had learned about Franklin was that he had passed his first winter at Beechey Island.

The Admiralty lords were aware that the criticisms directed at Austin by Penny might reflect upon the navy and upon themselves. It would not do if the public were led to believe that Austin, by refusing to commit one of his steamers to Wellington Channel, had unfairly thwarted Penny's initiative, thereby limiting the effectiveness of the search. They certainly did not wish the whaling captain to emerge victorious from the fracas at the expense of the naval captain. To open an investigation into the conduct of the two commanders and let Penny air his grievances involved certain risks, but these could be minimized by a careful selection of committee members. When its composition was announced, it was clear that the vested interests of the Admiralty and the Royal Navy would be protected. No civilian members were appointed, only five naval officers: Rear-Admiral Sir William Bowles (chairman), Rear-Admiral Arthur Fanshawe, and captains Sir William Edward Parry, Sir George Back, and Frederick William Beechey. Writing to Back in confidence, chairman Bowles assured him, "It will not be a long and troublesome affair." He went on to say, "I think we should begin with Penny, who in real truth is the only accuser, and how he will get out of all his previous assertions. I cannot devine, but we shall soon see."[3]

London newspapers welcomed the idea of an inquiry but some expressed concerns about its composition. The *Nautical Standard* (25 October) commented, "Such a committee should throw its doors open as wide as possible. There are many men of most unquestionable experience in Arctic matters, such as Dr. SCORESBY, Col. SABINE, and others, who should be invited to furnish the valuable information which they possess upon the subject." Lady Franklin and Sophia Cracroft were fully alert to the challenge facing Penny. Writing to Scoresby on 23 October, Sophia said that Penny "will hardly have a chance of legitimate indulgence if he be *tried*, as it were, by a jury of naval officers."[4]

In an undated letter, Sophia wrote to Margaret Penny, saying that she and Lady Franklin hoped to see her and Penny in the evening. She asked, "What can be said on Captain Penny's side by Lieut Osborn[,] Lieut Aldrich[,] Mr McDougall[,] Mr Allen[,] Lieut Elliot[,] Dr Pickthorne."[5] Although these officers had been on Austin's ships, they had been supportive of Penny's plan to take a steamer north through Wellington Channel. So obviously were the odds stacked against Penny that some of their naval friends believed that he should refuse to appear before the tribunal.

The Admiralty's first secretary, John Parker, wrote to Penny on 23 October requesting his presence before the Arctic Committee four days hence at 11 a.m. This left him little time to get his arguments organized, but any discomfiture at being rushed into a confrontation may have been overcome by the satisfaction he felt at being granted the opportunity to put matters straight. As he considered himself to be the innocent victim of Austin's machinations, he was glad to have the chance to present his own case in what had become a rather acrimonious dispute that threatened his reputation. As the relevant letters and dispatches had already been published in the major newspapers, the proceedings of the Arctic Committee were certain to be of considerable interest to the reading public.

Penny travelled from Aberdeen to London by train and took a room in the Ship Hotel on the Strand. On the morning of Monday, 27 October, he walked along the Strand, across Trafalgar Square, down Whitehall and through the elegant stone screen enclosing the Admiralty. Crossing the small courtyard, he entered the imposing building, which exuded authority and importance. As he walked through corridors that had seen generations of famous sea officers making their way to and fro on the country's business, and paced over marble floors that had felt the tread of Rodney, Hood, Nelson, and other heroes of the nation, he was very much in awe of the surroundings.

As the person whom the chairman of the committee had called "the accuser," Penny was the first witness on the agenda. When he was called into the committee room at 11 a.m., he faced five senior naval officers, all at least a decade or two older than himself, impressive in their uniforms, stern in their demeanour. The arctic veterans Back, Beechey, and Parry were household names, and Penny was familiar with their achievements. During his boyhood he may have seen Parry disembark at Peterhead after his remarkable traverse through the Arctic Archipelago to Melville Island. Beechey had served under Parry on that voyage and had later sailed HMS *Blossom* to Bering Strait. Back had undertaken four arctic expeditions, three of them overland and two of them with Franklin. Bowles and Fanshawe were more senior in rank and age but lacked arctic experience. A clerk, Frederic James Fegen, was on hand to record the proceedings in shorthand, transcribe his notes into longhand, and submit them each day to the committee chairman.

Penny was out of his element. The formal surroundings and the glittering array of naval officers confronting him were intimidating. What was worse, there was another person in attendance, one who was not a member of the committee at all. Seated in the room, ready to listen to every word of Penny's testimony, was the very man whose conduct he had criticized – Captain Horatio Austin, RN.

The chairman began by asking Penny whether he thought that the ice blocking Wellington Channel in the summer of 1851 would have broken up before winter. He replied that when the ships departed for home, they could not know whether the ice would go out or not, but if they had waited another month at the ice edge and if strong northeast winds had blown during that period, then perhaps the ice would have cleared out, enabling a steamer to reach Queen Victoria Channel. When in response to another question Penny stated that Austin had passed the ice barrier later than he had, the chairman rapped his knuckles. "You had better not remark upon Captain Austin, as we shall hear him during the course of the proceedings." Penny described how winds from the right quarter could affect great changes in ice distribution in a short period of time. On 9 September 1850, he had seen about fifteen miles of the barrier break away in Wellington Channel. He was certain that the northern part of the ice had never cleared out, however, and believed that a sailing ship could not have penetrated the ice that remained. As for the 1851 season, he admitted that "we left too early to judge," but he guessed that the ice probably went out completely once every two years or so. So far, his testimony was somewhat self-incriminatory, for if he truly believed that the ice went out every two years and that extensive changes in the ice cover could occur in a short time, why had he left for home a month before freeze-up instead of waiting for the breakup that might occur?

The questioning then turned to his experiences further north during the boat voyage in Queen Victoria Channel. Asked whether he thought Franklin had traversed this waterway, he stated, "I have no doubt on my mind that he went through it in clear water; my conviction has always been so. Nobody has heard me say anything else." This was a key point in Penny's testimony, and it had an impact on the committee. The chairman asked if he had taken soundings in Queen Victoria Channel. "No, we did not sound," Penny replied, whereupon Parry – who must have been either very inattentive or very hard of hearing – asked, "How many soundings did you get?"

The committee then brought up a highly sensitive issue: the apparent contradictions between Penny's account and Austin's regarding the decision to return home. "There is another question that you must expect to be asked," warned Bowles before striking right to the heart of the matter. "How was it you wrote a letter to Captain Austin stating that Wellington Channel was thoroughly searched and that nothing more could be done?" Penny replied, "Wellington Channel was searched but not beyond Wellington Channel." Bowles read out the badly worded note that an exasperated Penny had written to Austin in Assistance Bay, which stated that "Wellington Channel requires no further search." Penny now insisted that by "Wellington Channel" he had certainly not meant the region *beyond it*, through which he believed Franklin had gone, and that he had made this perfectly clear in conversations with Austin. "You mean then . . . to separate Wellington Channel from the Upper Channel?" asked Bowles. Penny confirmed that this was so: "In that remark I confined myself to Wellington Channel."

Although Penny had named many geographical features during and after his sledge and boat trips, the rough chart that he laid before the Arctic Committee contained no place names. Earlier in the morning, however, Bowles had questioned him on this point, saying, "I think it would be convenient if you were to distinguish the two channels, Queen Victoria's Channel from the Wellington Strait, or some confusion may ensue." Parry had agreed, saying, "It is a very important thing to make the distinction." Penny had probably used his own chart to show them the two connecting water bodies and the names he had applied, but Bowles appears to have forgotten this.

The question then arose as to whether Austin had clearly understood Penny's sentiments on 11 August. The proceedings went as follows:

> *Chairman*. Very good. Did you explain to Captain Austin at that time that a search should be made higher?
> *Captain Penny*. I asked him for a steamer, and would pilot her up the channel to the ice, and wait till the ice should clear away. I would have waited there a month.
> *Chairman*. That was your reason for asking for a steamer?
> *Captain Penny*. Yes; most undoubtedly.
> *Chairman*. And are you still of opinion that if the ice had cleared away you could have gone up?
> *Captain Penny*. Most unquestionably.
> *Chairman*. Did you explain that to Captain Austin at the time?
> *Captain Penny*. Yes, and he refused me that steamer to go up with.
> *Chairman*. Did you explain to Captain Austin at that time that Sir John Franklin had gone up that channel?
> *Captain Penny*. Most unquestionably.
> *Chairman*. That is a sufficient answer.
> *Chairman*. You meant by asking for a steamer that it would be the best vessel to prosecute the discoveries if the ice had cleared away?
> *Captain Penny*. Yes, the only vessel.
> *Chairman*. And your offer to pilot the steamer was on account of your own experience?
> *Captain Penny*. Yes.

Parry then asked, "Did you ask for a steamer to command it yourself?" to which Penny repeated that he had merely asked to pilot it. Parry's retort was rather curious: "Piloting through the ice I hardly understand. We are so much at the mercy of the ice that we cannot be secure." If anyone on the committee understood what piloting a ship among ice floes entailed, it should have been Parry, who on all of his arctic voyages had carried experienced whaling captains as ice masters. And to point out to Penny (who had made five times as many arctic voyages as Parry) that ships were never entirely "secure" among ice floes seems almost insulting. Penny had said nothing about being "secure" anyway. He had proposed acting as pilot for one of Austin's steamers because he had travelled along the west coast of Wellington Channel twice in each direction and had sailed an open boat for a month in Queen Victoria Channel. He considered that this, in addition to his long career in the arctic whale fisheries, would enable him to be of service to a young inexperienced steamer captain such as Osborn or Cator navigating through pack ice in unfamiliar waters.

Penny explained that when Austin had asked him in writing whether further searching by way of Wellington Channel was practicable or necessary, he had been "very angry at being written to at all by Captain Austin when we had conversed so long upon the very same subject, and I had told him all in my power, giving him my chart, and begged him to let me pilot one of his steamers up Wellington Channel." Austin's insistence on getting Penny's opinion *in writing* had gotten under his skin. "I am not accustomed to write official despatches, or probably I should have been more cautious," he admitted.

But the damage had been done. Conversations could be conveniently forgotten or misinterpreted, but a signed written statement – notwithstanding a possible ambiguity about whether or not Queen Victoria Channel was part of Wellington Channel – was impossible to deny. Admiral Bowles made this very point: Penny's note existed, he said, and "anyone on reading it would suppose you intended to convey that it was useless to wait for the breaking up of the ice, and that you were convinced that no further search was necessary." To this unfortunate truth Penny could only protest weakly that "Captain Austin could not have put that interpretation upon the letter, as he well knew my previous opinions. That letter was written under strong feelings of irritation."

Penny had been pushed to provide Austin with a written statement that justified an immediate return home. Why had he not simply said "Yes, I *do* feel that further searching by way of Wellington Channel *and* Queen Victoria Channel *is necessary*"? The reason may be that Austin had cleverly backed him into a corner. Austin did not wish to keep four large ships and 180 men out for another winter. They had done enough to *satisfy the public mind*. The drawing rooms and tea salons of London were calling, and it was time to go home. If Penny were to state unequivocally that he believed another effort should be made to penetrate Wellington Channel and Queen Victoria Channel, he would be the cause of their disappointment. If Penny stayed to resume the offensive, Austin could not afford to retreat; he would have to remain too or risk censure by the Admiralty and severe criticism in the newspapers.

Penny may have wondered what would happen if Austin did remain to assist in the search and subsequently one or more ships were crushed by ice, or if the entire squadron was inextricably beset in some remote and inaccessible locality. Who would be the scapegoat then? Penny, of course! At one point in the proceedings, he said, "I could not undertake the fearful responsibility of saying whether beyond the channel was searched." Austin had cleverly shifted the final decision into Penny's hands.

Penny's reliance on spoken words exchanged face-to-face between men, his apparent disdain for "official" written records, and his quickness to take offence had gotten him into a fix. At one point Parry asked him if he had ever advised Austin *in writing* to go up Wellington Channel. "Certainly not," he replied hotly. "What was the use of writing when I was telling him what I had done?" But, Parry asked, did he not think now that it would have been better to give a more "specific" answer to Austin? To which Penny replied, "Any one can see under what circumstances it was written, and the intention of it" – which not only evaded the question but assumed far too much, because in fact not everyone *could* understand the note's intention. When Admiral Bowles pressed him for a direct answer, Penny stubbornly stuck to his guns; "I think I pursued the right course." He refused to concede that he had acted incorrectly or unwisely. Later, however, he grudgingly admitted, "Had I been cool I should have used better words."

In view of the ambiguity of Penny's note to Austin, the committee then asked about the opinions he claimed to have expressed orally. He stated that on 11 August he had told Austin that more searching should be carried out by way of Wellington Channel, and he had volunteered to take a steamer north if Austin would provide one. As they parted, he had said "Now, Sir, go up there, and do the cause some service." The next morning he had repeated these sentiments: "Go up into the Wellington Channel, and you will do good service to the cause."

During the questioning, Penny was rattled by the presence of Austin, who not only listened to the proceedings but on at least three occasions took the extraordinary liberty of interrupting with his own remarks.

Austin testified before the committee on the third day. Penny was allowed to be present, and the chairman described this as "rather a stretch of fairness," a remark he had not made to Austin when he sat in on Penny's testimony. Evidently what was an *accepted right* for Austin was a *special privilege* for Penny. When the interview turned to the opinions that Penny had allegedly expressed orally about further searching, Austin told a different story. When asked if Penny had advised him on 11 August to go up Wellington Channel and do the cause some service, he stated, "No such thing, or any thing like it, was ever expressed." When asked if Penny on the following morning had again told him to serve the cause by going up Wellington Channel, he replied, "I declare he did not do so." When asked if Penny had requested a steamer to penetrate Wellington Channel, he said, "No, never." When Back later questioned him further on this point, he again denied that Penny had ever asked to pilot a steamer. How, then, could he explain Penny's statement that they *had* discussed these matters, Admiral Bowles wanted to know. Austin then admitted that they had talked, but it had merely been "a sort of rambling conversation . . . a rambling conversation that I never thought worthy of consideration."

At issue was whether two Admiralty expeditions comprising six vessels and more than 220 men should remain to prosecute the search through Wellington Channel or return immediately to Britain. It is difficult to see how remarks on such a vital matter could be considered *unworthy of consideration* by one of the two commanders. It seems more likely that Penny's request for a steamer during this *rambling conversation* had fallen on ears made deaf by intent. Or perhaps Austin had heard Penny's request but had chosen to ignore it and later deny that it had ever been made.

During the proceedings, Austin missed no opportunity to inflate his own accomplishments and denigrate those of Penny. After he stated that he and Penny had examined the graves on Beechey Island, the chairman said to him, "In fact you discovered undoubted traces of the missing expedition?" Austin did not deny it or bother to point out that the graves and other evidence of Franklin's wintering had already been discovered by men from the ships of Penny, Ross, and De Haven before he arrived at Beechey Island on HMS *Resolute*. "I saw that there were traces of the expedition having wintered there," was his only response, leaving the committee to think that he had played a crucial role in discovering Franklin's wintering place – patently false. And he made no mention of the fact that a party under Ommanney, his second-in-command, had been the first to land on Beechey Island but had departed in haste without even noticing the graves or any of the countless relics scattered about.

Explaining his decision to leave Barrow Strait and investigate Jones Sound on the way home, Austin managed to put in a sly dig at Penny: "There appeared to be something undone," he said, "Captain Penny not having searched Jones's Sound for which service he was equipped I proceeded there." In neglecting to point out that Penny had been prevented from entering the sound in the previous summer by impenetrable ice, he left the impression that Penny had been derelict in his duty. He did not think it necessary to admit that his own two ships, both steam-powered, had gone no further than Cone Island at the entrance to the Sound and could not have risked going further because he had sent most of their provisions away on board the two sailing vessels.

The first three days were allotted to interviews with Penny, Stewart, Sutherland, and Austin. On the fourth day, as the committee's attention turned to Ommanney, the chairman announced that he had received a note from Captain Penny objecting to Austin being present during the questioning of his own second-in-command. Penny felt that it was the right of persons who were testifying to do so without the intimidating presence of their superiors, so that they could express opinions frankly without jeopardizing chances for promotion and future postings. The committee had anticipated such an objection, Bowles told Penny, and they had decided "that Captain Austin had a right to be present during the examination of his officers." Austin continued to sit in on the testimony of Ommanney, and later he attended that of other officers under his command. Penny, on the other hand, discreetly withdrew whenever one of his subordinates appeared.

Ommanney stated that he had accompanied Austin on board the *Lady Franklin* to take leave of Penny shortly before the ships started for home, but he had never heard Penny urge Austin to "go up Wellington Channel." He had never heard Penny request a steamer, nor had Austin ever mentioned to him that Penny had done so. These denials are surprising because later on the same day, in response to a question from Parry, Ommanney testified: "I said I was ready to go up with steamers as soon as Penny gave his opinion. I was ready to go at any time Captain Austin wished. Captain Austin and I had frequently talked about going up with the steamers when we met Penny, should it become desirable." Furthermore, on 11 August at Assistance Bay, Ommanney had written *in his own journal* that Penny "expressed an anxiety to act in concert requesting one of his brigs should accompany a steamer if considered requisite."[6] Either he had a very short memory or his allegiance to his commanding officer had taken precedence over the truth.

During subsequent testimony, it became clear that Penny's idea of sending a steamer up Wellington Channel had often been discussed at Assistance Bay. Stewart said he had heard Penny ask Austin for a steamer after he had returned from the sledge and boat trip in late July. Commander Phillips of Ross's *Felix* said that Penny had said to him, "I shall ask Captain Austin to put a steamer under my orders." Dr Goodsir testified that Penny had spoken often of his intention to ask Austin for a steamer and that on 12 August there had been a rumour among the crews of Penny's two ships that he had done so. Even the officers on Austin's ship *Resolute* admitted they had been aware of the idea. Robert

Dawes Aldrich, the first lieutenant, stated that he had personally discussed with Penny the advisability of taking a steamer up the channel, and that Penny's application for a steamer had been a topic of conversation among the officers of the ship. The *Resolute*'s master, Robert Allen, admitted that he had heard people say that Penny had asked for a steamer, and Thomas Pickthorne, additional assistant-surgeon of the ship, declared that on the night of 11 August, after Austin had visited the *Lady Franklin*, Penny had told him, "I asked Captain Austin if he would let me go up Wellington Channel in one of his steamers to prosecute the search there; but that has not been agreed to." Cator, commander of the *Intrepid*, and Allen, master of the *Resolute*, had in fact volunteered to go north with Penny!

The fact that Penny's idea of piloting a steamer up Wellington Channel was common knowledge not only among his own officers but also among those serving under Austin throws suspicion on the protestations of Austin and Ommanney that they had never heard Penny make such a request.

Alexander Stewart, commander of the *Sophia*, returned twice to the hearings and, like Penny, he made some alterations to his earlier evidence for clarity or elaboration. He added that in late June Austin himself had pointed out that steamers could be very effective in the region beyond Wellington Channel. More importantly, he presented a detailed description of the crucial conversation between Penny and Austin on 11 August. Earlier in the proceedings, when the chairman had asked him to "repeat, as nearly as you can recollect, what he [Penny] said on that occasion," Stewart had had no chance to reply because Parry immediately interjected another question. But this time Stewart was not to be put off by interruptions. He testified as follows:

> After the "Resolute" came into Assistance Bay, I went on board with the chart, and Captain Penny followed immediately afterwards. After they had looked over the chart, and Captain Penny had explained to Captain Austin where he had been, he asked Captain Austin for a steamer. He said to him "You say we have been acting in concert. Let us prove the sincerity of that concert. Give me a steamer, and with the little 'Sophia' I will go up 500 miles further?" Captain Austin did not say "No;" but he drew himself up, – refused. I do not recollect the exact words he said; but it was a refusal. He said something, and Captain Penny said, "Then I know the truth of your sincerity, and I will have nothing more to do with you." Then Captain Penny left immediately, both displeased and disappointed. This took place on the 11th of August, on board the "Resolute."

Penny and Austin gave their final statements on 17 November. Penny insisted that he had requested a steamer and Austin denied that he had done so. Which of the two men was lying?

What the committee members said among themselves after the doors had closed is unfortunately not a matter of record. Mr Fegen's transcriptions of his shorthand notes, which are now in the National Archives at Kew, give only the evidence of the thirty-odd witnesses heard by the Arctic Committee, evidence that was subsequently published in the Parliamentary Papers on Arctic Expeditions (Arctic Blue Books). The deliberations of the five officers who comprised the committee occurred in camera, and if they were recorded, they have never come to light.

The members of the committee – all naval officers – perceived Penny as an outsider. Austin was one of their elite band and Penny was not. Parry, the doyen of arctic explorers, had a very high regard for Austin and had recommended him for leadership of the naval expedition with a remarkably extravagant testimonial: "I confidently believe there is no man like Austin in the whole Navy, and therefore in the whole world, for the service."[7] Penny, on the other hand, was a mere whaling captain, and the ill-mannered man did not seem to know his place! He had had the impudence to criticize the naval officer commanding four of Her Majesty's ships. Had the man no respect?

The Arctic Committee submitted its anxiously awaited report to the Admiralty three days after hearings had closed. Their judgment was that "both Captain Austin and Mr. Penny were fully justified by the tenor of their Lordships [*sic*] instructions in not risking their detention in the ice during another winter." They decided that Penny's sledging journals showed that he believed that Franklin had proceeded well beyond reach and that no search by way Wellington Channel could ever reach him; he had indicated his intention to return home and had never asked Austin for a steamer. Thus they accepted Austin's statement that Penny had not asked for a steamer and rejected Penny's testimony to the contrary, despite Stewart's declaration that he had heard Penny make that very request and the testimony of several of Austin's own officers that the idea was common knowledge on the naval ships.

The report of the Arctic Committee was widely seen as a whitewash for Austin. The *Nautical Standard* (20 December) called it an "inconsequential and puerile production." In his thoughtful analysis of the work of the Arctic Committee, Clive Holland concluded that the committee's judgment was "perversely hostile to Penny." Penny was "an utterly loyal, dedicated servant of Lady Franklin and the search; who was energetic, full of enthusiasm, highly talented and regarded by nearly all of his colleagues in the search, who presented a perfectly good argument for continuing the search by steamer, yet who, for his pains, was in consequence of the committee's insensitive judgement, maligned and, ultimately, excluded from future participation."[8]

Although Penny "lost his case badly,"[9] he emerged from the committee hearings with an enhanced reputation, while respect for Austin diminished. Holland attributed this paradoxical outcome to widespread public disappointment at the way the issue had been handled. Many resented that Penny had been an outsider facing a board of naval officers, and that Austin's presence in the room while his own officers were testifying had intimidated them. The public were sympathetic towards Penny because he was a common man whose unpolished manner and awkward statements had obviously worked greatly to his disadvantage, and they suspected that Austin had in fact treated Penny unfairly during the voyage, a point for which there is persuasive evidence.

Penny was not an articulate man, and he knew it. At the outset of the hearings, sensing that his response had been confused, he had asked the chairman, "Do you understand? As they say I sometimes put in a word that completely changes the sense of a sentence." What is curious is that Austin, whom one would expect to have superior skills in communication, was often more muddled than Penny. At one point, attempting to describe his outlook after all his sledge parties had returned to the ships and Penny's boat party was the only one still out, he made the following bewildering statement:

> During this period I had occupied my mind with plans for going overland in case of hearing from parties coming in from Captain Penny, so that in the event of anything being found I might send a party across the land at the head of the bay; my calculations were that, taking the best people I had, I must devote three weeks in order to do it. I mean to say I would send across the bay surveyed by Mr. M'Dougall. I had reports, and I suppose that I must have had something from Sir John Ross or some of the parties connected with Captain Penny who had gone to the westward. Perhaps I better tell exactly what I felt. If it had not been for delicacy in the one case, and having a doubt about land completed at the time of the year, I might have made an attempt to cross there; but that would have been mere hydrography, and would not have accomplished my object. I thought, also, that it would be going over ground my colleague was doing, and that I had better wait until he returned, having the full determination in my mind that I would never allow

a word to pass by which it might be inferred that I had any intention of returning to England.

From the moment of their first meeting in London in 1850, Penny and Austin had clashed. They were from different backgrounds, disliked one another, and often had contrary views about a situation. But it was more than that. Both men were clumsy with words, intolerant of views contrary to their own, and extremely sensitive to criticism. And both were "loners" who were reluctant to take subordinates into their confidence.

At the inquiry, Captain Stewart was asked if Penny consulted his officers about the necessity of further searching. "The matter was mentioned," Stewart replied, "but I cannot say he consulted the officers on the subject." Dr Goodsir said, "He was not much in the habit of communicating his intentions." Penny himself did not disagree. When asked if he had sought the opinion of his officers about remaining out to continue the search, he replied without hesitation, "Certainly not; the masters of our vessels take all that responsibility upon themselves."

Austin dealt even more swiftly with the same question; he simply replied, "No." Both men commanded from lofty pedestals. While this was far from unusual, it meant that their testimony could not always be confirmed by their shipmates. Penny could have made a stronger case if his officers had participated in the analysis of the situation and had been fully aware of his real intentions regarding the search through Wellington Channel.

* * *

After the conclusion of a government expedition, it was normal procedure for the commander and his officers to submit their journals to the Admiralty for inspection, following which they would be returned. Penny had done so shortly after his arrival. In mid-October, he wrote to the Admiralty requesting the return of "the reports and journals of my officers and myself as I am anxious that some account of what has been done should appear in print as soon as possible."[10] The Admiralty's second secretary, Captain W.A.B. Hamilton, acknowledged receipt of his letter and wrote, "I am commanded by my Lords Commissioners of the Admiralty to acquaint you that the documents will be returned as soon as the public service will admit"[11] – a response Penny must have found vague and unsatisfactory. Time passed, but no documents arrived, and when he appeared before the Arctic Committee, he still had not received the journals and was forced to prepare his case without them, a severe disadvantage. William Hogarth, the Aberdeen shipowner who had helped get Penny's ships ready for the search voyage, received a letter from Hamilton asking if he had Penny's journal, "as it was wanted at the Admiralty & *could not be found there.*"[12]

While the first secretary of the Admiralty served mainly as its spokesman in Parliament, the second (or permanent) secretary handled most of the routine business of the Admiralty Board. It was his responsibility to read incoming post and assign documents to appropriate persons on the Admiralty Board. If Hamilton had received the journals sent or delivered by Penny (which probably included those of Stewart and the three medical officers as well as his own) and had passed them on to the committee, how could he not know their whereabouts? Who had taken possession of them? Perhaps the journals had not been submitted to Hamilton at all but directly to a member of the committee. (Goodsir testified that he had delivered his meteorological journals not to the secretary but directly to Parry.)

On the first morning of the inquiry, the chairman, Admiral Bowles said to Penny, "You wrote to the Admiralty for your journals; did they return them to you?" He answered, "No, they did not." Bowles then asked, "Have you any other memorandum of your proceedings?" He replied, "The journals are here (in the Admiralty)." Several minutes later, Admiral Fanshawe came to life and asked, "Do I understand you that your journals from the Admiralty have not been returned to you." Penny confirmed, "They have not been returned."

The journals written on board the *Lady Franklin* and *Sophia* had certainly been examined by the committee members. On that same morning, Beechey said to Penny, "I find *from your journal* [italics added] that you make the following remarks," after which he quoted an entry of almost two hundred words. Soon after, Parry read a long paragraph from Stewart's journal. Later in the proceedings, Bowles also quoted from Penny's journal, and some extracts from it were appended to the committee's report. Only Bowles and Fanshawe expressed surprise that Penny and his officers had not received their journals back. Could the other three officers have managed to keep the journals to themselves throughout the proceedings without the knowledge of the two admirals or the permanent secretary? Whatever the case, Penny and his officers were under a serious disadvantage during the inquiry by not having their journals at hand.

When Austin took the stand on the third day, he had laid down conditions: "I am perfectly ready to answer any questions that may be proposed to me, reserving the privilege which I may hereafter have occasion for, of referring to my notes and to documents, some of which I shall refer to with great reluctance, if it become necessary. With that privilege

I am ready to go into the examination." Penny, on the other hand, was unable to benefit from such a privilege because his journals were still in the hands of one or more members of the committee (who remained silent on the matter).

Evidently Austin, unlike Penny, had not been required to submit his journals to the committee. On the eighth day of the proceedings – several days after Penny, Austin, and other major figures had testified – the chairman, seeking to redress the balance, asked Austin to produce his journals. Austin's reply to this request is revealing:

> Mr. Chairman. – With regard to a journal which you have called for, I beg to state that I have no journal, but that I have with care made the log a journal from which all may be obtained of any necessary interest required. It is necessary to remind you that the transmission of journals is not called for in my orders . . .
>
> Under these circumstances I have only in my possession notes that were made from time to time. These notes are my private thoughts, and never intended, in their present state, for other eyes than my own. I place them before the Committee, who will perceive that they are so; and the Committee I am sure will agree with me in opinion, that from their nature being private, in some measure sacred, I ought not to be asked to put forward their contents, further than so far as they may bear upon the subject of the search, or matter connected with the Wellington Channel.

By insisting that his "private" and allegedly "sacred" notes were meant for his eyes only, Austin was fortifying the high ground in the conflict with Penny. Apparently the committee deferred to his plea for privacy, for the content of his notes has never been revealed. As for the journals of Austin's officers, it appears that they were not requested by the committee either. Sixteen of his officers were interviewed, eight from the *Resolute*, four from the *Assistance*, and two each from the steamers *Pioneer* and *Intrepid*, but no journals written by them are quoted or even mentioned in the proceedings.

Like Penny, Dr Goodsir had delivered his journals to the Admiralty after the voyage. On 29 October, he wrote to the permanent secretary, "I will be very much obliged to their Lordships if they can find it convenient to instruct you to transmit the Journals to me as I am very desirous to prepare my narrative for publication as soon as possible."[13] His request brought forth a terse reply, "Nothing can be returned until after the proceedings of the Arctic Committee have terminated."[14]

Dr Sutherland had also relinquished his journals to the Admiralty and had to do without them during the inquiry. He needed them afterward because he also intended to write a book about the expedition (and did so). In early December, he wrote to John Barrow Jr, keeper of records at the Admiralty, "My journal & completed report have not been returned yet. May I express a hope it will not be long until I receive them."[15] Barrow promised to send them, but two weeks later, when they had still not arrived, Sutherland had to appeal to Barrow again: "The journal and report I hope will soon reach me if they are no longer required."[16] This was almost a month after the committee had completed its report.

Was Penny's journal ever returned? On 24 November, it was among the papers that were "delivered over to Mr. Hay Chief Clerk of the Admiralty" by Fegen, the committee's clerk.[17] Unfortunately the documents pertaining to this issue had been "weeded," which in the lingo of the National Archives means intentionally destroyed. But the digest of Admiralty correspondence provides the following terse entry: "Journals of Cap. Penny's Arctic Expedition. Returned to him."[18] The date is 20 January 1852, *four months* after he had provided them to the Admiralty and more than two months after the Arctic Committee had concluded its hearings. Whether they ever reached him is unknown. Much of Penny's correspondence has survived in family hands and is now safely preserved, but no shipboard journal has ever come to light.

* * *

Since the return of Penny and Austin in September, it had become perfectly clear that the relationship between the two commanders had been atrocious. Shortly after the conclusion of the Arctic Committee's proceedings, a London newspaper – probably the *Times* (9 December 1851) – pointed to "incompabilities of temper" between two "angry men." The root cause of their mutual dislike, the newspaper said, was their different backgrounds. "The rough merchant-seaman and the punctilious officer of the royal navy could not act in harmony."

27

An Open Polar Sea?

A geographical problem of first-rate importance.[1]

The discoveries made by Penny and his companions north of Cornwallis Island, and the charts that were subsequently published, brought an extensive unknown territory into the sphere of European awareness. More importantly, his suggestion that HMS *Erebus* and *Terror* might have proceeded north in Wellington Channel, northwest through Queen Victoria Channel, and out into the alleged Open Polar Sea radically altered prevailing ideas about the strategy of the search. Initially it had been assumed that Franklin would attempt to follow his orders, which directed him to sail west past Cape Walker, south to the continental mainland, and west to Bering Strait. In 1850, a panel of experienced arctic officers had reached a different conclusion – that he had struck directly westward to Melville Island along the route pioneered by Parry in 1819.

Wellington Channel had been cited in Franklin's instructions only as an alternative route to try in the event that he found Barrow Strait and passages leading south blocked by ice and saw nowhere else to turn. At the time, nobody knew whether Wellington Channel was a dead end or a route way to some unexplored region beyond, but Penny had revealed that it connected with a broad ice-free zone that extended to the horizon. The *Athenaeum* (7 February 1852) called this "the most important of modern hydrographical discoveries."

Sledge parties from the wintering ships of James Ross had travelled southward in Peel Sound without finding any evidence of Franklin's passage. Forsyth's probe into Prince Regent Inlet had revealed no signs. Austin's search parties had found no clues south of Griffith Island. Overland expeditions to the mainland coast had returned with no information; ships in Alaska had not encountered him. Apparently Franklin and his men had been unable to follow their intended route south to the continental coast and west to Bering Strait.

In three years of searching from the Atlantic, the Pacific, and across the vast expanse of barren tundra in between, Penny's discovery of an open channel leading northward into the polar basin was the only encouraging development. When Beechey, Parry, and James Ross visited Penny's ships in late September, interviewed his officers, and listened to the accounts of open water north of Cornwallis Island, they became convinced that Franklin had indeed passed through Wellington Channel and continued toward the northwest, and they – like Penny – concluded that he should be sought in that direction.

Penny's map of the newly discovered region appeared in newspapers within a few weeks of his arrival home, and there was widespread rejoicing that Franklin's path had finally been located. The *Aberdeen Journal* reported that Penny would take the *Lady Franklin* and *Sophia*, accompanied by a steamer, on a new search expedition in 1852 to follow the trail. The *Athenaeum* (11 October), after asserting that "Capt. Penny's energy, heroism, and Arctic experience eminently fit him to conduct a steamer to the scene of his discoveries," declared, "The nation will be disappointed should the Admiralty pass him over in their selection of Arctic officers to command the renewed searching Expeditions next spring." But although the Admiralty fully intended to renew the search in that region, they had not the slightest intention of employing Penny again.

* * *

Penny's discovery of open water at high latitude reawakened dormant ideas about an Open Polar Sea. The belief that there might be a navigable circumpolar ocean north of Eurasia, North America, and Greenland was an old idea that kept on recurring, fuelled by the desire to discover a direct route from Europe to the Pacific over the North Pole. Its adherents had cited a wide diversity of arguments to support the hypothesis, including the following:

- The North Polar region must be an ocean, because if it were land, the Northern Hemisphere would overbalance the Southern.
- As the earth is not a perfect sphere, its flattened polar regions are closer to the intense heat of the interior, so the polar seas stay warm enough to remain unfrozen.
- There cannot be ice in the polar sea because the heat transported to that region by the Gulf Stream and rivers would cause it to melt.
- In the circumpolar region, the sun shines continually for almost half the year, which provides significant warmth.

- The slanting solar rays of high latitudes reflect off water surfaces and heat the atmosphere.
- As ice floes are known to be carried out of the polar basin by currents, open sea must be left behind in their place.
- Wind, waves, and currents prevent the open ocean from freezing over, so ice floes must break off from the fixed ice adjacent to coasts.
- Animals in the sea and air are observed to be heading north in summer, presumably to take advantage of a milder climate in high latitudes.

Opponents pointed out that whaling ships in the Spitsbergen sector had always been stopped by pack ice around 81° N. If they encountered impenetrable ice at that latitude year after year, did that not refute the seductive stories of an ice-free sea? No, said the proponents; it merely showed that the ice-free polar sea was surrounded by a circumpolar ring of pack ice that was difficult to penetrate. If ships could somehow manage to locate a favourable route through that annular ice barrier, they could criss-cross the polar basin at will. A few reports circulated about Dutch whaling ships that had succeeded in breaking into the ice-free sea and in one case even advancing to the pole. Although none of these claims could be substantiated, they were often cited as "evidence."

Biological observations were often brought forward as well. Navigators in high latitudes often saw great numbers of sea birds, whales, narwhals, walrus, and seals migrating northward. As birds needed to find fish in the ocean, and sea mammals needed to breathe air, these animals must be heading toward feeding grounds in a sea that was unencumbered by ice. This argument overlooked the possibility that sea mammals might be abundant in an area of open water because ice prevented them from going any further, and that sea birds might congregate there because they could not secure fish beneath the ice cover further north. Such high-latitude zones of open water might be the limit of their poleward migrations.

Also persuasive were the claims of explorers who had reached the northern coast of Siberia in summer and looked out over a large expanse of unfrozen water. Because they could see no land in the distance, men who were anxious to discover the fabled Open Polar Sea could be deceived into thinking they had reached its margin. "The wish was usually father to the thought," observed John Kirtland Wright, a president of the American Geographical Society and author of the definitive study of the Open Polar Sea hypothesis.[2]

As humans were incapable in the mid-nineteenth century of travelling beyond the northern coasts of the continents, they could only reach the southern margin of the polar sea ice and consequently could only *speculate* about the nature of the vast region that extended over the North Pole. They were unable to obtain reliable facts either to support the hypothesis of an Open Polar Sea or to refute it. Although Fridtjof Nansen's bold polar voyage in the vessel *Fram* in 1893–96 showed the appealing notion of an ice-free Arctic Ocean to be a fiction, the spatial distribution and characteristics of land, water, and ice north of the continents could not be truly determined until the era of powered flight, air photography, and remote sensing by satellite. The establishment of scientific stations on drifting ice islands and the penetration of the polar ice by powerful icebreakers permitted reliable meteorological and oceanographic observations to be obtained. As we now know, the Arctic Ocean has a permanent ice cover that expands in winter and shrinks in summer. (In recent years, the ice has diminished in area and thickness in response to climatic warming.)

The existence of polynyas – areas in which the formation of sea ice is inhibited by local factors such as winds, currents, tides, or upwelling – was recognized by the time of Franklin's expedition. The North Water in Baffin Bay, vitally important in the itinerary of Davis Strait whaling ships, was a particularly well known one. But in order to know whether a large expanse of open water was a vast unfrozen sea or merely a polynya, one would have to determine its full extent, and that was seldom possible. The "newly discovered sea" around part of which the searching parties of Stewart, Sutherland, and Goodsir had hauled their sledges, and on whose surface Penny and his companions had explored by sledge and boat, was widely assumed to be part of the Open Polar Sea; in fact, it was merely a polynya, one that still exists today. During the warm part of the year, small patches of open water among the islands of Queen Victoria Channel and in Penny Strait expand and coalesce. The shape and maximum extent of the polynya vary from year to year.

At a meeting of the Royal Geographical Society in early November 1851, a new search expedition – "a very daring plan," the *Athenaeum* called it (15 November) – was announced by Bedford Clapperton Pim, a young naval lieutenant who had participated in the search for Franklin on the Pacific side of the continent. After penetrating Bering Strait on HMS *Herald* commanded by Henry Kellett, the party had crossed the Chukchi Sea in a northwesterly direction, sailed along the Siberian coast for four hundred miles, and discovered a small island, which they named after their ship and claimed for Great Britain.

Three months after returning to England in June 1851, Pim was promoted to lieutenant. Inspired by Penny's discoveries, and doubtless influenced by his own experience on the *Herald*, he believed that the *Erebus* and *Terror* had reached the Open Polar Sea by way of Wellington and Queen Victoria Channels. The ships, he believed, had continued westward but then had found their way southward through Bering Strait to the Pacific Ocean blocked – perhaps by ice or perhaps by land that extended from Banks Island to the Siberian coast somewhere beyond Herald Island. They had therefore maintained their westerly course along the roof of the Asian continent toward Europe, and Pim intended to look for them there. In a few weeks, he would sail from England to Saint Petersburg, present letters of recommendation to the Russian authorities, obtain permission to undertake his mission, and proceed overland across Siberia to the mouth of the Kolyma River to begin the search. He would use native people to help establish depots. He would be accompanied by only one man – "a gentleman of congenial habits would be sufficient" – and the venture would cost less than £3,000.[3]

When Pim submitted his search proposal to the Admiralty in late September, he mentioned an alleged sighting of two boats off the north coast of Alaska and suggested that the *Erebus* and *Terror* might have sent them in to shore while they remained in deep water to avoid the risk of running aground. Alaska was alive with such rumours at the time. Pim himself had been sent to a Russian fur-trade post to investigate rumours of strange white men – possibly from Franklin's ships. He also called the attention of the Admiralty to the existence of shipwreck wood near the mouth of the Kolyma River and other places along the Siberian coast, as revealed three decades earlier by the Russian explorer Ferdinand von Wrangel. He suggested that Franklin's ships might have come to grief along this shallow and uncharted Asian coast,

Pim sent the Admiralty a reminder two weeks later and was eventually notified that the lords commissioners did not accept his proposal. He was not ready to give up, however. He approached Lady Franklin, who was understandably eager to support any further effort in the search, and next appealed to the Royal Geographical Society. Penny was in London at the time for the hearings of the Arctic Committee, and he attended the society's meeting. According to the *Illustrated London News* (15 November 1851), he "spoke highly of Lieutenant Pim, and thought his plan was in the right direction, both agreeing that Sir John Franklin must have advanced far toward Behring's Straits."

Not everyone accepted the existence of an ice-free circumpolar sea or Pim's opinion about Franklin's course. The *Nautical Standard* (15 November) ruthlessly exposed the false premises underlying Pim's search plan, called his assumptions "utterly unsupported," his reasoning "inconsequential and inconclusive," and his speech "an entertaining goblin story for Christmas." The Open Polar Sea was "apocryphal," and there was "not a tittle of evidence" that Franklin had reached the western end of the Northwest Passage. "We should seek for evidence rather than run after shadows," the newspaper declared, with good reason.

Many in official positions were hesitant to express opposition to the notion of an Open Polar Sea because to do so might lead to a curtailment of the search for Franklin. One who dared to voice his disbelief was William Scoresby. In 1806, he had sailed on his father's whaling ship *Resolution* to the latitude of 81°30' N in the Greenland Sea, where they believed themselves to be further north than anyone in the world. They were approximately six hundred miles from the North Pole, and after examining historical records, they concluded that no ship had ever been closer. A decade later, Scoresby read a paper at a meeting of the prestigious Wernerian Society entitled "On the Greenland or Polar Ice," in which he said, "From the pretended excursions of the Dutch, many have believed that the sea at the Pole is free from ice . . . but I consider it too improbable to render it necessary to hazard any opinion concerning it."[4] Several years later, in his monumental work on the arctic regions, he wrote, "we can have no reasonable ground, I conceive, for doubting the continual presence of ice in all the regions immediately surrounding the Pole."[5]

He did not dispute reports of a few open water areas at high latitude but merely pointed out the obvious: the existence of open water here and there did not prove that it existed everywhere. Now, thirty years later, Penny had discovered another ice-free area, and veteran explorers, Admiralty officials, armchair geographers, and newspaper editors were rushing eagerly to embrace the Open Polar Sea hypothesis. Scoresby was still standing firm against it, but few were listening. It was not what they wanted to hear.

Despite the *Nautical Standard*'s harsh criticism of Pim's plan, and Scoresby's continued scepticism, the increasingly popular belief in a ice-free circumpolar ocean surrounded by a ring of pack ice altered the nature of the Franklin search. A merchant captain named Donald Beatson also announced plans for an expedition inside the Open Polar Sea. He proposed to take a steam-powered vessel through Bering Strait into the Chukchi Sea. Then, according to the *Aberdeen Herald* (6 December 1851), he would head "westward and northward" into open water and "work his way eastward, year by year, if necessary, in a high latitude, toward the northern entrance of Wellington Strait." He hoped to encounter Franklin's ships.

The maritime search expeditions sent by way of Baffin Bay and Bering Strait, and the overland parties that crossed the Barren Grounds, had been looking for Franklin *within* the Arctic Archipelago that lies poleward of the North American continent. Now Pim and Beatson, and the Admiralty as well, intended to seek him *beyond* the archipelago in the alleged Open Polar Sea. As a result of Penny's discoveries north of Cornwallis Island, the customary perception of a more or less latitudinal passage from the Atlantic westward to the Pacific, mainly in coastal waters adjacent to the continental coast, had evolved into a circumpolar outlook, or at least one in which the unknown region lying north of all the known continents was regarded as a possible route way, and therefore a place where Franklin should be sought. The new – although erroneous – view of the geography of northern lands and seas was changing the strategy of the search and vastly increasing its spatial extent.

In the summer of 1852, a new Admiralty expedition would follow Franklin's assumed path through Wellington Channel into the alleged Open Polar Sea. Meanwhile, the two private expeditions under Pim and Beatson would attempt to intercept that path from the Russian coast and Bering Strait. Surely this pincer movement would solve the mystery.

* * *

Three months after returning home in 1851, Penny had written from Aberdeen to the president of the Royal Geographical Society. As reported by the *London Chronicle* (9 December), he said, "I beg to lay before you and the scientific body to which you belong, a few observations which came under my notice during my late voyage in search of Sir John Franklin, which I consider of great importance at the present critical moment." He outlined the achievements of his expedition: the discovery of Franklin's first winter quarters; the examination of coasts bordering Wellington Channel; the addition of a large, hitherto unknown region to the charts; and the finding of a sizable expanse of open water abounding with marine life. During his boat journey, he had seen "norwals [*sic*], walrus, and white whales making their way down channel, seeking the protection of the ice" as these species customarily did in Davis Strait. He cited this phenomenon as "convincing proof that a sea does exist beyond Queen Victoria Channel, comparatively open and free from ice." Although it was proof of no such thing, the seductive image of an arctic oasis teeming with animal life and connected to an ice-free polar sea was compelling. He ended with a powerful plea for further efforts: "Humanity demands that this search should not be given up until the searchers shall have passed through the Arctic basin and out into the Pacific Ocean, and until such a course shall be pursued, the fate of our missing countrymen can never be ascertained."

The society's president, Sir Roderick Murchison, intended to have Penny's letter read out at the next general meeting but strong objections were raised by Captain Sir Edward Belcher RN. Everyone else was in favour of hearing it, however, so Belcher was overruled and the letter was read. Afterward, according to the *Morning Herald* (9 December), Belcher "rose to make some observations on the conduct of Captain Penny, but they were of so personal a nature, and were not participated in by the majority of the meeting, that the chairman called him to order, and wished the gallant officer to avoid personalities but to confine himself to facts." After contradicting some of Penny's conclusions (and in doing so betraying his own ignorance about the origin and distribution of icebergs) and then firing off a final criticism of Penny, Belcher finally relinquished the floor, following which the other members passed a unanimous vote of thanks to Penny.

Sherard Osborn later wrote to Penny, "I was present the other evening at the reading of your documents, it was all that could be wished and most ably drawn up – Capt. Belcher's remarks only exposed professional weakness & made others as well as myself blush."[6] Osborn, open-minded and a good judge of character, regarded Belcher's protestations as intolerant and unreasonable. Why should a society devoted to geography refuse to hear what Penny had to say about his remarkable discoveries?

One might also wonder how an officer whose only arctic experience had been a few months in Alaska a quarter-century earlier had presumed to criticize the remarks of a whaling captain who had been sailing to the Arctic year after year. Belcher's arrogance, and his utter rejection of the observations of a "mere whaling master," were as blatant as those of Austin a few years earlier, after which the *Daily News* (9 December 1851) had written, "The officers of the royal navy ought to be given to understand that they are engaged on an exceptional service, in which all mere *martinet* considerations must be left out of account."

Penny had discovered what appeared to be Franklin's path into the polar sea, but HMS *Erebus* and *Terror* had now been gone for more than six years. Could any of their officers, sailors, and marines still be alive? On this point, opinions differed, but in late December Penny revealed startling information that inspired hope. Writing to the editor of the *Times*, he said that he had learned a vital fact from a brother whaleman, Robert Martin, who had commanded the whaler *Enterprise* in 1845. Martin had been alongside HMS *Erebus* in Baffin Bay, had spoken to Franklin and some of his officers, and had been told that the discovery expedition had enough food for five years and it could be stretched to seven. They had salted down enough sea birds to fill five casks, and the hunters were still active. When Martin reported his encounter to the Admiralty at the conclusion of his voyage, he had not considered his information about provisions important enough to include. The Admiralty received his brief notice but did not request further details. Martin saw Lady Franklin a few years later when she visited Peterhead but was reluctant to approach her with the information.

Martin's information, made public by Penny six years after the event, was formalized in a notarized deposition. It reached the Royal Geographical Society, whose secretary, C.R. Weld, sent it to the *Times*, where it was published on 3 January 1852. Martin's former first mate, Robert Walker, also submitted a deposition attesting to hearing similar remarks about provisions, and on 6 February, Esther Blenky (sometimes spelled Blanky), wife of the *Terror*'s ice master, sent a letter to the editor of the *Morning Herald* saying that in 1845 her husband had written from Disko saying not to worry if they were not home in three years. "It may be from five to six years – it might even be into the seventh – before we return," he had told her.[7]

These optimistic statements were contradicted by Lieutenant Edward Griffiths, commander of Franklin's transport *Baretto Junior*, who had supervised the transfer of supplies to the *Erebus* and *Terror* at Disko before returning to England. He wrote to the *Times* (12 January 1852), "However unpleasant, the truth should be known. They had only three years', at full allowance; of course if they went on half rations this would give them six years." The officers, he admitted, carried as a perquisite of rank some "private stock," which included "at least six dozen fowls, some pigs, and a few sheep"; there was meat hanging in the rigging from three bullocks recently slaughtered on the transport; hunting and fishing would probably bring in some additional food. But he dismissed the idea that Franklin had full rations for seven years. Griffith's rebuttal agreed with a statement made by Commander James Fitzjames of HMS *Erebus* in 1845. Writing home before parting from the supply ship *Baretto Junior* at Greenland, he had said, "We shall start with three years provisions."[8]

Martin may have been misled by the optimism of Franklin and his officers, but despite Griffith's contrary opinion, his information was widely regarded as valid. It gave hope to relatives and friends of the missing men, reinforced Penny's view that survivors were still alive, and strengthened his determination to take part in another search effort. If he could follow up his efforts with a new expedition, one utilizing the power of steam, he might solve the Franklin mystery. He expected opposition from the lords commissioners of the Admiralty, but the support of Barrow Jr (keeper of records), Beaufort (hydrographer), and Hamilton (permanent secretary) was a great consolation.[9]

The Admiralty had rejected Penny's impetuous application for command of a search expedition in the autumn on the reasonable grounds that it was too late in the season, but surely they would now look favourably upon an application for an expedition to be sent out in the spring of 1852, for which there was ample time to make preparations. The Arctic Committee had recommended that ships be sent through the very region of open water that Penny and his men had discovered and explored, and Penny felt that he was the man to command the new expedition. In mid-January 1852, he applied again to the Admiralty, stating that he had earned the right to undertake another search. The Admiralty, however, had had enough of the outspoken whaling captain whose critical comments about Austin had forced them to expose some soiled laundry to public view. A week later, the secretary of the Admiralty thanked him for his offer and coldly informed him that they had no intention of hiring private individuals to command Admiralty expeditions. The abrupt, almost insulting dismissal greatly offended Penny. But as he remarked to Barrow, he had "never yet showed the white flag in the moment of danger."[10]

Two weeks passed before he managed to compose a response. "I have been too much hurt to reply," he wrote. "It never entered my mind for one moment that your Lordships would have cast me off in the contemptuous manner you have done." He and Alexander Stewart, his second-in-command on the recent search expedition, and their former crews, had rejected offers of employment for the forthcoming season, he said, because they expected to participate in another search. "Your Lordships must know that I had a claim to be employed when the next expedition was sent out. The country knows that it is to the whaling captain they owe their present hopes of the safety of Sir John Franklin, for it had not been for him their Lordships would be as ignorant of the way he has gone as they were when he came back to say what he had done."[11]

The lords commissioners of the Admiralty were probably not pleased to be called ignorant about any subject, particularly the search for Franklin, and would not accept Penny's claim that he had earned *a right* to be appointed. An internal memo a few days later reads, "It would be very injurious to employ Captain Penny."[12] They did not change their minds.

Clive Holland has revealed that although Penny's letter was written in the first person and signed by him, it had actually been composed by Lady Franklin and Sophia Cracroft, imitating Penny's style to the best of their ability, and taking advice from John Barrow and Sherard Osborn, both of whom read sections and made suggestions. When they completed the final draft, Penny travelled to London, copied it in his own hand, and affixed his signature.[13]

Penny's letter was much more than a personal complaint. It was a formidable salvo fired by a cadre of persons who disagreed with Admiralty policies in the search and with the treatment accorded Penny by the Arctic Committee, and who strongly supported his candidature for leadership of a new expedition through Wellington Channel, strengthened by steam power. Lady Franklin, the commander-in-chief of this undeclared protest movement, cannily consigned a copy of the letter and its appendices to a printer. If the Admiralty lords rejected Penny's offer again – as they almost certainly would do – his detailed summary of grievances against Austin, the Arctic Committee, and the Admiralty would be published and circulated. After waiting for a week, Penny sent a reminder in which he said if the lords commissioners had not yet considered his letter, would they please send it back. They wasted neither time nor words in replying, stating merely that they could not "avail themselves" of his offer of service.[14] Lady Franklin's counter-attack now got under way and Penny's letter was soon rolling off the presses at the London publishing house of Chapman and Hall.

Penny's published letter was long, comprising about seven thousand words with an additional three thousand words in seven appendices. Its published version (identical to the original except for a few small and inconsequential changes) ran to fifty-eight pages. Most of the text consisted of a well-reasoned, point-by-point rebuttal of the committee's report. He presented sound arguments to show that he had been judged unfairly while Austin had received preferential treatment: the committee had "sacrificed the Whaling-master to the Post-captain."[15]

Included as appendices were affidavits from three crew members of the *Lady Franklin*. In Assistance Bay, the steward had heard Austin promise Penny that he would not use in any prejudicial way his written statement that no further searching was required (but in fact he had done so). The second mate and a seaman had heard Penny urge Austin and Ommanney to go up Wellington Channel and do the cause some good – which both officers had flatly denied.

The Report of the Arctic Committee, which had been appended to a Parliamentary Paper shortly after the conclusion of proceedings in November, was available to the public in a so-called "Blue Book" by mid-February. It gave an unfavourable view of Penny, but the publication of his letter with its supporting appendices provided an effective rebuttal. It was too late to change the Admiralty's plans, however. They had already selected a commander for the forthcoming search expedition, and by late February he was at the Admiralty making initial preparations. The officer who would lead a five-ship squadron into Lancaster

Sound and take two of them north through Wellington Channel and Penny's "newly discovered sea," was Captain Sir Edward Belcher, the very man who only a few months earlier had attempted to prevent Penny's letter being read before the Royal Geographical Society and had rudely questioned the reliability of his observations.

Belcher, a capable surveyor, was highly regarded by the navy's hydrographer, Sir Francis Beaufort, whose powerful influence had probably helped to secure his appointment as commander of the new search expedition. Although he had been to the Bering Strait region in 1826 and 1827 as a lieutenant on HMS *Blossom*, he had never wintered in the Arctic. More importantly, he possessed personality traits that made him totally unsuited to a winter's confinement in the ice with absolute authority over a ship's crew. An officer who knew him called him "a devil incarnate with his officers."[16] The arctic historian Leslie Neatby compared him to Captain Bligh and said, "He was coarse, meddling, tactless, and fearfully egotistical."[17] Other writers have called him "a nit-picking, bullying martinet,"[18] "a notoriously heavy-handed disciplinarian,"[19] and "one of the most controversial and detested figures in the Royal Navy."[20] According to naval historian Andrew Lambert, Belcher was "able but pathologically irascible." Even the *Dictionary of Canadian Biography* describes him as "irritable, quarrelsome, and hypercritical."[21]

In 1831, during his first voyage in command of a ship, his dictatorial attitude and harsh punishments had stirred up the antagonisms of his officers and men. The voyage ended with Belcher bringing charges against two of his officers and their levying charges against him. Midshipman Henry Backhouse stated that when his boat was sent off from HMS *Aetna* for a week to explore the coast of Sierra Leone, Belcher had provided its crew with so little fuel that they had to eat their meat raw and so little fresh water that one man nearly died. Furthermore, Belcher had ordered one or more muskets to be fired at Backhouse's boat (possibly as a recall).[22] Two years later, several of Belcher's officers laid charges against him, and after another voyage a ship's crew submitted an official complaint of excessively harsh treatment. Although he was acquitted or exonerated in every case – by fellow officers – the danger signals were clearly visible. Recommending Belcher for command of the largest expedition in the entire Franklin Search was Beaufort's "worst mistake," his biographer has concluded.[23] True to form, Belcher would generate more bad will during the voyage and face another court martial after his extraordinary decision to abandon four of his five ships in the Arctic.

In addition to composing Penny's letter to the Admiralty, Lady Franklin lobbied members of Parliament on his behalf. Her friends Sir

Robert Inglis and Chisholm Anstey raised questions about an apparent inequity in the government's treatment of the commanders of the two official expeditions. While all but one of Austin's officers had received promotions (the exception being Osborn, who had expressed support for Penny), Penny and his officers had received no rewards. Could the Admiralty secretary explain this? And did the Admiralty intend to accept Penny's offer to participate in the next search? The responses by the present and former first secretaries of the Admiralty were annoyingly evasive, and the dismissive statement by Admiral Berkeley, one of the lords commissioners, was especially irritating: "If Captain Penny had been sent out with the present expedition it would not have been for the interest of the object in view."[24]

Penny responded vigorously to the statements made by the Admiralty spokesmen in the House. In a letter published in the *Times* (29 March 1852), probably written by or with the assistance of Lady Franklin, he declared, "I have maintained silence in the face of much provocation ... [but] there is a limit to forbearance, and ... very probably I have allowed that limit to be transgressed with respect to myself and my men." He then reminded the Admiralty of the achievements of his recent expedition: he and his men had proved Adam Beck's massacre story to be false, had tried to persuade Saunders to land his supplies within reach of the search expeditions instead of at Disko, had discovered Franklin's winter quarters and a watch station up the coast, had kept Austin and Ommanney informed of developments, had wintered in a secure position without health problems, and had discovered new territory and a large expanse of open water. He would not, he said, "allow my name to be used to cover the inefficiency of others ... I am not willing to be badgered and to have evidence cooked, to have my language perverted, and to have my tongue tied by respect to etiquette, which those who most require it are the last to observe toward me."

It was an angry letter, somewhat on the boastful side. To Sir John Ross, glancing through the *Times* in his "North-west Castle" at Stranraer, Penny's remarks were like a red flag waved in front of a bull, and Ross's response was suitably bullish. In a letter to the editor of the *Times* (10 April), he derided Penny for "the good opinion he appears to have formed of his superior abilities" and accused him of attempting to "gull the public" with his "statements and misrepresentations." Beck's report that Franklin and his men had been massacred somewhere north of Cape York (which Penny's interpreter Petersen had discredited) was in fact true, Ross insisted, and it was believed by "the Greenland authorities as well as our own." (He neglected to say which "authorities.")

Penny, he said, was actually preventing the search from being directed to the very region in which Franklin was "most likely to be found," by which he meant somewhere north of Cape York, Greenland.

Without giving reasons, Ross faulted Penny for having delivered to Saunders his new Admiralty orders in Lancaster Sound – which Penny had been instructed to do. He asserted that Penny could not claim to have discovered the graves on Beechey Island because they had actually been found by Phillips and Abernethy from his own expedition (patently false). He contradicted Penny's boast that none of his men had been on the Sick List during the winter because he knew that several of his sledge-haulers had experienced some frostbite (which had been merely superficial). He insisted that Penny's sledge parties had travelled only half as far as Penny claimed. As for Penny's request to be sent back to Wellington Channel in the fall, Ross declared that his motive was obvious: he wanted to "enjoy himself in some snug harbour in Danish Greenland" over the winter while continuing to receive a generous salary from the government. Ross concluded with a personal complaint. He had "spent 14 winters in the frozen regions" (by which he meant Stockholm and southern Sweden) and had "examined the records of Russia, Sweden, and Denmark for 40 years" (he did not specify which records or even which repositories). Yet he had been "excluded from every committee, every discussion, and every consultation on the subject." It was perfectly clear to him that Franklin was being sought in the wrong place, but no one would pay attention.

Ross's sarcastic and libellous remarks brought a brief, dismissive response from Penny in the *Times* (23 April): "I will not condescend to contradict Sir John Ross's wholesale misstatements in detail, but I give them all my most unqualified contradiction." Referring back to the winter of 1850–51 at Assistance Bay, when he had provided antiscorbutics for the crew of the *Felix*, he confessed that he felt sympathy "for a man whose want of foresight had placed himself and 16 of his fellow-creatures in a position which, had no aid been near, would have cost half of them their lives by scurvy, and placed him before a jury of his countrymen." Lady Franklin wrote in late April to tell Henry Grinnell about the rancorous exchange between the two men: "You will see a justly angry letter of Penny's in answer to one of old John Ross's. I believe Penny could unfold many a discreditable tale of the hoary-headed & alas! most mendacious veteran, whom he holds in ineffable contempt. Indeed, nothing saves him from more frequent exposure but the old man's hardihood and weight of years."[25]

* * *

Since his return from the Arctic, Penny's spirits had oscillated between elation and despondency. He was pleased when the Admiralty concurred with his conclusion that Franklin's ships had passed through Queen Victoria Channel but resentful when the Arctic Committee criticized his performance. He felt vindicated when the Admiralty decided to dispatch a new search expedition to the region he had discovered but furious when they rejected his own application for command, and instead appointed – of all men – Belcher! He was appreciative when members of Parliament spoke on his behalf but angry when Admiral Berkeley, the first lord of the Admiralty, stated that employing him would not "achieve the object" of the next search expedition. And then, of course, there were the venomous comments by Ross in the press. To Barrow he exclaimed in frustration, "I entered upon this search with the ardors of a generous hearted seaman. My God what is the return that I have met with?" The only thing they could not rob him of, he said, was his integrity.[26]

Through the winter of 1851–52, he felt overlooked, unappreciated, rejected, and excluded unfairly from the search. "The time will come when I will get justice done me but it is certainly hard for me in many ways in the meantime," he told Barrow.[27] He was still bitter about the attitude of Ross and Austin during the search expedition, and in June he told Barrow in confidence, "It was my misfortune to be brought into contact with two commanders whom I considered unworthy to bear her Majesty's commission the one a bustling deceitful fool the other a coward: they have injured me and those who have served me faithfully."[28]

It doubtless gave Penny considerable satisfaction that the proposals of Pim and Beatson to carry the search through Bering Strait into the polar sea were based on his own discoveries and convictions. In the end, however, their plans were not put into effect. Pim was turned back from St Petersburg by the Russian authorities, who probably feared intelligence-gathering along the Siberian coast, and he returned to England. Beatson was still intending to go, however, and in March he outlined his plan before the Royal Geographical Society. His speech was enthusiastically received, reported the *Illustrated London News* (27 March), and was followed by "loud and continued cheering," but his screw yacht *Isabel* never left the Thames. Despite having gained the approval of the Admiralty and the Russians (neither of which had supported Pim's proposal), as well as financial support from Lady Franklin and many private sponsors, Beatson ran up expenses that he could not pay and was arrested for debt. According to historian Andrew Lambert, he was "either a crank or a crook."[29]

Penny's discoveries were of great interest to the German geographer Augustus Petermann, an extraordinary man. After graduating in 1844 from the Geographische Kunstschule in Potsdam (established by Alexander von Humboldt and Heinrich Berghaus), Petermann had worked in Edinburgh and London. He cooperated with Alexander Keith Johnson in the production of a physical atlas (the one used for geography classes on board Penny's wintering ships) and established his own company for printing maps. He joined the Royal Geographical Society and was elected its under-secretary at the age of only twenty-eight. Two decades later, he would be awarded the Founder's Medal of the society. Petermann was a prolific writer who greatly stimulated polar research and exploration, especially in Germany. Although not an explorer in his own right, he was a "devoted armchair geographer"[30] who exerted an enormous influence upon cartographic methods, geographic thought, and the search for Franklin.

Unfortunately, Petermann's bold hypotheses were not always supported by observed facts. He published a letter in the *Athenaeum* (17 January 1852), setting out his ideas about the nature of the unexplored region around the Pole and suggesting a new approach to locating Franklin. It was *a fact*, he said, that "a sea open in all seasons" existed north of the Siberian coast, and there was *no doubt* that "a similar open sea exists on the American side to the north of the Parry group." Probably, "these two open seas form a large navigable Arctic Ocean." There were, he said, only four entrances into the navigable polar sea in which Franklin must be sought: Bering Strait, Wellington Channel, the sea between Greenland and Spitsbergen, and the gap between Spitsbergen and Novaya Zemlya. (Smith Sound was not yet known to be another entrance.) As the first three of these were characterized by very difficult ice conditions, Petermann suggested sending a search expedition through the fourth – "*the easiest and most advantageous entrance into the open navigable Polar Sea*" (Petermann's italics). For somewhat obscure reasons, he thought that access to the supposed ice-free sea, and navigation within it as well, would be easier in winter than in summer. Petermann's views, based precariously on theoretical arguments rather than observed data, failed to garner enough support for a search expedition in that direction, but they impressed Penny.

Despite cancellation of the searches by Pim and Beatson, and the lack of momentum behind Petermann's proposal, it seemed almost inevitable that some news of Franklin would come to light during the summer of 1852. In the previous summer, William Kennedy had sailed into the Arctic Archipelago in command of Lady Franklin's *Prince Albert*, intending to turn south into Prince Regent Inlet, pass the winter, and send out search parties. Richard Collinson and Robert John LeMesurier McClure were believed to be searching eastward from Alaska in HMS *Enterprise* and *Investigator*. HMS *Plover* was still on watch at Bering Strait and would be resupplied by HMS *Amphitrite* during the summer so that she could remain on station. John Rae of the Hudson's Bay Company was expected back from an overland voyage to Victoria Island. Belcher's squadron was making final preparations for departure, intending to follow up Penny's information, and Frederick Augustus Inglefield, sponsored by Lady Franklin, would be leaving for Baffin Bay in July on Beatson's former screw yacht *Isabel*. At least ten ships would be operating throughout the enormous region that stretched three thousand miles from Greenland to Bering Strait and northward toward the Pole.

Although Penny was devastated to be excluded from this extensive searching effort, he would eagerly await news of the various expeditions, in particular the one directed to the open water he had discovered. Belcher would have resources that Penny had lacked, notably two steam-powered tenders; he should be able to carry on where Penny had left off and follow Franklin into the presumed open sea beyond – if his ships succeeded first in breaking through the ice barrier in Wellington Channel. His expedition would comprise the same four ships that Austin had commanded in 1850, along with HMS *North Star*. The *North Star* would take up winter quarters in Erebus and Terror Bay adjacent to Beechey Island to serve as a store ship and fallback refuge.

Shortly before Belcher's departure, a curious story came to light. A year earlier, only thirty miles east of Cape Race, Newfoundland, the crew of the merchant vessel *Renovation* had allegedly sighted two square-rigged ships perched on the flank of an iceberg, According to the ship's mate (when he was located and examined later), the ships were high and dry on the berg, with no sign of human life. His report was confirmed by the man who had been at the helm at the time.

Could the ships be *Erebus* and *Terror*? If so, where and when had the crews abandoned them and where could the survivors be now, if indeed any were still alive? It was, as a contemporary historian expressed it, "a fruitful subject for speculation."[31]

Or had the ships been merely an optical illusion in those fog-shrouded, berg-crowded seas? Writing to the *Times* (21 April 1852), Penny said, "I think they were 'country ships,' as we whalers call them – formations upon an iceberg which deceive even practiced eyes." This gave John Ross another opportunity to square off against Penny. In the *Times* (27 April), he wrote, "Many of my brother officers, as well as myself, who have navigated Arctic seas, are of quite a different opinion. There is no such illusion, as he speaks of, near the Banks of Newfoundland." (Had he ever been there?) Penny's real motive, Ross insisted, was to sabotage the investigation of the story so that the open water he had discovered would remain at the top of the search agenda. If Franklin's crews had been lost after his ships had drifted out of Lancaster Sound, there would be no justification for anyone to renew the search north of Wellington Channel, which Penny wanted to do, Ross said.

The alleged sighting of ships on a berg brought forth various far-fetched hypotheses. One man signing himself "Arctic" wrote to the *Times* (16 April) suggesting that after entering the Open Polar Sea by way of Wellington Channel, the *Erebus* and *Terror* had been frozen in and then abandoned by Franklin and his men. The ships, elevated somehow onto an iceberg, had then drifted south into the Atlantic, either along the east coast of Greenland or between Spitsbergen and Novaya Zemlya. He did not attempt to explain how an iceberg could occur in the Open Polar Sea, how the ships could get on top of it, or how the berg could drift contrary to the prevailing current direction between Spitsbergen and Novaya Zemlya to reach the Atlantic Ocean.

The Admiralty took the report of ships on the ice seriously. Supposing the ships to have been real, they had to determine whether they were *Erebus* and *Terror*, whaling ships, fishing or sealing vessels, merchant transports, or some other type of commercial craft. And if they were in fact Franklin's ships, what had happened to their crews? To answer these basic questions was far from easy, because in the intervening year the ship *Renovation*, from which the sightings were said to have been made, had sailed vast distances, and most of the alleged witnesses had gone ashore or transferred to other ships that were now crossing the world's oceans or loading cargo in foreign ports.

The Admiralty's official investigation of the matter extended over three months and involved at least four departments of government. Queries were sent to shipping interests, customs authorities, and diplomatic representatives in several countries, and men were interviewed at various locations at home and abroad, but the sighting was never satisfactorily explained. The possibility that it was authentic and that the ships were in fact *Erebus* and *Terror* could not be confidently eliminated, and this influenced the search. Although the various expeditions proceeded into the Arctic Archipelago as intended, Belcher was ordered to look out for survivors from the wrecked ships along the coast of Baffin Island on his homeward voyage, and Inglefield was instructed to search as far south as Labrador on his way home.

Unlike Penny, the ex-whaling captain William Scoresby thought that there might be some truth in the report. In a letter to the *Athenaeum*, reprinted in the *Morning Herald* (15 June 1852), he wrote, "I do not think that the observers could have been deceived by visions of ships, 'country ships,' or resemblances of ships, in forms or fractures of ice, nothing of this kind having ever come within my own extensive observation, except refracted images, generally inverted, with which these could not be confounded."

Another intriguing report reached England shortly before Belcher's departure. During a voyage on behalf of the Hudson's Bay Company, John Rae had picked up two pieces of worked wood, one of oak and the other of pine, each several feet long, on the southeastern coast of Victoria Island, about fifty miles west of King William Island. They were just outside the area in which he had previously predicted Franklin's ships would probably be found – a rectangular zone bounded by the parallels of 71° N and 74° N and the meridians of 103° W and 115° W. Evidently the fragments were parts of a stanchion and a flagstaff, and as Rae reported in a letter to the Hudson's Bay Company in February 1852 (quoted later in the *Leader*, 27 March 1852), the stanchion "had certainly belonged to one of her Britannic Majesty's ships, as there was a piece of line and two copper tacks attached to it, all of which bore the Government mark."

This discovery should have brought the Admiralty, the Royal Geographical Society, and all other parties interested in Franklin's fate to full alert. The wood was from a naval ship and had been found close to the continent's north coast, which Franklin had intended to follow to Bering Strait. Only one Royal Navy ship had ever been wrecked among the arctic islands – Parry's *Fury* in Prince Regent Inlet a quarter-century earlier. But that inlet had no known connection with the waters around Victoria Island other than a lengthy and circuitous route north to Barrow Strait and south through Peel Sound, a highly unlikely drift path – or perhaps an *impossible* one, if Peel Sound was really blocked by land or ice as search parties had (erroneously) reported. The existence of Bellot Strait, which cuts through the Boothia Peninsula, making its northern part into an island, was not yet known.

Rae returned to England with the end parts of the long piece of pine, a sketch of its central portion, and the entire piece of oak. At Woolwich dockyard, Belcher, whose expedition was preparing to depart, and several officers and petty officers who had served on Admiralty expeditions, gathered in the superintendent's office to examine the wood. Among those present was James Ross, who had commanded HMS *Erebus* and

Terror in a four-year cruise to Antarctica, made two arctic voyages on HMS *Fury* (including the one on which she was driven ashore and wrecked in Prince Regent Inlet), and subsequently wintered with HMS *Enterprise* and *Investigator* at the entrance to the inlet. Also present was Austin, who had been on the *Fury* when she was lost.

According to the *Times* (3 April), the men agreed that the stanchion and flagstaff had *not* belonged to "the Enterprise or Investigator, or any of the vessels of the recently returned Arctic expedition" – the one commanded by Austin. Only three possible sources remained: Parry's *Fury* or Franklin's *Erebus* or *Terror.* Austin was certain that the wooden pieces had never belonged to the *Fury*, and therefore must have come from the *Erebus* or *Terror*. Ross thought the fixtures looked familiar and he felt they had come from a ship on which he had served, but he disagreed with Austin and insisted that they were in fact from the *Fury* wreck (in which case Franklin's ships might still be intact).

Although James Ross had commanded HMS *Erebus* and *Terror* over a four-year period, he did not recognize that the articles of pine and oak had once been parts of deck fixtures on one of his two ships. Doubtless he had had more important things to do on that voyage than commit to memory the appearance of all the stanchions and flagstaffs on each vessel, and the relatively small size of the pieces made their identification difficult. If shipwrights who had been involved in refitting HMS *Erebus* and *Terror* before Franklin's departure had been consulted, they might have recognized that the wooden pieces picked up by Rae on Victoria Island had once belonged to one of Franklin's missing ships, in which case the search might have had an entirely different outcome.

Murchison, president of the Royal Geographical Society, dismissed the idea that the wood might have come from one of Franklin's ships. According to the *Illustrated London News* (27 March), "He expressed a decided opinion, founded on the last intelligence heard of the expedition, that this conclusion was wholly erroneous, an opinion in which he was supported by the best geographers." (He did not reveal who the "best geographers" were or how they had reached that conclusion.) So sharply were the eyes of the society focused on the Wellington Channel route way and the exciting prospect of ascertaining the geography of the circumpolar region that evidence pointing elsewhere was not given the consideration it deserved. As for the Admiralty, it was heavily committed to a renewed search by way of the open water discovered by Penny, and Belcher's ships were only two weeks away from departure. With Rae's wooden fragments confidently attributed to the wreck of HMS *Fury*, the search for Franklin's ships could continue as scheduled – toward the North.

Some wondered whether there could be a connection between the Inuit report to Captain Kerr of four ships frozen into the ice west of Pond Inlet, the discovery of Franklin's first winter quarters at Beechey Island, the fragments of ship wood found by Rae on Victoria Island, and the sighting of two ships on an iceberg near Newfoundland. Perhaps after wintering at Beechey Island, the *Erebus* and *Terror* had been blown out of harbour and badly damaged by drifting ice, after which some wooden fragments had drifted west to Victoria Island, while the ships were carried by floes into Baffin Bay and southward to the Grand Banks, by which time their crews had abandoned them and made for land. The *Nautical Magazine* urged "the strictest search along the west shores of Davis Straits . . . where they may find the very sails unbent from the *Terror*'s yards forming houses over our poor unfortunate countrymen."[32]

The search for Franklin had started with a simple three-pronged strategy: first, send an expedition to follow his track westward from Baffin Bay; second, station ships at Bering Strait in case he emerged from the Passage; and third, search the continental coast in between these two extremes. Then came Penny's discovery of an expanse of open water at high latitude stretching to the horizon. It caused what the historian John Brown called "a perfect infatuation in favour of Wellington Channel and the north,"[33] spawning proposals to search the northern coast of Asia from both east and west. But just as momentum for the investigation of the vast and unknown circumpolar region was building, Rae had discovered wood from a naval ship *five hundred miles south* of Penny's open water – close to the route that Franklin had been ordered to follow. And two abandoned ships had been reported on an iceberg *more than two thousand miles to the southeast* near Newfoundland! Meanwhile Sir John Ross was still stubbornly insisting that the ships had been destroyed and their crews massacred in northwest Greenland, and a succession of rumours about sightings of white men, ships, and boats had been emanating from Alaska. "It is certainly remarkable," observed the *Nautical Magazine*, "that in the case of the Arctic ships *Erebus* and *Terror*, almost every step in the measures for their discovery has heaped mystery on mystery."[34]

28

A Dispute over Place Names

Shove the American flag aside.[1]

Normally it was the responsibility of a commander to write the account of his expedition for publication, but Penny probably considered this more of a chore than a privilege, and in the months following his return he was too busy with other matters.

In January, John Stuart, former assistant-surgeon on the *Lady Franklin*, published a pamphlet of sixty-seven pages entitled *The Relief of the Franklin Expedition: What Has Been Done and What May Yet Be Done*. In it, he made a cynical comment about "the unseemly and undignified race of nine vessels – their different Commanders less actuated by any anxiety for the rescue of Franklin than eager to outstrip his neighbours in a passage to the westward."[2] After a review of the publication appeared in the *Aberdeen Journal*, Peter Cormack Sutherland, the former surgeon of the *Sophia*, submitted a long letter to the editor in which he objected strenuously to Stuart's remark. A few months later, he presented his own account of events in a two-volume book entitled *Journal of a Voyage in Baffin's Bay and Barrow Straits, in the Years 1850–1851, Performed by H.M. Ships "Lady Franklin" and "Sophia," under the Command of Mr. William Penny.* More than a thousand pages in length, his work was the most detailed account of any of the expeditions to the eastern Arctic in 1850–51. (Sherard Osborn's excellent narrative of the Austin expedition was a slim volume of just over two hundred pages, and William Parker Snow's colourful account of Forsyth's expedition was about four hundred.) Sutherland's familiarity with the natural history of arctic regions and the challenges of navigation among ice floes, which he had acquired on whaling voyages with Penny, contributed to the authenticity and maturity of his narrative. His book added significantly to public awareness about the expedition and greatly enhanced Penny's reputation.

The lengthy sledging trips carried out around the coasts of Wellington Channel and Queen Victoria Channel from Penny's icebound ships, and his month-long voyage in an open boat, had resulted in many names being assigned to features in the region. The travelling reports of the various parties published in Sutherland's book contained more than a hundred new place names. Many of these were probably assigned during the expedition using lists made in advance. (Before departing in 1850, Penny had asked Barrow to send him a list of names of Franklin's friends and people interested in the search expedition.) Other place names may have been added after Penny returned to Britain.

To have a place or a geographical feature named after oneself or members of one's family conferred a great honour. Names inscribed on the official charts achieved permanence. They would be immortalized in books and on maps for all successive generations. Among the many names Penny laid down were those of his wife, Margaret, his daughter, Helen, and his Admiralty friend W.A.B. Hamilton.

As keeper of records at the Admiralty, John Barrow was in a position to influence the assigning of names in the Arctic. He made suggestions not only to Penny but to others as well. In a letter to William Kennedy, commander of the *Prince Albert* in 1851–52, he listed the names of forty-one individuals, most of them involved somehow in the arctic enterprise, some of them personal friends. He also included his own father and his two sisters, who had no connection with the search other than their relationship to him. Then he confessed to Kennedy, "I have a sort of fancy I should like to have some Cape, or Hill, or Bay, or whatever there may be in *Lat.* 71 named after me, and should be very proud of the spot on the globe which really bore my name [as] connected with your Expedition."[3] The *Gazetteer of Canada* (1980) contains five Barrow names, all in northern latitudes.

Assigning place names to geographical features in newly discovered lands was regarded as a prerogative of the first explorers to reach the region. The names might record remarkable characteristics of prominent landforms, indicate noteworthy events during the voyage, or commemorate familiar places in their homeland. It was taken for granted that in selecting names, an expedition commander would also honour famous individuals, prominent statesmen, the sponsors of his expedition, and possibly a few of his own family and friends. After place names appeared on international maps and charts, they generally attained permanence. Naming was not merely a matter of national prestige; territorial claims were often based on evidence of discovery, and what better evidence could there be than a network of names clearly associated with the nation whose explorers had been first on the scene?

But after Penny's expedition, a problem arose. Unknown to anyone in Britain, De Haven's American expedition had preceded Penny's into the northern part of Wellington Channel. In September 1850, after Penny's ships had taken up winter quarters in Assistance Bay, the *Advance* and *Rescue* had been trapped in the ice floes and carried into Wellington Channel. As they drifted slowly northward, De Haven had assigned names to several features on the adjacent coasts. From their northernmost position at 72°25' N, the Americans sighted mountainous land beyond the termination of Wellington Channel. De Haven named this region Grinnell Land in honour of his expedition's principal sponsor. Half a year later, knowing nothing of the Americans' ice drift and name-giving, Penny would refer to the same region as Prince Albert Land.[4]

While De Haven's ships drifted helplessly through Baffin Bay during the winter on their thousand-mile voyage among the ice floes, no news of the American sighting of land beyond Wellington Channel could reach the outside world, or even the seven British ships wintering in Assistance Bay and near Griffith Island. It was not until after the ice released its captives in Davis Strait in June that the Americans were able to sail to the Greenland coast, tell their story to Danish administrators, and leave dispatches to be carried to England by commercial vessel. Further up the coast, they spoke several whaling ships and consigned messages to them as well. De Haven's primary intention was to inform the British Admiralty that Franklin's first winter quarters had been found – vital information for the prosecution of the search – and in the various messages he consigned to Danish administrators and British whalers, he said nothing about the new land his ships had sighted or the place names they had laid down to mark their discoveries.

Early in October, shortly after the *Advance* and *Rescue* reached New York, De Haven submitted his official report to the Secretary of the US Navy. It included six place names assigned by him in northern Wellington Channel during their ice drift. Apparently, neither the report nor any extract relating specifically to discovery or nomenclature was sent to the British Admiralty.

Before leaving Assistance Bay for home, Penny had delivered to Austin an outline chart (without place names) of the vast territory he and his men had searched north of Barrow Strait. On arrival in Britain, he delivered a copy to the Admiralty. He also sent a copy to Scoresby containing a dozen place names, including that of Scoresby himself, tactfully assigned to a headland on one of the islands. In late October, the *Illustrated London News* published a small-scale version of Penny's outline chart with about sixty place names. Sutherland's book, which contained the travelling reports of all the search parties, contained more than a hundred place names. Before any news had reached England of the American sighting of new land, British map-makers began preparing new charts to incorporate Penny's recent discoveries and the place names he had laid down.

Information about the American ice drift eventually crossed the Atlantic by ship to Britain. An important conduit for information was the steady stream of correspondence flowing between the two principal civilian sponsors of search expeditions – Lady Franklin in England and Henry Grinnell in the United States. Between the time of Penny's return in early September and the end of December, Grinnell wrote to Lady Franklin every two weeks or so. After she sent Grinnell a new chart by John Arrowsmith showing the discoveries of Penny and Austin, Grinnell sent her a copy of De Haven's chart. "You will notice," he pointed out, "that some of these discoveries [by Penny] were previously made by De Haven."[5] He expressed the hope that the British government would make the corrections necessary to reflect the American discoveries and nomenclature. On Lady Franklin's behalf, Sophia Cracroft sent a copy of the American chart to Scoresby, remarking that it showed "the claim of the Am[ericans] to priority of discovery."[6] Undoubtedly Lady Franklin informed the hydrographer and the Admiralty lords as well. Soon afterward, Grinnell wrote to say that he hoped Beaufort, Parry, and the map-maker John Arrowsmith would review the facts and assign the "proper names" to the geographical features in question.[7]

At about the same time, the *Times* (21 October 1851) and the *Nautical Standard* in London reprinted an item from the *New York Tribune* that included a letter Kane had written to Grinnell after his return from Greenland. In his letter, Kane described what they had seen a year earlier from their northernmost position in Wellington Channel: "We saw at this time high and dome like ranges of land trending north-westwardly ahead of us, in the latitude of 76° at least, and probably much further." He said nothing about assigning a name to the mountainous land, however.

After recently issued British maps were seen in the United States, concern arose that De Haven's discoveries might be permanently overlooked. On 18 November, the nautical publishing firm of E. & W. Blunt sent the British Hydrographic Office a chart showing De Haven's discoveries in Wellington Channel, with a tracing of the tracks of his two vessels during their ice drift in September 1850. When Beaufort presented the information to the Admiralty Board, the lords commissioners expressed their conviction that the nomenclature on British and American and charts should be identical, but in order to see where conflicts existed, they would need to see De Haven's official report. Beaufort wrote to the Blunts and said that as soon as they could examine a copy of his report, "Not an hour shall be lost in transmitting the results to you."[8]

* * *

The place-name issue would soon fire up the interest and the ire of Peter Force, a former printer, politician, and newspaper man in Washington. An avid collector of rare documents relating to the American Revolution, his collections would later form an important part of the Library of Congress. He was also interested in the search for Franklin, and in January 1850, a few months before the various search expeditions set off from Britain and the United States, he had written to William Scoresby, whose book *The Franklin Expedition: Or Considerations on Measures for the Discovery and Relief of Our Absent Adventurers in the Arctic Regions* had recently been published. Receiving no reply, he wrote again on 28 March: "In your late very valuable publication on the Franklin Expedition, you cover the whole ground of research, save one point . . . the Gulf of Boothia."[9] In ten pages of detailed argument, Force explained why he believed that Franklin had gone south in Prince Regent Inlet, west through the Boothia Peninsula in some strait not previously known, and on toward Bering Strait.

Although his ideas on the searching strategy do not appear to have impressed Scoresby or influenced any of the expeditions that departed soon afterward, Force continued to follow the progress of the search. After the return of all the ships, he took up arms against what he regarded as an intentional British suppression of the American discoveries. In May 1852, he read an acerbic paper at a meeting of the National Institute in Washington. "On the authority of the Lords of the Admiralty," he said, "De Haven's discoveries of 1850 have been set aside. The name of 'Grinnell' has, there, been erased, to make room for that of Prince Albert."[10]

Force claimed that the London map-maker John Arrowsmith had intentionally omitted the American place names on his map *Discoveries in the Arctic Seas*, published on 21 October 1851. Whereas Penny's sketch map of the northern regions had been compiled before there was any knowledge of the American ice drift, Force said, Arrowsmith could claim no such excuse. "There could be no pretence of a want of information; as full accounts of De Haven's discovery in 1850, had then reached England."[11] Arrowsmith's map showed "the first certain public demonstration in England to rob him [De Haven] of the credit of the discovery." In publishing it, the map-maker had committed "a discreditable act."[12]

According to an inscription on Arrowsmith's map, Albert Land had been sighted from HMS *Assistance* in the southern reaches of Wellington Channel on 26 August 1850, a month before the Americans saw it, and the event had been recorded in Ommanney's journal. The newly discovered land seen far to the north was named Albert Land because the sighting coincided with the birthday of Prince Albert. Force, however, declared this claim to be false and presented several facts to refute it. Ommanney was too far away to see land; the visibility was poor; he had never mentioned the sighting in any official report; he had said nothing about it when he appeared before the Arctic Committee in November. Most damning was Force's assertion that, contrary to Arrowsmith's inscription, Ommanney's journal contained no reference to sighting land further north. "Whether the statement was made on the authority of Captain Ommanney, or was invented by Mr. Arrowsmith, must be settled by the gentlemen themselves."[13]

Had Force really examined Ommanney's journal? Evidently he had not. He must have assumed that the *journal* cited by Arrowsmith was the same as the *report* Ommanney submitted to his superior officer. In one place he quoted from Ommanney's "Report to Captain Austin of 10th September, 1850" and declared that the extracts provided "abundant proof that no discovery was made by him on the 26th Aug. 1850."[14] But Ommanney's *report* and *journal* were separate documents – the former an official account and the latter a personal day-to-day record of events. Ommanney's manuscript *journal* exists today in the archives of the Glenbow Museum in Calgary, and his entry for 26 August 1850 reads as follows: "Lt. McC reported that he saw land this morning in a Northly direction, mid-channel . . . It being the birthday of H.R.H. Prince Albert I proposed naming the land seen this morning Albert Islands."[15] Although this passage reveals that it was McClintock rather than Ommanney who first sighted land far to the north, and that it was at first thought to be group of islands rather than a continuous land mass (a logical mistake if only mountain summits were visible), it shows that Arrowsmith had a valid reason to believe that the British had been the first to sight the region that the Americans saw a month later.

As the land sighted on 26 August 1850 was reported to lie in mid-channel toward the north, it was almost certainly the part of Devon Island to which the names Albert and Grinnell were separately given. Although the distance from HMS *Assistance* was great, the land was high, and summits may have been visible either directly or through miraging. Men on other British ships had seen it too, although there were some differences of opinion: some officers on HMS *Assistance* and *Pioneer* said they saw land, but others did not. On the *Sophia*, Manson saw land, but Stewart did not. Perhaps the mountains were only visible from time to time, owing to changing conditions of cloud cover and visibility.

To Force, however, it was nothing less than a British conspiracy. Ommanney's sighting on 26 August was "fabricated;" Penny, under pressure from his superiors, had lied about it; Sutherland "intended to mislead." Clearly, the British objective was to "shove the American flag aside, that their own might appear to be in front."[16]

Penny and his men had sledged hundreds of miles, sailed an open boat among the ice floes for a month, and charted a vast unmapped territory. Yet their achievements went virtually unnoticed in Force's diatribe, which denigrated most of Penny's observations: his latitudes and longitudes were wrong; his bearings and distances were "absurd and preposterous"; his hydrography was "guess-work"; he was thirty-three days in the boat but "got no farther west than Baring Island, about twenty miles;" he extended the coast on his map "to the utmost limit of credence" in order to make his discovery "as large as possible"; his remarks on ice conditions before the Arctic Committee were incorrect.[17] Force ruthlessly belittled everything Penny and his men had done while saying nothing about the hardships they had faced or the impressive extent of their discoveries.

Having revealed all the supposed deficiencies of Penny's methods, Force wondered how he could have dared to correct "the observations of an American naval officer, an experienced and accomplished seaman, thoroughly versed in all the branches of nautical science."[18] What corrections Penny was alleged to have made, he did not bother to reveal, but in any case, his inference that De Haven was an officer of extensive experience and ability whereas Penny was not was very far from the truth. Penny had nine more years of nautical experience behind him than De Haven; he had been sailing to the Arctic for almost three decades, while it was De Haven's first voyage to the region. He had commanded ships for fourteen years, but it was De Haven's first command. All Penny's whaling voyages had been in the proximity of sea ice, but De Haven had been on only one voyage into ice-filled waters, as a junior officer near Antarctica.

Force sent his *Remarks* to the principal individuals involved in the matter in England, but in a "Supplement" published later, he complained that "none had the candor to admit that the truth had been fairly stated. The Admiralty, the Hydrographer, Captain Ommanney, Mr. Arrowsmith, Captain Penny, &c., were all silent."[19] It would not be surprising if Force's scurrilous tone, unfounded accusations, and false statements had made them reluctant to reply directly to the author or to comment publicly. In the United States, Henry Grinnell, whose own name was vying with that of Prince Albert for a prominent place on the charts, remarked to Kane, "Although I believe the facts he states are true, still I do not like the spirit of the work, evidently he has strong prejudices against the English."[20] In his second pamphlet, Force emphasized that the sole authority for the British claim to the discovery of Albert Land was Ommanney's journal – "an authority that was proved to be entirely fabulous."[21] This was strong language from a man who had never actually seen the journal! After repeating many of his previous points and pointing out that books were now perpetuating the myth of British discovery and suppressing the truth, he ended on a dramatic nationalistic note: "America is too proud to claim a discovery that may not be claimed with honour, and honesty, and truth. AMERICA NEITHER STRUTS NOR FLAUNTS IN BORROWED OR STOLEN PLUMES."[22]

Reviewing Force's address to the National Institute, the Washington correspondent of the *Baltimore Sun* wrote, "The impudence of the English Government is astounding . . . they have actually stolen the American discovery of 'Grinnell Land," claim it as British, and call it 'Albert Land.'" The Admiralty and British map-makers were guilty of "the grossest falsehood and most unblushing robbery."[23]

Penny's opinion on Force's comments is not known. At the time he was trying to obtain command of a search expedition, form the Royal Arctic Company, obtain a land grant in Cumberland Sound, introduce steam power to the whaling fleet, expand whaling into the supposed Open Polar Sea, act as consultant for the Atlantic Telegraph project, and sit for a portrait by Stephen Pearce. The reaction of Jane Franklin and Sophia Cracroft is on record, however. The former wrote to Scoresby, "Have you seen the most bitter and unjust pamphlet by Col. Force upon the [American] discoveries in Well^{n} Channel, versus our own. The Lords of the Admiralty, the Hydrographer, Arrowsmith & Captn Ommanney are [said to be] the veriest rogues & cheats."[24] (Why she referred to Force as a colonel is unclear. He had enlisted as a private in the US army in the War of 1812 and had risen to the rank of lieutenant but no higher. Perhaps it was an honorary title awarded later, or even a rank he had assumed on his own.)

The controversy over place names on the charts persisted for several years. The British retained the name Albert Land, presumably because they believed that it had first been sighted from Ommanney's *Assistance* – which it had. The Americans called it Grinnell Land because they believed it had first been sighted by De Haven – which it had not.

A few years later, Sir John Ross could not resist stirring up the waters with his own oar. He declared that when the *Felix* had remained in Union

Bay in September 1850 after all the other ships had advanced westward, the ice had opened to the northward, and "we were enabled to see the land to the north of this barrier, which I named 'North Victoria,' but which was subsequently claimed by Mr. Penny as his discovery, and named 'Albert Land.'"[25] In fact, Penny never claimed to have discovered Albert Land, and Ross's alleged sighting was not mentioned in his diary or anywhere else.

Among the few voices raised in England in favour of the recognition of De Haven's names was that of Scoresby, who appears to have been unaware of the prior sighting from HMS *Assistance*. After reading Kane's book about the American expedition a few years later, he wrote to the editor of the *Morning Chronicle* (25 August 1854) about "the injustice done, in our authoritative charts of the far north, to the claims of Captain De Haven, of the Grinnell United States search expedition, to priority of discovery of the coast immediately northward of Wellington Channel." Scoresby rejected the argument that as the American ships had reached their northernmost position by accident rather than intent – through happenstance rather than skill and effort – their place names should not be accepted. How they got there was irrelevant, he said. He also rejected the argument that the priority of discovery probably belonged to Franklin, who appeared to have passed that way in 1846. Unless this could be proven, Scoresby declared, the names laid down by De Haven and Penny must stand. It bothered him that three years after the return of the search expeditions, the matter had still not been resolved. "Why, indeed, the remonstrances from America have not before this time obtained their object, is difficult to account for."

In 1854, Grinnell told Lady Franklin that he did not intend to protest to Beaufort at the Hydrographic Office because he had "not the least doubt he will do just the right thing when he has all the proper evidence before him."[26] The issue was unresolved until almost four years had elapsed since Penny and De Haven returned from their voyages. The Admiralty's hydrographer decided to adopt the name Grinnell Land, either because he concluded that the British claim of sighting Albert Land from HMS *Assistance* on 26 August 1850 was too tenuous, or simply because he wanted to pay tribute to the American involvement in the search and restore amicable relationships. On British charts, Albert Land became – and still is – Grinnell Land, and all but two of the six names assigned by De Haven in Wellington Channel were placed on the charts as well. (The name Mount Franklin was dropped because its location was too uncertain, and Murdaugh's Island did not exist.)

In New York, Grinnell expressed satisfaction with the revised British chart. After displaying it at the American Geographical Society, he wrote to Lady Franklin. Everyone had been pleased, he told her, "to see what the Admiralty had done, in placing the names given to Lands, Capes &c. by De Haven on the chart."[27]

The perceived injustice toward De Haven's American expedition did not go quietly into the night, however. As nobody had examined Ommanney's private journal, the prior British sighting of the mountainous region north of Wellington Channel was never revealed in the United States. A few years later, when William Elder published his biography of Kane, he wrote, "our 'friendly allies' in the search for Franklin did not behave handsomely, nor fairly, nor respectfully, nor justly, in this matter, which so nearly touched the honor of the American wing of that service. It is all settled now rightly, but it was not done gracefully, by the Lords-Commissioners of the British Admiralty."[28]

29

Penny's Portrait

An expression of great firmness. . .[1]

Penny's unfair treatment at the hands of the Admiralty and the Arctic Committee following his return from the Arctic had not diminished the public perception that he was a skilled and determined navigator who had done his utmost in the search for Franklin despite restrictive orders and a lack of cooperation from Austin. In the autumn of 1851, newspapers had generally extolled his boldness and expressed their conviction that he was the man to command the next search expedition. The Admiralty's rejection of his offers to serve again in the search only reinforced the widely held view that he was being improperly denied the opportunity to contribute to the great humanitarian venture. In the eyes of the British public, Penny was a champion. He also had some loyal supporters in the corridors of power, including Lady Franklin and Sophia Cracroft, Osborn, and several other naval officers, Barrow and Hamilton at the Admiralty, and Beaufort at the Hydrographic Office.

What Penny and his men had achieved during their search expedition was impressive by any standards. Their search of Beechey Island resulted in the discovery of Franklin's first winter quarters, and their explorations eastward and northward of the island revealed two outlying campsites used by Franklin's men for hunting or monitoring ice conditions. They climbed to the summit of Beechey Island to re-examine the summit cairn inspected earlier by Ommanney. In the absence of Austin's four ships, it was Penny who suggested a plan of action for the small-ship expeditions at Union Bay. On two occasions, he communicated with Ommanney on the opposite coast of Wellington Channel, and when Stewart's boat party intercepted HMS *Resolute* and *Pioneer* as they were arriving from Pond Inlet, Stewart gave Austin the first news of developments at Beechey Island and alerted him to navigational hazards on the island's west coast. When the *Pioneer* ran aground on the way to

Union Bay (despite Stewart's warning), it was Penny rather than Austin who came to assist.

More than the other expedition commanders, Penny had been conscious of the need for everyone to share information about what they had done and what they intended to do, so that resources could be used effectively without wastefully duplicating the work of others or leaving gaps in the search. Not all the commanders felt the same.

In addition to being the only commander to employ sledge dogs, Penny used boats to good effect. Using open boats less than thirty feet in length among ice floes was usually considered too risky, and with good reason. Penny, however, knew that whaling crews fleeing from ships that had been crushed by ice along the Greenland coast had sometimes been forced to make lengthy voyages to Lievely or other settlements and had coped successfully with drifting floes. This knowledge, and his own experience in handling boats, gave him the confidence to dispatch Stewart to search eastward from Beechey Island as far as Gascoyne Inlet, and to send Stuart across Wellington Channel to contact Ommanney. Both parties travelled in open whaleboats. His most enterprising move was to haul a boat a hundred miles north from his wintering ships and undertake an extensive voyage during June and July. With half a dozen men, he had explored the region north of Cornwallis Island for a month, contending with jostling ice floes and swift inter-island currents. This dangerous but successful venture – unprecedented in arctic exploration – would inspire others, including Elisha Kent Kane.

Penny and his men had contributed in an important way to Euro-American knowledge of the Arctic. They mapped the northern coasts of Wellington Channel and much of Queen Victoria Channel, identified several islands within the latter area, and drew attention to the existence of open water at high latitude, a phenomenon that influenced the search for Franklin.

John Barrow, the Admiralty's keeper of records and confidant of many of the officers involved in arctic exploration, was an ardent admirer of Penny. He was also a friend of Stephen Pearce, a London artist who had made a name for himself painting equestrian subjects and had exhibited at the Royal Academy. In 1850, Barrow had commissioned Pearce to paint a group portrait of the ten men who were most actively involved in the planning of the search for Franklin. They were collectively referred to as the Arctic Council, although they rarely, if ever, met together as a group or functioned as a committee. The subjects (including Barrow himself) sat individually for the artist and were assembled later into a composite portrait. In his autobiography, the artist would recall, "As soon as the 'Arctic Council' was finished, it was sent by command to Buckingham Palace for the Queen and Prince Albert to see; and then Messrs. Graves & Co. exhibited it in Pall Mall, and at the principal places in England, Ireland, and Scotland, for the purpose of obtaining subscribers' names for the excellent steel engraving executed by James Scott."[2] The popularity of the work led Barrow and Lady Franklin to commission or purchase individual portraits of more than two dozen prominent arctic explorers.

Penny was one of the first arctic explorers to sit for an individual portrait by Pearce, and one of the few non-naval men to be so honoured. When the sittings began is not clear, but in March 1852, shortly before the departure of Belcher's flotilla, Pearce wrote to acknowledge receipt of the carpetbag in which Penny had sent his arctic travelling clothes for the portrait. The sittings must have been already under way, for he scolded Penny for not sending the clothes sooner: "Mr. Barrow has been in quite a fidget about the [sealskin] dress not coming, for both he and Lady Franklin are most anxious that your Portrait should if possible be sent to the Royal Academy Exhibition. They say it will do the cause some good, & you also – your Portrait there will be seen by thousands & your name thus kept before the public."[3]

Evidently Pearce was not *au courant* with the garb normally worn by explorers in the Arctic, for he asked, "When you wore the Esquimaux jacket, did you wear under it a shirt, black neckerchief, waistcoat, & frock coat? If you did not wear all these under the Esquimaux jacket, what did you wear?"[4] Having lost ten days' work owing to the absence of the sealskin parka and trousers, Pearce was not sure if he could complete the work in time, and he asked Penny if he would be in London during the next few weeks. In mid-June, the painting was still unfinished, and in an letter sometime later, Pearce told Penny that he had finished with his "seal skin dress &c" but would require him for two more sittings. As artist and subject lived four hundred miles apart, the meetings were not easy to arrange. However, the painting was finished by the autumn. In early October 1852, Lady Franklin told Penny that she had seen his portrait at Pearce's studio and thought it should be placed in the National Gallery.

Penny's portrait, which is now at the National Portrait Gallery, measures approximately forty by fifty inches. It is a three-quarter-length view of Penny standing, bareheaded and glancing slightly to his left with an expression of calm resolution. He has brown eyes, dark brown hair, and a neatly trimmed beard. His left hand rests jauntily on his hip clutching a fur cap with earflaps, while his right fist rests upon a long, tapering sledge flag bearing a Union Jack with the motto "God Aiding Us We Will Do Our Duty." He wears a sealskin parka, the fur facing outwards, trimmed around the bottom with coloured cloth, and the hood folded down behind his neck. Underneath the parka, he appears to be wearing a white shirt with a collar (no sign of a frock coat). A leather pocket sewn over the left breast may have contained a compass. His trousers, like the parka, appear to be of sealskin.

In the background, a man is sitting on a dog sledge to which several dogs are harnessed, near a building apparently made of snow blocks with wooden rafters, and in the far distance two vessels can be seen, banked with snow and housed in for the winter. The scene is clearly Assistance Bay, with HMS *Lady Franklin* and *Sophia* securely frozen in, while Carl Petersen, Penny's Danish dog handler and travelling companion, waits with his team beside the expedition's storehouse on the snow-covered shore of Cornwallis Island. Penny, dressed in fur clothing of Greenland Eskimo style, is about to employ native travel methods in the search for Franklin.

In the summer of 1853, Pearce wrote to Penny, "You will be pleased to hear that very many of the London newspapers have given most flattering notices of your Portrait in the Exhibition of the Royal Academy, & I hear on all sides from friends, acquaintances, & even strangers that it is one of the most striking Portraits there."[5] After seeing it, some of Margaret Penny's friends told her they had been "startled with the striking likeness & much admired the exquisite manner in which it is painted."[6] Penny's "head and figure coincide in an expression of great firmness," stated the Academy's catalogue.[7]

Penny received additional exposure to the public through the sale of engravings made from the painting. In April 1853, a handbill announced that Henry Graves and Company, "Printsellers and Publishers in Ordinary to Her Majesty the Queen and His Royal Highness Prince Albert," were preparing to publish the portrait of "Captain Penny, late of the discovery ship, 'Lady Franklin,' represented in the costume he wore in the Polar Regions while on his sledge journies [*sic*] in search of Sir John Franklin, engraved in the finest style of mezzotinto, by Mr. James Scott, from the original picture, in the possession of John Barrow, Esq., F.R.S. Painted by Stephen Pearce."[8]

Penny's portrait would be admired by many at an exhibition in Manchester in 1857, displayed along with two other arctic paintings by Pearce – his famous group picture of the Arctic Council, and his portrait of Robert McClure, the man credited (but not in the eyes of Lady Franklin!) with the discovery of the Northwest Passage. Later Penny's portrait would hang in the Arctic Navigators Room of the National Portrait Gallery in London along with portraits of Franklin, McClintock, McClure, and Nares – arguably the most famous British arctic explorers of the nineteenth century.

Of all the men who served on the two Admiralty search expeditions to the eastern Arctic in 1850–51, Penny was the first to have his portrait painted by Pearce. His second-in-command, Alexander Stewart, followed in 1854, but it was not until a few years later that any officers of Austin's expedition were painted – McClintock in 1856, Osborn in 1857, Austin in 1860, and Ommanney in 1861.[9] Undoubtedly the patronage of Lady Franklin and John Barrow had much to do with the priority given to Penny.

Henry Grinnell, the sponsor of the American search expedition under De Haven, told Lady Franklin that Penny's portrait hung in his dining room along with ones of Wellington, Napoleon, her husband, and the Arctic Council. In this American home, Penny was in very lofty company! John Barrow's admiration of the whaling master may even have exceeded of Grinnell. He had reproductions of Penny's portrait hanging in his drawing room, dining room, bedroom and dressing room. "I never look at it without feelings of the strangest attachment & admiration," he confessed.[10]

30

A Gratifying Acknowledgment

Mr. Penny urges his plans – not always wisely; still they have a national spirit.[1]

Shortly after arriving back in Scotland, Penny had received a brief note from the Admiralty acknowledging the "praiseworthy conduct" of his officers and men and the "untiring and praiseworthy exertions of the travelling parties," but nothing more than that. Through the winter months, as Austin and his men garnered honours, awards, and promotions, Penny came to feel that he and his crews had been slighted.

He was determined to obtain a decent gesture of appreciation for his services. He told Barrow that he wanted *justice*, and public opinion backed him up. The *Aberdeen Journal* (17 March 1852), after receiving a copy of his published *Letter to the Lords Commissioners of the Admiralty*, expressed a view that was widespread: "His explanations will be found so far satisfactory as to show that not only has his conduct been in every respect honourable towards his colleagues, and zealous in the momentous cause in which he took so important a part, but that he has not been treated by the Admiralty with that consideration which his services merited."

In the autumn of 1852, when he succeeded in obtaining an interview with the Duke of Northumberland, first lord of the Admiralty, Penny pointed out that he and his men had not only searched to the best of their abilities but had also made important geographical discoveries. Surely their services deserved to be officially recognized. Soon afterward, the *Herald* reported that Penny had been *honoured* by the duke with an interview and had received "*a gratifying acknowledgement*" of his services and those of his men. The magazine *Punch* could not refrain from unleashing some scathing sarcasm:

> Mr. Penny, being only Merchant-captain PENNY, having no epaulets upon his shoulders, but simply brains schooled by long experience

> in his head, was treated with the most contemptuous neglect, all his counsel disregarded – which was afterward proved to be so wise and true – by the officers of "the service." CAPTAIN AUSTIN returns home, having done nothing except "pooh-pooh" the merchant skipper, and is thereupon appointed to a most lucrative post at Southampton. At length, however, the skipper is rewarded. Yes; he is dignified for life, "with an interview" of the First Lord of the Admiralty; and receives "a gratifying acknowledgement of his services!" By what means, may we ask, comes the acknowledgement? By cold word of mouth, or has skipper PENNY anything tangible to show for it?[2]

Sherard Osborn's brother Edward clipped the item out of the magazine and sent it to Penny, remarking that it was "the most honest compliment you have ever had paid you. *Punch* often jests without cause; but is never serious without strong and good reason."[3]

In their *gratifying acknowledgement* of his services, the lords commissioners of the Admiralty assured Penny of their "entire concurrence in the opinion expressed by the late Arctic Committee" and thanked him for his "humane and arduous exertions."[4] In fact, the Arctic Committee had ignored or rejected much of his testimony, and he and his men had done much more than merely exert themselves. What the *Herald* called a *gratifying acknowledgement* was really a backhanded compliment, given reluctantly more than a year after his return. Still, it provided closure of a sort, after which Penny launched energetically into plans for a new search expedition and once again applied to the Admiralty for command. Whether or not he really thought that they would seriously consider hiring him is not known, but they did not. He passed the summer of 1852 at home, anxious to learn whether any of the various expeditions had achieved success.

* * *

Following the disappointing voyage of the *Prince Albert* under Forsyth in 1850, Lady Franklin had sent the small ship back to the Arctic in the following year, this time under the command of William Kennedy. She and Sophia Cracroft had been in Orkney when the little vessel had passed through in the summer of 1851 before heading west to Greenland and beyond. Sophia wrote to her mother and sisters, saying that they had gone on board, met several of the crew ("on the whole a remarkably fine set of men"),[5] and admired the seven message-carrying homing pigeons, particularly "Lady Ross," now setting forth on her second arctic voyage.

They said emotional farewells to their "dear little French friend," the keen, young Lieutenant Bellot, who was serving as a volunteer on the expedition, and watched the *Prince Albert* fade into the distance. They remarked on Kennedy's lack of business skills; after arriving with a broken jib boom, he had seemed reluctant to make the necessary arrangements to have it repaired. "We have therefore had to deal with this work ourselves," Sophia Cracroft wrote, "& very unpleasant work it is."[6]

Now, a year and a half later, the *Prince Albert* was back from the search, and when Kennedy arrived in Aberdeen, his lack of business skills was again apparent. He did not have enough money to pay the crew's wages. He got in touch with Lady Franklin in London, and she in turn wrote Penny in Aberdeen to ask for help in settling the matter. She promised to send him £70 immediately ("all I have in the house") and probably another £300 as soon as the banks opened, and she would reimburse Penny for the money he had already lent to Kennedy.[7] It was certainly the only time Penny was able to bail out another expedition commander and help the wealthy Lady Franklin out of financial difficulties. No doubt he enjoyed relating the incident later.

From their wintering vessel in Batty Bay on the west coast of Prince Regent Inlet, Kennedy and his men had performed impressive man-hauling sledge trips, even in the coldest months of winter. Instead of travelling out and back along the same line, according to the method devised by McClintock for Austin's expedition, they followed a circular track, thus searching a much larger territory. Lady Franklin and Sophia Cracroft were pleased with the efforts and expressed admiration for Kennedy. He is the "mildest of heroes,"[8] Sophia told Scoresby, while Lady Franklin considered him "one of the worthiest beings I ever knew."[9] Penny, on the other hand, resented the fact that a fur trader with no maritime experience had been placed in command of one of Lady Franklin's search expeditions, while he was excluded from the search.

Kennedy did not think much of Penny either. When he wrote to Kane on 19 November 1852, he said that Penny had been "putting himself foremost not only as the most worthy but as the greatest discoverer ... to the utter disregard of *all others.*" During the dispute about place names in Wellington Channel, Kennedy declared, "I actually heard him ... deny the very existence of Murdaugh Island,"[10] which had been placed on maps of Wellington Channel by De Haven's expedition as his ships were carried north in the pack. (In fact, Penny was right: the island did not really exist.)

Other career sailors may have shared Penny's opinion that a fur trader – a mere *landsman* – should not be commanding a ship in the search for

Franklin. McClintock remarked to James Ross, "Kennedy is now known to be an imposter unworthy of belief."[11] Despite his lack of maritime qualifications, however, Kennedy completed a successful voyage, accomplished some notable searching trips by sledge, and made a significant geographical discovery. In the recesses of Brentford Bay, he had found a strait cutting westward through the mountains of North Somerset, which had hitherto been regarded as a peninsula extending all the way to Barrow Strait. As a result of this discovery, the northernmost tip of the continental mainland withdrew about 150 miles southward, and an alternative route toward the west was revealed. Kennedy named it Bellot Strait.

At the western extremity of the strait, Kennedy saw what appeared to be "a continuous barrier of land"[12] crossing Peel Sound father north. Even if it consisted of a maze of islands rather than unbroken land, he was certain that it could not be navigable. He decided to cut across to Prince of Wales Island and circle northward via Barrow Strait and back to Batty Bay. Kennedy's unfavourable report of Peel Sound appeared to confirm the opinion of W.H. Browne, whose sledge party had entered the waterway from Austin's ships at Griffith Island a year earlier, and also that of McClintock in 1849 during James Ross's search expedition. It now seemed clear that Franklin's ships could not have gone south through Peel Sound, and this strengthened the widespread feeling that they had turned north into Wellington Channel instead and had continued through the open water discovered by Penny.

If Kennedy and Bellot had turned north instead of south in Peel Sound, they would have discovered that the strait was navigable. (It was in fact the path that Franklin had taken.) And if they had travelled south through what is now called Franklin Strait and continued as far as King William Island, they would have discovered evidence of the Franklin disaster. A sledging voyage from their winter quarters in Batty Bay to King William Island and back would have been approximately the same distance as their futile circular trek by way of Prince of Wales Island and Barrow Strait.

Kennedy returned to Britain with important news from Beechey Island. Belcher and Osborn had taken HMS *Assistance* and *Pioneer* north through Wellington Channel without any difficulty and had confirmed the existence of Penny's open water! This was delicious irony, because Belcher was the obnoxious fellow who had objected to the reading of Penny's letter at the Royal Geographical Society meeting and had rudely cast aspersions on the reliability of his information. Now, having found Penny's observations to be perfectly correct, he would be obliged to eat humble pie. Penny was vindicated, and the news excited Lady Franklin, who wrote to him, "What a triumph for you & all of us who have our [hearts] in the cause that is the news that Mr. Kennedy brings us of Wellington Channel. So Penny was right after all! is what the fiercest opponents must now say. But even without this conclusive testimony to the value of your discovery, every body speaks of you & all you have accomplished with respect & admiration – I speak of naval officers with whom I have lately been in communication."[13]

In August 1850, the ships from all five search expeditions had been restricted to the southern part of Wellington Channel by an ice barrier extending from one coast to the other, and when they departed homeward in the following summer the impediment was still there. But a year later, in the same month, Belcher's *Assistance* and the steamer *Pioneer* had passed easily up the channel and continued for a considerable distance through Queen Victoria Channel and further toward the northwest. Like Parry in his famous traverse from Baffin Bay to Melville Island in 1819 and back in the following summer, Belcher had been blessed with unusually favourable ice conditions that no one could have predicted. Instead of acknowledging his good fortune, however, he chose to malign the explorers who had preceded him. "How the imagination of the croakers had conjured up the 'barrier of ice,' I am totally at a loss to conceive," he wrote later, "for here not a particle of ice, to be seen from the land, about eighty feet above the [sea] level, could warrant the idea of obstruction."[14] Penny, one of the so-called *croakers*, must have been furious. The ice barrier that he and the other commanders had encountered in the two previous seasons had been perfectly real.

The idea that arctic climate and ice conditions were constant year after year was a persistent misconception. Within the living memory of many Londoners, the River Thames, although normally open through the entire winter, had occasionally frozen over from bank to bank, enabling shopkeepers to hold fairs on the ice near Westminster Bridge. Yet many found it difficult to realize that the phenomenon of environmental variability extended to arctic regions. If one explorer managed to cover a greater distance than his predecessors, newspapers usually credited the feat solely to his superior navigation or zeal. Important as those characteristics were, they seldom constituted a complete explanation.

Lady Franklin and Sophia Cracroft understood the real situation. Writing to Osborn about the barrier of ice in Wellington Channel and Barrow Strait, Sophia asked rhetorically, "Can we in the face of facts assure its permanency?" She pointed out that although both waterways were frozen up in 1850, both had been open in 1819. "We at home do not think that either the one or the other is always closed," she wrote.[15] To Scoresby, she said that although "the barrier of ice in Wellington

Channel is made much of as an impediment," there was no reason to assume that it was permanent.[16]

The officers of Franklin's ships had also been aware of the variability of ice conditions. Writing from Greenland in 1845, Fitzjames had expressed it nicely: "I can see, however, that even if there be a good passage, that it is a perfect lottery what sort of season we have and whether we happen to be at the particular spots at the most favourable moments."[17] A lottery it truly was, and Belcher had drawn the winning number in 1852. Would he strike it lucky again when he attempted to return southward?

Shortly after Kennedy returned on the *Prince Albert*, another of Lady Franklin's expeditions reached England. She had sent the screw yacht *Isabel* out under Edward Augustus Inglefield to search the west coast of Baffin Bay and Davis Strait. Before proceeding to the West Side, he had sailed north along the Greenland coast and, as the *Illustrated London News* reported (11 December 1852), he now claimed that Whale Sound in northwest Greenland was "a great strait passing into the Arctic Basin" and that he had sailed further north in Smith Sound than any other navigator – right to the very margin of the Open Polar Sea! Later, when the inevitable book was published, he described the elation he had experienced at that moment: "We were entering the Polar Sea, and wild thoughts of getting to the Pole – of finding our way to Behring Strait – and most of all, of reaching Franklin and giving him help, rushed rapidly through my brain. A few hours and we should either be secure in our winter quarters, or else plying onward in the unfreezing Polar Basin."[18]

Like John Ross in Lancaster Sound, Penny in Queen Victoria Channel, Austin in Jones Sound, and McClintock, Browne, and Kennedy in Peel Sound, Inglefield had made unwarranted judgments about uncharted bodies of water without confirming their full extent and nature. Whale Sound was in fact merely a fiord leading into Greenland's ice-covered mountains, and Smith Sound was only a part of the North Water polynya (although it does lead eventually to the Arctic Ocean by way of Kane Basin, Kennedy Channel, Hall Basin, and Robeson Channel). With open water extending northward into the distance, Inglefield supposed that he was on the edge of the Open Polar Sea, but, as subsequent explorations would reveal, the Arctic Ocean was still three hundred miles away and it was covered by ice. Nonetheless, his report of open water in Smith Sound was seized upon as additional evidence for the existence of the Open Polar Sea, and that waterway would soon attract other explorers, including De Haven's former surgeon, Elisha Kane, whose ambitions became focused on the North Pole.

Inglefield also reported that he had later sailed the *Isabel* westward into Jones Sound to a longitude of 84° W roughly fifty miles beyond Cone Island. Whereas Gravill, Penny, and Austin had been unable to penetrate the ice-choked waterway, he had apparently succeeded, aided – it may have seemed – by the same unusually open ice conditions that had enabled Belcher to proceed northward through Wellington Channel. Inglefield had not seen HMS *Erebus* and *Terror*, or the mysterious cairn allegedly seen by Captain Lee a few years earlier, or any other evidence of Franklin's passage. His report did not entirely eliminate Jones Sound from the Franklin search, however. The longitude he claimed to have attained was still a hundred miles from the western end of Jones Sound (the full extent of which was still unknown); he had remained at this position only briefly; and he had not been able to search for evidence along the north or south coasts because they were icebound. Furthermore, a recent examination of Inglefield's original track chart by Brian Graville (a descendent of the whaling master John Gravill) has concluded that his position-finding was faulty and that he had reached "no further west into Jones Sound than Austin did the year before."[19]

As another winter descended over the arctic regions, the hopes of finding Franklin were pinned on the two expeditions still out searching – Belcher's four ships to the west of Baffin Bay, and Richard Collinson's two ships intending to search eastward from Bering Strait.

* * *

In early November 1852, Penny attended a meeting of the Royal Geographical Society. Responding to an invitation from the president to outline his views on the search, he rose and delivered an impassioned speech on its status and future direction, which was summarized in the *Illustrated London News* (13 November 1852). Belcher, Penny noted, was already "at the door," but it looked as if Collinson and his second-in-command, McClure, commanding HMS *Enterprise* and *Investigator* somewhere in the western Arctic (no one knew precisely where), had become "second Franklins." It would be a severe setback if more ships were marooned somewhere in the arctic ice, so Penny suggested that a small, shallow-draught, steam-powered vessel be sent through Bering Strait to find and assist them, and that Rae (already famous for his travelling exploits) be sent back overland in case the crews decided to abandon their ships and head south toward fur-trade posts on the Mackenzie River. (He did not explain what assistance Rae and a small party of men could provide.)

Adopting the essence of the unrealized plans of Pim and Beatson, which in turn had been based on his own discovery of open water at high latitude, he called for a renewed search in which two steamers would go through Bering Strait and along the Asiatic shore "into the Arctic Basin, where an extensive sea must exist." The Open Polar Sea had now become a fundamental part of Penny's perception of arctic geography. He also urged the government to adopt Petermann's suggestion of entering the polar sea by way of Novaya Zemlya. In this, Penny was keeping in mind his own vocation of pursuing whales; Petermann had suggested that a probe in that direction could reveal new whaling grounds.

Penny then made an interesting announcement. He had learned that the crew of an American whaling ship had wintered ashore in Cumberland Sound and had succeeded in killing thirteen whales – "an amount of food upon which Franklin and his whole crew might have subsisted during the whole time he has been absent," He said. But it was now seven and a half years since HMS *Erebus* and *Terror* had sailed, and although Penny's news was encouraging, there was growing feeling that the solution to the Franklin mystery would not be a pleasant one. He later wrote to tell a friend about the success of his "maiden speech" at the society: "When I sat down, you would have thought they would have brought the house about our eyes, and Sir Roderick [Murchison] said he had that night heard more valuable facts than had yet been brought to bear upon the subject."[20]

Penny's presence at the meeting was not altogether disinterested. He still wanted command of a search expedition, and Lady Franklin and others were supporting his cause. Gaining the backing of the Royal Geographical Society was an essential step. Sophia Cracroft wrote to Scoresby, "Penny is in town & his friends are striving hard to get him in command of an Exp. either by way of Bering St. or by Nova Zembla upon the theory of W. Petermann."[21] Lady Franklin asked for Scoresby's support: "We are now most anxious for Penny getting employment in a new exped^{n}. There is nobody equal to him – Cannot you write him up a little? It would go a great way to help."[22] She and Sophia hoped that Penny could lead an expedition into the polar sea in the summer of 1853 to meet Belcher's ships as they advanced westward from the Wellington Channel entry point.

The women reminded Scoresby that there might be commercial as well as humanitarian benefits: "The Americans are reaping a golden harvest w^{h}. might have been ours," Sophia wrote.[23] For the past few years, whaling ships had been sailing from San Francisco through Bering Strait to pursue bowhead whales in the Chukchi and Beaufort Seas. Those lucrative whaling grounds, she thought, could have been exploited by British whaleships sailing eastward from Europe along the Siberian coast or directly across the allegedly ice-free circumpolar ocean.

* * *

Without definite prospects of securing command of another expedition in search of Franklin in 1852, Penny turned his thoughts back to his real vocation of whaling and reconsidered an idea he had expressed a decade earlier after bringing Cumberland Sound back into the sphere of European awareness and initiating whaling there.

The problem facing whalemen in the sound was that, as the whales were smaller than the ships and capable of swimming under water for a limited distance, they were able to enter the sound long before the sailing ships could penetrate the massed remnants of the winter ice cover. Although the whales were numerous and conveniently concentrated near the retreating margin of the landfast ice, the whale hunters could not get to them through the floes. By the time the ships were able to make their way into the sound, the whales had dispersed and were therefore harder to catch.

Penny had considered wintering on board ship or establishing a year-round shore station somewhere within the sound so that the whalers could be on hand when the whales advanced into the sound and could pursue them in boats. His expedition in search of Franklin had provided experience in wintering and insight into the effectiveness of steam power. It was clear that screw propulsion could be applied to great advantage in the whaling fleet, whereas vessels had previously depended solely on wind and the muscle power of their crews.

Penny also considered the future of the arctic whaling in relation to the growing belief in an Open Polar Sea. If the circumpolar region were really free of ice in summer, it was reasonable to expect to find whales there. The geographer Petermann had suggested that the gap between Spitsbergen and Novaya Zemlya was the easiest route into the polar sea, and he had elaborated on this theme in relation to marine resources, first in a letter to the *Times* in early November and later when addressing the Royal Geographical Society. According to a summary of his speech in the *Times* (12 November 1852), he declared that the Greenland whale fishery (between Greenland and Spitsbergen) was "fished out," and he urged British whalemen to exploit the waters east of Spitsbergen. The prevailing belief that the circumpolar region was barred by a ring of impenetrable ice he dismissed as "groundless . . . resting upon prejudice and imagination." The whales there were very numerous, he said, because they had not been disturbed by hunting, while "farther to the north, it was well known that a great many whales occurred along the New Siberian Islands, and prodigious must be the number found in the great unknown Polar Sea." But in the absence of any reliable evidence about the polar sea – which Petermann himself admitted was *unknown* – his statements about the absence of ice and the abundance of whales were mere conjecture in the garb of certainty.

In fact, the Greenland fishery had not been totally "fished out," but voyages to the Davis Strait fishery had declined by a third since 1840, which was of special concern to Penny because he had placed high hopes on the development of the industry in Cumberland Sound. In Aberdeen he had been discussing with ship captains and owners ways to improve the prospects for arctic whaling. After he described the important benefits that could accrue from introducing steam power, wintering ships, and year-round shore stations, a new company was formed to adopt and benefit from these new methods. Known at first as the Arctic Screw Steam Whaling Company, it evolved into the Royal Arctic Company, and Penny was appointed general superintendent. After an encouraging interview in November with the colonial secretary, Sir John Pakington, Penny applied for a charter and a grant of land in Cumberland Sound. Although the government fell shortly afterward, the outlook remained favourable for the incorporation of the company and the grant of land on which to erect facilities.

What he and the other organizers of the proposed Royal Arctic Company envisioned for the Davis Strait whaling region was extraordinary. They requested a grant of land on Baffin Island extending from Frobisher Bay in the south to Home Bay in the north, and from the Davis Strait coast westward across the island to Foxe Basin – a territory larger than England and Wales together. A permanent whaling colony in Cumberland Sound would contain residences for whalemen, sailors, company officers, a surgeon, and miners to extract plumbago (graphite). While men at the settlement were pursuing whales and digging out ore, two five-hundred-ton steam-auxiliary ships would proceed in spring from Aberdeen to the vicinity of Spitsbergen to hunt seals on the ice floes, continue toward Novaya Zemlya for whales, return north of Scotland and across the Atlantic to Cumberland Sound, unload supplies at the settlement, and take on blubber, baleen, and plumbago to transport back to Aberdeen. Inuit attracted to the year-round colony would supply sealskins, walrus hides, and the tusks of walrus and narwhals; they would provide labour for tasks such as manning whaleboats, flensing whales, and loading and unloading the ships; and they would hunt animals to provide fresh meat for the wintering whalemen. For the first time, Inuit would be fully integrated with European whalemen in a mutually beneficial economic system.

Penny was well aware that contact with whalemen might be detrimental to the Inuit, but the presence of a missionary should alleviate its worst impacts, as it had under Danish administration on the opposite coast of Davis Strait. If a Moravian brother from Greenland, fluent in the Eskimo language, could accompany his ships, he could instruct the Inuit in the principles of Christianity during the winter. In the words of the *Aberdeen Journal* (23 February 1853), he could awaken them "from the torpor of long ages of darkness."

Owing to careless wording in the prospectuses issued to potential investors and in newspaper summaries of the scheme, whaling interests in other British ports became alarmed that the government was about to grant exclusive privileges to the new company. They objected strongly. "This charter," read a petition from Hull to the Admiralty, "would be giving to Captain Penny & his party a monopoly" and "fettering a trade hitherto free."[24] Powerful opposition from the whaling fraternity killed the project.

The failure of the Royal Arctic Company was a blow to Penny's aspirations, but he resolved to implement at least part of the company's overly ambitious plan. Investors backed him enough to purchase the veteran brigs *Lady Franklin* and *Sophia* from the government, and he began to make preparations for a wintering voyage to Cumberland Sound. Although some men from an American whaler had wintered ashore in the sound a few years earlier, this would be the first time that whaling crews had intentionally spent the winter on board their ships anywhere in the Arctic. Penny still intended to take a missionary from Greenland, and in April he was invited to attend a meeting of friends of the Moravian Brethren in London. "You might have an opportunity of saying a word on behalf of your friends the Esquimaux," the director told him, before dangling a carrot: "Our gallant & estimable friend Sir John Franklin attended such a meeting and addressed us only a few weeks before he sailed."[25]

The *Lady Franklin* and *Sophia* left Aberdeen in early August 1853 and crossed the Atlantic to Davis Strait. Unable to reach the Moravian station at Lichtenfels to collect the missionary, they continued across to Cumberland Sound and took up winter quarters close to its southwest coast, on the island of Nuvujen. After killing several whales in the autumn, the crews rendered the oil from the blubber during the winter in try pots erected ashore. In the spring, as the margin of the landfast ice retreated, whales arriving from their wintering zone in the Atlantic advanced toward the head of the gulf as the area of open water increased. In May, when the floe edge was about twenty miles from the winter harbour, Penny and his men commenced an impressive whaling operation. With the assistance of Inuit hunters and their dog teams, they hauled three whaleboats to the floe edge on the hunters' komatiks and set up tents for the crews. Whenever whales appeared, the men launched the boats and gave chase. After a kill, they towed the whale to the ice edge, flensed it in the water, and stacked the blubber on the ice. Inuit dog teams shuttled back and forth between the floe-edge camp and the harbour, carrying blubber in one direction and supplies and crew replacements in the other. Penny later stated that twenty-two dog sledge teams had covered altogether a distance of some fourteen thousand miles. Whaling at the floe edge yielded a total of eighteen whales, one of which could not be retrieved. It was a highly profitable voyage.

* * *

In late April 1854, as Penny's crews were preparing for whaling at the floe edge in Cumberland Sound, half a dozen men led by the Hudson's Bay Company's hardy explorer John Rae were approximately seven hundred miles to the west, trudging over the tundra near Pelly Bay. Rae's mission was to fill in the last remaining blank space on the map of the continent's northern coast and determine whether King William Land was really attached to the Boothia Peninsula, as James Ross had suggested two decades earlier. If, on the other hand, it turned out to be an island, the intervening strait might form an important link in a Northwest Passage (which, in fact, it did).

As they hauled their sledges over snow-covered tundra and sea ice, they met Inuit hunters from time to time. One of them, whose native attire of animal skins was accented by an incongruous gold Royal Navy cap band, told them that some years earlier a group of about forty white men had died of starvation several days' travel to the west as they were walking south, and that some had resorted to cannibalism to prolong life. He claimed (possibly in fear of retribution) that he had never been to the place himself but had heard the stories and obtained the cap band from other Inuit who had been there. Rae purchased the cap band on the spot.

His party then continued west down the Murchison River to salt water. Across the narrow strait before them lay King William Land. They turned north and traced the coast of Boothia Peninsula for forty miles or so, far enough to see that King William *Land* – hitherto assumed to be a peninsula – was in fact King William *Island*. Franklin had lacked this useful geographic information. (Half a century later, Roald Amundsen's sloop *Gjoa* would go through this strait and continue westward to Bering Strait, becoming the first vessel to complete a Northwest Passage.)

Rae and his companions began the return journey to their winter quarters in Repulse Bay. On the way and after arriving at the base, they purchased some additional relics of Franklin's expedition from Inuit they encountered. There was no doubt at all that the relics had come from the missing expedition, as several pieces of silver plate bore the initials or family crests of officers of HMS *Erebus* or *Terror*. One Inuk possessed a handsome medal, an ornate Maltese cross surmounted by a crown, with a gold ball at the tip of each of the points. In its centre was a white horse on an enamelled blue field, encircled by a gold wreath of oak leaves and the Latin motto *Nec aspera terrent* (Nor do difficulties daunt). It was the badge of the Royal Hanoverian Guelphic Order that had been awarded to Sir John Franklin a decade before he embarked on his final expedition in search of a Northwest Passage. He had worn it at his throat when he sat for Richard Beard's daguerreotype portrait before departure.

Also among the articles purchased from the Inuit was a silver table fork bearing the initials of Alexander McDonald, Penny's former whaling surgeon and protégé, who had sailed on HMS *Terror* in search of a Northwest Passage. This was the second macabre souvenir of the young man; in the fall of 1850, Penny had found a note he had written at an abandoned campsite overlooking Wellington Channel.

In August, Rae's party travelled by boat to Churchill. After sending a report of his discoveries to his employers, the Hudson's Bay Company in Winnipeg, he continued to York Factory and sailed for England, accompanied by the grisly souvenirs of Franklin and his men.

A few years earlier, when Rae had found pieces of wood from a naval ship – almost certainly one of Franklin's – his information had been largely ignored. But this new evidence could not be pushed aside so easily. The numerous relics and the Inuit stories of a large party of ill and starving men were unmistakable indicators of a major catastrophe. On arrival in London in late October, Rae went directly to the Admiralty and presented his evidence. The following day, his report appeared in the *Times*. Reprinting the story, the *Aberdeen Free Press* (27 October 1854) wrote that Rae's oral information and relics "place the fate of a portion, if not all, of the survivors of Sir John Franklin's long-lost party beyond a doubt – a fate as terrible as the imagination can conceive." The *Illustrated London News* (28 October 1854) proclaimed, "The veil that obscured the fate of Sir John Franklin has been unexpectedly lifted."

Many blamed Rae for not having confirmed the oral reports of Inuit, saying that he should have continued further west to the scene of the alleged disaster to look for evidence before returning home with the dreadful news. Some even asserted that he had hurried home in order to secure the £10,000 reward for information about the fate of the expedition. As the messenger bearing unwelcome news, Rae was raked over the coals in a number of prominent newspapers. His critics included Lady Franklin, Penny, and many officials at the Admiralty. It was easy enough in Britain to suppose that he could have easily travelled to the place to which the Inuit had referred; knowledge about the geography of the region was sketchy, and few people understood the seasonal limitations on travel techniques and the difficulty of transporting provisions or obtaining food by hunting.

Rae had not been aware of the exact location of the tragedy until after he returned to his winter quarters at Repulse Bay in late May and obtained more information from Inuit. By that time it was too late to travel westward on foot. Snow was turning into slush, and sea ice was beginning to melt. During and after breakup, it would be impossible to cross to King William Island or reach the estuary of the Back River without a boat. A boat would have to be transported three hundred miles from Repulse Bay, but this could not be accomplished in summer because no suitable vehicle existed to haul it over bare ground. Rae would therefore have had to keep his party at Repulse Bay over another winter in order to travel westward with sledges in the fall or late winter, and that would require intensive hunting to obtain enough meat to feed his party for an additional year. Meanwhile, his discovery of the location of the disaster would remain unknown in Britain, and new expeditions might be setting off to search other regions, needlessly risking more ships and men. His decision to return was the correct one.

Rae's situation in 1854 was not unlike that of Penny in June and July 1851. While returning to the ships after their boat voyage in Queen Victoria Channel, Penny and his men had been forced to abandon their boat and proceed along the coast of Cornwallis Island by sledge. When the snow ended, they had to leave the sledge and walk. When they came upon a boat that had been left for them, they took to the water again, but when their progress was halted by sea ice they had no choice but to resume walking, carrying sleeping gear and food on their backs, wading through ice-cold streams in full flood.

Once the optimum season for sledge travel ended with the melting of snow and ice, no entirely suitable overland travel method was available until the next winter. No one knew this better than Rae, who had accomplished remarkable feats of travel by adopting native techniques and living off the land. In departing for York Factory and England in August 1854, he had done the right thing. On his way south, he sent a full report from Churchill to the headquarters of the Hudson's Bay Company, which enabled them to dispatch a small overland expedition to the mouth of the Back River the following summer. It would be the first expedition to see Franklin relics there in situ, thus confirming the Inuit reports.

* * *

Rae's shocking revelation proved the experts wrong and showed that many alleged reports had been unfounded rumours or – sad to say – intentional hoaxes. Franklin and his companions had not been massacred in northwest Greenland or carried to the Grand Banks on an iceberg. They had not sailed into an iceless circumpolar sea abounding with whales or been wrecked on the coast of Siberia. The missing crews had not sent a note from Victoria Island to England by balloon, or a message from the Beaufort Sea to Ireland by drift bottle; they had not released glass balls at the mouth of the Ob River in Siberia, or dispatched reindeer with specially notched ears from Novaya Zemlya to Norway; they had not built a cairn in Jones Sound, fired canons near the mouth of the Mackenzie River, made their way overland to Lake Superior, or been shipwrecked in Hawaii; their frozen bodies had not been found in the Arctic.[26]

Clairvoyants and spiritualists had put Franklin all over the map – in Hudson Bay, at various places within the Arctic Archipelago, and in the alleged Open Polar Sea. Two of them, however, appear to have been on the mark. In 1849, a dead Irish girl, Weesy Coppin, communicating with her family from the afterlife, is said to have indicated four sequential positions of the *Erebus* and *Terror* from Lancaster Sound into Victoria Strait, the very region in which the ships were in fact abandoned and probably crushed. Her father kept Lady Franklin apprised of the information, which is said to have influenced her decision to send two successive search expeditions, Forsyth's and Kennedy's, in that direction.

Kennedy – himself a believer in paranormal phenomena – travelled to Ireland to meet the Coppin family and attended séances there. Forsyth's second-in-command, Parker Snow, also claimed to have experienced a vision in which the two ships appeared off King William Island.

* * *

Rae's evidence dealt a severe blow to Penny, who saw himself as the man who had discovered Franklin's path northward through Wellington Channel and had turned the search in the right direction. The notion that Franklin had reached the Open Polar Sea now appeared to be wrong. It looked as if his expedition, or at least a large part of it, had actually travelled southward as instructed and had met its end within striking distance of the continental mainland – precisely where he had been ordered to go! Conscious of the impact that Rae's news of the deaths would have on Lady Franklin, Penny wrote to console her. "I can easily suppose what must have been your feelings when you learned the melancholy tidings brought over by 'Dr Rae,'" he said.[27]

He then offered a different interpretation of Rae's evidence. The ships, he suggested, might have attained the polar sea by way of the open water he had discovered north of Wellington Channel but had then been "pressed upon an island" and wrecked, after which the crews retreated by boat and sledge, stopping at Beechey Island for a short time to rest and replenish supplies from a depot they had left there, before pressing on southward toward the continental coast. To Penny this sequence of events explained the sledge tracks, tent rings, and crude hearths they had found in the vicinity of Beechey Island and the apparent signs of a hasty departure. To Lady Franklin, he pictured the men leaving Beechey Island: "See the lean sufferers again starting after a short stay . . . to carry out their Lordships' instructions to the letter in their retreat by passing across Barrow Strait to that channel which is formed by Cape Walker and Cape Bunny to the American shore which they have reached, but too much exhausted to enable them to procure food for the winter."[28] Writing to Barrow a few days later, he said, "Much must remain conjecture until we recover their journals . . . My heart bleeds for Lady Franklin."[29] Two months later, he mentioned another possibility: perhaps Franklin had never left his ships, but had sent some of his possessions south from the polar sea with men who were leaving, as "memorials" for his wife.

It came as a shock to most of the individuals involved in the search for Franklin to learn that the evidence of the tragic end of the missing expedition had been recovered nowhere near the entrance to the alleged Open Polar Sea but far to the south in an area believed to have been thoroughly searched. The expeditions commanded by James Ross, Forsyth, Austin, and Kennedy had not found the remains of Franklin's expedition simply because they had not travelled far enough south. Yet because their published accounts emphasized their impressive feats of travel and the hardships they had heroically endured, while neglecting to point out the geographical *limitations* of their searches, it was easy to assume that they could not have missed anything important. Thus the very region in which a large number of starving white men had been observed by Inuit, and in which the entire expedition had probably perished, had evaded all searching efforts, while the seductive lure of Penny's ice-free water and the hopeful hypothesis of an Open Polar Sea diverted the search in the opposite direction.

The arctic officers who had concluded that the pieces of a naval ship's stanchion and flagstaff found by Rae on Victoria Island had come from the wreck of HMS *Fury* and not from the *Erebus* or *Terror* had been mistaken. The Arctic Committee, which had concluded that no further exploration southward toward the continental coast was required, had made a serious error.

Rae's wooden fragments were not the only clues overlooked on the south coast of Victoria Island. Richard Collinson's ship *Enterprise*, after advancing eastward from Bering Strait and passing a winter on the west coast of Victoria Island, wintered at Cambridge Bay on the south coast in 1852–53. Inuit who visited the ship showed him a piece of metal bearing the navy's broad arrow, reported that ships were frozen into the ice to the east, and indicated their position on a sketch map. Lacking an interpreter, Collinson failed to comprehend their report, or even its vital point – *ships in the ice further east* – and because the Inuit map (sketched freehand from memory) contained what he perceived as errors, he chose to ignore it.

In late winter, Collinson's three sledge parties travelled more than a hundred miles to the island's southeastern tip. King William Island was only sixty-five miles away, and the Royal Geographical Society Islands in mid-channel divided the traverse over sea ice into two relatively short sections. But the sledgers did not continue in that direction. Instead, they turned northward along the east coast of Victoria Island (which, unknown to them, had already been examined by Rae two years earlier) for a hundred miles or so. Geographer William Barr, who has consulted unpublished first-hand narratives to examine Collinson's expedition in detail, states, "Had he sent one party east across Victoria Strait to King William Island . . . it would have discovered the traces of the Franklin expedition."[30] Collinson and his officers had considered such a course of action, but rough sea ice off the southeastern end of Victoria Island apparently dissuaded him from searching in that direction. He observed "a most confused jumble of angular pieces, many of which were upwards of 20 feet high, while the snow between them laid so loose that we frequently sank up to the middle, and only extricated ourselves by our hands."[31] Whether sledge parties could have crossed the zone of rough ice, or circumvented it by travelling around the south coast of Queen Maud Gulf, is uncertain.

So close and yet so far! Collinson could have discovered the tragic trail of Franklin's men a year before Rae obtained the first relics and stories near Pelly Bay.

After the sledging journeys of Collinson's expedition, another piece of evidence came to light near Cambridge Bay – a wooden object more than four feet in length bearing the queen's mark and containing a copper hasp, several metal screws, and a residue of paint. It was probably part of a ship's door frame or hatch, and Collinson realized that it might have been carried from the missing ships by coastal currents. By this time, however, he was headed for home.

The voyage of HMS *Enterprise* had already extended more than three and a half years, including two winters in the Arctic (there would be one more), and the relationship between Collinson and his officers had been – and would continue to be – nothing less than atrocious. He left Cambridge Bay with his four executive officers under arrest for various trivial offences and later tried (unsuccessfully) to have them all court-martialled. After the ship passed through Hong Kong, a naval officer there wrote, as reported in the *Aberdeen Herald* (13 January 1855), "I am sorry to say that misery and disunion have long been the tenants of the ship while among the ice." The *Enterprise* was what geographer William Barr has called an "arctic hell-ship."[32] In view of the voyage's long duration and the dismally low morale on board, it is perhaps just as well that Collinson left Cambridge Bay in August 1853 and started for England.

* * *

While Rae was retrieving evidence of Franklin and his men west of Repulse Bay, and Collinson was preparing to set off from Alaska for home, Penny was completing his floe whaling operations in Cumberland Sound. None of these groups was aware of extraordinary events occurring only a few hundred miles further north.

In 1850, Collinson's subordinate, Robert McClure, commander of HMS *Investigator*, had sailed eastward from Bering Strait, penetrated almost to the northern tip of Prince of Wales Strait, wintered, and then circled westward around Banks Island to its northern coast, facing Viscount Melville Sound. His ship was brought to a halt in Mercy Bay, and he and his men were held in the ice for another two years. Nobody knew they were there.

While Collinson and McClure were still thought to be somewhere in the western part of the archipelago, Penny had warned that they might become "second Franklins" unless some action was taken to locate them or at least to deposit emergency supplies for them in the region. Yet when the Admiralty drew up plans for Belcher's 1852–54 search expedition, they ordered all four of his searching ships to proceed northward through Wellington Channel and Penny's open water. Fortunately, the Admiralty altered these orders before departure and split Belcher's force, detaching two of his ships, the *Resolute* (Kellett) and *Intrepid* (McClintock), to search westward toward Melville Island for Collinson and McClure, while HMS *Assistance* (Belcher) and *Pioneer* (Osborn) proceeded northward in search of Franklin. The Admiralty is said to have changed its mind in response to a letter from the father of Samuel Gurney Cresswell, a lieutenant serving on the *Investigator*, but other people – among them Lady Franklin and William Penny – had voiced a similar appeal.

The western division, wintering near Melville Island, came upon a cairn message left by a party from HMS *Investigator*, then hopelessly icebound in Mercy Bay. Soon after, they made contact with McClure and his men, who were on the verge of abandoning the ship and attempting to walk south to the Fort McPherson fur-trade post, approximately six hundred miles away. It was an escape plan similar in essence to that conceived by Crozier on King William Island half a dozen years earlier after the entrapment of the *Erebus* and *Terror* and the death of Franklin – and a plan just as likely to fail.

Kellett's men transferred the entire crew to the *Resolute* and *Intrepid*, where the men endured yet another arctic winter (their fourth) when both ships were beset and held in the ice off Bathurst Island. The final stage of this convoluted journey occurred when Belcher ordered Kellett and McClintock to abandon their ships and convey all the men, including those of the abandoned *Investigator*, over the ice to Beechey Island, where HMS *North Star*, the expedition's fallback vessel, awaited. They arrived there in mid-May 1854.

Belcher's northern division had failed to locate Franklin by way of the open water Penny had discovered, not because their searching techniques were ineffective but simply because they were looking in the wrong place. The bones and belongings of Franklin and his companions lay five hundred miles to the south. As the *Assistance* and *Pioneer* were returning through Wellington Channel toward Barrow Strait, they became trapped in the same barrier of ice that in Belcher's opinion had merely been *conjured up* by the *imagination* of Penny and other *croakers*. Unfortunately for Belcher, the imaginary ice barrier had proved to be quite real. After clearing out conveniently to allow his ships to proceed north in the first summer, it had closed behind them like a steel trap. As there appeared to be no prospect of penetrating the ice barrier, Belcher and his men abandoned the *Assistance* and *Pioneer*, trudged over the ice to Beechey Island in August, and joined the other refugees.

Beechey Island suddenly became a very populous place. McClure's translator, Augustus Miertsching, who with his shipmates and the men of the *Resolute* and *Intrepid* had travelled more than five hundred miles through the Arctic Archipelago from Mercy Bay by ship, sledge, and on foot, wrote that "on one ship [HMS *North Star*] are the crews of six: the total number is 278 men."[33] In late August, HMS *Phoenix* and *Talbot* arrived from England with supplies for Belcher's expedition, bringing the total on the island to well over three hundred men. Not since Frobisher's mining venture in the sixteenth century – Canada's first gold rush – had so many Europeans been brought together at one place in the North American Arctic.

* * *

The officers of Belcher's four searching ships returned to England as mere passengers on board HMS *North Star*, *Phoenix*, and *Talbot*. Writing to inform Scoresby of Belcher's arrival, Sophia Cracroft said, "We have immense quantities of letters from the officers of both divisions of the Squadron . . . All the letters tell the same tale – the writers are indignant & disgusted at returning again unsuccessful, but all declare that Sir Edward would consult no one." In addition, he had "acted up to his long standing reputation for oppressive cruelty to those under him."[34]

Two and a half years earlier, on the occasion of Belcher's appointment, the *Athenaeum* (7 February 1852) had remarked, "It is well known that Sir Edward Belcher is not the man to turn back while the smallest prospect remains of his being able to carry out his instructions." But turn back he had, and in a most dramatic way. For his unprecedented decision to leave four intact and serviceable Royal Navy ships in the Arctic, described by historian Glyn Williams as "one of the most controversial episodes in the whole convoluted story of the Franklin searches,"[35] Belcher was criticized, ridiculed in the press, and court-martialled but, once again, not convicted.

After the government announced that it still held a claim over the four vacant ships left encased in the arctic ice, Penny wrote to the Admiralty asking whether they intended to "give salvage to any vessel or vessels who may pick them up." If so, he said, "I should consider it my duty to recommend to my partners to aid their Lordships in recovering the valuable property."[36] Unfortunately Penny was given no encouragement. Salvage claims would be considered "as circumstances arise," the first secretary wrote dismissively.[37]

The opportunity thus denied to Penny and other British whalemen who had gone out of their way to deposit stores and messages at various places in Lancaster Sound and look for evidence of Franklin was eagerly seized by others who had done nothing to assist in the search.

HMS *Resolute*, one of the abandoned ships, was carried in the ice into Baffin Bay and south through Davis Strait to a position off Cape Mercy, near the northern flank of the entrance to Cumberland Sound. Penny was whaling in the sound with his two ships at the time, intending to spend another winter there. How he would have loved to retrieve the *Resolute* and sail her back to England, but he was unaware of the presence of the unmanned ship outside.

It was an American whaling captain, James Buddington of the *George Henry*, who won the prize. He and his men sighted the derelict and boarded her. After three years unattended in the ice, the vessel was shabby and forlorn; yet below deck everything seemed strangely normal, as if the officers and men had left only moments before. Buddington decided to salvage the ship. After pumping out the bilges, overhauling the running rigging and hoisting the mouldy sails, he sent his own vessel homeward and followed with a skeleton crew on the *Resolute*. They soon encountered the British whaling ship *Alibi*, on her way to join Penny in Cumberland Sound, and Buddington gave her captain a pair of epaulettes that had belonged to Henry Kellett, the *Resolute*'s former commander, to take back to England and return to their owner.

Earlier in this season, as the whaler *Anne* of Hull was attempting to force a passage westward through the Middle Ice of Baffin Bay, her crew had sighted a spar protruding from the ice, apparently part of the wreckage of a ship beneath. William Barron, then one of the ship's harpooners and later a famous whaling captain, recounted the episode in his memoirs half a century later. His shipmates reckoned that the ice holding the spar was "Sound ice" – floes that had drifted out of Lancaster Sound.[38] The wreckage underneath may have been from another of Belcher's ships, possibly the *Intrepid*, which had been close to the *Resolute* when both were abandoned. Later the *Anne* sighted a rigged ship apparently embedded in pack ice. "At first we thought it was one of our unfortunate friends who was beset," Barron recalled. "Later on we heard that it was the Resolute, belonging to Sir Ed. Belcher's expedition, which was afterward picked up by Captain Buddington."[39] Thus the whaler *Anne* may have been not only the first to sight the *Resolute* but the only one to see part of the wreck of her consort, the steam tender *Intrepid*.

The recovery of the *Resolute* by American whalemen did nothing to improve Belcher's reputation. On learning of the event, Sophia Cracroft wrote to Scoresby, "I wonder how Sir Edward feels? Surely he is in the way of being punished for his misdeeds."[40] The pinnacle of his humiliation was attained when the American government refitted the *Resolute*, sailed her across the Atlantic, and presented her – shipshape and Bristol fashion – to the queen. After all her adventures, the *Resolute* was once again Her Majesty's Ship. Penny may have considered the episode a fitting retribution for Belcher's remarks at the Royal Geographical Society a few years earlier.

Lady Franklin and supporters on both sides of the Atlantic expected that the ship would be used in the search for Franklin. Sophia, writing to Scoresby, expressed what many people felt: "To let her rot in some harbour would be a bad compliment indeed to the munificent donors."[41] Regrettably, the Admiralty had a different view. Having tired of the Franklin search and turned its attention to the Crimean War, the Admiralty let HMS *Resolute* sit idle for the next quarter-century before sending it to the breaking-up yard. The generous Americans had given Britain a beautifully restored full-rigged ship. In return they got back two desks made of wood from the wreckage.

As Buddington was sailing his prize back to New London, the Hudson's Bay Company was attempting to corroborate the alarming evidence reported by its employee, John Rae, in the previous summer. From its subarctic fur-trade post at Fort Resolution on Great Slave Lake, the company sent a small overland expedition down the Great Fish River (now Back River) to the alleged scene of disaster. Travelling by canoe, James Anderson, James Stewart, and their boatmen became the first non-Inuit to reach the place that Rae had been told about. Although they had no Inuit interpreter and had insufficient time to search the estuary thoroughly or cross to King William Island, they did succeed in retrieving more relics of Franklin's expedition from Montreal Island and adjacent coasts, thus confirming Rae's report of a major tragedy in that region. No longer could sceptics dismiss the stories of starving men as "the vague babble of savages," as Charles Dickens had contemptuously written,[42] for these men were *British*, and they had seen and collected relics from the scene of the disaster.

31

Persistence

The matter is not solved to my satisfaction.[1]

The fact that Belcher had met impenetrable ice beyond the open water of Queen Victoria Channel did not end Penny's belief in an Open Polar Sea, or at least a sea open enough in summer to permit navigation by steam-powered vessels. He told Barrow in 1855 that he still wanted to cross the Arctic Ocean and determine whether Franklin had gone there. He would name a steamer *John Barrow* and take her "across the 'Pole in commercial enterprise."[2]

His belief in a navigable Arctic Ocean led him to consider new and imaginative ways to exploit cetacean resources. The British government, he believed, should establish a base called "Port Franklin" on Vancouver Island as a memorial to the man who had contributed so much to European knowledge of the North American Arctic. He envisaged whaling ships proceeding from there to Britain by way of Novaya Zemlya through what is now known as the Northeast Passage – along the northern coast of Asia. As late as 1875, seven years after his last command, Penny suggested that whaling ships should proceed from Britain around Norway's North Cape and follow the Asian coast eastward to the New Siberian Islands. "By the time they reach this position" he wrote, "the Polynia will extend to a high latitude. The whales here will be so numerous that it baffles description and what will pleasure them more[,] fine light floes to cross and recross the Pole making the best of their way to Behring Straits which in a favourable season they will reach by the middle of August and less trouble than in rounding Melville Bay."[3] Penny was still holding fast to ideas expressed by the geographer Petermann and others two decades earlier, for which no concrete evidence had ever been produced.

Long before this, the historian John Brown had remarked, "Mr. Penny urges his plans – not always wisely; still they have a national spirit. We cannot help admiring him, even in his wildness."[4] If Penny's views about the geography of the circumpolar region appear today to be naïve, one should remember that the naval and scientific communities held similar opinions at the time. A bandwagon had picked up speed and many had jumped on board. William Scoresby – always a voice of reason – had tried to throw himself beneath the wheels, but its momentum had been too great and it had kept right on rolling.

The notion of an Open Polar Sea persisted because it had not been disproved, and it had not been disproved because men had never been there. They had seen only a few ice-free places around its periphery. In 1875, the year in which Penny wrote confidently about crossing and recrossing the polar sea in a steamer, an official British expedition under George Strong Nares, who had served on HMS *Resolute* during Belcher's expedition, pushed northward through Smith Sound in an attempt to reach the North Pole. Elisha Kent Kane had already failed; so had Isaac Israel Hayes and Charles Francis Hall. Yet it was confidently believed that Nares would succeed. Like Franklin, he had two engined ships with screw propellers, but his ships were twice as large and their engines were twelve times as powerful – not second-hand railway locomotives but dedicated marine engines. There seemed little doubt that such ships could break through the ring of ice thought to encircle the Open Polar Sea and could continue to the North Pole with little trouble.

After HMS *Alert* attained the highest latitude reached by a ship to that time (82° 28' N), ice blocked further progress. Nares and his crews wintered safely on board their ships in two separate locations on the coast of Ellesmere Island, and in spring a party under Albert Hastings Markham set off northward with sledges and a boat. After struggling over thick, rough, multi-year ice for about forty miles beyond the northern tip of Ellesmere island, they turned back, weak from the combined effects of strenuous physical exertion and scurvy, one man already dead. The pole was still four hundred miles away, but they could see nothing but hummocky ice ahead. Two decades earlier, Hayes had called the narrative of his expedition *The Open Polar Sea*. Markham's book about the Nares expedition was pointedly entitled *The Great Frozen Sea*.

Although Nares's expedition did not prove that the entire circumpolar zone was a frozen sea – any more than Penny's discovery of an expanse of open water had proven it to be an *unfrozen* sea – it did cool the ardour of Open Polar Sea proponents. The real proof would have to await Nansen's drift voyage in the *Fram*, the over-ice treks of Cook and Peary, the polar flights of dirigibles and airplanes, and the view from above with air photography and satellite imagery. The first vessel to reach the pole – the nuclear-powered USS *Nautilus* in 1958 – did not travel across the surface of an Open Polar Sea but through tranquil waters underneath the ice of a permanently frozen sea.

* * *

After his government search expedition of 1850–51 and the failure of his grandiose plans for a Royal Arctic Company, Penny resumed his whaling career, commanding ships for almost two more decades. Although he was no longer actively involved in the search for Franklin, he could not forget about the missing expedition. Rae's information revealed much, but what was the full story? Penny followed developments in the search closely and continued to correspond with some of those who had been actively involved in its planning and execution, including Barrow, Osborn, Lady Franklin, and Sophia Cracroft. When he encountered Inuit on his whaling voyages, he routinely questioned them about unfamiliar ships, boats, and men.

Wintering in Cumberland Sound on four occasions, he had an opportunity to collect information relating to an extensive area. Many Inuit travelled from distant places to take up temporary residence near his ships for employment and trade, and some were able to pass on stories from regions farther afield. During Penny's voyage of 1853–54, one of his Inuit workers mentioned "much fighting in the year 1850 and of four of his friends being killed" somewhere to the west, as Penny later told Lady Franklin.[5] Could there have been armed conflict between Inuit and Franklin's men? Two months after Penny returned to Scotland, Rae's shocking revelations were published in major newspapers. The graphic Inuit stories of ill and starving men in their final desperate moments of life, and the accompanying sketches of various personal belongings of Franklin and his officers, were difficult to refute. Even so, some sceptics insisted that, as Rae had not actually reached the alleged site of the disaster, it was entirely possible that hostile Inuit had massacred Franklin's men, stolen the most useful of their belongings, and invented a fictitious cover story. Inuit in Cumberland Sound told Penny they did not believe the information communicated to Rae. Two years later, Penny told Sophia Cracroft, "Dr. Rae's story had reached my fishing station the same year as he brought it to England 1854 but they contradicted it most distinctly."[6]

Of the authenticity of Rae's relics there could be no doubt, but the recovery of a silver spoon or plate inscribed with the initials of a particular officer did not necessarily mean that the officer had died. He might have left it behind to save weight, or bartered it with Inuit in return for food. As Rae himself had not seen any bodies, his information and relics did not really *prove* the death of anyone. Furthermore, Inuit who had described the starving men and dead bodies claimed they had not seen these things themselves but had merely heard stories from others.

Even if the accounts that Rae heard from Inuit were true, some important questions remained. If only thirty or forty men had died on King William Island and around the estuary of the Back River, what had happened to all the others? Had they perished too, or could they still be alive somewhere, perhaps among bands of roving Inuit? If the ships had been abandoned, some desperate men might have reoccupied one or both later. Where were the ships located now? Had any records of the expedition, such as logbooks, order books, private journals, or letters, survived to reveal what had happened after leaving Beechey Island? Questions such as these continued to bother Penny.

On his return from Cumberland Sound in 1854, he sent Barrow a sealing harpoon head that his men had found embedded in the body of a dead seal on the coast of Baffin Island somewhere between Pond Inlet and Cape Dyer. It appeared to be quite different from those customarily used by the Inuit, and he suggested that it might have been used by Franklin's men. "The moment I put my eye upon it," he said, a "voice seems to say to me, 'desert us not we still exist on an island in the Arctic Ocean.'"[7] The harpoon head was examined at Woolwich dockyard and a sketch of the type of harpoon normally supplied to Royal Navy ships was sent to Penny. No records of harpoons provided specifically for the *Erebus* or *Terror* could be found, however. Barrow told Penny that the harpoon head resembled ones made in Liverpool for sealing ships, but Penny insisted that no British sealers carried ones of this sort.

* * *

While Penny was developing a whale fishery in Cumberland Sound based on wintering ships and Inuit labour, the question of who had discovered the Northwest Passage was being debated. As McClure's *Investigator* had reached Viscount Melville Sound from the west, and Parry had reached it three decades earlier from the east, a sea route between the Atlantic and the Pacific – a Northwest Passage – was now known to exist, and McClure could claim to have forged its last link. Furthermore, he claimed that he and his men were the first ever to traverse a Northwest Passage from one ocean to the other. By the time they reached England on other vessels after their ship had become hopelessly beset in Mercy Bay, they had circumnavigated North and South America by sailing south in the Atlantic, around the Horn, north through the length of the Pacific to Bering Strait, along the north coast of Alaska, around Banks Island to Viscount Melville Sound, and then eastward through the Arctic Archipelago and home by ship to Britain. It was an odyssey of nearly five years' duration.

But had they *discovered* the Northwest Passage? Some pointed out that their ship had been frozen in for more than two years at Mercy Bay with no hope of getting out, that they were about to abandon ship and attempt to walk southward to subarctic fur trade posts, that the men were no longer in adequate physical shape for such a journey, and that if they had not had the extreme good fortune to be rescued by Kellett's expedition, they would probably all have died. McClure, eager for fame, denied being hopelessly trapped and insisted that he could have extricated the *Investigator* and sailed her east to Baffin Bay and home. But it was an idle and unprovable boast, one that has been called "a monstrous assertion."[8] Even if he had seriously contemplated such a course of action (rather than the long trek to a trading post that they were about to attempt), ice conditions would probably have made it impossible. HMS *Investigator* has never emerged from Mercy Bay. It was found by Parks Canada archeologists in July 2010, lying serenely on the sea bottom under the ice in about eleven metres of water.[9]

In the opinion of many, McClure had shown great initiative in reaching Mercy Bay, spending three winters in the region, and intersecting Parry's track of three decades earlier. Others, however, pointed out that he had departed from Hawaii before his scheduled rendezvous with Collinson, commander of the two-ship expedition, in order to strike off by himself toward the Arctic. Was this *initiative*, or was it what naval historian Andrew Lambert has called a "foolhardy lunge for glory"?[10]

In April 1855, McClure laid claim to the Admiralty's reward of £10,000 for the discovery of the Northwest Passage. There was no denying that after entering the Arctic from the Pacific he had intersected Parry's track from the Atlantic, thus proving the existence of one maritime passage through the archipelago. However, a decade earlier, the Admiralty had sent Franklin to find a link between Barrow Strait and the well-known route along the continental coast to Bering Strait. Rae's oral information from Inuit, confirmed by James Anderson and James Stewart, showed that some of Franklin's men had crossed the ice from King William Island to the mainland, thereby connecting with the familiar coastal route to Alaska, half a dozen years before McClure's transit. Therefore Franklin's men, rather than McClure and his crew, had been the *discoverers* of a Northwest Passage.

Penny was concerned that the achievement of Franklin's expedition was going to be overlooked if Parliament decided to recognize McClure as the discoverer of the passage. In a letter he was composing to Lady Franklin, he wrote,

> I have just learned that McClure has laid claim to the North West Passage[.] You still have another painful duty to perform to the memory of your husband and his brave companions[,] that is to enter your protest against such a claim and when a committee of the house shall have leasure [*sic*] I shall lay such facts before them as I have no doubt will justify your claim. McClure was ice bound. The wife of Franklin and the Whaling Master obliged the Lords of the Admiralty to do their duty which saved a McClure and now . . . he would rob the memory of her husband.[11]

In July, Lady Franklin wrote to W.A. Mackinnon, chairman of the Select Committee on Arctic Expedition, which was evaluating McClure's claim. She acknowledged that McClure had discovered one Northwest Passage but pointed out that "the 'Erebus' and 'Terror' under my husband had previously, though unknown to Captain M'Clure, discovered another and a more navigable passage; that passage, in fact, which if ever ships attempt to push their way from one ocean to another, will assuredly be the one adopted."[12]

Her prescient prediction about the relative importance of the two passages was absolutely correct. Franklin's expedition and much of the subsequent search for it occurred during "one of the least favourable climatic periods within the last 1000 years."[13] But since the end of the so-called "Little Ice Age," which ended in mid-nineteenth century, global warming has made arctic transits comparatively easy. From 1906, when Roald Amundsen completed the first successful voyage from the Atlantic through the Arctic Archipelago to the Pacific, up to the end of the 2015 season, there were no fewer than 238 transits in one direction or the other by surface vessels.[14] The vessels ranged from large icebreakers down to small yachts.

Of the seven different passages used, the one from Barrow Strait southward through Peel Sound and Victoria Strait to the continental coast, then west to Bering Strait (or the reverse direction) was chosen more than any other – sixty-four times. This is precisely the route that HMS *Erebus* and *Terror* were attempting to follow in 1846 when they became trapped in the ice. Although the ships did not reach the mainland, some of their men did, thus closing the last link in what has become – as Lady Franklin predicted – the most popular Northwest Passage. In contrast, during the same span of slightly more than a century, McClure Strait was navigated only three times.

Despite Lady Franklin's protest, however, the committee recognized McClure and his men as the discoverers of the Northwest Passage and awarded them the prize.

* * *

In 1850 the British government had announced a series of rewards: £20,000 for the rescue of Franklin and his men; £10,000 for the rescue of any of them, or for providing information leading to their rescue; £10,000 for ascertaining their fate. As Penny was preparing for another whaling voyage in 1855, he told Barrow that he would send Inuit westward from his winter station in Cumberland Sound to search for survivors and records. Barrow applauded this intention. "Who knows but Captain Penny may yet be the man to save the lives of a remnant . . . I know of nothing that would more please me and to hear that you had got the reward of £10,000," he wrote. "Bring back but one man and the reward is yours."[15]

Penny misinterpreted Barrow's sentiments and tersely pointed out that he was motivated not by money but by a desire to save men's lives. Barrow hastened to assure him that he was perfectly aware that the reward had never been foremost in Penny's mind; nonetheless he would be glad to see Penny receive the money if – through God's grace – he did succeed in rescuing any survivors. Shortly before Penny's departure from Aberdeen with his favourite brigs *Lady Franklin* and *Sophia*, Barrow wrote to wish him luck. He hoped Penny would be the one to clear up "the mystery of the lost ones," and "if any survive, to restore them to their homes."[16]

As Penny and his two ships were establishing winter quarters in Cumberland Sound, he was unaware of two interesting developments that had occurred in the search for Franklin – the finding of relics around the estuary of the Back River by James Anderson and James Stewart, and the recovery of Belcher's abandoned ship *Resolute* by Captain Buddington of the American whaleship *George Henry*. He would not learn of these events for another year. While wintering in Cumberland Sound, he kept alert for any reports that might relate to the Franklin expedition. On returning to Scotland in the autumn of 1856, he told Lady Franklin and Sophia Cracroft what he had heard. An Inuk named Toutou – a brother of Eenoolooapik – had told him that Franklin's men had been killed by Inuit. Toutou's tongue, Penny admitted, had been loosened by brandy and his story of a massacre might simply have been the liquor talking. The report appeared to be nothing more than a resurrection of Adam Beck's fabulous story from Cape York five years earlier.

Penny had also learned that some Inuit from Cape Searle had travelled northward to Lancaster Sound. He told Sophia Cracroft that he found this "remarkable," because they must have been aware that his own ships were wintering in Cumberland Sound, only seventy miles away, "& more especially me who they had known for 28 years & given them so much. The temptation must have been far greater to induce not less than 8 sledges to go some 500 miles."[17] Penny did not believe their long journey was related to Franklin, however. He thought it probably had some connection with the depot established at the Wollaston Islands by HMS *North Star* shortly after it separated from the *Lady Franklin* and *Sophia* in 1850. The cache of supplies was said to have been plundered by Inuit, and the people of Cape Searle probably wanted a share of the wealth.

Another story, of much greater interest, concerned a tent. The *Aberdeen Journal* (17 September 1856) carried an account:

> In the course of the winter he [Penny] saw many of the native Esquimaux, and was informed by some of them that they had heard from some other tribes, of their having seen, a long distance in a north-westerly direction from Hogarth [Cumberland] Sound, and probably about the year 1850, a circular white tent, erected on the ice, and that, in the absence of all the inmates, they had stolen from it articles made of bright metal. On a second visit, several moons after, two white men were seen at the tent. The natives also brought the story of a party of white men, in the locality indicated, having been compelled by hunger to devour each other. These reports are quite consistent with that brought by Dr John Rae. The articles of "bright metal" stolen may have been the silver forks, medals, and nautical instruments, &c., recovered from the natives in a broken state by Dr Rae.

The story of a tent and two white men certainly captured Lady Franklin's attention. What if Rae's hearsay evidence of starving men, dead bodies, and cannibalism referred to only a portion of the crews, and some other men had still been alive in 1854? If so, could Sir John have been among them? Writing to Penny, she asked if the Inuit with whom he had spoken had actually seen any of the articles allegedly stolen from the tent. Had they noticed "anything like my husband's large silver star of the Guelphic order?"[18] If so, she may have reasoned, Sir John might have been one of the occupants of the tent, and the medal might have been stolen in his absence by Inuit and later bartered to Rae. She pointed out to Penny that, despite the stories Inuit told to Rae, there was still no proof that any deaths had occurred. Anderson, in the first search party to reach the scene, she said, had "found nothing whatever to confirm that part of Rae's story. He found no graves, no bodies, nor parts of bodies, no clothes, no guns, no ammunition, but only the remains or part of the material of a large boat."[19]

In late September 1856, Rae himself called on Penny in Aberdeen. The two Scotsmen, one a renowned traveller in subarctic and arctic regions, and the other a leader in the arctic whaling trade, had much in common. Both were tough men of action, inured to hardship, single-minded when pursuing an objective, and ready to adopt native methods of travel and survival. But the meeting did not go well. Describing it to Sophia Cracroft a week later, Penny said, "I had a long visit from Dr. Rae . . . I suppose his visit was because he thought I agreed with him. He seemed much disappointed I did not do so. I said I regretted much when he was within so short a distance he had not solved the bulk of his report, at the same time stating I did not believe [it was] the truth and he could not have believed it himself or he would have done so."[20]

A few months earlier, the Admiralty had awarded Rae and his companions the reward of £10,000 for ascertaining the fate of Franklin and his men. Yet in the eyes of Penny and Lady Franklin, Rae had *not* ascertained their fate, because he had not travelled to the alleged scene of the disaster to confirm the truth of the Inuit stories. Penny and Lady Franklin were reluctant to believe the oral reports; if they were true, they would mean for Lady Franklin a serious erosion of hope regarding the survival of her husband, and for Penny, a contradiction of his opinion that Franklin had traversed the open water north of Cornwallis Island and continued into an ice-free sea. Although a year had passed since Anderson and Stewart had seen relics with their own eyes near the estuary of the Back River, Lady Franklin and Penny were still not ready to face the implications. They continued to hope that HMS *Erebus* and *Terror* would be found.

In addition to blaming Rae for not having continued on to the reported location of the dying white men, Lady Franklin looked upon him with extreme disfavour because he had kept to himself some of the relics he had obtained from Inuit. Wherever this was known, Sophia Cracroft told Grinnell in July 1856, "it excites the deepest disgust. This while my aunt has to go to Greenwich & look with the multitude into the glass case containing my uncles Guelphic Order & his own plate bearing his crest (she has not a morsel or scrap of anything brought home by Rae). He parades ab^{t} at dinner parties with his relics, making a show of them."[21]

Later Penny wrote to Lady Franklin with exciting news. He had just talked with Captain Paterson, whose whaling ship *Pacific* had stopped near Pond Inlet during the 1856 season. Paterson had obtained from Inuit "three copper bolts and a galvanized rod about 3½ feet long" – articles that could only have come from a large ship. Penny continued,

> I have seen them. The bolts are of pure copper the iron rod quite bright . . . The Esquimaux made signs it was from a ship which had been pressed on shore in a gale of wind by the Ice[,] that she slept there still (which means she was still lying there). Capt Paterson also states that the Esquimaux made signs that the ship had come up Barrows Straits and turned down I have no doubt he meant Peel Sound. He also saw two new sledges made of hard wood of the colour of mahogany[.] He mentioned that Capt Deuchars of Dundee saw a patent-copper scoop of some description with one of the ships names upon it. He wished to barter it but the Esquimaux had made off.
>
> This is a very important fact which will prove that the Franklins has [*sic*] not launched a boat over land as some would have it but has made the N.W. passage with one of his ships at least.[22]

Penny's statement in this letter (or draft) about a completion of the Northwest Passage is somewhat obscure, and the Inuit reports of an unmanned ship in Lancaster Sound or Pond Inlet did not eliminate the possibility that some men from the *Erebus* and *Terror* had headed south up the Back River with boats, as indicated by Rae's Inuit reports and the discoveries of Anderson and Stewart. Yet there was enough in Paterson's information to inspire hope that some men from the ships – perhaps including Sir John himself – had sailed one of the ships back out through Lancaster Sound.

As soon as Lady Franklin received Penny's letter, she rushed off a reply, urging him to contact Captain Deuchars and try to find out what ship's name had been on the copper scoop. In addition, she suggested that if Penny could obtain some of the bolts, they could be examined in England by naval shipwrights familiar with the *Erebus* and *Terror*. There was also a report circulating about some native sledges near Pond Inlet made of hardwood containing holes for treenails. Captain Parker of the *Truelove*, she added, had learned that *two ships*, not merely one, had been aground near the mouth of Lancaster Sound.

Derelict ships, copper bolts, a metal scoop, sledge wood containing holes for treenails! "I think they may have belonged to Franklin's ships," an excited Barrow wrote to Penny.[23] By this time, however, there was another possibility. The mysterious vessels – if in fact the reports were true – might have been from Belcher's expedition. Of the four vessels he had abandoned in the ice two years earlier, only the *Resolute*, and possibly some wreckage from the *Intrepid*, had reappeared after drifting out of Lancaster Sound, the *Resolute* into the welcoming grasp of an American whaling master. The other two abandoned ships were still unaccounted for.

Reports such as this were tantalizing. In the absence of concrete facts about place and time, the names of ships involved, and the identification of witnesses, they could not be verified or acted upon. Furthermore, in their transmission from one Inuk to another, then across a linguistic boundary from Inuit to whalemen, and later from whalemen to newspapers and individuals in Britain or America, stories were often hopelessly distorted. Different interpretations of such reports were always possible.

It is *conceivable* that several of the events referred to in preceding pages were connected. They *might* have occurred in the following order:

May 1854: HMS *Resolute* and *Intrepid* are abandoned in the ice of Melville Sound and Barrow Strait. Their crews travel over the ice to Beechey Island and depart for England on other vessels.

Summer 1854: After the ice breaks up, the two abandoned ships drift eastward among the floes, through Barrow Strait and into Lancaster Sound. Still held within pack ice, they are carried close to the south coast of Lancaster Sound and perhaps are grounded for a time. Inuit go on board and retrieve useful objects and materials, including worked wood, metal bolts, and a copper scoop.

Winter 1854–55: Both ships drift into Baffin Bay and southward along the coast of Baffin Island. HMS *Intrepid* is crushed by the pack.

Summer 1855: The whaler *Anne* sights the top of a spar sticking through "Sound ice" in Davis Strait – the only visible part of the crushed *Intrepid*. Later, the *Anne* sights the *Resolute* within the pack, apparently a derelict.

Summer 1855: The American whaling ship *George Henry* (Captain Buddington) retrieves the *Resolute* and sets off for New Bedford with his prize. On the way, the *Resolute* speaks the whaling ship *Alibi*. Buddington consigns Kellett's epaulettes to her captain to return them to their owner in England.

This sequence of events is highly conjectural but not impossible. It makes unwarranted assumptions about place, date, and identity, and it fails to consider the other two ships abandoned by Belcher, the *Assistance* and *Pioneer* in Wellington Channel. But the speculative reconstruction may serve to underline the difficulty of utilizing the vague and often contradictory reports that originated among nomadic Inuit who did not speak English, were communicated by itinerant whalemen who did not speak Inuktitut, and were often published in newspapers and journals whose personnel were not familiar with places or conditions in the Arctic. Many interpretations of such stories were – and still are – possible.

* * *

The expeditions by Rae in 1854 and then Anderson and Stewart in the next year had identified the general area of the Franklin disaster, but much more information was needed. As neither had reached King William Island, and the Hudson's Bay men had found no bodies around the estuary of the Great Fish (Back) River, a close examination of the region needed to be carried out.

In June 1856, three dozen leading scientists pleaded with the prime minister, Lord Palmerston, to send an expedition to the region in order "to satisfy the honour of our country, and clear up a mystery which has excited the sympathy of the civilised world." Such a venture would "dispel the obscurity in which the voyage and fate of our countrymen are still involved."[24] It could be accomplished with little risk, they believed, because the destination was small in scope. But by the time the government addressed the request, it was too late in the season to send out a vessel.

In the autumn, Bedford Pim, whose proposal several years earlier to search for Franklin along the Asiatic coast had been rejected, suggested sending a ship through Bering Strait and eastward along the continental coast to King William Island, where it would winter and send out search parties. There were a few problems with this idea: the voyage through the Atlantic and Pacific Oceans would take several months; unpredictable ice conditions in the Beaufort Sea could halt, imprison, or crush the ship; and only one vessel – HMS *Enterprise* – had ever reached as far east as Victoria Strait, still 150 miles short of King William Island. Pim's suggestion was not followed up.

In December Lady Franklin wrote to the prime minister, pointing out that the government's delay had made it impossible for her to dispatch her own ship; an entire season had been wasted. She began making preparations for a private expedition in the summer of 1857.

* * *

In June 1857, six weeks after returning to Aberdeen from a sealing voyage to Spitsbergen waters, Penny was preparing to set off for Cumberland Sound on another wintering whaling voyage. This time he would be accompanied by his wife, Margaret, their twelve-year-old son, Willie, and a missionary, Matthäus Warmow. Penny intended to demonstrate that the Arctic was not the forbidding hostile place usually depicted in exploration narratives but rather a place in which European men and their families could live comfortably, either on board their ships or ashore, provided they adopted reasonable measures. His common-sense outlook, based on his own experience, was similar to Vilhjalmur Stefansson's famous concept of a *friendly Arctic*[25] but preceded it by two-thirds of a century.

In Cumberland Sound, Inuit hunters could provide fresh meat and fish to keep scurvy at bay. The missionary could satisfy spiritual needs and, if necessary, curb the rowdy behaviour occasionally manifested by seamen. On two previous voyages, Penny had been unable to pick up a Moravian brother from a mission settlement on the Greenland coast, so this time the Moravians had arranged to have Warmow join the expedition before the ships left Britain.

Also in port was the steam-powered yacht *Fox*, about to depart for Prince Regent Inlet on Lady Franklin's final search expedition. She had hoped that the Admiralty would employ HMS *Resolute* in the search, as the Americans had urged, but despite strong support from newspapers, petitions to the House of Commons from cities such as Liverpool, and personal appeals to influential people in Britain and the United States to put pressure on the prime minister, the government had done nothing.

Lady Franklin had therefore seized the initiative once again. In command of her expedition was the veteran sledger McClintock, on leave from the Royal Navy. "The rescue of any possible survivor of the 'Erebus' and 'Terror,'" she told him, "would be to me, as it would be to you, the noblest result of our efforts." Failing that, she hoped that he would be able to recover "the unspeakably precious documents of the expedition, public and private, and the personal relics of my dear husband and his companions."[26]

Even two weeks after the departure of the *Fox*, Sophia Cracroft expressed the hope that the *Resolute* could still be of assistance. Writing to Grinnell on 12 July, she said that she and Lady Franklin hoped "to make the Govt. yield us the Resolute at last, & send her to meet the Fox by way of Behring St."[27] But it was not to be.

The *Fox*, a private luxury yacht built in Aberdeen, with "velvet hangings and splendid furniture," large skylights, spacious companionways, and commodious cabins, was transformed into an efficient polar vessel with an iron-encased stem ("a ponderous chisel set up edgeways"), a network of strengthening beams inside, a sheathing of stout planks outside, new boilers, a new propeller, and an altered rig.[28] With McClintock were two men who had sailed on Penny's search expedition seven years earlier – the Danish interpreter and dog driver Carl Petersen, and the Scotsman Alex Thompson who had assisted in caring for the dogs. Like Penny in 1850–51, McClintock intended to use dog traction as well as man-hauling. The crew numbered twenty-five all told, and of these, seventeen had participated in the search before.

Before the *Fox* departed, Penny made many practical suggestions about the outfitting of the vessel. He told Barrow that he saw McClintock every day and that his officers were "first-rate."[29] McClintock had served on two search expeditions on naval ships that were up to three times the tonnage of the *Fox*, and he may have found the advice of the commander of the small *Lady Franklin* useful. (In his book, however, he did not mention that Penny's ships were also in port preparing to leave.)

As Penny and McClintock prepared to depart, there was a special intensity of emotion around the harbour. Aberdonians were accustomed to seeing whaling ships come and go on their voyages to and from Spitsbergen and Davis Strait, and within the past decade they had witnessed the departure of Franklin search expeditions commanded by Forsyth, Penny, and Kennedy. Now what appeared to be the final search venture was about to leave, and it might well return with the answer to the lingering question of whether any members of Franklin's expedition were still alive. Lady Franklin and Sophia Cracroft were on hand to see the ships off. Lady Franklin met and chatted with Brother Warmow, soon to become the first missionary to work among the Inuit of the Arctic Archipelago. (He later said that at the last minute she had tried to persuade him to shift his gear to the *Fox* and accompany McClintock on her own expedition, instead of going with Penny.)

The three ships left on the last day of June. The *Fox*'s departure was marred by a pilot's error that put the ship aground on a bar and delayed its departure for a night, but the *Lady Franklin* and *Sophia* (the latter now commanded by John Cheyne) got away cleanly. The two expeditions headed separately around mainland Scotland's northern tip and westward across the Atlantic.

A month later, the *Lady Franklin* and *Sophia* made landfall at Resolution Island, the lofty guardian of Hudson Strait. Turning north, they crossed the mouth of Frobisher Bay where, almost three centuries earlier, a huge English expedition of fourteen ships and four hundred men, commanded by Martin Frobisher, had extracted more than a thousand tons of worthless ore, believing it to contain gold. Continuing north along the coast of what is now called the Hall Peninsula, Penny's ships entered Cumberland Sound. After a brief stop at Niantelik, they reached Nuvujen and set up the buildings they would need for whaling operations and winter residence.

During the twelve months that his ships spent in Cumberland Sound, Penny and Warmow interviewed Inuit whenever the opportunity arose, seeking information that might add to the incomplete picture of Franklin's expedition. In mid-October, they encountered eighteen Inuit in an umiak and a kayak. One of them, nicknamed Captain Pakak (a corruption of *Parker*), had worked for British whalers before and was well known to Penny. He told of shipwrecked white men arriving in four boats at Pond Inlet several years before, staying for two weeks, and then departing. Pakak had traded a large salmon for one of their guns. He had noticed that one of the men was left-handed; another, a red-headed man, had set off on his own for Greenland. (How he intended to get there was not explained.) Warmow, who interpreted Pakak's statements about the boats manned by white men, later wrote, "May not this have been Sir John Franklin's party?"[30] Penny, on the other hand, thought the boats had probably belonged to whaling ships, which in retrospect appears more likely.

* * *

While the *Lady Franklin* and *Sophia* were wintering securely in Cumberland Sound, McClintock's *Fox* was trapped among ice floes in Baffin Bay. After proceeding northward along the Greenland coast in order to cross the North Water to Lancaster Sound, the ship had been beset in heavy pack ice in Melville Bay and condemned to a winter within the pack. Encased within a desolate frozen seascape, the vessel drifted slowly southward with the current, day after day, week after week, through the gloom of the long arctic winter. It was not until the end of April 1858 that the little steam yacht was liberated, having been carried almost fourteen hundred miles.

Eight months in the ice did not deter McClintock and his two dozen shipmates. They returned resolutely to the Greenland coast, retraced their path northward through Melville Bay, crossed the North Water to Lancaster Sound, and proceeded southward into Prince Regent Inlet. After putting the *Fox* into a secure harbour at the eastern end of Bellot Strait, they settled in for their second arctic winter on board their small ship.

In February 1859, three sledge parties, assisted by almost two dozen dogs, set off. When they encountered Inuit, they learned more details about the starving men seen several years earlier and obtained additional relics of Franklin's unfortunate expedition. Among the articles they recovered was the John Argyll Robertson silver medal that had been awarded to Alexander McDonald for high achievement in medical studies at the University of Edinburgh in 1838. When he was appointed assistant-surgeon on Franklin's expedition seven years later, he had brought the medal from his home in Laurencekirk, Scotland, to London by train, then across the Atlantic and into the Arctic Archipelago on board HMS *Terror*. After wintering safely at Beechey Island, the ship had continued on her quest for a Northwest Passage with McDonald and his medal still on board. Somehow, the precious academic prize had ended up among the Inuit in the vicinity of King William Island.

How could the medal have reached the hands of nomadic hunters? Had McDonald kept it with him when the men abandoned their ships? If so, for what purpose? Was it a nostalgic souvenir from which he could not bear to part? Did he hope that in the event of his own death the medal might be found and eventually returned to his family? It may have appeared to Penny that his former whaling surgeon, who had published a book about Eenoolooapik's winter in Scotland, and who had written so proudly when he had been appointed to Franklin's great history-making expedition, had laid a trail to follow. First there had been the paper bearing his name, found by Penny at the abandoned campsite overlooking Wellington Channel, then the silver cutlery bearing his initials retrieved from Inuit by Rae, and finally a medal he had won at medical school two decades before, obtained by McClintock. Where would the trail end?

In April, McClintock began a second round of journeys, heading south with five sledges and seventeen dogs. When he and his men met up with Inuit again, they secured more information and more relics. From what the Inuit said, it appeared that both of Franklin's ships had been lost, one crushed by the ice and the other driven ashore. The latter had provided a treasure trove for nomadic hunters inhabiting a treeless land. Among their possessions, McClintock and his companions saw wooden tent poles, spear handles, bows, arrows, kayak paddles, and snow shovels, all made from the oars, thwarts, and bottom boards from one or more of the ships' boats.

With a wintering ship and two dozen men, McClintock was able to organize a much more thorough search than the small canoe party under Anderson and Stewart had four years earlier. Utilizing snow and ice surfaces for sledge travel, they could cover far more ground and collect more relics and, having an interpreter, could learn what the Inuit knew about the disaster. Search parties travelled around the entire coast of King William Island and across to the continental mainland and the estuary of the Back River. What they saw removed any lingering doubts about the reliability of the Inuit stories told to Rae five years earlier, for the telltale detritus of the disaster was scattered over the barren tundra for hundreds of miles along the west and south coasts of the island – sledges, uniforms, blankets, a clothes brush, a comb, spectacles, sun goggles, tools, rope, scientific and navigational instruments, a telescope, a medicine chest, pocket chronometers, watches, guns, gun cases, gunpowder, ammunition, stoves, canteens, flasks, silver cutlery, clay pipes, commemorative medals, coins, buttons, Bibles, prayer and hymn books, a novel, and assorted other items.

Many of these articles were found along the western coast of the island, some in or near a twenty-eight-foot open boat mounted on a sledge. McClintock later recalled that his men had been "transfixed with awe"[31] when they saw inside the boat two partially clothed skeletons, one propped in the bow and the other lying beneath a thwart near the stern. Neither body had a head.

Beside one of the headless skeletons was a beautifully ornamented pair of slippers "lined with calf-skin, with the hair left on, and edges bound with red silk ribbon."[32] These clearly indicated that the owner had been an officer – but why would he carry ornate slippers on a thousand-mile survival trek through an unknown wilderness? And why had Crozier's officers considered silk handkerchiefs, towels, scented soap, toothbrushes, combs, hair brushes, and half a dozen pairs of spare boots to be essential equipment on this desperate trek? McClintock described the baggage as "a quantity of articles of one description and another truly astonishing in variety, and such as, for the most part, modern sledge-travellers in these regions would consider a mere accumulation of dead weight, but slightly useful, and very likely to break down the strength of the sledge-crews."[33]

Some of the articles, particularly knives and guns, would have been essential on their journey, but others seemed to be unnecessary, even frivolous. The five watches and the twenty-six silver spoons and forks embossed with the crests of Franklin and a few of his officers would certainly have been regarded as valuable in England – but what was the cost of getting them there? Perhaps Crozier and his men thought the items could be traded with Inuit for food, but to what extent would nomadic hunters covet heavy metal utensils?

British naval officers were neither trained nor equipped for overland marches in treeless country, and the gear available for travel, hunting, and camping was by today's standards notoriously heavy. McClintock reported "two double-barrelled guns – one barrel in each loaded and cocked – standing muzzle upwards against the boat's side"[34] – apparently shotguns for hunting birds and small game. Each gun probably weighed at least eight pounds and required subsidiary items. He noticed a waterproof gun cover, powder, bullets, cartridges, cartridge cases, wads, and two rolls of sheet lead for moulding bullets, altogether a heavy burden.

McClintock estimated that the sledge and boat together weighed about fourteen hundred pounds, and that a party of twenty or more men must have been involved in dragging the cumbersome craft over the snow, presumably in shifts. But where had all the men gone? Although graves were almost impossible to dig in the permafrost and the snow cover was thin, no bodies could be seen. Food was another mystery: all they could find in the boat was a little tea and forty pounds of chocolate – hardly adequate rations for hard-working men.

Among the many unanswerable questions facing McClintock was a fundamental one: what had been the destination of the boat party? Instead of heading south toward Simpson Strait and the Back River – the general direction of Crozier's retreat – the sledge and boat were pointing northward, back toward the place where the survivors had earlier come ashore from the ships. Perhaps they had been returning in order to obtain more food, or possibly they had given up entirely on the idea of striking out for a distant fur-trade post. Had the companions of the headless men intended to come back to the boat later for the chocolate and various other items?

McClintock's description of the pathetic relics of Franklin's failed expedition evoked intense feelings of sorrow everywhere. The *New York Journal* (5 November 1859) wrote:

> Here lie the arms, the sail, the oar,
> Dark with the storms of winters ten,
> And by their inexhausted store
> The bones that once were stalwart men.

* * *

On 7 May 1859, McClintock's second in command, William Hobson, retrieved two messages from a cairn at Victory Point on King William Island – the only official records of the expedition ever found, other than a bottle note cast overboard in Baffin Bay in 1845. The first message, written on an official naval form, said the expedition had wintered at Beechey Island. "All well," it concluded. But a second message, written in a shaky hand around the margins of the text, revealed that Franklin had died on 11 June 1847. The surviving men had now abandoned their ships and were making their way on foot to "Back's Fish River."

The two messages recounted in the briefest of terms what had happened. In 1845, the *Erebus* and *Terror* had turned north into Wellington Channel, proceeded to Queen Victoria Channel, and continued northwestward between Devon Island and Bathurst Island to a latitude of 77° N before turning back. They returned to Barrow Strait west of Cornwallis Island (thus circumnavigating it), then wintered at Beechey Island. In the summer of 1846, they had resumed their mission and made their way south through the allegedly *impassable* waterway of Peel Sound toward the continental mainland. Their intention was to follow the mapped coast westward to Bering Strait. Unfortunately, before attaining that coast, they became embedded in a mass of thick, hard, multi-year ice from which they never escaped.

The existence of the treacherous stream of polar ice that drifts from the Arctic Ocean through Viscount Melville Sound and McClintock Channel had been noted sixteen years earlier, so Franklin must have known about it. Had he boldly attempted to work his ships through it? Or had they been carried into it unwittingly by currents and drifting floes? The ships appeared to be heading west of King William Island, which was then thought to be a peninsula connected to Boothia. Once the *Erebus* and *Terror* were embedded in the drifting ice, they were trapped. During nineteen months from mid-September 1846 to mid-April 1848, they are said to have drifted less than twenty miles,[35] only a mile per month, or an average speed – if one can call it that – of about seven feet per hour. For the men on board the imprisoned vessels, it must have been dispiriting beyond description.

Out of the 126 officers and men on board HMS *Erebus* and *Terror* when the ships departed from Beechey Island in the summer of 1846, 105 were still alive in April 1848 when Crozier, who had assumed overall command on the death of Franklin a year earlier, made the decision to abandon the icebound ships and lead the survivors out of the Arctic. He intended to head for the subarctic fur-trade post of Fort Resolution on Great Slave Lake. To reach it, the men would walk southward along the west coast of King William Island, cross the ice of Simpson Strait to the continental mainland, travel up the valley of the Back River, and cross the height of land to Great Slave Lake. But the post was *more than a thousand miles away* – an impossible journey for starving, scurvy-ridden, poorly equipped men who had no experience in overland travel, camping, and hunting. Once they left their ships, "they were incompetent to live in Arctic conditions."[36]

As their intended journey would take place over different seasons, more than one travel technique would be required. Sledges would be of use only as long as an adequate snow cover remained on land and ice, and boats would be of use only after river ice melted but before it froze again. Furthermore, the Back River was known to have scores of rapids and falls, and the availability of game in the region was uncertain. The venture was a desperate gamble, the odds unfavourable.

The men fleeing the ships had little food and found wildlife to be scarce. As they came face to face with starvation, some resorted to cannibalism in a last desperate attempt to prolong their miserable lives. Many died on King William Island, and others got as far as the estuary of the Back River before the last dregs of life drained away.

Crozier had decided to attempt this hopeless voyage in preference to travelling north to Fury Beach, only a quarter as far, where they would have been able to exploit the food cache left by Parry two decades earlier. The depot was familiar to Crozier, who had been present when it was established, and to Thomas Blanky, his ice master on the *Terror*, who had been with John Ross several years later when he and his men consumed some of the provisions before taking to their boats to reach a whaling ship and escape from the Arctic. The provisions were still well preserved. If Crozier and his companions had reached Fury Beach during the summer of 1848, they might have encountered the first expedition sent out to look for them, for in September James Ross's two ships *Enterprise* and *Investigator* took up winter quarters at Port Leopold, less than a hundred miles away, and in the following spring his sledge parties set out to search the adjacent coasts.

After Rae's Inuit information about dead or dying men in the vicinity of King William Island was made public in 1854, Penny had assured Barrow that "Fury Beach would have been visited" if Franklin and his men had been trapped in that region.[37] But they had not.

* * *

Fourteen years after HMS *Erebus* and *Terror* left England in search of a Northwest Passage to the Pacific, the researches of McClintock and his men had revealed in dramatic fashion the tragic fate of the expedition. The Inuit stories told to Rae in 1854 were now confirmed, and many relics could be added to the few collected by Anderson and Stewart in 1855. The long search was over. The *Daily News* revived the popular light/dark metaphor, and proclaimed on 23 September 1859, "The dark veil is lifted and the very secret of these treeless solitudes is given up."

Even after McClintock returned with the gruesome evidence, there were tenuous hopes that some men had managed to survive and might be living on one of their ships or among nomadic bands of Inuit – or at least that some further written record of the tragic voyage might be found. Penny shared in these hopes and did not abandon his inquiries. In 1863–64, he commanded the whaler *Queen* on another wintering voyage, accompanied again by his wife, Margaret. His itinerary was unusual. He sailed through Hudson Strait and across Hudson Bay to Roes Welcome Sound, more than eight hundred miles in from the Atlantic Ocean. American whalers from New England ports had been exploiting this whaling ground for four years, but it was not familiar to Penny. After a short period of unsuccessful cruising, he retraced his path to the Atlantic and sailed around into the familiar Cumberland Sound to winter. He later told Barrow that if he had wintered in Repulse Bay (the place in northwestern Hudson Bay from which Rae had set off westward in 1854), he would have sent Inuit west to look for Franklin's ships. He was still not convinced that they had both been wrecked, and he believed that even if they had been, some records might have survived.

Almost two decades had passed since Franklin and his crews vanished into the labyrinth of arctic islands. One decade had expired since Rae reported hearsay evidence of a tragedy and returned with many relics of the missing expedition, their authenticity corroborated in the following year by the overland expedition of Anderson and Stewart. Half a dozen years had elapsed since McClintock carefully examined the area and found conclusive evidence of the disaster. Yet many questions remained unresolved, and Penny was still intent on completing the puzzle. As late as 1872, long after the Admiralty had given up the search and four years after Penny made his last sea voyage, he was still thinking about the tragic end of the Franklin expedition, in particular the disappearance of the ships. The matter, he told Barrow, is not "solved to my satisfaction."[38]

Nor was it solved to anyone's satisfaction after Franklin and his men sacrificed their lives in a futile quest for a sea passage of extraordinary geographical interest but no commercial value. Since then, many expeditions searched King William Island and adjacent regions, retrieved more relics, and subjected skeletal bones to rigorous scientific analysis.

Whereas much of the evidence on land was visible and easy to collect, the unseen evidence beneath the sea stubbornly resisted discovery. During the Franklin search, the historian John Brown wrote, "The *Erebus* and *Terror* may yet be in existence – yet unfold their own story."[39] A century and a half later, his prophecy came true.

In recent seasons, research vessels employing sophisticated underwater techniques methodically eliminated a succession of areas west of

King William Island from the search, steadily reducing the possibilities. This process led in September 2014 to the sensational discovery by Parks Canada marine archeologists of HMS *Erebus*, lying less than thirty-five feet below the surface in Queen Maud Gulf off Adelaide Peninsula.[40] Two years later, Canadian news media announced that on 2 September 2016 the well-preserved wreck of HMS *Terror* had been located about sixty miles to the north of the *Erebus*, south of King William Island in Terror Bay.

Postscript

In 1850, as the search expeditions were preparing to depart for Baffin Bay, the *Nautical Standard* (27 April) commented, "It is, therefore, desirable before the several expeditions leave this country, that some scheme or plan should be agreed upon by the several adventurers, so as to secure the largest amount of co-operation amongst themselves." In fact, Penny's brigs had left two weeks earlier and were already sailing past icebergs along the coast of Greenland, but the *Nautical Standard* had a point. No overall plan of operation existed.

Would such a plan have been feasible? One serious difficulty was that six of the ships were proceeding under orders drawn up by the British Admiralty, two were following instructions formulated by the US navy, one was sailing under orders from the Hudson's Bay Company, and another was governed by the directives of Lady Franklin. They were setting off from three different ports: one in England, two in Scotland, and one thousands of miles away in the United States. Among the sponsors and commanders, there was no unanimity about the searching strategy.

Communication at the time was primitive. Telegraph networks in Britain and in the United States were limited and not yet connected to each other by submarine cable. Postal service was efficient on land but slow by ship across the Atlantic. The only way to exchange messages directly between ships at sea, or between ships and the land, or between different locations on land in the Arctic, was by flags or voice, both of which had a very limited range. Notes in rock cairns were a haphazard way of disseminating information, and messages deposited in drift bottles were of no use other than contributing to the knowledge of ocean currents.

In view of the distances between the various sponsors and organizers, the lack of consensus on an appropriate searching plan, the crude communications of the time, the incomplete knowledge of the geography of the region, the short navigation season, and the variability of ice conditions, it would have been difficult, if not impossible, to design a realistic overall plan for one ship, let alone ten.

A year and a half later, after the searching expeditions had arrived back, the same newspaper (25 October 1851) complained that its "timely admonition" had not been heeded, and commented smugly, "To us it appears that the want of some original plan by which the several ships dispatched upon this service might have been distributed over the largest possible surface, was the radical vice of the plan which regulated the search." But the *largest possible surface* across which the ten ships could sail in 1850 had turned out to be a disappointingly small area surrounded by unbreachable ramparts of ice. The searchers had gained admittance to the vestibule of the archipelago but then found all its interior doors firmly shut. Had the ice been as open as it was during Parry's voyage to Melville Island three decades earlier, the ships of Penny, Austin, De Haven, Ross, and Forsyth could have dispersed into Wellington Channel, Prince Regent Inlet, Peel Sound, Barrow Strait, and Melville Sound, but the utterly unpredictable arctic ice had hemmed them all in. No plan, no matter how brilliant, could have avoided this.

The *Nautical Standard* called the search effort of 1850 "a mere scramble – a sort of helter-skelter race between the ships, to attain first the point in which they ultimately arrived together." Of course they were racing – racing to find Franklin and his 128 companions, racing to achieve as much as possible in a navigation season measured in mere weeks! And of course, when ice brought them all to a halt, the ships in front were overtaken by those in the rear, so that they all came together. The searchers had not been defeated by the lack of an overall plan but by the arctic ice.

In July, as the ships were advancing northward along the Greenland coast in fits and starts, Penny and Austin had often discussed search plans. But as wind direction and ice distribution could change suddenly, any plans were, in Sutherland's words, "reduced to mere or faintly probable outlines."[1] At Union Bay, Penny had attempted to unite the small-ship expeditions in cooperative action, including the search of Beechey Island that discovered the evidence of Franklin's first wintering. In early September, off Griffith Island, Austin had suggested areas of search for each expedition, and after going into winter quarters, he had designated separate spheres of responsibility for the various commanders. Because ad hoc plans such as these took into account the actual conditions encountered at a particular place and time, they were much more realistic than any advance plan could have been.

It is difficult to see how the search expeditions of 1850–51 could have accomplished much more. The ships, most of them dependent solely on wind power, criss-crossed the limited open water area and sent small parties ashore to search the accessible coasts, resulting in the discovery of Franklin's first winter quarters and various outlying campsites. The sledging operations during autumn and spring were very extensive (a point not mentioned by the *Nautical Standard*), and they failed to uncover any traces of Franklin and his companions only because none existed in the regions they searched. If the expeditions under Penny and Austin are to be criticized, it is for their cessation of searching activities after breaking out of winter quarters in August 1851. Yet even if Austin had allowed Penny to pilot one of his steamers into the open water beyond Cornwallis Island, or even if both expeditions had headed north and stayed out for another winter, they would not have solved the mystery of Franklin's course and location because they would have been searching in the wrong place.

Penny's discovery of open water north of Cornwallis Island turned the Franklin Search on its head for four years. What he at first considered a possibility that Franklin had gone through Queen Victoria Channel into a navigable ocean was magnified into a near-certainty by the Admiralty, the Hydrographic Office, the Arctic Committee, the Royal Geographical Society, the geographer Petermann, Penny's friends Osborn and Barrow, and even Lady Franklin. Writing before the return of McClintock's expedition on the *Fox*, the historian John Brown observed that there had been "a perfect infatuation in favour of Wellington Channel and the north," but this "monomania"[2] had not been based on facts. "In short," he wrote, "we knew nothing positive, for notwithstanding the multitude of opinions, written and expressed, they were all mere surmises – all merely the emanations of wandering imaginations, and without a tittle of positive evidence to prove that the Wellington Channel route had been adopted."[3]

When McClintock returned home the following year in September 1859, Brown would have been astounded to learn that in fact Franklin *had gone north through Wellington Channel* and had continued for a hundred miles beyond – but then had then returned south, presumably because he had met impassable polar ice. Nonetheless, Brown's point was valid. Franklin had left no evidence of his passage, not even a note in a cairn. Was it a death wish? Did he yearn to become a martyr?

In February 1847, before any official searches were initiated, Sir William Edward Parry, the dean of arctic navigators, had advised the masters of Davis Strait whaling ships to investigate "prominent positions for the piles of stones and flag-stones, under which, in former expeditions it has been customary to bury bottles, or copper cylinders, containing some account of their Proceedings."[4] In addition, John Ross claimed he had advised Franklin to deposit messages from time to time to summarize his progress and intentions. Yet it appears that, contrary to tradition and to his advice from Ross – indeed, contrary to common sense – Franklin never left any account at all, not even at his winter quarters at Beechey Island. If he had left a message there reporting that he had circumnavigated Cornwallis Island but had not found a passage leading north, and that in the forthcoming season he would attempt to go south from Beechey Island through Peel Sound toward the continental mainland, the search could have focused on that route way instead of being misled by the fanciful concept of an Open Polar Sea. One wonders whether Franklin really did not want to be found and rescued, but rather to become a martyr in the glorious cause of arctic exploration.

Even without a note from Franklin, the search would probably not have been misdirected toward the allegedly ice-free Arctic Ocean during the period 1850–54 if James Ross's sledge parties had been able to continue south to King William Island in 1849, or if sufficient attention had been paid to Rae's discovery of portions of a stanchion and a flagstaff from Franklin's ships on the shore of Victoria Island in 1851, or if Kennedy had turned south instead of north after sledging through Bellot Strait in 1852, or if Collinson had followed up the Inuit reports and wreckage on Victoria Island in 1853, or if three different searching parties had not jumped to the erroneous conclusion that Peel Sound was impassable.

The search for Franklin took place in a remote, vast, largely unknown, climatically severe region, in which ship navigation was limited to a small portion of the year and the distribution of sea ice could not be known in advance. As the various commanders could not communicate directly with each other unless they were in close contact, they often did not know where the other expeditions were, what they had accomplished, what problems they had encountered, or what they intended to do next, which made cooperation very difficult. They deserve great credit for operating under such conditions.

Unit Conversion Table

IMPERIAL	METRIC
Lengths and distances	
1 inch	25.4 millimetres
1 foot	0.3 metres
1 yard	0.9 metres
1 mile	1.6 kilometres
Weights	
1 pound (avoirdupois)	0.45 kilograms
1 ton (UK long)	1,016 kilograms
Liquid volumes	
1 cup (10 ounces)	284 cubic centimetres
1 pint (20 oz.)	568 cubic centimetres
1 quart (40 oz.)	1,137 cubic centimetres
1 gallon (160 oz.)	4,546 cubic centimetres
Temperatures	
Fahrenheit	Celsius
−60	−51
−40	−40
−20	−29
zero	−17
+20	−7
+32 (water freezes)	zero
+40	+4
+60	+15
+212 (water boils)	+100

Notes

Sources are given below for quotations in the text and for selected facts and works that may be unfamiliar to the general reader and worthy of note.

Authored works, whether published or unpublished, are identified by author surname and title. Full bibliographical details and a list of abbreviations are given in the bibliography. Official logbooks are listed under ships' names. Newspaper names and dates of publication are indicated in the text.

CHAPTER ONE

1 *Illustrated London News*, 24 May 1845.
2 Woodward, "William Penny," 253.
3 SPRI, MS 312/3, D.
4 SPRI, MS 313/3, D.
5 SPRI, MS 248/175, D.

CHAPTER TWO

1 Scoresby, *Journal of a Voyage*, 1823, 51.
2 Ross, Distant and Unsurveyed Country.
3 Spink and Moodie, "Eskimo Maps."

CHAPTER THREE

1 Mirsky, *Elisha Kent Kane*, 57.
2 SPRI, MS 116/45/1, D.
3 SPRI, MS 116/45/2, D.
4 SPRI, MS 116/45/3, D.
5 SPRI, MS 116/45/5, D.

6 SPRI, MS 312/1, D.
7 SPRI, MS 312/2, D.
8 SPRI, MS 116/45/5, D.
9 *Alexander*, Shipboard Journal, 8 June 1845.
10 Tillotson, *Adventures in the Ice*, 118–21.
11 *Alexander*, Shipboard Journal, 23 June 1845.
12 Ibid., 25 June 1845.
13 *Cumbrian*, Shipboard Journal, 14 July 1823.
14 *Alexander*, Shipboard Journal, 1 July 1845.
15 Ibid.
16 Ibid., 2 July 1845.
17 Ibid., 10 May 1845.
18 Quoted in Fluhmann, *Second in Command*, 91.

CHAPTER FOUR

1 Sutherland, *Whaling Years*, 25.
2 *Pacific*, Shipboard Journal, 7 June 1846.
3 Ibid., 1 July 1846.
4 Sutherland, "On the Esquimaux," 205.
5 Ibid., 204.
6 Ibid.
7 McDonald, *A Narrative*, 9.
8 Sutherland, "On the Esquimaux," 202.
9 Ross, *Rear Admiral Sir John Franklin*, 6.
10 SPRI, MS 116/27/2, D.
11 SPRI, MS 116/81/1, D.
12 TNA, ADM, 1/187, 14.
13 Ibid.
14 Brown, *North-West Passage*, following xi.
15 Cyriax and Wordie, "Centenary of the Sailing," 176.
16 TNA, ADM, 7/187.
17 Woodward, "William Penny," 260.
18 SPRI, MS 1100/3/1, D.
19 Ibid.
20 Ibid.
21 Kinnes Lists, Tay Whale Fishing Company, Minutes.
22 SPRI, MS 116/63/133, D.
23 Kinnes Lists, Tay Whale Fishing Company, Minutes; italics added.
24 Great Britain, *Additional Papers*, AB45228, 148.
25 SPRI, MS 1100/1/8, D.

CHAPTER FIVE

1 Letter from "An Observer," *London Times*, 1 January 1850.
2 Osborn, *Stray Leaves*, 35.
3 SPRI, MS 116/63/56, D.
4 Woodward, "William Penny," 264.
5 Ibid., 265.
6 SPRI, MS 1100/2/1, D.
7 SPRI, MS 1100/3/5, D.
8 SPRI, MS 1100/3/6, D.
9 Ibid.
10 Ibid.
11 Goodsir, *Arctic Voyage to Baffin's Bay*, 2–3.
12 Ibid.
13 Ibid., 51.
14 Ibid., 97.
15 Ibid.
16 Ibid., 75.
17 Ibid., 106.
18 TNA, ADM, 7/189 (13).
19 Goodsir, *Arctic Voyage to Baffin's Bay*, 107.
20 Ibid., 104.
21 Great Britain, *Additional Papers*, AB45228, 69.
22 SPRI, MS 248/104, BJ.
23 SPRI, MS 1100/3/9, D.
24 SPRI, MS 116/63/71, D.
25 WLPS, SA 2D-14.
26 SPRI, MS 248/240, BJ, 74–5.
27 Ibid.
28 SPRI, MS 1100/3/9, D.
29 TNA, ADM, 7/189.
30 SPRI, MS 248/104, BJ.
31 Ibid.
32 Ibid.
33 HL, Beaufort Papers, FB 33.

CHAPTER SIX

1 Great Britain, *Arctic Expedition*, AB45223, 134.
2 SPRI, MS 248/196/1, D.
3 Ibid.

4 SPRI, MS 116/63/84, D.
5 SPRI, MS 116/63/57, D.
6 Stamp and Stamp, *William Scoresby*, 67.
7 SPRI, MS 116/63/58, D.
8 SPRI, MS 116/63/59, D.
9 SPRI, MS 248/343/2, D.
10 Ibid.
11 SPRI, MS 116/63/74, D.
12 Ibid.
13 Great Britain, *Additional Papers*, AB45228, 134.
14 SPRI, MS 116/63/83, D.
15 TNA, ADM, 1/5606, 1–2.
16 SPRI, MS 248/454/4 D.
17 SPRI, MS 116/63/73, D.
18 Ibid.
19 SPRI, MS 248/406, D.
20 Kinnes Lists, Tay Whale Fishing Company, Minutes.
21 SPRI, MS 116/39/12, D.
22 Kinnes Lists, Tay Whale Fishing Company, Minutes.
23 Great Britain, *Additional Papers*, AB45228, 134.
24 SPRI, MS 116/59, D, quoted in Holland, "Arctic Committee of 1851," 9–10.
25 Ibid.
26 SPRI, MS 116/96, D.
27 Ibid.
28 SPRI, MS 116/63/30.
29 *Nautical Magazine* 19 (November 1850): 604.
30 SPRI, MS 248/454/9, D.
31 SPRI, MS 248/454/7, D.
32 Great Britain, *Additional Papers*, AB45228, 137.
33 Ibid.
34 SPRI, MS 116/81/3, D.
35 Great Britain, *Additional Papers*, AB45228, 136.
36 SPRI, MS 248/247/23, D.

CHAPTER SEVEN

1 SPRI MS 116/63/80, D.
2 Stuart, "Journal of an Unknown Officer," 13 April 1850.
3 BL, Barrow Bequest, ADD 35,306(2), 79.
4 Osborn, *Stray Leaves*, 13.
5 Ibid.

6 Sutherland, *Journal of a Voyage in Baffin's Bay*, 1: 6.
7 Stuart, "Journal of an Unknown Officer," 15 April 1850.
8 Ibid., 17 April 1850.
9 Ibid., 23 April 1850.
10 Sutherland, *Journal of a Voyage in Baffin's Bay*, 1: 20.
11 Stuart, "Journal of an Unknown Officer," 2 May 1850.
12 Sutherland, *Journal of a Voyage in Baffin's Bay*, 1: 35.
13 SPRI, MS 116/63/62, D.
14 SPRI, MS 116/24/7, D.

CHAPTER EIGHT

1 *Aberdeen Herald*, 26 January 1850.
2 Osborn, *Stray Leaves*, 73.
3 *Nautical Magazine* 19 (June 1850): 398.
4 SPRI, MS 248/344/2, D.
5 TNA, ADM, 7/190/4.
6 *Nautical Magazine* 21 (April 1852): 200.
7 Great Britain, *Papers Relative to the Recent Arctic Expeditions*, AB45241, 112.
8 Ross, *Rear Admiral Sir John Franklin*, 55.
9 Ibid., 53–4.
10 HBC, John Ross Fonds, E15/7, 125.
11 Ibid. The following quotations are also from the Ross Fonds.
12 WLPS, SA 2D.
13 SPRI, MS 248/455/2, D.
14 HBC, John Ross Fonds, 117.
15 Snow, *Voyage of the Prince Albert*, 171.
16 Quoted in Ross, *Polar Pioneers*, 327.
17 Woodward, "William Penny," 265.
18 *Nautical Magazine* 18 (July1849): 370.
19 Ibid., 370.
20 Ibid.
21 SPRI, MS 248/414/2, D.
22 Quoted in Friis, "Beginnings of Exploration," 10.
23 Wright, *Geography in the Making*, 30.
24 SPRI, MS 248/414/2, D.
25 Ibid.
26 Ibid.
27 Old Officer, *Observations on the Missing Ships*.
28 TNA, ADM, 7/190 (4).

29 RGS, "Arctic Special" card file, S/Div 2.
30 Stone, "William Parker Snow," 163.
31 Woodward, "William Penny," 272.
32 Loomis, *Weird and Tragic Shores*, 151.
33 Hazlewood, *Survival, Revenge*, 163.
34 Great Britain, *Additional Papers*, AB45228, 140.
35 Neatby, *Search for Franklin*, 125.
36 Jones, "Captain William Parker Snow," 370.
37 Great Britain, *Additional Papers*, AB45228, 139.
38 Ibid.
39 Ibid., 139–40.
40 Snow, *Voyage of the Prince Albert*, iv.
41 Skewes, *Sir John Franklin*.
42 Kennedy, *Short Narrative*, 27.
43 BL, Barrow Bequest, ADD 35,306(3), 163.
44 Snow, *Voyage of the Prince Albert*, 28.
45 BL, Barrow Bequest, ADD 35,306(3), 161–2.
46 Snow, *Voyage of the Prince Albert*, 31.
47 Ibid.
48 Mirsky, *Elisha Kent Kane*, 51.

CHAPTER NINE

1 Sutherland, *Journal of a Voyage in Baffin's Bay*, 1: 213.
2 SPRI, MS 248/ 454/ 11, D.
3 Great Britain, *Arctic Expeditions*, AB45227, 80.
4 Sutherland, *Journal of a Voyage in Baffin's Bay*, 1: 213.
5 Ibid., 45–6.
6 Ibid.
7 Ibid., 66.
8 Stuart, "Journal of an Unknown Officer," 17 May 1850.
9 Great Britain, *Arctic Expeditions*, AB45226, 80.
10 Sutherland, *Journal of a Voyage in Baffin's Bay*, 2: ccxxiii.
11 Ibid., ccxxvii.
12 Sutherland, *Journal of a Voyage in Baffin's Bay*, 1: 83.
13 Stuart, "Journal of an Unknown Officer," 22, 23 May 1850.
14 Sutherland, *Journal of a Voyage in Baffin's Bay*, 1: 65.
15 Stuart, "Journal of an Unknown Officer," 24–25 May 1850.
16 Sutherland, *Journal of a Voyage in Baffin's Bay*, 1: 75.
17 Stuart, "Journal of an Unknown Officer," 30 May 1850.
18 Sutherland, *Journal of a Voyage in Baffin's Bay*, 1: 100.

19 Great Britain, *Arctic Expeditions*, AB45226, 80.
20 Sutherland, *Journal of a Voyage in Baffin's Bay*, 1: 104.
21 Penny in Sutherland, *Journal of a Voyage in Baffin's Bay*, 2: cix–cx.
22 Stuart, "Journal of an Unknown Officer," 4 June 1850.
23 Great Britain *Arctic Expeditions*, AB45226, 80.
24 Snow, *Voyage of the Prince Albert*, 230.

CHAPTER TEN

1 Osborn, *Illustrated Arctic News*, 13, quoted in Behrisch, "'Far as the Eye Can Reach,'" 3.
2 Sutherland, *Journal of a Voyage in Baffin's Bay*, 1: 199.
3 Ibid., 134.
4 Stuart, "Journal of an Unknown Officer," 2 July 1850.
5 Ibid., 158.
6 Stuart, "Journal of an Unknown Officer," 17 June 1850.
7 Osborn, *Stray Leaves*, 39.
8 Ibid.
9 Ibid., 13.
10 *Nautical Magazine* 20 (February 1851): 92.
11 TNA, ADM, 7/190(2).
12 *Athenaeum*, 7 February 1852, 169.
13 Osborn, *Stray Leaves*, 41.
14 Ibid.
15 Ibid., 43.
16 Ibid., 50.
17 Ibid.
18 Ibid.
19 Ibid., 51.
20 Cator, Private Journal, 7 July 1850.
21 Ibid.
22 Ibid., 12 July 1850.
23 Ibid., 8 July 1850.
24 Stuart, "Journal of an Unknown Officer," 1 July 1850.
25 Ibid.,13 July 1850.
26 Sutherland, *Journal of a Voyage in Baffin's Bay*, 1: 225.
27 Ibid., 221.
28 Ibid., 225.
29 SPRI, MS 116/63/28.
30 Holland, "Arctic Committee of 1851," 1: 11.
31 SPRI, MS 116/63/66, D.

32 Osborn, *Stray Leaves*, 52.
33 Ibid.
34 Cator, Private Journal, 11 July 1850.
35 Stuart, "Journal of an Unknown Officer," 5 July 1850.
36 Ibid., 9 August 1850.
37 Ibid., 13 July 1850.
38 Sutherland, *Journal of a Voyage in Baffin's Bay*, 1: 202–3.
39 Ibid., 205–6.
40 Cator, Private Journal, 5 August 1850.
41 De Haven, Diary, 7 July 1850.
42 Ibid.
43 Carter in Gill and Young, *Searching for the Franklin Expedition*, 37.
44 Ibid., 41.
45 Ibid., 43.
46 Ibid.
47 De Haven, Diary, 25 July 1850.
48 Carter in Gill and Young, *Searching for the Franklin Expedition*, 43
49 Kane, *United States Grinnell Expedition*, 94.
50 Ibid., 83.
51 Ibid., 93.
52 Ibid.
53 De Haven, Diary, 4 August 1850.
54 Kane, *United States Grinnell Expedition*, 104–5.
55 Snow, "Remarks on the Prince Albert's Track," 172.
56 Kane, *United States Grinnell Expedition*, 77.
57 Harrison, "Lost Whaling Ships," 189–95.

CHAPTER ELEVEN

1 Sutherland, *Journal of a Voyage in Baffin's Bay*, 1: 255.
2 Ibid., 226.
3 Ibid., 237.
4 Ibid., 238.
5 Snow, *Voyage of the Prince Albert*, 200.
6 Ibid., 199.
7 Washington, *Esquimaux and English Vocabulary*, 5–6.
8 Carter in Gill and Young, *Searching for the Franklin Expedition*, 37.
9 Washington, *Esquimaux and English Vocabulary*, 102, 105.
10 Sutherland, "On the Esquimaux," 200.
11 Washington, *Esquimaux and English Vocabulary*, xvi.
12 LAC, RGS MF A-871, part 1E.

13 Ibid.
14 Snow, *Voyage of the Prince Albert*, 212.
15 Ross, John 1850–51, 13 August 1850.
16 LAC, RGS MF A-871, part 1E.
17 Ibid.
18 Ibid.
19 Cyriax, "Adam Beck," 38.
20 LAC, RGS MF A-871, part 1E.
21 Ibid.
22 Sutherland, *Journal of a Voyage in Baffin's Bay*, 1: 255.
23 LAC, RGS MF A-871, part 1E.
24 Sutherland, *Journal of a Voyage in Baffin's Bay*, 1: 259.
25 LAC, RGS MF A-871 part 1E.
26 Cator, "Private Journal," 14 August 1850.
27 Snow, *Voyage of the Prince Albert*, 196–7.
28 Ibid.
29 Ibid.
30 Darwin, *Voyage of the Beagle*, 207.
31 Moorehead, *Darwin and the Beagle*, 90.
32 Darwin, *Voyage of the Beagle*, 229.
33 Hazlewood, *Savage*, 174, 176.
34 Snow, *Voyage of the Prince Albert*, 224.
35 Cator, Private Journal, 15 August 1850.
36 Ibid.
37 Savours, *Search for the North West Passage*, 198.
38 Beattie and Geiger, *Frozen in Time*.
39 Cookman, *Ice Blink*.
40 Osborn, *Stray Leaves*, 72.
41 Cator, Private Journal, 17 August 1850.
42 Markham, Franklin's Footsteps, 60n.
43 Osborn, *Stray Leaves*, 72.
44 Woodman, *Unravelling the Franklin Mystery*, 59.
45 Mary-Rousselière, *Qitdlarssuaq*.

CHAPTER TWELVE

1 Kane, Notebook of Polar Exploration, 19 August 1850.
2 Sutherland, *Journal of a Voyage in Baffin's Bay*, 1: 263.
3 Osborn, *Stray Leaves*, 73.
4 Great Britain, *Arctic Expeditions*, AB43225, 3.
5 SPRI MS 248/247/24, D.

6 SPRI MS 248/247/26, D.
7 Sutherland, *Journal of a Voyage in Baffin's Bay*, 1: 272.
8 De Haven, Diary, 15 August 1850.
9 Ibid.
10 Kane, *The United States Grinnell Expedition*, 130.
11 Kane, Notebook of Polar Exploration, 19 August 1850.
12 Carter in Gill and Young, *Searching for the Franklin Expedition*, 55.
13 Kane, *The United States Grinnell Expedition*,152.
14 Sutherland, *Journal of a Voyage in Baffin's Bay*, 1: 277.
15 Stuart, "Journal of an Unknown Officer," 19 August 1850.
16 Carter in Gill and Young, *Searching for the Franklin Expedition*, 55.
17 Kane, *United States Grinnell Expedition*, 152.
18 Ross, Pocket Diary, 19 August 1850.
19 Ibid.
20 Cooke 1981, 58.
21 Kane, *United States Grinnell Expedition*, 153.
22 Cator, Private Journal, 19 August 1850.
23 Ibid.
24 Snow, *Voyage of the Prince Albert*, 236.
25 Osborn, *Stray Leaves*, 73.
26 Ibid., 74–5.
27 Sutherland, *Journal of a Voyage in Baffin's Bay*, 1: 282.
28 Great Britain, *Arctic Expeditions*, AB45226, 82.
29 Sutherland, *Journal of a Voyage in Baffin's Bay*, 1: 283.
30 Great Britain, *Arctic Expeditions*, AB45226, 82–3.
31 Ibid., 66–7. See also *Nautical Magazine* 19 (November 1850): 601.
32 Sutherland, *Journal of a Voyage in Baffin's Bay*, 1: 284.
33 Osborn, *Stray Leaves*, 104.
34 De Haven, Diary, 28 August 1850.
35 *Nautical Magazine* 20 (January 1851): 9.
36 Brown, *North-West Passage*, 157.

CHAPTER THIRTEEN

1 Osborn, *Stray Leaves*, 76.
2 Sutherland, *Journal of a Voyage in Baffin's Bay*, 1: 286, 287.
3 Cator, Private Journal, 21 August 1850.
4 Sutherland, *Journal of a Voyage in Baffin's Bay*, 1: 284.
5 Ibid., 290.
6 Cator, Private Journal, 23 August 1850.
7 Ibid.

8 McClintock, Journal Kept during the Voyage, 23 August 1850.
9 Ommanney, Extract from Rough Journal, 23 August 1850.
10 Ibid.
11 Cator, Private Journal, 23 August 1850.
12 Ibid.
13 Carter in Gill and Young, *Searching for the Franklin Expedition*, 57.
14 Ibid.
15 Ibid., 58.
16 Ibid.
17 Ibid.
18 De Haven, Diary, 25 August 1850.
19 Kane, *United States Grinnell Expedition*, 155.
20 Snow, *Voyage of the Prince Albert*, 249.
21 Potter, *Arctic Spectacles*.
22 Snow, *Voyage of the Prince Albert*, 249.
23 Ibid.
24 Ibid., 261.
25 BL, Barrow Bequest, ADD 35,306(3), 173.
26 Ibid., 168.
27 Ibid., 173.
28 *Nautical Magazine* 19 (November 1850): 612.
29 Stone 1993a, 141–2.
30 *Nautical Magazine* 19 (November 1850): 614–15.
31 LAC, RGS MF A-871.
32 Ibid.
33 SPRI, MS 1100/3/16, D.
34 Ross, *Rear Admiral Sir John Franklin*, 9.
35 *Nautical Magazine* 20 (January 1851): 24.
36 Ibid.
37 Ibid., 23.
38 Weld, *Arctic Expeditions*, 45.
39 Cator, Private Journal, 17 September 1850.
40 "Mr. Hampton, the Celebrated Aeronaut," handbill at Bloomsbury Auctions, London.
41 Ibid.
42 *Dundee Advertiser*, 29 September 1848.
43 *Nautical Magazine* 19 (November 1850): 606–15.
44 Kane, *United States Grinnell Expedition*, 176.
45 Carter in Gill and Young, *Searching for the Franklin Expedition*, 65.
46 Kane, *United States Grinnell Expedition*, 173.
47 Ibid., 177.

48 Cator, Private Journal, 25 August 1850.
49 Kane, *United States Grinnell Expedition*, 173.

CHAPTER FOURTEEN

1 Cator, Private Journal, 25 August.
2 Ibid., 24 August 1850.
3 Lovell, Journal of a Cruize, 27 August 1850.
4 Kane, *United States Grinnell Expedition*, 166.
5 McClintock 1850, 25 August 1850.
6 Cator, Private Journal, 25 August 1850.
7 *Morning Herald*, 9 December 1851.
8 Sutherland, *Journal of a Voyage in Baffin's Bay*, 1: 300–1.
9 Carter in Gill and Young, *Searching for the Franklin Expedition*, 60.
10 Kane, *United States Grinnell Expedition*, 161.
11 Cator, Private Journal, 29 August 1850.
12 Osborn, *Stray Leaves*, 90.
13 Kane, *United States Grinnell Expedition*, 165.
14 Osborn, *Stray Leaves*, 89.
15 Sutherland, *Journal of a Voyage in Baffin's Bay*, 1: 306.
16 Ibid., 307–8.
17 Osborn, *Stray Leaves*, 86.
18 Ross, *Rear Admiral Sir John Franklin*, 28, 29 August 1850.
19 Osborn, *Stray Leaves*, 95.
20 Parry, *Journal of a Voyage*, 50.
21 Osborn, *Stray Leaves*, 95.
22 Sutherland, *Journal of a Voyage in Baffin's Bay*, 1: 316.
23 Quoted in Beattie and Geiger, *Frozen in Time*, 118.
24 Ibid., 140.
25 Carter in Gill and Young, *Searching for the Franklin Expedition*, 61.
26 Ibid.
27 Ibid., 66.
28 Sutherland, *Journal of a Voyage in Baffin's Bay*, 1: 309.
29 Cator, Private Journal, 29 August 1850.
30 Ibid., 8 January 1851.
31 Stuart, *Relief of the Franklin Expedition*, 19.
32 Cator, Private Journal, 24 August 1850.
33 Ibid., 25 August 1850.
34 De Haven, Diary, 29 August 1850.
35 Stuart, "Journal of an Unknown Officer," 29 August 1850.
36 Cator, Private Journal, 29 August 1850.

37 Stuart, "Journal of an Unknown Officer," 30 August 1850.
38 De Haven, Diary, 30 August 1850.
39 Ross, Pocket Diary, 6 September 1850.
40 Sutherland, *Journal of a Voyage in Baffin's Bay*, 1: 340.
41 Ibid., 342.
42 Quoted in Cyriax, *Sir John Franklin's Last Expedition*, 46.
43 SPRI, MS 116/38/5, D.

CHAPTER FIFTEEN

1 Osborn, *Stray Leaves*, 100.
2 De Haven, Diary, 5 September 1850.
3 Cator, Private journal, September 1850.
4 Carter in Gill and Young, *Searching for the Franklin Expedition*, 69.
5 Kane, *United States Grinnell Expedition*, 176.
6 Ibid., 178.
7 Carter in Gill and Young, *Searching for the Franklin Expedition*, 70.
8 Cator, Private journal, 5 September 1850.
9 Ibid., 15 September 1850.
10 Ibid.,7 September 1850.
11 Ibid.
12 Ibid., 9 September 1850.
13 Osborn, *Stray Leaves*, 102.
14 Ibid.
15 Ibid., 102–3.
16 Carter in Gill and Young, *Searching for the Franklin Expedition*, 71.
17 Snow, *Voyage of the Prince Albert*, 300.
18 Cator, Private journal, 7 September 1850.
19 Ibid., 8 September 1850.
20 Ibid., 10 September 1850.
21 Ibid.
22 Parry, *Journal of a Second Voyage*, 117.
23 Ibid.
24 Great Britain, *Additional Papers*, AB45228, 63.
25 Osborn, *Stray Leaves*,105.
26 Great Britain, *Papers Relative to the Recent Arctic Expeditions*, AB45241, 113.
27 Ibid., 116.
28 Ibid., 123.
29 Ibid., 12.
30 Great Britain, *Arctic Expeditions*, AB45227, 29.

31 Stuart, Journal of an unknown officer, 11 September 1850.
32 Cator, Private journal, 11 September 1850.
33 Stuart, Journal of an unknown officer, 11 September 1850.

CHAPTER SIXTEEN

1 Kane, Notebook of polar exploration, 13 September 1851.
2 Carter in Gill and Young, *Searching for the Franklin Expedition*, 64.
3 Ibid., 66.
4 Osborn, *Stray Leaves*,106.
5 McClintock, Journal kept during the voyage, 23 August 1850.
6 Carter in Gill and Young, *Searching for the Franklin Expedition*, 64.
7 Snow, *Voyage of the Prince Albert*, 302.
8 Kane, *United States Grinnell Expedition*, 171.
9 Carter in Gill and Young, *Searching for the Franklin Expedition*, 66.
10 Osborn, *Stray Leaves*,108.
11 Cator, Private journal, 28 July 1850.
12 Ibid., 9 August 1850.
13 Carter in Gill and Young, *Searching for the Franklin Expedition*, 71, 73.
14 Ibid.,71.
15 De Haven, Diary, 11 September 1850.
16 Ibid., 12 September 1850.
17 Carter in Gill and Young, *Searching for the Franklin Expedition*, 73.
18 Ibid., 75.
19 De Haven, Diary, 13 September 1850.
20 Cator, Private journal, 13 September 1850.
21 Kane, *United States Grinnell Expedition*, 187.
22 Ibid., 187–8.
23 De Haven, Diary, 13 September 1850.
24 Ibid.
25 Carter in Gill and Young, *Searching for the Franklin Expedition*, 75.
26 Ibid.,77.
27 Ibid.
28 Ibid.
29 De Haven, Diary, 21 September 1850.
30 Ibid., 22 September 1850.
31 Ibid., 3 October 1850.
32 Carter in Gill and Young, *Searching for the Franklin Expedition*, 95.
33 De Haven, Diary, 31 October 1850.
34 Ibid., 13 December 1850.
35 Ibid., 20 December 1850.

CHAPTER SEVENTEEN

1 *Sophia*, Official logbook, 15 September 1850.
2 Sutherland, *Journal of a Voyage in Baffin's Bay*, 1: 369.
3 Ross, Pocket diary, 15 September 1850.
4 Great Britain, *Arctic Expeditions*, AB45227, 29.
5 Cator, Private journal, 13 September 1850.
6 *Assistance*, HMS, Logbook, 13–16 September 1850.
7 Stuart, Journal of an unknown officer, 5 September 1850.
8 Ibid., 21 September 1850.
9 Ibid., 23 September 1850.
10 Ibid., 26 September 1850.
11 *Arctic Miscellanies*, 275.
12 Cator, Private journal, 8 October 1850.
13 Stuart, Journal of an unknown officer, 4 October 1850.
14 Ibid.
15 Cator, Private journal, 17 October 1850.
16 Ibid., 18 October 1850.
17 Ibid., 27 November 1850.
18 Ibid., 13 December 1850.

CHAPTER EIGHTEEN

1 Osborn, *Stray Leaves*,73.
2 Sutherland, *Journal of a Voyage in Baffin's Bay*, 1: 382.
3 Ibid., 407–8.
4 Ibid., 495.
5 Sutherland, *Journal of a Voyage in Baffin's Bay*, 2: 14n.
6 Stuart, Journal of an unknown officer, 17 February 1851.
7 Ibid., 26 October 1850.
8 Sutherland, *Journal of a Voyage in Baffin's Bay*, 1: 428.
9 Stuart, Journal of an unknown officer, 17 February 1851.
10 Sutherland, *Journal of a Voyage in Baffin's Bay*, 1: 430.
11 Johnston, *Physical Atlas*, 101.
12 Ibid., 2.
13 Markham, *Arctic Navy List*, 7.
14 Cator, Private journal, 5 October 1850.
15 Honigsbaum, *Fever Trail*, 91.
16 Ibid., 90.
17 *Nautical Magazine* 19 (October 1850): 556.
18 Osborn, *Stray Leaves*,121.

19 Sutherland, *Journal of a Voyage in Baffin's Bay*, 1: 411.
20 Osborn, *Stray Leaves*,138–9.
21 Ibid.
22 Cator, Private journal, 17 December 1850.
23 McClintock, Journal kept during the voyage, 57.
24 Osborn, *Stray Leaves*,138.
25 Cator, Private journal, 5 November 1850.
26 McClintock, Journal kept during the voyage, 55–6.
27 Ibid., 56.
28 Cator, Private journal, 6 November 1850.
29 Ibid.
30 Ibid., 7 November 1850.
31 Stuart, Journal of an unknown officer, 12 November 1850.
32 Carpenter, *History of Scurvy*, 29.
33 Sutherland, *Journal of a Voyage in Baffin's Bay*, 1: 416.
34 Ross, Pocket Diary, 10 November 1850.
35 Sutherland, *Journal of a Voyage in Baffin's Bay*, 1: 426.

CHAPTER NINETEEN

1 Osborn, *Stray Leaves*,132.
2 McClintock, Journal kept during the voyage, 57.
3 Cator, Private journal, 24 December 1850.
4 Ibid., 25 December 1850.
5 Ibid.
6 McClintock, Journal kept during the voyage, 57.
7 Cator, Private journal, 26 January 1851.
8 Ibid., 1 January 1851.
9 Ibid.
10 Ibid.
11 Ibid., 15 January 1851.
12 De Haven, Diary, 27 November 1850.
13 Philbrick, *Sea of Glory*, 239.
14 De Haven, Diary, 10 December 1850.
15 Lovell, Journal of a cruize, 11 December 1850.
16 Kane, *United States Grinnell Expedition*, 267–8.
17 De Haven, Diary, 25 December 1850.
18 Kane, *United States Grinnell Expedition*, 269.

CHAPTER TWENTY

1 Motto on Penny's sledging flag.
2 Sutherland, *Journal of a Voyage in Baffin's Bay*, 1: 451.
3 Ibid., 462.
4 Stuart, Journal of an unknown officer, 4 February 1851.
5 Sutherland, *Journal of a Voyage in Baffin's Bay*, 1: 462.
6 McClintock, "Reminiscences of Arctic Ice Travel," 183.
7 Parry, *Journal of a Voyage for the Discovery of a North-West Passage*, 196.
8 McClintock, "How to Prepare Oneself to Enjoy the Winter," 55–6.
9 See Sutherland, *Journal of a Voyage in Baffin's Bay*, appendix 1.
10 Cator, Private journal, 3 April 1851.
11 Ibid., 4 April 1851.
12 Stuart, Journal of an unknown officer, 5 April 1851.
13 Ibid., 8 April 1851.
14 Sutherland, *Journal of a Voyage in Baffin's Bay*, 2: 30.
15 Stuart, Journal of an unknown officer, 10 April 1851.
16 LAC, RGS MF A-871, 1(F).
17 SPRI, MS 116/63/115.
18 Ibid.
19 Sutherland, *Journal of a Voyage in Baffin's Bay*, 2: 39.
20 Ibid.
21 SPRI, MS 116/63/116, D.
22 Sutherland, *Journal of a Voyage in Baffin's Bay*, 2: 48.
23 Ibid., 53.
24 Osborn, *Stray Leaves*,171.
25 Cator, Private journal, 28 April 1851.
26 Sutherland, *Journal of a Voyage in Baffin's Bay*, 2: 57.
27 SPRI, MS 116/63/116, D.
28 LAC, RGS MF A-872.
29 Cator, Private journal, 19 May 1851.
30 Ibid., 17 May 1851.
31 Ibid., 19 May 1851.
32 McClintock, "Reminiscences of Arctic Ice-Travel," 11, 12.
33 SPRI, MS 116/63/116, D.
34 Stewart in Sutherland, *Journal of a Voyage in Baffin's Bay*, 2, appendix, viii.
35 LAC, RGS MF A-871, part F.
36 Pack, *Nelson's Blood*, 69.
37 Milton-Thompson, "Two Hundred Years of the Sailor's Diet," 31.

CHAPTER TWENTY-ONE

1 Sutherland, *Journal of a Voyage in Baffin's Bay*, 2: 31.
2 Penny in Sutherland, *Journal of a Voyage in Baffin's Bay*, 2: 126.
3 Ibid., 128.
4 Ibid., 130.
5 Cator, Private journal, 23 May 1851.
6 Ibid., 25 May 1851.
7 Great Britain, Arctic Expeditions, AB45227, 31.
8 Ibid.
9 Ibid.
10 Penny in Sutherland, *Journal of a Voyage in Baffin's Bay*, 2: 144.

CHAPTER TWENTY-TWO

1 Sutherland, *Journal of a Voyage in Baffin's Bay*, 2: 250.
2 Penny in Sutherland, *Journal of a Voyage in Baffin's Bay*, 2: 149.
3 Ibid., 151–2.
4 John Ross, Pocket diary, 22 December 1850.
5 Penny in Sutherland, *Journal of a Voyage in Baffin's Bay*, 2: 158.
6 Ibid., 159.
7 Ibid., 160–1.
8 Ibid., 162.
9 Ibid., 163.
10 Ibid., 176.
11 Stewart in Sutherland, *Journal of a Voyage in Baffin's Bay*, 2: xviii.
12 Penny in Sutherland, *Journal of a Voyage in Baffin's Bay*, 2: 177.
13 Ibid., 178.
14 Ibid.
15 Ibid., 182.
16 Ibid.
17 Ibid., 184.
18 Ibid., 185.
19 Ibid.
20 Osborn, *Stray Leaves*,176.
21 Sutherland, *Journal of a Voyage in Baffin's Bay*, 2: 231.
22 Ibid., 245–6.
23 John Ross, Pocket diary, 1 August 1851.
24 Ibid.
25 Sutherland, *Journal of a Voyage in Baffin's Bay*, 2: 217.
26 Ibid., 219.

27 Ibid., 218.
28 Ibid., 219.
29 Ibid., 218.
30 Great Britain, *Arctic Expeditions*, AB45227, 49.
31 Ibid., 32.
32 Ibid., 33.
33 Ibid.
34 Ibid., 34.
35 Ibid., 112.
36 Penny in Sutherland, *Journal of a Voyage in Baffin's Bay*, 2: 178.
37 Sutherland, *Journal of a Voyage in Baffin's Bay*, 2: 245–6.
38 Great Britain, *Arctic Expeditions*, AB45227, 61.
39 Sutherland, *Journal of a Voyage in Baffin's Bay*, 2: 301.
40 Stuart, Journal of an unknown officer, 6 August 1851.
41 Ibid., 8 August 1851.
42 Great Britain, *Arctic Expeditions*, AB43225, 4.
43 Ibid.
44 Stuart, Journal of an unknown officer, 11 August 1851.
45 Ommanney, Journal written on HMS *Assistance*, 11 August 1851.
46 Ibid.
47 Great Britain, *Arctic Expeditions*, AB45227, 34–5.
48 Ibid.
49 Ibid., 36.

CHAPTER TWENTY-THREE

1 *Athenaeum*, 11 May 1850, 503.
2 Kane, *United States Grinnell Expedition*, 275.
3 De Haven, Diary, 18 January 1851
4 Kane, *United States Grinnell Expedition*, 305.
5 Ibid., 297.
6 De Haven, Diary, 29 January 1851.
7 Lovell, Journal of a cruize, 29 January 1851.
8 NYHS, MS 2783.
9 Carter in Gill and Young, *Searching for the Franklin Expedition*, 80.
10 Kane, Notebook, September 1850.
11 Ibid.
12 Kane, *United States Grinnell Expedition*, 271.
13 Ibid., 288.
14 Kane, Notebook, February 1851.
15 De Haven, Diary, 17 February 1851.

16 Kane, Notebook, letter to De Haven, 23 June 1851.
17 Kane, *United States Grinnell Expedition*, 310.
18 Ibid., 331.
19 De Haven, Diary, 5 March 1851.
20 Lovell, Journal of a cruize, 30 May 1851.
21 Carter in Gill and Young, *Searching for the Franklin Expedition*, 113.
22 Ibid., 135.
23 Ibid., 125.
24 De Haven, Diary, 5 June 1851.
25 Kane, *United States Grinnell Expedition*, 405.
26 De Haven, Diary, 6 June 1851.
27 Kane, *United States Grinnell Expedition*, 408.
28 De Haven, Diary, 7 June 1851.
29 Carter in Gill and Young, *Searching for the Franklin Expedition*, 145.
30 Kennedy, *Short Narrative of the Second Voyage*, 47.
31 Photostat copy of a notebook containing information and advice supplied Capt. Kennedy by Lady Franklin prior to his setting out for the relief of the Franklin party in 1851, AANDC, G660 F73.
32 Bellot, *Memoirs of Lieutenant Joseph-René Bellot*, 1: 225.
33 Kane, *United States Grinnell Expedition*, 465.
34 Bellot, *Memoirs of Lieutenant Joseph-René Bellot*, 1: 212.
35 Ibid., 230.
36 Ibid., 247–8.
37 Kane, *United States Grinnell Expedition*, 462.

CHAPTER TWENTY-FOUR

1 Sutherland, *Journal of a Voyage in Baffin's Bay*, 2: 305.
2 SPRI, MS 486/1/1, D.
3 SPRI, MS 486/5/2, D.
4 Osborn, *Stray Leaves*,181.
5 LAC, RGS MF A-871.
6 *Nautical Magazine* 20 (November 1851): 573.
7 Ibid., 578.
8 Ibid.
9 Osborn, *Stray Leaves*, 185–6.
10 Ibid., 184.
11 Ibid., 195.
12 Great Britain, *Additional Papers*, AB45228, 297–8.
13 Osborn, *Stray Leaves*, 204.
14 Ibid., 206.

15 *Nautical Standard*, 18 October 1859 (reprinted from the *Ayr Observer*).
16 Ibid.
17 TNA, ADM, 7/192: 445.
18 Stuart, Journal of an unknown officer, 20 August 1851.
19 Sutherland, *Journal of a Voyage in Baffin's Bay*, 2: 339–40.
20 TNA, ADM, 7/190.

CHAPTER TWENTY-FIVE

1 Traditional folk ballad "Lady Franklin's Lament."
2 TNA, ADM 7/190.
3 Ibid.
4 Ibid.
5 TNA, ADM 7/190.
6 SPRI, MS 1100/1/9, D.
7 TNA, ADM 7/192(28), 448.
8 TNA, ADM 7/193.
9 *Nautical Magazine* (January 1851): 7.
10 Great Britain, *Arctic Expeditions*, AB 45229, 120.
11 *London Chronicle*, 9 December 1851.
12 Ibid.

CHAPTER TWENTY-SIX

1 SPRI, MS 116/28/3, D.
2 All quotations in this chapter from the proceedings of the Arctic Committee are from its published report, Great Britain, *Arctic Expeditions*, AB45229, unless otherwise noted.
3 SPRI, MS 395/97, BL.
4 SPRI, MS 1100/1/10, D.
5 SPRI MS 116/16/9, D.
6 Ommanney, Journal written on HMS *Assistance*, 11 August.
7 TNA, ADM 7/190(2).
8 Holland, "The Arctic Committee of 1851," part 1, 118.
9 Ibid., part 2, 5.
10 *London Chronicle*, 9 December 1851.
11 Ibid.
12 SPRI, MS 116/38/3, D; italics added.
13 TNA, ADM 7/192(26), 426.
14 Ibid., 423.
15 TNA, ADM 7/190(3).

16 Ibid.
17 TNA, ADM 7/612.
18 TNA, ADM 12/554.

CHAPTER TWENTY-SEVEN

1 *Nautical Magazine* 20 (November 1851): 587.
2 Wright, "Open Polar Sea," 114.
3 TNA, ADM 7/192, 112
4 Scoresby, "On the Greenland or Polar Ice," 333.
5 Scoresby, *Account of the Arctic Regions*, 49.
6 SPRI MS 116/57/10, D.
7 WLPS, SA 2B-9.
8 *Nautical Magazine* 21 (April 1852): 200.
9 BL, BB, ADD 35,306(2), 83.
10 Ibid., 86.
11 Great Britain, *Arctic Expeditions*, AB45232, 1–2.
12 TNA, ADM 7/190.
13 Holland, "Arctic Committee of 1851," 2: 116.
14 Great Britain, *Arctic Expeditions*, AB45232, 13.
15 Penny, *Letter to the Lords Commissioners*, 29.
16 Quoted in Friendly, *Beaufort of the Admiralty*, 320.
17 Neatby, *The Search for Franklin*, 198.
18 Lambert, *Franklin: Tragic Hero*, 230.
19 Williams, *Arctic Labyrinth*, 319.
20 Berton, *Arctic Grail*, 244.
21 Belcher, Sir Edward, *DCB*, http://www.biographi.ca/en/bio.php?id_nbr=4819.
22 TNA, ADM 1/5477.
23 Friendly, *Beaufort of the Admiralty*, 319.
24 *Times* (London), 13 March 1852.
25 Kane Papers, Correspondence, Lady Franklin, box 3, B.K132.
26 BL, BB, ADD 35,306, 82.
27 Ibid., 87.
28 Ibid., 90.
29 Lambert, *Franklin: Tragic Hero*, 224.
30 Tammiksaar, Sukhova, and Stone, "Hypothesis versus Fact," 238.
31 Brown, *North-West Passage*, 245.
32 *Nautical Magazine* 21 (May 1852): 276.
33 Brown, *North-West Passage*, 250.
34 *Nautical Magazine* 21 (May 1852): 276.

CHAPTER TWENTY-EIGHT

1 Force, *Supplement to Grinnell Land*, 16.
2 Stuart, *Relief of the Franklin Expedition*, 19.
3 Barrow to Kenndy, AANDC, G660 F73, 81–3.
4 Penny in Sutherland, *Journal of a Voyage* 2: 131.
5 SPRI, MS 248/414/28, D.
6 SPRI, MS 1100/1/11, D.
7 SPRI, MS 248/414/30, D.
8 Force, *Supplement to Grinnell Land*, appendix B, 49.
9 WLPS, SA 2B-3.
10 Force, *Grinnell Land*, 5.
11 Ibid.,9.
12 Ibid.
13 Ibid., 11.
14 Ibid., 15.
15 Ommanney, Journal written on HMS *Assistance*, n.p.
16 Force, *Supplement to Grinnell Land*, 16, 18, 6.
17 Force, *Grinnell Land*, 5–10.
18 Force, ibid., 7.
19 Force, *Supplement to Grinnell Land*, 1.
20 Kane, Notebook of polar exploration, Grinnell to Kane, 8 September 1852.
21 Force, *Supplement to Grinnell Land*, 2.
22 Ibid., 48.
23 *Baltimore Sun*, n.d., Grinnell scrapbooks, part O.
24 SPRI, MS 1100/1/21, D.
25 Ross, *Rear Admiral Sir John Franklin*, 65.
26 SPRI, MS 248/414/35, D.
27 SPRI, MS 248/414/36, D.
28 Elder, *Biography of Elisha Kent Kane*, 159.

CHAPTER TWENTY-NINE

1 "Eighty-Fifth Exhibition," 148.
2 Pearce, *Memorials of the Past*, 59.
3 SPRI, MS 116/60/1, D.
4 Ibid.
5 SPRI, MS 116/60/2, D.
6 BL, BB, ADD 35,306(2), 97–8.
7 "Eighty-Fifth Exhibition ," 148.
8 TNA, ADM 7/190(3).

9 Stone, "Arctic Portraits of Stephen Pearce," 55–8.
10 SPRI, MS 116/9/33, D.

CHAPTER THIRTY

1 Brown, *North-West Passage*.
2 *Punch*, 13 November 1852, 217.
3 SPRI, MS 116/56/9, D.
4 SPRI, MS 116/27/9, D.
5 SPRI, MS 248/247/23, D.
6 Ibid.
7 SPRI, MS 116/24/1, D.
8 SPRI, MS 1100/1/21, D.
9 SPRI, MS 1100/3/30, D.
10 Kane Papers, B.K132.
11 SPRI, MS 1226/17/1, D.
12 Kennedy, *Short Narrative of the Second Voyage*, 133.
13 SPRI, MS 116/24/1; D.
14 Belcher, *Last of the Arctic Voyages*, 1: 83.
15 SPRI, MS 248/259/1; D.
16 SPRI, MS 1100/1/9; D.
17 *Nautical Magazine* 21 (April 1852): 200.
18 Inglefield, *Summer Search*, 65.
19 Graville, "Examination of Edward Inglefield's 1852 Voyage," 95.
20 SPRI, MS 116/63/125; D.
21 SPRI, MS 1100/1/21; D.
22 SPRI, MS 1100/3/30; D.
23 SPRI, MS 1100/1/21; D.
24 TNA, CO 42.
25 SPRI, MS 116/43/3; D.
26 Ross, "False Leads."
27 SPRI, MS 116/63/86, D.
28 SPRI, MS 116/63/64, D.
29 BL, BB, ADD 35,306(2), 109.
30 Barr, *Arctic Hell-Ship*, 161.
31 Quoted in ibid., 165.
32 Barr, *Arctic Hell-Ship* (book title).
33 Miertsching, *Frozen Ships*, 233.
34 SPRI, MS 1100/1/30, D.
35 Williams, *Arctic Labyrinth*, 327.
36 SPRI, MS 116/63/93, D.

37 SPRI, MS 116/27/12, D.
38 Barron, *Old Whaling Days*, 64–5.
39 Ibid.
40 SPRI, MS 1100/3/34, D.
41 SPRI, MS 1100/1/39, D.
42 Quoted in Williams, *Arctic Labyrinth*, 333.

CHAPTER THIRTY-ONE

1 Penny to Barrow, BL, BB, ADD 35,306(2), 153.
2 SPRI, MS 116/63/34, D.
3 BL, BB, ADD 35,306(2), 158.
4 Brown, *North-West Passage*, 270.
5 SPRI, MS 116/63/64, D.
6 SPRI, MS 248/454/2, D.
7 BL, BB, ADD 35,306(2), 99–100.
8 Neatby, "McClure and the Passage," 40.
9 Cohen, *Lost beneath the Ice*.
10 Lambert, *Franklin: Tragic Hero*, 244.
11 SPRI, MS 116/63: 85, D.
12 Great Britain, *Report from the Select Committee*, AB45248, 35.
13 Alt et al., "Arctic Climate," 91.
14 Headland, "Transits of the Northwest Passage," 1–11.
15 SPRI, MS 116/9/17, D.
16 SPRI, MS 116/9/22, D.
17 SPRI, MS 248/454/2, D.
18 SPRI, MS 248/196/5, D.
19 Ibid.
20 SPRI, MS 248/454/2, D.
21 SPRI, MS 248/250/4, D.
22 SPRI, MS 248/454/14, D.
23 SPRI, MS 116/9/28, D.
24 McClintock, *Narrative of the Discovery*, appendix 2, 361.
25 Stefansson, *Friendly Arctic*.
26 McClintock, *Narrative of the Discovery*, 13.
27 SPRI, MS. 248/250/15, D.
28 McClintock, *Narrative of the Discovery*, 6.
29 BL, BB, ADD 35,306(2), 134.
30 Quoted in Ross, *This Distant and Unsurveyed Country*, 89.
31 McClintock, *Narrative of the Discovery*, 294.
32 Ibid.

33 Ibid., 296.
34 Ibid., 294–5.
35 Cyriax, *Sir John Franklin's Last Expedition*, 133.
36 Cooke, "A Gift Outright," 52.
37 BL, BB, ADD 35,306(2), 111.
38 BL, BB, ADD 35,306(2), 153.
39 Brown, *North-West Passage*, 389.
40 See Geiger and Mitchell, *Franklin's Lost Ship*; Watson, *Ice Ghosts*.

POSTSCRIPT

1 Sutherland, *Journal of a Voyage in Baffin's Bay* 2: 213–14.
2 Brown, *North-West Passage*, 250, 251.
3 Ibid., 283.
4 TNA, ADM 7/187(14).

Bibliography

ABBREVIATIONS

AANDC	Aboriginal and Northern Development Canada
ADM	Admiralty, Royal Navy, Whitehall
APS	American Philosophical Society, Philadelphia, PA
BB	Barrow Bequest
BL	British Library, London
GM	Glenbow Museum, Calgary, AB
HO	Hydrographic Office, Taunton, UK
HL	Huntington Library, San Marino, CA
ILN	*Illustrated London News*
LAC	Library and Archives Canada, Ottawa
LC	Library of Congress, Washington, DC
MF	Microfilm
NMM	National Maritime Museum, Greenwich, UK
NYHS	New-York Historical Society, New York
PRO	Public Record Office, Kew (now The National Archives of the United Kingdom)
RGS	Royal Geographical Society, London
UI	University of Illinois, Urbana–Champaign, IL
SA	Scoresby Archive, Whitby, UK
SPRI	Scott Polar Research Institute, Cambridge, UK
STEF	Stefansson Collection, Baker Library, Dartmouth College
TNA	The National Archives of the United Kingdom, Richmond, UK
USNA	United States National Archives, College Park, MD
WLPS	Whitby Literary and Philosophical Society, Whitby, UK

USEFUL SOURCES

The Arctic Blue Books are forty-seven British parliamentary papers on expeditions in the Canadian Arctic during the nineteenth century. They are all included in Marie Tremaine's *Arctic Bibliography*, with detailed summaries. The bibliography below lists those consulted under Great Britain, followed by the specific number assigned by Tremaine (e.g., AB45227). The Arctic Blue Books are now available online, accessible via Andrew Taylor's extensive index, from the University of Manitoba Archives and Special Collections in Winnipeg.

Statistics on British arctic whaling voyages presented in this book are taken from the Kinnes Lists unless otherwise noted.

ARCHIVAL SOURCES

Alexander. Shipboard journal, 1845. Dundee: Museums and Art Galleries.

Assistance, HMS. Logbook, 1850–51. LAC, RGS, MF A-872, part 4.

Brown, John. Arctic scrapbook (newspaper clippings), 1836–69, compiled by John Brown, FRGS, and continued by his son J. A. Brown. 2 vols. SPRI.

Cator, John B. Private journal kept whilst commanding HMS *Intrepid* on H.T. Austin's Franklin search expedition, 1850–51. 2 vols. Quotations in this book are from a typed transcript made in 1979. SPRI, MS 1183/1-2, BJ.

Cracroft, Sophia. Journal kept in the Orkneys and Shetlands, 29 July–27 September 1849. SPRI, MS 248/240, BJ.

Cumbrian. Shipboard journal, 1823. Lubbock notebook O, vol. 9. National Maritime Museum, Greenwich, UK.

De Haven, Edwin Jesse. Diary, 1850–51. De Haven Papers, 536D/10. Library of Congress, Washington, DC.

Grinnell, Henry. Scrapbooks (newspaper clippings), 1848–61. 8 vols. American Geographical Society Collection. University of Wisconsin–Milwaukee Library.

Kane, Elisha Kent. Notebook of polar exploration, 1850–51. Kane Papers, EKK1. Includes Kane's partial copy of the logbook of the *Advance*. American Philosophical Society, Philadelphia.

Kane Papers. Papers of Elisha Kent Kane, MSS B.K132, American Philosophical Society, Philadelphia.

Lady Franklin. Official logbook, kept by John Marshall, 1850–51. TNA: PRO, ADM 55/85.

Lovell, W.S. Journal of a Cruize [*sic*] in the Artic [*sic*] Regions in Search of Sir John Franklin, 1850–51. MMC 2383, Library of Congress, Washington, DC.

McClintock, Francis Leopold. Journal kept during the voyage of HMS

Assistance in search of Franklin, 22 March–31 December 1850. SPRI, MS I, MF; original in NMM.
Ommanney, Erasmus. Journal written on HMS *Assistance*, 25 April 1850 to 31 August 1851. Erasmus Ommanney Collection, M925, Glenbow Museum, Calgary.
– Extract from rough journal, 1850–51. Erasmus Ommanney Collection, M925, Glenbow Museum, Calgary.
"On the baptism of Erasmus Augustine Kallihirua, at St. Martin's Church, Canterbury, on Advent Sunday, 1853." LAC, RGS, MFA-871.
Pacific. Shipboard journal, 1846, by Alexander Hamilton, surgeon. Dundee: Museums and Art Galleries.
Ross, John. Pocket diary, 1850–51. SPRI, MS 1413, D.
Sophia. Official logbook, 1850–51, kept by Donald Manson. TNA: PRO, ADM 55/127.
Stuart, John. Journal of an unknown officer on Penny's Ship Lady Franklin, 1850–51 , MS X919.8 J826. Archives of the University of Illinois.

PUBLISHED WORKS

Alt, B.T., R.M. Koerner, D.A. Fisher, and J.C. Bourgeois. "Arctic Climate during the Franklin Era, as Deduced from Ice Cores." In Sutherland, *The Franklin Era in Canadian Arctic History, 1845–1859*, 69–92.
Arctic Miscellanies: Souvenir of the Late Polar Search; By the Officers and Seamen of the Expedition. London: Colburn 1852.
Auerbach, Jeffrey. *The Great Exhibition of 1851*. New Haven: Yale University Press 1999.
Baird, P.D. "Expeditions to the Canadian Arctic." Pts. 1–3. *The Beaver* 279 (March 1949): 44–7; 280 (June 1949): 41–7; 280 (September 1949): 44–8.
Barr, William. *Arctic Hell-Ship: The Voyage of* HMS *Enterprise, 1850–1855*. Edmonton: University of Alberta Press 2007.
Barron, William. *An Apprentice's Reminiscences of Whaling in Davis's Straits; Narrative of the Voyages of the Hull Barque Truelove, from 1848 to 1854*. Hull: Waller 1890. Reprint, Allerthorpe, York: K Book Editions 1980.
– *Old Whaling Days*. Hull: William Andrews 1895. Reprint, London: Conway Maritime Press 1970.
Beattie, Owen, and John Geiger. *Frozen in Time: Unlocking the Secrets of the Franklin Expedition*. Saskatoon: Western Producers Prairie Books 1988.
Behrisch, Erika. "'Far as the Eye Can Reach': Scientific Exploration and Explorers' Poetry in the Arctic, 1832–1852." *Victorian Poetry* 41, no. 1 (2003): 73–91.
Belcher, Edward. *The Last of the Arctic Voyages: Being a Narrative of the*

Expedition in H.M.S. Assistance under the Command of Captain Sir Edward Belcher . . . 2 vols. London: Lovell Reeve 1855.

Bellot, Joseph-René. *Memoirs of Lieutenant Joseph-René Bellot* . . . 2 vols. London: Hurst & Blackett 1855.

Benson, Arthur Christopher, and Viscount Escher, eds. *The letters of Queen Victoria: A Selection from Her Majesty's Correspondence between the Years 1837 and 1861*. 3 vols. London: John Murray 1907.

Berton, Pierre. *The Arctic Grail: The Quest for the North West Passage and the North Pole, 1818–1909*. Toronto: McClelland & Stewart 1988.

Brown, John. *The North-West Passage, and the Plans for the Search for Sir John Franklin: A Review*. London: S. Stanford 1858.

Carpenter, Kenneth J. *The History of Scurvy and Vitamin C*. Cambridge: Cambridge University Press 1988.

Cooke, Alan. "A Gift Outright: The Exploration of the Canadian Arctic Islands after 1880." In Zaslow, *A Century of Canada's Arctic Islands*, 51–60.

Cooke, Alan, and Clive Holland. *The Exploration of Northern Canada, 500 to 1920: A Chronology*. Toronto: Arctic History Press 1978.

Cookman, Scott. *Ice Blink: The Tragic Fate of Sir John Franklin's Lost Polar Expedition*. New York and Toronto: John Wiley & Sons 2000.

Corner, George W. *Doctor Kane of the Arctic Seas*. Philadelphia: Temple University Press 1972.

Cyriax, Richard J. "Adam Beck and the Franklin Search." *Mariner's Mirror* 48, no. 1 (1962): 35–51.

– "Arctic Sledge Travelling by Officers of the Royal Navy, 1819–49." *Mariner's Mirror* 49, no. 2 (1963): 127–42.

– *Sir John Franklin's Last Expedition: A Chapter in the History of the Royal Navy*. London: Methuen 1939.

– "The Voyage of H.M.S. *North Star*, 1849–50." *Mariner's Mirror* 50, no. 4 (1964): 307–18.

Cyriax, Richard J., and J.M. Wordie. "Centenary of the Sailing of Sir John Franklin with the *Erebus* and *Terror*." *Geographical Journal* 106 (1945): 169–97.

Darwin, Charles. *The Voyage of the Beagle*. Edited by Leonard Engel. Originally published in 1839. New York: Doubleday 1962.

Davis, Richard C., ed. *Lobsticks and Stone Cairns: Human Landmarks in the Arctic*. Calgary: University of Calgary Press 1996.

Day, Alan Edwin. *Search for the Northwest Passage: An Annotated Bibliography*. New York and London: Garland Publishing 1986.

"The Eighty-Fifth Exhibition of the Royal Academy." *Art-Journal, London* 15 (1 June 1853): 148.

Elder, William. *Biography of Elisha Kent Kane*. Philadelphia: Childs & Peterson 1858.

Fluhmann, May. *Second in Command: A Biography of Captain Francis Crozier R.N., F.R.S., F.R.A.S.* Yellowknife: Government of the Northwest Territories 1976.

Force, Peter. *Grinnell Land: Remarks on the English Maps of Arctic Discoveries, in 1850 and 1851, Made at the Ordinary Meeting of the National Institute, Washington, in May, 1852.* Washington: R.A. Waters [1852].

– *Supplement to Grinnell Land.* Read at the Ordinary Meeting of the National Institute, July 1853. Washington: R.A. Waters [1853].

Forsyth, Charles Codrington. "Arctic Voyage of the Prince Albert." *Nautical Magazine* 19 (1850): 606–15.

Friendly, Alfred. *Beaufort of the Admiralty: The Life of Sir Francis Beaufort, 1774–1857.* New York: Random House 1977.

Friis, Herman R. "The Beginnings of Exploration and Discovery in the Arctic by the United States and the Role of Matthew F. Maury, Henry Grinnell, and the American Geographical and Statistical Society: 1842–1861." Paper read to the Society for the History of Discoveries in New Bedford, MA, 25 October 1968.

Geiger, John, and Alanna Mitchell. *Franklin's Lost Ship: The Historic Discovery of HMS Erebus.* Toronto: Harper Collins 2015.

Gill, Harold B., and Joanne Young, eds. *Searching for the Franklin Expedition: The Arctic Journal of Robert Randolph Carter.* Annapolis: Naval Institute Press 1998.

Goodsir, Robert Anstruther. *An Arctic Voyage to Baffin's Bay and Lancaster Sound, in Search of Friends with Sir John Franklin.* London: Van Voorst 1850. Facsimile edition, Plaistow and Sutton Coldfield: Arctic Press 1996.

Graville, Brian. "An Examination of Edward Inglefield's 1852 Voyage into Jones Sound." *Polar Record* 49, no. 1 (2013): 91–6.

Great Britain. *Additional Papers Related to the Arctic Expeditions under the Orders of Captain Austin and Mr. William Penny.* House of Commons, Sessional Papers, Accounts and Papers 1852, 50 (1436), AB45228. London: Eyre & Spottiswoode.

– *Arctic Expeditions: Further Correspondence and Proceedings Connected with the Arctic Expedition.* House of Commons, Sessional Papers, Accounts and Papers 1852, 50 (1449), AB45229. London: Eyre & Spottiswoode.

– *Arctic Expeditions: Report of the Committee Appointed by the Lords Commissioners of the Admiralty to Enquire into and Report on the Recent Arctic Expeditions in Search of Sir John Franklin.* House of Commons, Sessional Papers, Accounts and Papers, 1852, 50 (1435), AB45227. London: Eyre & Spottiswoode.

– *Arctic Expeditions: Return to an Address of the Honourable the House of Commons, Dated 10 May 1850.* House of Commons, Sessional Papers,

Reports and Papers, 1850, 35 (397), AB43225. Ordered by the House of Commons to be printed, 31 May 1850.

– *Arctic Expeditions: Return to an Address of the Honourable the House of Commons, Dated 7 February 1851.* House of Commons, Sessional Papers, Accounts and Papers, 1851, 33 (97), AB45226. Ordered by the House of Commons to be printed, 7 March 1851.

– *Arctic Expeditions: Return to an Address of the Honourable the House of Commons, Dated 20 February 1852; for Papers in Connection with the Late Arctic Expeditions, or with Any Which May Be in Preparation.* House of Commons, Sessional Papers, Accounts and Papers, 1852, 50 (115), AB45232. Ordered by the House of Commons to be printed, 27 February 1852.

– *Arctic Expedition: Return to an Order of the Honourable the House of Commons, Dated 5 February 1850.* House of Commons, Sessional Papers, Accounts and Papers, 1850, 35 (107), AB45223. Ordered by the House of Commons to be printed, 5 March 1850.

– *Papers Relative to the Recent Arctic Expeditions in Search of Sir John Franklin and the Crews of HMS* Erebus *and* Terror. House of Commons, Sessional papers, Accounts and Papers 1854, 42 (1725), AB45241. Ordered by the House of Commons to be printed, 27 February 1852.

– *Report from the Select Committee on Arctic Expedition; Together with the Proceedings of the Committee, Minutes of Evidence, and Appendix.* House of Commons, Sessional Papers, Accounts and Papers, 1855, 7 (409), AB45248. Ordered by the House of Commons to be printed, 20 July 1855.

Great Britain. Admiralty. £20,000 Sterling Reward. To Be Given by Her Majesty's *Government* . . . Handbill, 23 March 1849. WLPS, SA 2B-3.

The Great Exhibition: London's Crystal Palace Exposition of 1851. The Art-Journal Illustrated Catalogue of the Industry of All Nations Exposition, 1851. Facsimile, New York: Gramercy Books 1995.

Harrison, George. "Lost Whaling Ships: Extract of a Letter from Mr. George Harrison, of Thirty Years Experience in the Command of Whalers." *Nautical Magazine* 21 (1852): 189–95.

Hazlewood, Nick. *Savage: Survival, Revenge and the Theory of Evolution.* London: Hodder & Stoughton 2000.

Headland, R.K. "Transits of the Northwest Passage to End of the 2014 Navigation Season." Scott Polar Research Institute, University of Cambridge, 2014, 1–11. https://docplayer.net/41079582-Transits-of-the-northwest-passage-to-end-of-the-2014-navigation-season-atlantic-ocean-arctic-ocean-pacific-ocean.html.

Historic Tinned Foods. International Tin Research and Development Council publication no. 85. Middlesex, UK [1939].

Hobhouse, Christopher. *1851 and the Crystal Palace; Being an Account of the Great Exhibition and Its Contents* . . . London: John Murray 1937.

Hodgson, Maurice. *On Site with Maurice Haycock, Artist of the Arctic: Paintings and Drawings of Historical Sites in the Canadian Arctic.* Campbellville, ON: Edgar Kent 2007.

Holland, Clive. "The Arctic Committee of 1851: A Background Study," Part 1, *Polar Record* 20, no. 124 (1980): 3–17.

– The Arctic Committee of 1851: A Background Study," Part 2, *Polar Record* 20, no. 125 (1980): 105–18.

– *Arctic Exploration and Development c. 500 B.C. to 1915: An Encyclopedia.* Garland Reference Library of the Humanities, vol. 930. New York and London: Garland Publishing 1994.

– ed. *Manuscripts in the Scott Polar Research Institute, Cambridge, England: A Catalogue.* New York and London: Garland Publishing1982.

– "William Penny, 1809–92: Arctic Whaling Master." *Polar Record* 15, no. 94 (1970): 25–43.

Honigsbaum, Mark. *The Fever Trail.* London: Pan Macmillan 2001.

Inglefield, Edward Augustus. *A Summer Search for Sir John Franklin* . . . London: Thomas Harrison 1853.

Johnston, Alexander Keith. *The Physical Atlas of Natural Phenomena.* Edinburgh and London: William Blackwood & Sons 1850.

Jones, A.G.E. "Captain William Parker Snow." In *Polar Portraits: Collected Papers*, 366–71. Whitby: Caedmon of Whitby 1992. Originally published in the *Falkland Islands Journal* 1979.

– "The Halkett Boat and Other Portable Boats." *Mariner's Mirror* 44, no. 2 (1958): 154–8.

– "The Voyage of H.M.S. *Cove*, Captain James Clark Ross, 1835–36." *Polar Record* 5, no. 40 (1950): 543–56.

Kane, Elisha Kent. *The United States Grinnell Expedition in Search of Sir John Franklin: A Personal Narrative.* Boston: Phillips, Sampson; Philadelphia: Childs & Petersen 1857. Originally published 1854.

Kemp, Peter, ed. *The Oxford Companion to Ships and the Sea.* London: Oxford University Press 1976.

Kennedy, William. *A Short Narrative of the Second Voyage of the Prince Albert, in Search of Sir John Franklin.* London: W.H. Dalton 1853.

Kinnes Lists. "An Account of the Number of Fish and Produce of Oil Brought by Each Ship from the Greenland and Davis's Streights Whale Fishery for the Year 1814." Annual lists with similar titles and various printers from 1814 to 1911, collected by Robert Kinnes & Sons, Dundee. London: Devereux and Lambert. Microfilm copy now at University of Dundee.

Lambert, Andrew. *Franklin: Tragic Hero of Polar Navigation*. London: Faber & Faber 2009.

Littell, Eliakim. *Littell's Living Age*. Vol. 24 (January–March 1850): 435. T.H. Carter and Co. 1844–96.

Lloyd, Christopher. *The British Seaman, 1200–1860: A Social Survey*. London: Collins 1968.

Lloyd-Jones, R. "The Paranormal Arctic: Lady Franklin, Sophia Cracroft, and Captain and 'Little Weesy' Coppin." *Polar Record* 37, no. 200 (2001): 27–34.

Loomis, Chauncey. *Weird and Tragic Shores*. Lincoln and London: University of Nebraska Press 1991.

Lubbock, Basil. *The Arctic Whalers*. Glasgow: Brown, Son & Ferguson 1955. First published 1937.

Mackinnon, C.S. "The British Man-Hauled Sledging Tradition." In *The Franklin Era in Canadian Arctic History, 1848–1859*, edited by Pat Sutherland, 129–40. Mercury Series. Ottawa: National Museums 1985.

Markham, Albert Hastings. *The Great Frozen Sea: A Personal Narrative of the Voyage of the "Alert" during the Arctic Expedition of 1875*. London: Daldy, Isbister 1878.

Markham, Clements R. *The Arctic Navy List; or, a Century of Arctic and Antarctic officers, 1773–1873*. London and Portsmouth: Griffin 1875.

– *Franklin's Footsteps*. London: Chapman & Hall 1852.

Mary-Rousselière, Guy. *Qitdlarssuaq: Histoire d'une migration polaire*. Montreal: Presses de l'Université de Montréal 1980.

McClintock, Francis Leopold. "How to Prepare Oneself to Enjoy the Winter." In *Arctic Miscellanies*, 55–6. London: Colburn 1852.

– *A Narrative of the Discovery of the Fate of Sir John Franklin and His Companions*. Spine title: *The Voyage of the 'Fox' in the Arctic Seas*. London: John Murray 1859.

– "Reminiscences of Arctic Ice-Travel in Search of Sir John Franklin and His Companions." *Journal of the Royal Dublin Society*, February 1857.

McDonald, Alexander. *A Narrative of Some Passages in the History of Eenoolooapik* . . . Edinburgh: Fraser & Hogg 1841.

Middleton, Lewis. *Whaling Recollections, 1818–1888*. Aberdeen, n.d.

Miertsching, Johann. *Frozen Ships: The Arctic Diary of Johann Miertsching, 1850–1854*. Translated with an introduction and notes by L.H. Neatby. Toronto: Macmillan 1967.

Mills, William James. *Exploring Polar Frontiers: A Historical Encyclopedia*. 2 vols. Oxford: ABC-CLIO 2003.

Milton-Thompson, G.J. "Two Hundred Years of the Sailor's Diet." In *Starving Sailors: The Influence of Nutrition upon Naval and Maritime History*, edited by J. Watt, E.J. Freeman, and W.F. Bynum, 27–34. Greenwich: National Maritime Museum 1981.

Mirsky, Jeanette. *Elisha Kent Kane and the Seafaring Frontier*. Edited by Oscar Handlin. Westport, CT: Greenwood Press 1971. Originally published 1954.

Moorehead, Alan. *Darwin and the Beagle*. Harmondsworth: Penguin Books 1971.

"Mr. W. Parker Snow – Sailor, Explorer, and Author." *Review of Reviews* 7 (1893): 371–86.

Neatby, Leslie H. *In Quest of the North West Passage*. London: Constable 1958.

– "McClure and the Passage." *The Beaver* 291 (1960): 33–40.

– *The Search for Franklin*. Edmonton: M.G. Hurtig 1970.

Nuttal, Mark. *Encyclopedia of the Arctic*. 3 vols. London and New York: Routledge 2005.

Old Officer. *Observations on the Missing Ships of Sir John Franklin; with Some Propositions and Considerations for Their Relief and Extrication from the Ice: By an Old Officer of Rank in the Royal Navy*. London: Trelawney Saunders 1849. Reprinted in Littell, *Littell's Living Age*, 24: 453.

Osborn, Sherard. *Stray Leaves from an Arctic Journal; or, Eight Months in the Polar Regions, in Search of Sir John Franklin's Expedition, in the Years 1850–51*. New York: George G. Putnam 1852.

– "A Tender Strain." *Illustrated Arctic News*. Published on Board HMS *Resolute*, Captain Horatio T. Austin, CB, 1850, 13.

Pack, A.J. *Nelson's Blood: The Story of Naval Rum*. Annapolis, MD: Naval Institute Press 1982. First published by Kenneth Mason, Homewell, Havant, Hampshire, 1982.

Parry, William Edward. *Journal of a Voyage for the Discovery of a North-West Passage from the Atlantic to the Pacific* . . . London: John Murray 1821. Reprinted in facsimile, New York: Greenwood Press 1968.

– *Journal of a Second Voyage for the Discovery of a North-West Passage from the Atlantic to the Pacific* . . . London: John Murray 1824.

Pearce, Stephen. 1903. *Memorials of the Past*. Edinburgh: Riverside Press.

Penny, William. *Letter to the Lords Commissioners of the Admiralty*. London: Chapman & Hall 1852.

Petersen, Carl. 1857. *Grindringer fra Polarlandene*. Copenhagen: Philipsens.

Philbrick, Nathaniel. *Sea of Glory: America's Voyage of Discovery, the U.S. Exploring Expedition, 1838–1842*. London: Penguin Books 2004.

Pim, Bedford Clapperton T. *An Earnest Appeal to the British Public on Behalf of the Missing Arctic Expedition*. London: Hurst & Blackett 1856.

Porsild, A.E. 1957. *Illustrated Flora of the Canadian Arctic Archipelago*. Bulletin no. 146. Ottawa. Department of Northern Affairs and Natural Resources.

Potter, Russell A. *Arctic Spectacles: The Frozen North in Visual Culture, 1818–1875*. Montreal and Kingston: McGill-Queen's University Press 2007.

Report of the Committee Appointed to Inquire into the Expediency of Diminishing the Present Quantity of Spirits Served Out Daily to the Seamen in the Royal Navy. Presented to both Houses of Parliament by Command of Her Majesty. London: Printed by William Clowes and Sons 1850.

A Review of the Proceedings of the Arctic Searching Expeditions, under the Command of Captain H.T. Austin, C.B., and Captain Wm. Penny. Extract from *Nautical Magazine*, October 1851.

Richards, Robert L. *Dr. John Rae*. Whitby: Caedmon of Whitby 1985.

Ross, John. *Rear Admiral Sir John Franklin, a Narrative of the Circumstances and Causes Which Led to the Failure of the Searching Expeditions Sent by Government and Others for the Rescue of Sir John Franklin*. London: Longmans, Green, Brown, & Longmans 1855.

Ross, M.J. *Polar Pioneers: John Ross and James Clark Ross*. Montreal and Kingston: McGill-Queen's University Press 1994.

Ross, W. Gillies. "Clairvoyants and Mediums in the Search for Franklin. *Polar Record* 39, no. 208 (2003): 1–18.

– "False Leads in the Franklin Search." *Polar Record* 39, no. 209 (2003): 131–60.

– *This Distant and Unsurveyed Country: A Woman's Winter at Baffin Island, 1857–58*. Montreal and Kingston: McGill-Queen's University Press 1997.

– "The Type and Number of Expeditions in the Franklin Search, 1847–1859." *Arctic* 55, no. 1 (2002): 57–69.

Savours, Ann. *The Search for the North West Passage*. London: Chatham Publishing 1999.

Scoresby, William. *An Account of the Arctic Regions, with a History and Description of the Northern Whale Fishery*. 2 vols. Edinburgh: Archibald Constable 1820. Reprinted in facsimile, Newton Abbot, Devon: David & Charles 1969.

– *The Franklin Expedition: or Considerations on Measures for the Discovery and Relief of Our Absent Adventurers in the Arctic Regions*. London: Longman, Brown, Green, & Longmans 1850.

– *Journal of a Voyage to the Northern Whale-Fishery: Including Researches and Discoveries on the Eastern Coast of West Greenland, Made in the Summer of 1822, in the Ship Baffin of Liverpool*. Edinburgh: Archibald, Constable & Co. 1823.

– "On the Greenland or Polar Ice." In *The Polar Ice (1815); The North Pole (1828)*. Whitby: Caedmon of Whitby 1980. First published in *Memoirs of the Wernerian Society* 20 (1915): 261–338.

Skewes, J.H. *Sir John Franklin: The True Secret of the Discovery of His Fate; A "Revelation."* London: Bemrose & Sons 1889.

Smith, Michael, and Bruce Rigby. "Distribution of Polynyas in the Canadian Arctic." In *Polynyas in the Canadian Arctic*, edited by Ian Stirling and Holly Cleator, 7–28. Canadian Wildlife Service occasional paper no. 45. Ottawa: Minister of Supply and Services 1981.

Snow, William Parker. "Remarks on the Prince Albert's Track through the Ice of Baffin Bay." *Nautical Magazine* 20 (1851): 169–79.

– *Voyage of the Prince Albert in Search of Sir John Franklin: A Narrative of Every-Day Life in the Arctic Seas.* London: Longman, Brown, Green, and Longmans 1851.

Spink, John, and D.W. Moodie. "Eskimo Maps from the Canadian Eastern Arctic." Cartographica Series, monograph no. 5. Toronto: B.V. Gutsell 1972.

Stamp, Tom, and Cordelia Stamp. *William Scoresby, Arctic Scientist.* Whitby: Caedmon 1976.

Stefansson, Vilhjalmur. *The Friendly Arctic: The Story of Five Years in Polar Regions.* New York: Macmillan 1921.

Stone, Ian. "The Arctic Portraits of Stephen Pearce." *Polar Record* 24, no. 148 (1988): 55–8.

– "Charles Codrington Forsyth (ca. 1810–1873)." *Arctic* 38, no. 4 (1985): 340–1. Reprinted in Davis, *Lobsticks and Stone Cairns*, 84–6.

– "An Episode in the Franklin Search: The *Prince Albert* Expedition, 1850." Pts. 1 and 2. *Polar Record* 29, no. 169 (1993): 127–42; no. 170 (1993): 197–208.

– "William Parker Snow, 1817–1895." *Polar Record* 19, no. 119 (1978): 163–5.

Stuart, John. *The Relief of the Franklin Expedition: What Has Been Done, and What May Yet Be Done.* Edinburgh: R. Grant & Son 1852.

Sutherland, Gavin. *The Whaling Years: Peterhead* (1788–1893). Centre for Scottish Studies, University of Aberdeen 1993.

Sutherland, Patricia D., ed. *The Franklin Era in Canadian Arctic History, 1845–1859.* Archeological Survey of Canada paper no. 131. Ottawa: National Museum of Man 1985.

Sutherland, Peter Cormack. *Journal of a Voyage in Baffin's Bay and Barrow Straits . . .* 2 vols. London: Longman, Brown, Green, & Longmans 1852.

– "On the Esquimaux." *Journal of the Ethnological Society of London* 4 (1856): 193–214.

Tammiksaar, E., N.G. Sukhova, and I.R. Stone. "Hypothesis versus Fact: August Petermann and Polar Research." *Arctic* 52, no. 3 (1999): 237–43.

Taylor, Andrew. *Geographical Discovery and Exploration in the Queen*

Elizabeth Islands. Geographical Branch memoir no. 3. Ottawa: Department of Mines and Technical Surveys, 1955.

– "A Preliminary Guide to the Arctic Blue Books and Parliamentary Papers of the Nineteenth Century." *Arctic Circular* 8, no. 3 (1955).

Tillotson, John. *Adventures in the Ice: A Comprehensive Survey of Arctic Exploration, Discovery, and Adventure. Including Experiences of Captain William Penny, the Veteran Whaler* . . . London: James Hogg [1869].

Tremaine, Marie, ed. *Arctic Bibliography*. Vol. 8: 317–52. British Parliamentary Papers on Arctic Expeditions in the Nineteenth Century. Washington: US Government Printing Office 1959.

Warmow, Mathias. "Extract from Br. M. Warmow's Journal of His Residence in Cumberland Inlet, during the Winter of 1857–1858." In *Periodical Accounts Relating to the Missions of the United Brethren, Established among the Heathen*, vol. 23, 90. London: Brethren's Society for the Furtherance of the Gospel 1858.

Washington, John. *Esquimaux and English Vocabulary, for Use of the Arctic Expedition*. Published by order of the Lords Commissioners of the Admiralty. London: John Murray 1850.

Watson, Paul. *Ice Ghosts: The Epic Hunt for the Lost Franklin Expedition*. Toronto: McClelland & Stewart 2017.

Weld, Charles Richard. *Arctic Expeditions: A Lecture Delivered at the London Institution, February 6, 1850*. London: John Murray 1850.

Williams, Glyn. *Arctic Labyrinth: The Quest for the Northwest Passage*. Toronto: Viking Canada 2009.

"A Winter in the Arctic Regions Forty Years Ago." *Nautical Magazine* 47, no. 2 (1878): 117–27.

Woodman, David C. *Unravelling the Franklin Mystery: Inuit Testimony*. Montreal and Kingston: McGill-Queen's University Press 1991.

Woodward, Frances J. "William Penny, 1809–92." *Polar Record* 6, no. 46 (1853): 808–11.

– *Portrait of Jane*. London: Hodder & Stoughton 1951.

Wright, John Kirtland. *Geography in the Making: The American Geographical Society, 1851–1951*. New York: The Society 1952.

– "The Open Polar Sea." *Human Nature in Geography: Fourteen Papers, 1925–1965*. Cambridge, MA: Harvard University Press 1966. Originally printed in *Geographical Review* 43 (1953): 338–65.

Zaslow, Morris, ed. *A Century of Canada's Arctic Islands, 1880–1980*. Ottawa: Royal Society of Canada 1981.

Index of Names

Index of Places

Index of Ships